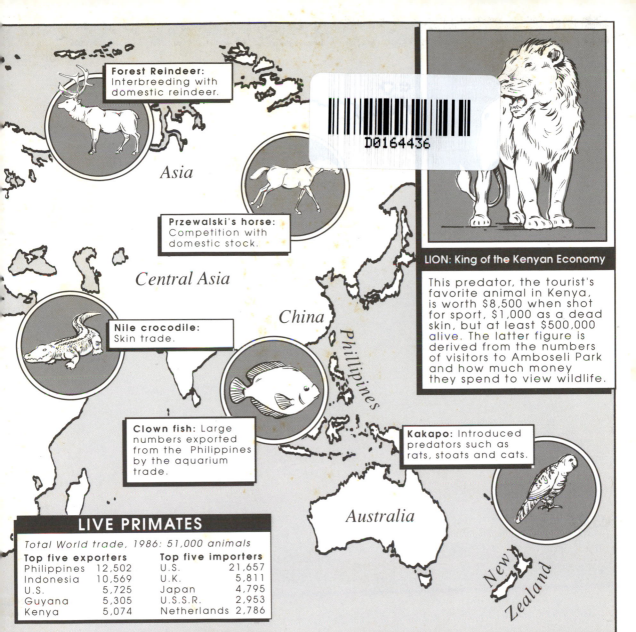

Forest Reindeer: Interbreeding with domestic reindeer.

Asia

Przewalski's horse: Competition with domestic stock.

Central Asia

China

Phillipines

Nile crocodile: Skin trade.

Clown fish: Large numbers exported from the Philippines by the aquarium trade.

Kakapo: Introduced predators such as rats, stoats and cats.

Australia

New Zealand

LION: King of the Kenyan Economy

This predator, the tourist's favorite animal in Kenya, is worth $8,500 when shot for sport, $1,000 as a dead skin, but at least $500,000 alive. The latter figure is derived from the numbers of visitors to Amboseli Park and how much money they spend to view wildlife.

LIVE PRIMATES

Total World trade, 1986: 51,000 animals

Top five exporters		Top five importers	
Philippines	12,502	U.S.	21,657
Indonesia	10,569	U.K.	5,811
U.S.	5,725	Japan	4,795
Guyana	5,305	U.S.S.R.	2,953
Kenya	5,074	Netherlands	2,786

IVORY

Total World trade, 1988: 430,000 kilograms

Top five exporters		Top five importers	
Hong Kong	135,938	Hong Kong	179,608
Singapore	100,568	Japan	100,985
Japan	33,149	China	50,879
Belgium	24,034	Belgium	38,730
Somalia	22,638	Singapore	17,722

CAT SKINS

Total World trade, 1986: 192,000 skins

Top five exporters		Top five importers	
China	68,274	F.R. Germany	82,240
U.S.	65,419	Canada	21,109
Canada	24,104	U.S.	17,543
Switzerland	7,622	U.K.	16,400
Bolivia	6,124	Japan	14,321

LIVE PARROTS

Total World trade, 1986: 6-700,000 parrots

Top five exporters		Top five importers	
Argentina	177,992	U.S.	305,997
Tanzania	84,228	F.R. Germany	60,499
F.R. Germany	60,499	U.K.	34,520
Indonesia	58,832	Netherlands	27,822
Guyana	30,324	Japan	27,790

REPTILE SKINS

Total World trade, 1986: 10,481,000 skins

Top five exporters		Top five importers	
Indonesia	3,081,313	Singapore	2,856,468
Thailand	1,650,317	U.S.	1,265,018
Singapore	1,637,061	Italy	1,026,928
Argentina	1,153,967	Spain	1,009,054
U.K.	473,519	Japan	802,397

Introductory Ecology

About the Cover Photo. Purple loosestrife (*Lythrum salicornia*) is an emergent aquatic plant that was introduced into eastern North America from Europe at the turn of the century. It probably arrived independently many times, either accidentally in ships' ballasts or lodged in the wool of sheep or purposefully as a valued herb. By 1985, it had become present in every state north of the 35th parallel, except Alaska and Montana. It grows rampantly in riparian habitats such as roadside ditches in the northeastern United States, delighting commuters from Massachusetts to Minnesota. Unfortunately, this purple haze signals disaster to wildlife biologists and land managers because it chokes out native vegetation and makes life tough for waterfowl, songbirds, and muskrats, hindering their search for food and shelter. It is very difficult to eradicate purple loosestrife. An old plant can produce more than two million seeds in one summer. Mowing or cutting makes more plants because new stalks grow from cut roots and the cut shoots lying on the ground may grow their own roots and send up new sprouts. It is too labor intensive to pull plants up entirely and the use of herbicides kills off native plants. Help may be on the way, from Europe, in the form of a root-eating weevil and two leaf-eating beetles that attack and eat only purple loosestrife. But not everyone wants the plant destroyed; bee keepers say it makes delicious honey and local nurseries sell hundreds each season.

Introductory Ecology

Peter Stiling
University of South Florida

Prentice Hall, Englewood Cliffs, NJ 07632

Library of Congress Cataloging-in-Publication Data

Stiling, Peter D.
    Introductory ecology / Peter Stiling.
      p.   cm.
    Includes bibliographical references and index.
    ISBN 0-13-223942-6
    1. Ecology.   I. Title.
QH541.S675   1992
574.5—dc20   91-36494
                                  CIP

Acquisitions Editor: David Kendric Brake
Editorial/Production Supervision: Tom Aloisi and Technical Texts
Interior Design: Sylvia Dovner, Technical Texts
Cover Design: Susan Slovinsky
Cover Photo: Frank Siteman
Prepress Buyer: Paula Massenaro
Manufacturing Buyer: Lori Bulwin
Editorial Assistant: Mary DeLuca

 © 1992 by Prentice-Hall, Inc.
A Simon & Schuster Company
Englewood Cliffs, New Jersey 07632

Printed in the United States of America
10 9 8 7 6 5 4 3 2 1

I S B N 0-13-223942-6

Prentice-Hall International (UK) Limited, *London*
Prentice-Hall of Australia Pty. Limited, *Sydney*
Prentice-Hall Canada Inc., *Toronto*
Prentice-Hall Hispanoamericana, S.A., *Mexico*
Prentice-Hall of India Private Limited, *New Delhi*
Prentice-Hall of Japan, Inc., *Tokyo*
Simon & Schuster Asia Pte. Ltd., *Singapore*
Editora Prentice-Hall do Brasil, Ltda., *Rio de Janeiro*

We acknowledge permission to use the following figures and tables: **Figure 2.1,**
from V. Grant, *The Origin of Adaptations* (New York, Columbia University Press,

continues on page 580

To Don and Dan

Contents _____

Preface

What Is Ecology?

The word ecology, defined by Ernest Haeckel in 1869, is formed from the Greek words *oikos* (house) and *logos* (study). It has become a household word as people have become more aware of their environment. In local and in general elections, "ecology" is often said to be a major issue, and indeed, in Germany the Green party is an appreciable political force. To most people, ecology is associated with the broad problems of the human environment. To others, ecology is synonymous with conservation and saving the whales or the forests. Yet, ecology does not simply deal with environmental pollution of human origin or conservation of species we find intriguing; it is related to environmental science much as physics is to engineering. Ecology provides the scientific framework upon which conservation programs can be set up. The scientific framework is often erected, however, not from studies of rare, exotic animals but from studies of insects or other small, relatively unappealing plants and animals because these best can be manipulated experimentally to test ecological theories. For example, it is rather difficult to experiment with blue whales or Florida panthers, of which there are relatively few. This book is, thus, peppered with studies of the teeming hordes of common life.

The biosphere is composed of some 1.8×10^{12} metric tons of living tissue and covers the surface of the globe in a layer proportionately thinner than the skin of an apple. If all the material were evenly distributed, it would be less than 1 cm thick and weight about 3.6 kg m^{-2} (Anderson 1981), but the Earth's biomass is not evenly distributed. It shows great variation, from about 45 kg m^{-2} in tropical forests to 0.003 kg m^{-2} in the oceans. Ecology is concerned with explaining this variation; it asks why plants and animals are found in certain areas and what controls their numbers. The study of ecology can be divided into two areas. *Functional ecology* asks how populations are maintained at a particular size and touches on behavior, competition, predation, and other contemporaneous factors. *Historical ecology* asks how populations have come to be this way and places great emphasis on biogeography (the study of global distribution patterns and shared characteristics), which presupposes a knowledge of continental drift and of evolution.

Text Development

In 1983, I began teaching ecology to students in the Department of Zoology at the University of the West Indies, Trinidad. Most had little prior ecological background, so I started from scratch. I wanted to give the students a full overview of ecology and not specialize in one particular area; I needed to present an introductory ecology course. This course should encompass population biology, the regulation of plant and animal numbers, which is the core material of most texts, and a knowledge of communities. Of course, a full understanding of ecology requires some knowledge of historical ecology and, thus, of evolution, so, to begin, I introduced a little evolutionary ecology. Also, I wanted to include some ecology of animal behavior, a fascinating area and one enjoyed by most students. And finally, I thought there should be some reference to applied ecology. For most of these students, a knowledge of applied ecology would be essential in the real world to which they would soon belong. Having set this outline for my introductory course, I found that there was no textbook incorporating evolutionary ecology, behavioral ecology, population ecology, community ecology, and applied ecology. This book is an attempt to remedy that situation. It assumes only an elementary knowledge of college-level biology, perhaps that offered by an introductory biology course in which some knowledge of genetics and heredity, physiological adaptation of plants and animals, and the diversity of life have been touched upon.

Themes

This book has several main themes. First, it is broad in coverage. No other ecology text attempts as diverse a treatment in explaining the distribution and abundance of many plants and animals. Take for example the distribution patterns of the polar bear. Why is it not found in Antarctic? Clearly, climate and topography are not sufficient answers. For this species, a knowledge of biogeography and evolutionary ecology is vital. For many others species, particularly those with good dispersal abilities, such as plants, physiological factors are the main constraints on observed distribution patterns. Within these distribution patterns, the abundance of species can be affected by food supply, competitors, predators, and parasites and, for animals, territorial behavior. Thus, a sound knowledge of population ecology is essential. Ultimately population fluctuations may be influenced by higher-order community interactions—hence, the value of examining ecological processes at these large spatial scales. Finally, so many ecological processes are impacted by people, especially in these times of exploding population growth, that it is worthwhile to examine ecological phenomena as they are affected by humans—applied ecology. No other ecology book is as broad. Breadth, of course, has its drawbacks. By combining many elements into an introductory course, the text could become overly long. I have resisted this temptation by presenting ideas as succinctly as possible. The result is, I hope, a streamlined introductory ecology text.

A second main theme of this book is to present both sides of an argument. Many texts are content to tell a "just so" story. This is often true when ecological processes are modeled. There is a great temptation either to present a model and then the evidence to back it up or to spend a long time concocting a model to fit available data. It is very easy to present an ecological principle with a "happy ending," that is, a nice tidy explanation with no loose ends that appeals to the student and is intellectually satisfying. Unfortunately, as most field ecologists know, the actual data are rarely neat and often fit no clearcut hypothesis. Even if the data do fit a hypothesis, the next set of data taken in another system is just as likely to show the opposite trend. To combat this, I have tried, wherever possible, to present a variety of evidence, both for and against proposed theories. For example, in Chapter 17, I show how two different indices of diversity give different conclusions about which of two communities is most diverse. This is an important issue given the present value attached to maintaining biological diversity.

Chapter Format

It is difficult, in such a wide-ranging book as this, to stick to one particular format for each chapter. There are, however, some generalities of approach that are worth bringing out. In each chapter, the ideas and theory are discussed first, then some examples follow, both for and against the proposed hypotheses. I have tried to keep these examples to a minimum. Often, modern examples are used but sometimes older work is referred to. There are many older pieces of work that will forever remain standards. In the field of competition theory, for example, the displacement of one species of parasitic *Aphytis* wasp by another in Southern California will probably forever remain a classic example of competitive displacement in the field. Nevertheless, references are kept as updated as possible; over half are from the 1980s. This is, therefore, an unusually current treatment of ecology. Following the examples, I discuss any recent overviews of the subject. Ultimately, one is swayed either for or against a theory by the weight of the evidence. I cannot present all the pertinent data for and against one theory in this book, but I can refer the reader to the research papers that do. For example, in discussing the frequency of density dependent parasitism (Chapter 15), I point out that two independent reviews (Stiling 1987; Walde and Murdoch 1988) have shown that in nature density dependence does not occur very frequently despite the fact that many ecologists have based much of their work on modeling density dependence. Similar reviews are referred to in the section on competition (Connell 1983; Schoener 1983). It is my hope that in the not-too-distant future all subject areas of ecology will be methodically reviewed in this way so that ecologists will no longer be content to pull contrasting examples out of a hat and that there will be a review that will show a preponderance of the evidence going one way or the other.

Chapter Elements

In determining the actual layout of each chapter, I have relied heavily on the use of tables, figures, and photographs to present the evidence. As Connor and Simberloff (1979a) noted, ''You can't falsify ecological hypotheses without data.'' I personally chose each table, figure, and photograph that is used in the book (and had to write all the letters of permission to use them!). Each piece of illustrative material is directly related to a point raised in the text. To maintain the brevity of style, however, the text does not go over the same ground. Each table, figure, or photograph is intended to be self-explanatory. Other more superficial learning tools, such as boxed essays and asides, are absent. Material that could be presented in this way is simply integrated into the text. I have tried to lessen the effects of ecological jargon by providing a glossary. Terms that are boldface, plus some that are not, are defined at the end of the book. Though the book tries to be brief and not burden the student with unnecessary baggage, which they would have to sift through with a highlighter, it is well referenced so that the interested student can always find where to read further on a subject. Despite its size, there are more references in this text than any of its competitors.

Key Content

There is a necessary tradeoff between classical material and recent developments. In ecology, subjects such as conservation, restoration ecology, bioengineering, acid rain, and global warming get much of the press. These are all important issues. But it is worth remembering that predicting the effects of such things is often based on more traditional disciplines. Thus, conservation ecology relies heavily on island biogeography theory. To predict the effects of genetically engineered organisms, many experts have used as analogs the results of releasing exotic species into novel environments. Physiological ecology can best give us the likely answers to questions of how the distribution patterns of species will change in the event of climatic alterations. This book attempts to integrate new concepts with new and older theory.

It has often been said that biology only makes sense in the context of evolution. I begin this book with a treatment of evolutionary ecology, the end point being to discuss the effects of continential drift on the distribution patterns of plants and animals we see today. This is followed by Section Three, Behavioral Ecology, a common theme of which is to examine whether animals behave optimally, so as to maximize their numbers of offspring, or whether their behavior is merely designed to enable them to persist, albeit suboptimally, for as long as possible, in an evolutionary sense. In Section Four, Population Ecology, I discuss the multitude of effects of the environment, competition, predation, herbivory, parasitism, and other factors on the abundance of plants and animals and tie these together by comparing their effects in the final chapter entitled ''The Causes of Population Change.'' Section Five, Community Ecology, completes the study of natural populations. With this background established, we are in a position to discuss the impacts of humans. The last section, Applied Ecology, documents the main effects of people: habitat destruction, exploitation of wildlife for its own sake, pollution, and the introduction of exotic organisms.

Acknowledgments

This book could not have been completed without the help of many friends and associates. I would like to thank Dana Bryan, T. S. Carter, K. R. McKaye, A. Murie, R. H. Reeves, P. M. Room, D. Simberloff, D. A. Sutton, and J. O. Wolff for kindly lending me their own photographs. The following people provided invaluable help in locating photographs: Lavonda Walton, Mae Goff, Robert Hailstock, and Nancy Chedester of the United States Department of Agriculture; Clark Frazier of the Florida Game and Fresh Water Fish Commission; Joan Morris of the Florida Archives; Barbara Mathe of the American Museum of Natural History; Raymond Rye of the Smithsonian Institution; R. W. Paugh of the United States Coast Guard: Giuditta Dolci-Favi of the Food and Agricultural Organization of the United Nations; Tracy Hornbein of the Florida Department of Natural Resources; and Dale Connelly of the United States National Archives. I am also grateful to Caroline Reynolds and Elizabeth Fairley for tracking down some hard-to-find references. All the authors whose tables or figures are reproduced here freely gave permission for their use.

I am indebted to the following reviewers for their useful comments and suggestions: Stanley H. Faeth, Arizona State University; Nicholas J. Gotelli, University of Oklahoma; Robert P. McIntosh, University of Notre Dame; David M. Gordon, University of Massachusetts—Amherst;

Richard Tracy, Colorado State University; Mark A. Hixon, Oregon State University; Thomas H. Kunz, Boston University; William Rowland, Indiana University.

Finally, I should like to offer my sincere appreciation to Anne Thistle for a meticulous job of typing and editing this book and for making me look much better grammatically. Heartfelt thanks to Sharon Strauss, University of Illinois, who had the fortitude to read the entire thing. My editors, Betty O'Bryant at Technical Texts and David Brake at Prentice Hall, provided many helpful suggestions. Students at the University of West Indies, Trinidad, and at the University of South Florida have been quick to point out any inconsistencies. However, I would be pleased to hear about errors in fact or interpretation, omission of material, further examples, or other relevant ideas.

A Contemporary View

The New York Times and Prentice Hall are sponsoring *A Contemporary View,* a program designed to enhance student access to current information of relevance in the classroom.

Through this program, the core subject matter provided in the text is supplemented by a collection of time-sensitive articles from one of the world's most distinguished newspapers, *The New York Times.* These articles demonstrate the vital, ongoing connection between what is learned in the classroom and what is happening in the world around us.

To enjoy the wealth of information of *The New York Times* daily, a reduced subscription rate is available. For information, call toll-free: 1-800-631-1222.

Prentice Hall and *The New York Times* are proud to cosponsor *A Contemporary View.* We hope it will make the reading of both textbooks and newspapers a more dynamic, involving process.

Section One

Introduction

How can we know how much of a natural resource can be harvested without impairing its renewal for generations to come? What measures are necessary to preserve natural beauty for the future? The study of ecology addresses these and similar questions of importance to us all. (Top photo: U.S. Department of Agriculture. Bottom photo: P. Stiling.)

Slobodkin (1988, p. 337) has commented, "there is no dearth of problems for applied ecologists. Red tides litter beaches with dead fish, greenhouse gases are changing the climate, habitats are being destroyed and species are becoming extinct." There is a widespread belief that people of preindustrial civilizations did far less damage to their environment than do their modern industrial counterparts. This belief supposes that hunter-gatherers lived in harmony with nature, practicing a conservation ethic and somehow avoiding short-sighted, destructive exploitation. They did not. For example, on every oceanic island for which we have adequate knowledge, the first arrival of humans was quickly followed by extermination of all or most large animals (Diamond 1986*a*). Easter Island, home of the famous monolithic statues, was once covered with palms, trees, and shrubs. Polynesians reached the island around A.D. 400. By A.D. 1500, 7,000 people lived there; they had deforested the island so completely that its tree species are now extinct. The **deforestation*** had serious implications: No logs were available to be made into canoes, so offshore fishing was curtailed, and the huge statues could not be erected without log levers. Once the population exceeded the carrying capacity of the island, warfare was rampant, as were chronic cannibalism and slavery. Spear points were manufactured in enormous quantities, and people reverted to living in caves for defense.

The scale on which destruction occurs has now shifted from small islands to whole continents, from which no escape is possible. In the deserts of the U.S. Southwest stand huge, empty communal houses or pueblos, relics of the Anasazi, one of the most advanced pre-Columbian civilizations in North America. When construction began, the cliffs were covered with pinyon-juniper woodland. Collection for firewood and construction had denuded the area completely to a radius of 40–70 km by the time the site was abandoned.

---

*Boldfaced terms are defined in the Glossary at the end of the book, where other useful terms are also defined.

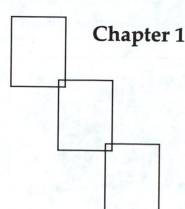

Chapter 1 _____

Why and How to Study Ecology

A side from the concerns expressed over general development, specific projects in themselves can have dire consequences. In 1970, after 11 years and an expenditure of $1 billion, construction of the Aswan High Dam, depicted in Photo 1.1, was completed. It is the largest dam of its kind in the world, located in southern Egypt on the world's longest river, the Nile. It contains more than four times the capacity of Lake Mead, the lake behind Hoover Dam, the largest dam in the United States. The Aswan High Dam was projected to provide several years of irrigation reserve, to add 1.3 million acres to the arable lands of Egypt, and to produce 10 billion kilowatts of electric power annually. Today the dam does produce more than 50 percent of Egypt's electrical power. The reservoir of water saved the rice and cotton crops during the droughts of 1972 and 1973. It has facilitated the cultivation of two or three crops annually rather than one (Azim Abul-Atta 1978), thus increasing productivity by 20 percent for some crops and by 50 percent for others and also increasing governmental and national annual income from agriculture by 200 percent. One million hectares (ha) of additional farm land can be irrigated year round, and 380,000 ha of desert are being irrigated for the first time (Moore 1985).

In many ways, however, the dam has been argued to stand as a monument to ecological ignorance. First, loss of water to seepage through bedrock meant that the dam was still not full by 1988, though it had been predicted to fill by 1970. This problem may have been compounded by 20 years of below-average rainfall in the area (Wright 1988). Until unusually heavy rains fell in August 1988, the volume of water in the impoundment had fallen so low that serious conservation measures were about to be implemented, and the electric turbines were about to stop, which would have caused widespread power cuts. The volume of the reservoir had dropped to around 3 billion m^3, out of a capacity of 110 billion m^3.

Second, the incidence of the tropical flatworm causing schistosomia-

Photo 1.1   The Aswan High Dam, Egypt. Economic boon or ecological nightmare? The dam produces much of Egypt's electrical power and counteracts the crippling effects of drought on agriculture, but it is also thought to reduce silt deposition and, thus, fertility renewal in the valley, to reduce fish catches in the Mediterranean Sea, and to increase the frequency of schistosomiasis, a severe parasitic disease of humans. (WFP photo by J. Van Acker. Courtesy of FAO.)

sis (a debilitating parasitic disease) in the area is said to have increased from 47 to 80 percent because the parasite's secondary hosts, snails, reproduce year-round in the reservoir and thus are no longer reduced by drought (van der Schalie 1972). This problem may since have abated (Moore 1985) and, according to some sources, may never have been as severe as was thought (Walton 1981). The variety of schistosomiasis now prevalent, however, is a much more severe one.

Third, reduced flow into the Mediterranean reduced phytoplankton blooms and fish harvest in the discharge area. Sardine catches alone dropped from 15,000 tons annually to 500 tons (George 1972), and yields from the new fishing areas behind the dam have been low. Reduced silt deposition along the floodplain (Petts 1985) has increased the need for commercial fertilizers to the tune of $100 million annually. The new fertilizer plants use much of the hydroelectric power from the dam.

Many of the adverse opinions of the dam are based on "Western" standards; the Egyptians themselves have chosen to look more positively upon the dam (Fahim 1981). If such costly errors are to be avoided in the future, however, a much deeper ecological knowledge is needed.

1.1 Ecology and Diversity

Ecology comes repeatedly to the economic foreground. During the latter half of the 1980's, the reduction of the Earth's biological diversity emerged as a critical issue and was rightly perceived as a matter of public policy (U.S. Congress 1987). A major concern was that loss of plant and animal resources would impair future development of important products and processes in agriculture, medicine, and industry. For example, *Zea diploperennis*, an ancient wild relative of corn, could be worth billions of dollars to corn growers around the world because of its resistance to seven major diseases plaguing domesticated corn. Two species of wild green tomatoes discovered in an isolated area of the Peruvian highlands in the early 1960's have contributed genes for a marked increase in fruit pigmentation and soluble-solids content currently worth $5 million per year to the tomato-processing industry (U.S. Congress 1987). Loss of tropical forests could mean loss of billions of dollars in potential plant-derived pharmaceutical products. About 25 percent of the prescription drugs in the United States are derived from plants, and as long ago as 1980 their total market value was $8 billion per year. Even in the United States, many habitats are threatened and need "saving." North American wetlands, for example, provide breeding, feeding, and overwintering grounds for millions of economically important waterfowl, and coastal wetlands also provide spawning grounds and nursery habitats for about two-thirds of the major U.S. commercial fish, crustaceans, and molluscs. On a smaller scale, individual species often thought worthless can actually be very valuable for research purposes. Armadillos, for example, are the only known species, other than humans, that can be used in research on leprosy (Photo 1.2). Desert pupfishes, found in the U.S.

Photo 1.2   *The nine-banded armadillo. This is the only animal other than humans to contract leprosy and is, therefore, valuable to medical researchers. (Photo © copyright by Florida Game and Fresh Water Fish Commission, A.V. Department, Neg. No. 4365.)*

Southeast, tolerate salinity twice that of seawater and are valuable models for research on human kidney diseases. Sea-urchin eggs are commonly used in studies of cell structure and fertilization and in tests of the
teratological effects of drugs. The technology does not exist to recreate
ecosystems or even individual species. Once a species or a system is
gone, it is lost forever. Some techniques for maintaining existing ecological habitats and plant and animal species are listed in Tables 1.1 and 1.2.
Prominent among these are the preservation of natural areas. The role of
the U.S. government in passing legislation and providing funds to help
maintain diversity is potentially substantial. As of 1987, 29 federal laws
impinged on the maintenance of biological diversity (U.S. Congress
1987), but gaps still existed in the laws, the national policies, and the data
needed to address these issues. A specific edict regarding diversity and
public policy was lacking.

Warren and Goldsmith (1983) listed arguments for the conservation
of one wood in Yorkshire, England: a source of timber, a good shoot, a
reservoir of natural enemies of pests, a useful field experiment, a good
open-air teaching laboratory, a retreat for renewal from urban areas, a
place to overcome alienation from nature, an important influence on local
property prices, a place to employ youth, a historic landscape, an example of ancient land use, an essential part of the beauty of the area, a place
to paint pictures or to photograph, a part of a romanticized past, a part of
the local heritage, the habitat of rare species, a species-rich ecosystem,
and a source of species being made extinct elsewhere. Perhaps one of the
soundest arguments for conservation is that, at the least, it keeps our options open for the future.

Finally, good arguments can be made against ecological mismanage-

Table 1.1   *Management techniques for maintenance of biological diversity.(After U.S. Congress 1987.)*

Onsite		Offsite	
Ecosystems Maintenance	*Species Management*	*Living Collections*	*Germplasm Storage*
National parks	Agroecosystems	Zoological parks	Seed and pollen banks
Research natural areas	Wildlife refuges	Botanic gardens	Semen, ova, and embryo banks
Marine sanctuaries	In-situ genebanks	Field collections	Microbial culture collections
Resource development plans	Game parks and reserves	Captive breeding programs	Tissue culture collections

Increasing human intervention ⟶

⟵ Increasing emphasis on natural processes

Table 1.2 *Management goals and conservation objectives to be maintained.(After U.S. Congress 1987.)*

| Onsite | | Offsite | |
Ecosystems Maintenance	Species Management	Living Collections	Germplasm Storage
Reservoir or "library" of genetic resources	Genetic interaction between semidomesticated species and wild relatives	Breeding material that cannot be stored in genebanks	Convenient source of germplasm for breeding programs
Evolutionary potential	Wild populations for sustainable exploitation	Field research and development on new varieties and breeds	Collections of germplasm from uncertain or threatened sources
Functioning of various ecological processes	Viable populations of threatened species	Offsite cultivation and propagation	Reference or type collections as standard for research and patenting purposes
Vast majority of known and unknown species	Species that provide important indirect benefits (for pollination or pest control)	Captive breeding stock of populations threatened in the wild	Access to germplasm from wide geographic areas
Representatives of unique natural ecosystems	"Keystone" species with important ecosystem support or regulating functions	Ready access to wild species for research, education, and display	Genetic materials from critically endangered species

ment and loss of biotic diversity on moral and ethical grounds. We simply have no right to destroy species and the environment around us. About 95 million U.S. citizens participate every year in nonconsumptive recreational uses of wildlife (observation, photography, and so on). In addition, 54 million people fish and 19 million hunt. These figures may have more impact when it is considered that $32.4 billion per year are spent on such activities (U.S. Department of the Interior 1982). Recreational hunters in North America pursue some 90 species (Prescott-Allen and Prescott-Allen 1986). In other countries, money from the observation of wildlife can be extremely important to the economy. In 1985, Kenya netted about $300 million from almost 300,000 visitors, making wildlife tourism the country's biggest earner of foreign exchange (Achiron and Wilkinson 1986).

1.2 *Scientific Methods*

Can science in general and the study of ecology in particular help overcome the loss of biotic diversity and all the problems it entails? Yes it can.
Although acid rain, toxic wastes, air and water pollution, and nuclear radiation are often seen by the public as direct results of scientific "progress"—and science is seen not as the "hero" but as the "goat"—one
must bear in mind that these ecological problems are the results of the
misuse of scientific knowledge. Solutions to ecological problems can also
be found through scientific methods.

The scientific method is at the heart of the acquisition of scientific
knowledge. Generally, the first step in the scientific method is sound observation, repeated to determine the frequency of an event and verified
and confirmed by independent observers. After observations, the next
step is construction of a hypothesis to explain the observed events. Good
hypotheses can be tested by further observations or experiments. If a hypothesis stands up to repeated testing, it may reach the status of a theory
or law. Popper (1972*a*) has emphasized the important point that science
progresses not by trying to confirm theories but by attempting to falsify
them. It is usually possible to find at least some confirmatory evidence for
any hypothesis; one piece of negative data, on the other hand, refutes the
hypothesis absolutely. Unfortunately, manuscripts that report the non-
occurrence of something are often distrusted—reviewers apparently regard it as easier to overlook a real phenomenon than to find a spurious
one. As a result, it is often easier to get confirmatory papers accepted
than to publish "negative" results. Mahoney (1977) submitted two sets
of contrived research papers differing only in the results they reported.
He found that papers reporting "negative" results were less likely to be
accepted for publication and less likely to be rated methodologically
sound than were those reporting confirmatory results.

The scientific method, of course, seems rather formal and tedious to
the layperson and sometimes even to the scientist. Roughgarden (1983)
and Quinn and Dunham (1983) argue that sometimes hypotheses can be
developed less formally, by the type of commonsense logic that we use in
everyday life, which usually involves the research for confirmatory evidence. Simberloff (1983) has replied that, although this type of approach
is most seductive, it is likely to be wrong; at the very least, formalization
should help us become more efficient in using our research time. Besides,
Simberloff points out, even perception and the search for confirmatory
evidence itself imply the tacit mental construction of some part of a hypothesis. In practice, scientific advances probably occur by several
routes. Isaac Newton claimed that scientists work from the particular to
the general, first observing phenomena and only later deriving generalizations from them. Popper argued that imagination comes first, then

hypotheses are tested by experimentation. Both avenues have undoubtedly proved valuable in research.

Popperian science has its drawbacks, of course. Popper (1972*b*) has denied that Darwin's theory of evolution is a valid scientific theory; it seems untestable as a whole and is best formulated from many separate lines of evidence (see Section Two). The same arguments apply to the science of astronomy. Creationists have exploited this view in an attempt to refute the entire theory of evolution. For this reason, some philosophers (for example, Suppe 1977) and ecologists (Dunbar 1980) have abandoned Popper's type of approach. Fagerström (1987) has even made the controversial suggestion that it would be better to erect a theory and throw out the data if they are not in agreement. He argues that this approach is not as absurd as it at first sounds. For example, ecological data are not always "hard facts"; they are produced, digested, and interpreted with the aid of theories and apparatus and by people, who are constrained by prejudices and previous experiences. Data are, in actuality, theory-laden. Ecological theories are judged more often by their simplicity, beauty, and intuitive appeal than by strict agreement with data. Even the simplest ecological statement assumes more about nature than can be concluded from observations alone. There can never be a complete match between theories and data no matter how much empirical evidence is gathered. Murray (1986) suggests that, in fact, most ecologists interpret their data by formulating the question, "How can I explain my data in terms of theory X," a procedure that often leads to a rash of "me too" papers in the wake of publication of a new theory and that can slow scientific progress by leading to the uncritical acceptance of any plausible hypothesis. It would be preferable simply to ask, "How can I explain my data?"

Dyson (1988) has suggested that among scientists in all fields there are "diversifiers," who are content to explore the details of phenomena, and "unifiers," who strive to unearth general principles. He believes that biology is the natural realm of diversifiers, much as physics is the realm of unifiers. As a result, biology lacks general themes. It has been further argued that biologists think inductively and that physicists and chemists think deductively. For example, from a series of observations, a physicist might predict the next element in the periodic table. Biologists group facts together, draw conclusions from them, but are loath to make subsequent predictions.

Null Hypotheses

Often, the working hypothesis that is set up in ecological research is that of "no effect"—the hypothesis that observed ecological patterns arise by chance and not through any effect of the forces of nature. This hypothesis of "no effect" is called the *null hypothesis* (Connor and Simberloff 1979b; Strong, Szyska, and Simberloff 1979), and there exists a wide range of statistical tests designed to challenge it (Conover 1980; Fleiss 1981; Zar 1984). Of course, this type of hypothesis is not the only one that is useful in ecology. For example, one might have reason to erect the hypothesis that the disjunct distribution of two species in a particular habitat is due to competition and that they therefore live in mutually exclusive zones. However, the null hypothesis is often the most logical place to start (Simberloff 1983).

With the right type of observation, experimentation, and attempts at falsification, hypotheses can be solidified into ecological theories. J.B.S. Haldane (1963) has observed, rather cynically, that this process is normally reflected in four stages through which the hypothesis passes in the regard of the scientific community:

1. This is worthless nonsense.
2. This is an interesting, but perverse, point of view.
3. This is true but quite unimportant.
4. I have always said so.

Ecological Questions

What types of ecological questions should we be asking? Slobodkin (1986a) has suggested that the "big questions" in ecology and evolution (what is life, how do higher taxa evolve, what determines the number of species in one locality?) may be too large to be amenable to theoretical formulation. He maintains that the most useful approach is to confine ourselves to more specific questions, such as, "How does meat dissolve in the stomach of a kite?" He suggests that in such minimal systems it may be possible to see meaning and derivations more clearly and to examine the criteria for theoretical quality. Colwell (1984), Slobodkin (1986b), Bartholomew (1986), and others have all urged a return to focusing on organisms and real problems rather than on the external criteria of philosophers and mathematicians. Those "pesky biological details" matter a great deal; they often violate the assumptions of mathematical theory, and theorists who ignore them risk wasting their efforts.

There is no doubt that for just about any ecological problem more data are needed. Simon (1986) and other proponents of exploitation of natural resources take advantage of this lack of information, saying, for instance, that the data are insufficient to show that deforestation is causing a great reduction in the numbers of plant and animal species on earth. To take another example, Parker and Douglas-Hamilton have used the same data to discuss elephant conservation and the ivory trade. Parker concludes that a substantial portion of harvested tusks come from natural mortality and that hunting for profit is a serious threat in only a few areas. He further claims that increases in hunting deaths are due to the elephants' competition with Africa's rapidly expanding human population. Douglas-Hamilton disagrees with these contentions on every point. He claims that Parker considerably exaggerates the number of deaths due to natural mortality, that elephants are overhunted in all but the most inaccessible areas, and that the high price of ivory is to blame. When such disparate positions can be reached from identical information, more data are needed. Parker and Douglas-Hamilton's work is discussed in a paper by Pilgram and Western (1986).

1.3 Experiments

As outlined in discussions of the scientific method, ecological hypotheses are best tested by experiments. Experiments can be classified in several ways. Diamond (1986*b*) distinguishes three main types: laboratory experiments, field experiments, and natural experiments. In practice, these types form a continuum. Diamond further divides the natural experiments into two categories: *Natural trajectory experiments* are comparisons of an ecosystem or species before and after some dramatic perturbation like a storm, a volcanic eruption, or the introduction of another species. *Natural snapshot experiments* compare natural areas that differ from one another in only one or two characteristics, for example, presence or absence of certain predators. Such differences have often been maintained throughout recent history. Strengths and weaknesses of these different types of experiments are outlined in Table 1.3. For example, the spatial scale of laboratory experiments is likely to be limited to the size of a constant-temperature laboratory room, around 0.01 hectare, and that of field experiments to usually less than 1 hectare. Natural experiments, however, may be virtually unlimited in scale and often use large islands or continents.

Laboratory experiments can regulate exactly all **abiotic factors,** from light, temperature, and moisture to available nutrients. They are valuable in investigations of these factors. The biotic community represented in a laboratory experiment, however, is likely to be limited at best. Laboratory experiments are best used to study the physiological responses of individual animals rather than the population dynamics of reproducing populations.

Field experiments are conducted outdoors and have the advantage of operating on natural rather than synthetic communities. The most commonly used manipulations include local elimination of a species, local introduction of a species, and erection of a fence or cage. Darwin used a field experiment to demonstrate that either mowing or the introduction of grazing animals increases plant species diversity on a lawn (by pre-

Table 1.3   *The strengths and weaknesses of different types of experiments in ecology.
(After Diamond 1986b.)*

Factor	Laboratory Experiment	Field Experiment	Natural Trajectory Experiment	Natural Snapshot Experiment
Regulation of independent variables	Highest	Medium/low	None	None
Site matching	Highest	Medium	Medium/low	Lowest
Ability to follow trajectory	Yes	Yes	Yes	No
Maximum temporal scale	Lowest	Lowest	Highest	Highest
Maximum spatial scale	Lowest	Low	Highest	Highest
Scope (range of manipulations)	Lowest	Medium/low	Medium/high	Highest
Realism	None/low	High	Highest	Highest
Generality	None	Low	High	High

venting some species from outcompeting others). Field experiments commonly manipulate systems through use of phenomena (like cages or fences) unlikely to be generated by nature itself.

Natural experiments are usually the sole technique for following the trajectory of a perturbation beyond a few decades. Simberloff (1976) was able to examine defaunation and recolonization on mangrove islands for several years, but only on the island of Krakatau has the process been followed in the long term, for over 100 years (see chapter section 8.2). One of the few exceptions is the experimental fertilization of selected experimental plots at Rothamstead Experimental Station, England, for over 100 years, beginning in fact in 1843 (Williams 1978). The weather is frequently shown to be vital in influencing the population densities of many species, but we cannot manipulate the weather. Natural experiments involving drought situations or floods commonly provide the best types of data on this subject. Furthermore, natural experiments often have general implications for the ecological system, because they sample from a wider range of natural variation among sites than do field experiments. In summary, it is apparent that there is no best type of experiment; the choice depends on what one is investigating. This point has ramifications relating to the preceding section on the scientific method—no one "right" type of experiment has inherent superiority over others.

Krebs (1988) has suggested that, of the three types of experiments, field experiments are preferable. He has detailed how laboratory experiments in the 1940's and 1950's failed to provide any useful insights into the population dynamics of rodent populations but that field experiments, begun in the 1950's, have been successfully used to test single- and multifactor hypotheses of population regulation. That experiments must be correctly set up should be obvious. A simple example illustrates the point. Otto Korner (cited by Sparks 1982) performed experiments in the early 1900's to see whether fish could hear. He engaged a well-known opera singer to perform before his aquaria. He watched for the signs of enthusiasm from the fish that would surely result as their piscine hearts were uplifted. None was forthcoming. Korner concluded that fish could not hear. It was left to Von Frisch to show that if fish were "given a reason" to respond to sounds, perhaps by association of sound with food, they would respond.

Experiments should be planned for greatest usefulness. For example, simple removal of one species to examine possible competitive effects on another might document a phenomenon, say the elevation of density of one species, and a hypothesis might then be developed that species *a* and *b* do compete, but no idea of the mechanism involved is provided. In such cases, the ability to predict the outcome of other pairwise or multispecies interactions is limited because no idea of mechanisms has been gleaned. Experiments that take the phenomenological approach and merely document a phenomenon (such as competition) are less useful than those taking mechanistic approaches, which attempt also to explain why the experimental results are obtained (in this case, say, resource competition, allelopathic effects, or effects of shared predators), because the level of prediction obtained in the latter is greater (Tilman 1987).

1.4 *The Effects of Scale*

Just as laboratory experiments, field experiments, and natural experiments are preferred on different scales, it has become obvious that the effects of scale on ecological research and conclusions are staggering (Wiens et al. 1986). What is the "right" spatial scale over which to look for a phenomenon? Ecologists must address this question in planning their experiments.

The size of the study area and the duration of an investigation can limit what one can see of an ecological system (Dayton and Tegner 1984). Wiens et al. (1986) have made the analogy that studying ecology is comparable to what it would be like to study chemistry if the chemist were only a few angstroms long and lived for only a few microseconds. If the chemist were no larger or longer-lived than the molecules and processes under study, the overall course of chemical reactions would be difficult to distinguish from the random collisions of molecules. Wiens et al. emphasize that scale is, of course, a continuum but that five major points on that continuum can be recognized:

Table 1.4   *The effects of scale on ecological investigations.*

Scale	Type of Investigation
Individual space	Physiological ecology, sociobiology, foraging ecology, reproductive biology (special problems arise for migratory species, for which important behavior may occur at another location entirely)
Local patch/ecological neighborhood	Predation, herbivory, parasitism, and pollination
Regional scale	Immigration, emigration, outbreaks, habitat preference
Closed system	Nutrient cycling, ecological energetics
Biogeographical scale	Climatic limits, evolutionary ecology

□ A space occupied by a single individual sessile organism or a space in which a mobile organism spends its entire life
□ A local patch occupied by many individuals
□ A region large enough to include many patches or populations linked by dispersal
□ A space large enough to contain a closed ecosystem (one receiving no migration)—an unlikely scenario in practice
□ A **biogeographic** scale, large enough to encompass different habitats and climates

Investigations on these different scales yield answers to different types of questions, and it is important to realize this point (see Table 1.4). Addicott et al. (1987) discuss this matter further and suggest that the correct scale depends exactly on the question being asked, so there will never be a single ecological neighborhood for a given organism, but rather a number of neighborhoods, each appropriate to a different process.

Just as with spatial scale, the appropriate choice of temporal scale depends on the phenomenon and the species to be studied. One would choose a relatively short time over which to study behavioral responses, a longer one for population dynamics, and a still longer scale for studies of genetic change and evolution (Wiens et al. 1986).

Section Two ——————

Evolutionary Ecology

What caused the extinction of some species millenia ago? Why did some species, such as the horseshoe crab, continue almost unchanged? The study of evolutionary processes seeks to answer these questions and to elucidate the patterns and processes involved. (Top photo: U.S. Department of Agriculture. Bottom photo: P. Stiling.)

I f ecology is concerned with explaining the distribution and abundance of plants and animals and with control of their numbers, then evolutionary ecology is an important part of the discipline. For example, one may argue about what controls penguin numbers in the Southern Hemisphere, but a nagging question remains—Why are there no penguins in the Northern Hemisphere? The answer is not insufficient food or too many predators. Penguins simply evolved in the Southern Hemisphere and have never been able to cross the tropics to colonize northern waters. A knowledge of evolution and historical ecology is clearly of paramount importance to contemporary ecology.

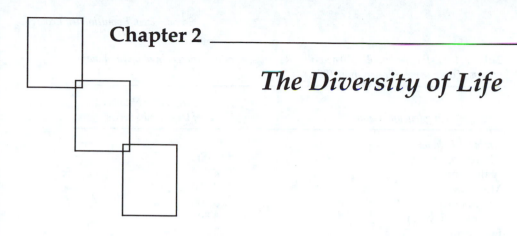

Chapter 2

The Diversity of Life

As Grant (1977, p. 3), said, "The world of living organisms exhibits several general features that have always aroused feelings of wonder in mankind." First, there is a tremendous diversity of life forms—over 3,700 species of mammals, 8,600 species of birds, and about 20,000 species of fish, which together with other organisms give a total of 42,000 known vertebrates. Diversity of other groups, especially the arthropods and molluscs, is even more staggering (Table 2.1). It has been tentatively estimated that at one point or another during the history of life there have been in the neighborhood of 1 billion species.

Second, among these organisms there is a great structural complexity and an apparently purposeful adaptation of many characteristics to the environment. Darwin's own example was that of the woodpecker, with its chisel bill, strengthened head bones and neck muscles, extensile tongue with barbed tip, and stout tail for balance while chiseling. Another classic example is provided by the variety of feet in birds, from the perching feet of warblers, to the grasping feet of eagles, the walking feet of quail, the wading feet of herons, the swimming feet of ducks, and of course the specialized feet of woodpeckers (see Fig. 2.1). How can we explain this diversity?

2.1 Historical Background

Lamarck

A number of naturalists and philosophers, beginning with the ancient Greeks, had supposed that many forms of life evolved from each other. The first actually to formalize and publish a theory of **evolution**—"transformism" as he called it—was Jean Baptiste Lamarck (1744–1829), though similar ideas had been thought of by others, including the Comte de Buffon (1707–1788) (see Greene 1959). (See Photo 2.1.) Whereas Buffon conceived variation as a process of degeneration or random deviation

Table 2.1 *Estimated numbers of described Recent species in various kingdoms and major groups of organisms. (From Grant 1985.)*

Kingdom and Group	Approximate Number of Described Species	
Animal Kingdom		
Chordates	43,000	
Arthropods	838,000	
Molluscs	107,250	
Echinoderms	6,000	
Segmented worms	8,500	
Flatworms	12,700	
Nematodes and relatives	12,500	
Coelenterates	5,300	
Bryozoans and relatives	3,750	
Sponges	4,800	
Miscellaneous small groups	2,100	
Total		1,043,900
Plant Kingdom		
Flowering plants	286,000	
Gymnosperms	640	
Ferns and fern allies	10,000	
Bryophytes	23,000	
Green algae	5,280	
Brown and red algae	3,400	
Total		328,320
Fungus Kingdom		
True fungi	40,000	
Slime molds	400	
Total		40,400
Protistan Kingdom		
Protozoans, plant flagellates, diatoms	30,000	30,000
Moneran Kingdom (prokaryotes)		
Blue-green algae	1,400	
Bacteria	1,630	
Total		3,030
Viruses	200	200
Grand Total		1,445,850

Note that Erwin (1982, 1983) has estimated the existence of about 30 million extant species.

from innumerable ancestral forms, Lamarck viewed it as evolution from simple beginnings. Lamarck's (1809) work, however, was largely ignored because the mechanism he proposed to explain how evolution works was based on the inheritance of **acquired characteristics.** For example, Lamarck supposed that giraffes, in their continual struggle to reach the

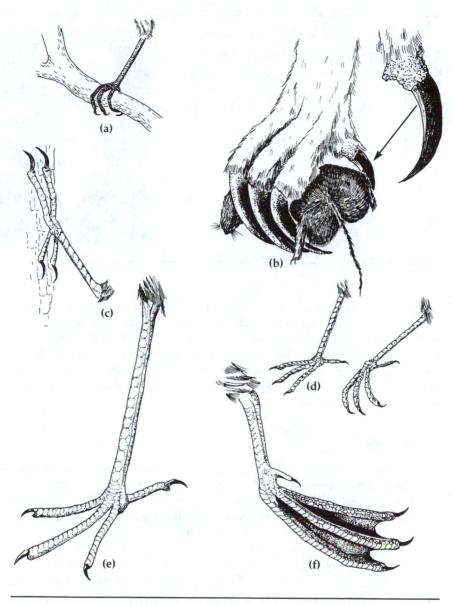

Figure 2.1 *Different adaptive types of bird feet.* **(a)** *Perching foot (Audubon warbler,* Dendroica audubon*).* **(b)** *Grasping foot with strong talons (horned owl,* Bubo virginianus*).* **(c)** *Climbing foot (acorn woodpecker,* Melanerpes formicivorus*).* **(d)** *Walking and scratching foot (California quail,* Lophortyx californicus*).* (e) *Wading foot (green heron,* Butorides virescens*).* **(f)** *Swimming foot (pintail duck,* Anas acuta*)* Drawings not to same scale. (Redrawn from Grant 1963.)

Photo 2.1   Jean Baptiste Lamarck, 1744–1829. (Photo Neg. No. 124768. Courtesy Department of Library Services, American Museum of Natural History.)

highest leaves on trees, stretched their necks by a few millimeters in the course of their lifetimes. This increase in neck length was passed on to their offspring, who continued the process until the necks of giraffes reached their current proportions. In a similar way, Lamarck explained racial variation in skin color by assuming that the suntan developed by ancestral races was transmitted to their descendants, who were in turn a little darker than their parents.

Darwin

Lamarck's theories were replaced by the ideas of Charles Robert Darwin (1809–1882), the founder of modern evolutionary theory (Photo 2.2). Darwin was born into a prosperous background. (His birthday, February 12, 1809, was the same as that of Abraham Lincoln.) His father was a successful doctor, and his mother and wife were both Wedgwoods, of china fame. His family's fortune enabled Darwin, educated at Edinburgh and Cambridge universities, to accept an unpaid job as scientific observer on board H.M.S. *Beagle,* which sailed on a five-year world survey from 1831 to 1836, concentrating on South America. In some ways, Darwin was "primed" to accept the theory of gradual biological change and evolution because he had with him a copy of *Principles of Geology (1830),* newly published by Charles Lyell, who had taken the unprecedented step of describing the physical world as one that changed gradually through physical processes, not through a few catastrophic events (the view held firmly up until that time by the clergy and the masses).

To be accurate, Darwin was strictly the captain's guest, a position of some social standing, which was intended to alleviate the boredom of the

Photo 2.2   Charles Robert Darwin, age 51, 1860. (Photo Neg. No. 326668. Courtesy Department of Library Services, American Museum of Natural History.)

captain during the five-year voyage (in those days captains were ''forbidden'' to talk to men of lower rank). The ship already had an official surgeon-naturalist, named McKormick, but he left the ship at Rio de Janeiro in 1832. The previous captain of the *Beagle* had shot himself in the third year of the voyage, and the family of Captain Fitzroy, his successor, had some history of mental illness (his uncle, Viscount Castlereagh, suppressor of the Irish rebellion of 1798 and foreign secretary during the defeat of Napoleon, had killed himself in 1822). Fitzroy knew he could certainly use some company to help retain his sanity. (He himself broke down temporarily, while on board the *Beagle,* and later he too committed suicide, feeling guilty that Darwin's heretic theory was originated on board his ship.)

During the voyage of the *Beagle* up and down both coasts of South America, Darwin was able to view diverse tropical communities, some of the richest fossil beds in the world in Patagonia, and the Galapagos islands, 600 miles west of Ecuador. The Galapagos contain a fauna different from that of mainland South America, exhibiting tortoises and other animals different in form on virtually every island. By the time he had finished the expedition, Darwin had amassed a wealth of data, described an astonishing array of animals, and built up a vast collection of specimens.

A year after his return, Darwin read a revolutionary book on human population growth by the English clergyman Thomas Malthus, written (anonymously) in 1798, which pointed out that populations had the capacity for geometric growth (that is, they could grow as rapidly as the series 1, 2, 4, 8, 16), yet food supply increased only arithmetically (that is, like the series 1, 2, 3, 4). Malthus proposed that—because the earth was not overrun by humans as it should be—disease, war, famine, or conscious control must limit population growth. Darwin quickly established that the **Malthusian theory of population** would apply to animal populations. He made the logical deduction that these factors would act to the detriment of weaker, less well-adapted individuals and that only the strongest would survive.

Darwin had formulated his theory of **natural selection,** survival of the fittest: a better-adapted plant or animal would leave more offspring. In this way, giraffes born, by genetic chance, with longer necks would be better fed and able to reproduce more successfully. This trait, because it was genetically determined, would be passed on to their offspring, and long necks would become common. Only rarely would such long-necked mutants evolve independently; much more commonly, distinct traits, such as neck length, are inherited unchanged.

Incredibly, Darwin did not immediately publish his theory. He waited for nearly 20 years, collecting data on a wide range of organisms. Eventually, he was pushed into publication (Darwin 1859) by the arrival of a manuscript by Wallace, who had independently and years later arrived at the same conclusions.

Wallace

Alfred Russel Wallace (1823–1913) was born into poverty and farmed out to an older brother in London at age 14, after only six years of schooling (Photo 2.3). He held down a succession of jobs until his brother died and left him some money. Wallace set sail immediately for the Amazon. A year later, a second brother joined him. Both men contracted yellow

Photo 2.3   Alfred Russell Wallace, age 46, 1869. (Photo Neg. No. 326812. Courtesy Department of Library Sciences, American Museum of Natural History.)

fever, and the brother died. After four years in the jungle, Wallace sailed for home with his precious collections. En route the ship caught fire and sank. After 10 days in the open sea in a small boat, Wallace was saved, but four years of labor went down with the ship. Back in England, Wallace began to prepare for a second voyage, this time to the Malay archipelago as a professional collector and naturalist in the company of W.H. Bates (for whom Batesian mimicry is named). It was there, during another bout of fever, that Wallace conceived the idea of natural selection. Wallace's one major advantage over Darwin was that he was persuaded before he left on his voyages that species evolve; Darwin did not abandon his creationist beliefs until after his return. Thus Wallace was able to gather data with an eye to his evolutionary hypothesis.

Unfortunately, Wallace's earlier papers had been ignored by the scientific community, and he was faced with the problem of lack of recognition. His solution was to send his manuscripts to Darwin, with whom he had previously corresponded. Darwin immediately sought the advice of friends (geologist Lyell and botanist Hooker) and, as a result of their suggestions, Darwin and Wallace presented their theories jointly at a historic meeting of the Linnean Society of London on July 1, 1858. One year later, Darwin at last published his *Origin of Species*, an abbreviated version of the manuscript based on his 20 years of work.

Although Wallace deserves full credit as a codiscoverer of the chief mechanism of evolution, Darwin garnered, and continues to garner, most of the credit because his subsequent books, although on a variety of subjects, all seemed to explore the ideas and principles inherent in the original work. Furthermore, Darwin has been credited as a pioneer of the hypothetico-deductive method, that is, the testing of hypotheses through use of observations to confirm deductions from them. In Darwin's day, most science was done by induction—conclusions were drawn from an accumulation of individual observations. Darwin's main conclusions about the origin of species were two.

First, all organisms are descended with modification from common ancestors. All the prominent scientists of the day were convinced of this point within 20 years, although the religious community of course was skeptical and still is. (For those wishing to read about the fascinating and ongoing creationism-evolution debate, an excellent entry into the literature and the rhetoric is provided by Numbers [1982], Godfrey [1983], Wilson [1983], and the National Academy of Sciences [1984].)

Second, the mechanism for evolution was natural selection. This conclusion convinced few people at the time and was not fully accepted until the late 1920's partly because of a widespread belief in blending inheritance, in which the traits of the parents were thought to be blended in the offspring, like the colors of two paints blending to produce an intermediate color. Natural selection would not work in such a system. For example, if a long-necked giraffe mated with a short-necked giraffe, the offspring would have a neck of medium length, and the advantage of a long neck would be lost. Furthermore, the belief that environmentally induced variation could be inherited was widespread and provided an alternative to natural selection.

Darwin's theory of natural selection had one serious flaw. He knew nothing of the causes of hereditary variation and could not well answer questions on that subject. The evidence of genetics and Mendel's laws of heredity were available but had passed into obscurity and were only resurrected in the early part of the 20th century.

Mendel

Gregor Johann Mendel (1822–1884) was an Austrian who entered monastic service at Brno in 1843. As the monastery was a center of learning, Mendel was sent to Vienna University and later taught at Brno Technical School during 1856–1864, where he performed his revolutionary work on peas. He became abbot in 1868 but still carried on his hybridization work on plants, although he never published the results.

Mendel crossed tall and dwarf strains of peas and found in the second generation a 3:1 ratio of tall to dwarf plants. He could therefore conclude that the parents differed with respect to a single gene (known today to be a single unit of DNA) that controlled size. The gene for height in pea plants existed in two different forms (or **alleles**), tall and short. We know today that these alleles can always be found at the same point on a chromosome, the locus of that gene. During meiosis, DNA is replicated (doubled in quantity), and the resulting two sets of chromosomes recognize each other and become precisely aligned on the meiotic spindle. During the first meiotic division, homologous chromosomes are drawn to opposite poles. Mendel's experiments formed the basis of a genetic understanding of the acquisition of inherited traits, although much controversy was still to exist for many years because his simple experiments did not work for all species. For example, the equivalent cross of tall and short humans does not yield a neat 3:1 ratio in the second generation because in more complex species features like height are often controlled by many genes, each of which alone has only a slight effect on size. Each of the individual genes controlling these so-called polygenic traits follows the principles Mendel outlined, however. Genes that contribute to a single character or to functionally related characters often show strong linkage or linkage disequilibrium; that is, they are located close to one another on the same chromosome and are often inherited together. Such clusters of genes are often called **supergenes.** There are only a few cases in which parental experience can influence offspring characteristics. These are usually due to the maternal origin of the cytoplasm of the egg (the sperm contributes almost nothing except DNA) and involve self-replicating bodies like mitochondria and chloroplasts.

Mendel's work showed that inheritance is generally particulate, with the exception of genes on sex-linked chromosomes. He showed that inherited factors did not contaminate one another and could be passed down from ancient ancestors in the same form. Furthermore, genes appeared exceptionally stable. As Weismann (1893) was to show, germ plasm is entirely separate from, and immune to any influences from, the soma (the rest of the body), so environment has no influence on heredity.

Yet some variation must occur in populations if selection is to work at all, and it is readily obvious that not all members of the same species are identical. This point is addressed in Section Three, but it is worth reconsidering here the many apparently purposeful adaptations discussed at the beginning of this section.

It is important to realize that the development of adaptations does not always proceed via natural selection for those adaptations. Feathers, for example, may have evolved as insulation for warm-blooded birds and only secondarily become useful in flight. Sometimes chance alone may influence evolution. Perhaps more commonly, the developmental biology and other morphological constraints may impose selection by coupling one trait with another. Because genes exert their effects on the **phenotype** through biochemical reactions, other reactants must be incorporated into the equation. A phenotypic effect at one locus often depends on the genotype at one or more other loci: this is the phenomenon of **epistasis.** Finally, some features may actually be neutral with respect to "fitness." For example, blue eyes and brown eyes confer equal visual ability on the majority of humans; they are merely different morphological solutions to the same problem.

2.2 *How Variation Originates*

It is genetic variation that can produce either an increase or a decrease in the variability of a population. In all organisms (except RNA viruses), the genetic material is DNA, deoxyribonucleic acid. In prokaryotes the DNA exists in the form of a single circular chromosome; in eukaryotes it is arranged into a set of linear chromosomes that reside in the cell nucleus. (The exact structure of the famous double-stranded DNA molecule is given in most introductory biology texts and should be familiar to biology students. Further detail will not be given here.) When the DNA code is copied for delivery into the gametes, mistakes are possible; these mistakes are the source of much genetic variation.

Increases in the amount of genetic variability present in the gene pool arise chiefly from **mutations** during copying, of which there are two kinds: gene or point mutations (the most important in enriching the gene pool) and chromosome mutation (the most important in rearranging it). Most mutations result in a loss of fitness; the effects of mutations are at random with respect to adaptiveness, and the chances are small that a random change to an already well-adapted system would result in an improvement.

Point Mutations

A point mutation results from a misprint in DNA copying (Fig. 2.2). Most point mutations are thought to involve changes in the nucleotide bases that make up the DNA base pairs (adenine, thymine, guanine, and cytosine) at single locations only (cistrons), for example from adenine to guanine or from adenine to thymine. Two such changes, which result in the change of the sequence GAA to CUA, combine to substitute the amino acid valine for glutamic acid, the change that causes the abnormal beta chains in sickle-cell hemoglobin. Because of the redundancy of the genetic code, about 24 percent of these codon (base triplet) substitutions do not change amino-acid sequences and thus do not alter phenotype. The genetic code is universal among prokaryotes and eukaryotes (Jukes 1983); the same triplets code for the same amino acids in all organisms.

More drastic changes in amino-acid sequences are caused by frameshift mutations. An addition to (or deletion from) the amino-acid triplet sequence alters the whole reading frame, leading to drastic and

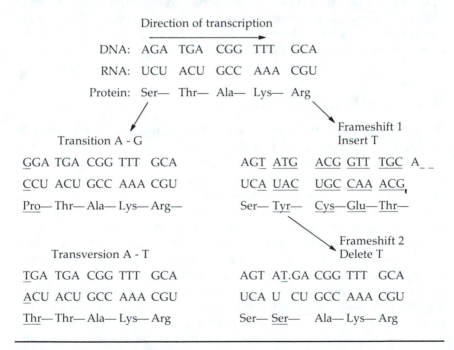

Figure 2.2 *Types of point mutation.*

often fatal mutations. A second such change at another site may rectify the original reading frame so that only a short nucleotide sequence is then read in altered triplets. Genes themselves consist of many long sequences of base pairs. A gene for collagen in chickens is 40,000 base pairs (bp) (40 kilobases, kb) long. The size of the complete genome (all the genes) varies greatly among organisms from less than 400 bp in some virus-like particles to more than 10^{11} bp in some vascular plants. It has been estimated that the average mammalian genome is sufficient to code for more than 300,000 genes, that of *Drosophila*, 100,000. In *Drosophila*, however, only about 10,000 genes are recognized. Much DNA therefore codes for repetitive sequences or codes for nothing; it is essentially junk DNA. In prokaryotes, there are many single copies of each DNA sequence.

Although mutations may be accelerated by man-made radiation, UV light, or substances such as colchicine, in nature such mutagens (usually in the form of weak cosmic rays) are too rare to produce many mutations. Nevertheless, it has been conservatively estimated that mutations occur in nature at the rate of one per gene locus in every 100,000 sex cells (Dobzhansky 1970; Neel 1983). Higher organisms contain at least 10,000 gene loci, so 1 out of 10 individuals carries a newly mutated gene at one of its loci. Of course most mutations are deleterious—they arise by chance, and the genes cannot know how and when it is good for them to mutate. Only 1 out of 1,000 mutations may be beneficial, and thus 1 in 10,000 individuals carries a useful mutation per generation. In actuality, every individual *Drosophila* fly, corn plant, and human being seems to carry at least one abnormal, mutant allele (Spencer 1957; Crumpacker 1967; Morton, Crow, and Muller 1956, respectively), some originating in that organism and some inherited from ancestors.

If we estimate 100 million individuals per generation and 50,000 generations for the evolutionary life of a species, then 500 million useful mutations would be expected to occur during this span. It has been estimated that only 500 mutations may be necessary to transform one species into another, so only 1 in 1 million of the useful mutations needs to be established in a population in order to provide the genetic basis of observed rates of evolution. The chief factor limiting the supply of variability is therefore not the rate of new mutations. In fact, variability is limited mainly by gene recombination and the structural patterns of chromosomes. Even more astonishing is that some scientists have argued, with good evidence, that pure genetic drift is probably sufficient to cause directional genetic change in species with effective population sizes of under 5 million (Charlesworth 1984), which is probably greater than the population size of many mammals. Lande (1976) showed mathematically how even very weak selection on horses' teeth, two selective deaths out of 1 million individuals per generation, would be sufficient to explain the dramatic evolution of horse teeth through the ages.

Chromosome Mutations

Chromosomal mutations do not actually add to or subtract from gene-pool variability; they merely rearrange it. Followed by natural selection, chromosomal mutations make certain adaptive gene combinations easier for the population to maintain.

Chromosomes can undergo four types of changes (deletions, duplications, inversions, and translocations) in which the order of base pairs within the gene is unaffected but the order of genes on the chromosome is altered (Fig. 2.3).

A deletion is the simple loss of part of a chromosome. A deletion is usually lethal unless, as in some higher organisms, there are many duplicated genes.

When two chromosomes are not perfectly aligned during crossing over, the result is one chromosome with a deficiency and one with a duplication of genes. The change may be advantageous in that greater amounts of enzymes may be coded for. In yeasts, for example, an increase of acid monophosphatase enables cells to exploit more efficiently low concentrations of phosphate in the medium.

An inversion occurs when a chromosome breaks in two places and then refuses with the segment between the breaks turned around so that the order of its genes is reversed with respect to that on the unbroken chromosome. Such breaks probably occur at prophase, during which the chromosomes are long and slender and often bent into loops. In translocations, two nonhomologous chromosomes break simultaneously and exchange segments.

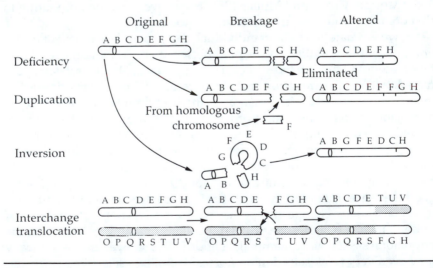

Figure 2.3 *Chromosome breakage and reunion can give rise to four principal structural changes.*

Together, gene and chromosomal mutation provide most of the genetic variability in a population. However, some other mechanisms for promoting genetic diversity do exist and will be mentioned briefly. If the first meiotic division fails to occur, unreduced **diploid** ($2n$) gametes are formed. Normally the union of a diploid gamete with a normal **haploid** (n) gamete forms a triploid ($3n$) zygote. Such individuals, if they can exist at all, are usually sterile; they cannot produce gametes with effectively balanced complements of chromosomes. However, if two unreduced ($2n$) gametes unite, the resulting tetraploid ($4n$) zygote may be fertile. Such a phenomenon can occur in some plants where, because of a doubled gene dosage, these polyploids are often bigger and more robust than diploids (Levin 1983). Tetraploidy is also found, rarely, in animals that are **parthenogenetic** (those in which females can produce daughters without fertilization by male gametes), like brine shrimp, *Artemia salina*, in which tetraploids, pentaploids, octoploids, and even decaploids have been found.

The existence of all this variation was something of a shock in the 1920's and 1930's, before which genetic uniformity had been taken for granted, and it led to a new ground swell of opinion, led by Theodosius Dobzhansky, that populations were in fact huge collections of diverse genotypes (Futuyma 1986). How much change in the amount of DNA, in the genome size, or in chromosome structure is necessary to cause speciation? There is no simple correlation. Two species of deer in the same genus, *Muntiacus reevesi* and *M. muntjac*, even have vastly different complements of chromosomes; $2n = 46$ and $2n = 6$, respectively (White 1978)!

2.3 How Variation Is Maintained

How Variation Is Maintained Without Selection

Although variation is created by mutation and chromosomal rearrangements, given simple Mendelian genetics and dominance, how is it that a dominant allele, responsible for a 3:1 numerical ratio of phenotypes, does not gradually supplant all other types of alleles, given that all alleles confer equal fitness on the organism? That is, if a gene pool in one generation consists of 70 percent A alleles and 30 percent a alleles, what stops the proportion of A alleles from increasing dramatically? What will the proportion of alleles be in the next generation? This very question was posed and independently answered by G.H. Hardy and W. Weinberg in 1908. They assumed three things: (1) populations are large, which thus negates sampling errors; (2) individuals contribute equal numbers of gametes; and (3) mating is random. In short, gametes carrying different alleles combine in pairs in proportion to their respective frequencies in the gamete pool.

The Hardy-Weinberg Theorem

Assume that a diploid (normal) population polymorphic for A contains the following proportions of genotypes: 60 percent AA, 20 percent Aa, and 20 percent aa. The genotype frequencies are 0.60 AA + 0.20 Aa + 0.20 aa = 1. The allele frequencies (q) must then be

$$qA = \frac{0.60 + 0.60 + 0.20}{2} = 0.70$$

$$qa = \frac{0.20 + 0.20 + 0.20}{2} = 0.30$$

The paired combinations of matings are then

Female Gametes		Male Gametes		Zygotic Frequencies	
0.70 A	×	0.70 A	=	0.49 AA	
0.70 A	×	0.30a	=	0.21 Aa	
0.30 a	×	0.70 A	=	0.21 Aa	= 0.42 Aa
0.30 a	×	0.30 a	=	0.09 aa	

The gene pool of the second generation now contains the two alleles in the following frequencies:

$$qA = \frac{0.49 + 0.49 + 0.42}{2} = 0.70$$

$$qa = \frac{0.42 + 0.09 + 0.09}{2} = 0.30$$

The allele frequencies are the same as they were in the first generation, which is why the proportion of alleles is unchanging.

In general if p is the frequency of allele A and q the frequency of allele a (q = 1 − p), then the combinations of A and a gametes will produce zygotes in proportions given by the expansion of the binomial square $(p + q)^2 = p^2 + 2pq + q^2$, that is, p^2 AA, $2pq$ Aa, and q^2 aa. This generalization is known as the *Hardy-Weinberg theorem*. For three alleles whose frequencies are p, q, and r, the equilibrium frequency of genotypes is given by the trinomial square $(p + q + r)^2$. In general, for an n-ploid organism, the genotype frequencies are given by $(p + q)^n$.

The Effects of Nonrandom Mating

Of course, genotype frequencies are not always constant over time. Species change and evolve as a result of deviations from the Hardy-

Weinberg assumptions. For example, not all mating is random or **assortative.** Bateson, Lotwick, and Scott (1980) have shown that individual Bewick's swans (*Cygnus columbianus*) can distinguish other individuals by their face markings. Swan families tend to have characteristic family markings, and young birds prefer to mate with partners whose faces are clearly different from their own. This preference reduces inbreeding, which can dramatically increase the proportions of genotypes in favor of homozygotes (Grant 1977). Furthermore, sexual selection (see chapter section 7.2) in many organisms ensures that individuals do not always contribute equal numbers of gametes. In territorial species some males contribute more genes than others; in those with harem-holding males, just one individual will fertilize the majority of the females. New genes can also often enter the population via the migration of new individuals from different areas. The result is that there are very few truly **panmictic** species, in which all the individuals from one species form a single randomly mating population. One example of such a species is the common eel, *Anguilla rostrata*, in which individuals from U.S. eastern-seaboard drainages and Europe as well migrate to one area of the ocean near Bermuda to breed.

Environmental Variance

In addition to disruption of the Hardy-Weinberg assumptions by nonrandom mating, each genotype may be phenotypically variable to some extent because its development is directly affected by the environment. Fly weight depends often on the amount of medium available to the larvae and whether or not they have been competing for it. This dependence is termed environmental variance, V_E, and affects variance of phenotypes in nature, the Hardy-Weinberg theorem notwithstanding. It is in contrast to the genetic variance V_G. Total phenotypic variance V_P would thus be

$$V_P = V_G + V_E$$

Very often, however, the reaction to a difference in environment differs among genotypes, so there is a genotype-by-environment interaction (V_{G+E}), which also contributes to the phenotypic variance. In other words, certain genotypes do better in some habitats, and other genotypes do better in others. Thus

$$V_P = V_G + V_E + V_{G+E}$$

In this case genetic variance is hard to distinguish from environmental variance because each depends on the other (Gupta and Lewontin 1982).

The heritability of a trait is the fraction of the phenotypic variance that is attributable to genetic variation and is denoted by h^2. Thus

$$h^2 = \frac{V_G}{V_P}$$

In most populations, it is hard to know whether a gene exists for a particular trait, and it is hard to study its properties. The reason is that alternative alleles at that locus are hard to identify; studies of variation have therefore focused on distinct polymorphisms such as red eye versus white eye in *Drosophila*. In nature, however, species are seldom found with two or more discrete phenotypes that lend themselves well to Mendelian crosses. The reason and studies of them are the subject of Chapter 3. In many of these cases, the two phenotypes or **morphs** were originally described as distinct species. The differential survival of these morphs in nature and how sources of mortality act differentially upon them fall into the realm of natural selection.

How Variation Is Maintained with Selection

When selection acts at a locus and a homozygous genotype is most fit, the less advantageous alleles should be eliminated. This is the concept of natural selection (which will be developed further in Chapter 3). This situation contrasts with the Hardy-Weinberg theorem, which showed how two different alleles are maintained in the same population, generation after generation, if one is not favored over the other. Natural selection is obviously a common and potent force in the real world. How then does variation continue to exist in nature? Much of the effort of population geneticists has been devoted to this question. There are three common answers:

1. Selection acts on the locus so as to maintain a stable polymorphism, in which different genotypes are most fit under different situations; the "wild type" allele is not always most fit (see Chapter 3).
2. Fixation by selection is counteracted by mutation (for example, mutation may explain the reoccurrence of albinos in populations from which they are eliminated by natural selection).
3. Fixation by natural selection is counteracted by **gene flow.**

2.4 *How Much Variation Exists in Nature*

Given the existence of so many mechanisms by which variation is produced, it is interesting to speculate on how much genetic variation exists in nature. For a long time, such speculation was difficult because it was impossible to tell how many loci exhibited no genetic variation—no clues were given by the phenotype. In the 1960's it was realized that most loci actually code for proteins, especially enzymes. Different forms of, say, alcohol dehydrogenase are coded for by different alleles. Two individuals with the same form of enzyme are presumably genetically identical at that locus. Therefore, invariant gene loci can be found through a search for invariant enzymes. The most common technique for distinguishing different genetic forms of enzymes (allozymes) is gel electrophoresis. In this technique, specially prepared samples of specimen tissue are placed in a porous gel, and an electrical potential is applied, causing the electrically charged enzymes to migrate through the gel along the lines of electrical force. Because slightly different forms of the same amino acid differ in charge, they migrate at different rates and separate into bands at different distances from the original specimen. These bands become visible when the gel is first flooded with a substrate on which the enzyme acts and then stained with a substance that colors the reaction products. Such enzyme differences have proved to be of genetic origin, and experienced workers have identified much genetic variation among individuals in populations whose phenotypes looked identical. For example, as Photo 2.4 shows, morphologically identical strains of bacteria have proved to be distinctly different when examined electrophoretically.

The first assays of genetic variation by protein electrophoresis were published in 1966 by Harris, who reported on variation in 10 enzyme loci in humans, and by Lewontin and Hubby (1966), working on *Drosophila*. The results were something of a surprise; no one had suspected just how much variation existed in nature. In *Drosophila* the average population was polymorphic at about 30 percent of its loci, with about two-to-six alleles per polymorphic locus. An average fly was likely to be heterozygous at 12 percent of its loci, so any two flies would, on average, differ at about 25 percent of their loci. In humans, Harris' data indicated 30 percent of the loci were polymorphic and about 10 percent were heterozygous. Moreover, it has since been shown that the enzyme products of some alleles have similar electrophoretic mobility, and variation would not be distinguished by the gel technique (Coyne 1976; Coyne, Felton, and Lewontin 1978). Therefore, existing estimates of variation in nature could err on the low side.

Genetic variation for a wide variety of taxa is summarized in Table 2.2. In general, vertebrates are less polymorphic than invertebrates; species that form small local populations or are inbred have reduced heterozygosity. It must always be noted, however, that some enzymes are not soluble and cannot be examined electrophoretically; whether the same degree of polymorphism exists in these is not known.

Very often, allele frequencies differ at loci from one population to another, so the variation that arises within populations becomes transformed into variation among populations. In plants, morphologically or

Photo 2.4   Genetic variation within a species. The upper photo is DNA from 19 different cultures of Xanthomonas maltophilia taken from a Tallahassee, Florida, hospital in 1989. The lower photo is the same DNA, subjected to an additional treatment. All the cultures in each photo were treated in the same way (according to a method that produces a DNA "fingerprint"). Despite the superficial similarity of the cultures, their DNAs are clearly different. The additional techniques used in the lower photo show that even some of those that appear identical in the upper photo (for example, 10, 25, and 32) are unique. (Photo courtesy of R.H. Reeves, Florida State University, Tallahassee.)

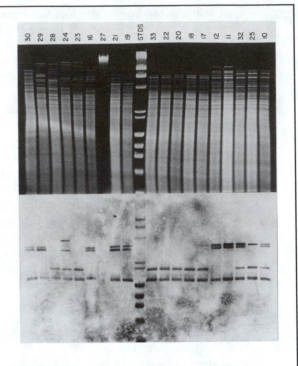

physiologically different forms, called **ecotypes,** are often found in a mosaic pattern in association with different microhabitats. A famous case was studied by Antonovics, Bradshaw, and Turner (1971), who showed that plants that occupied metal-impregnated soils on mine tailings were genetically distinct from plants of the same species on nontoxic soils only a few feet away.

Variation among populations exists on much larger geographic scales as well. Commonly, the farther apart populations are, the more different they are in allele frequencies and in phenotype characteristics. A gradual change along a geographic transect is called a **cline.** Clines may extend over the whole geographic range of a species. For example, body size in white-tailed deer (*Odocoileus virginianus*) increases gradually with increasing latitude over most of North America—this phenomenon is common in mammals and birds and has been termed **Bergmann's rule.** In the clover *Trifolium repens*, the proportion of plants that produce cyanide increases in warmer locations in Europe (Fig. 2.4). This difference in frequency is caused by a balance between the advantage cyanogenic plants

Table 2.2 *Genetic variation at allozyme loci in animals and plants.(After Selander 1976.)*

	Number of Species Examined	Average Number of Loci per Species	Average Proportion of Loci	
			Polymorphic per Population	Heterozygous per Individual
Insects				
Drosophila	28	24	0.529	0.150
Others	4	18	0.531	0.151
Haplodiploid wasps[a]	6	15	0.243	0.062
Marine invertebrates	9	26	0.587	0.147
Marine snails	5	17	0.175	0.083
Land snails	5	18	0.437	0.150
Fish	14	21	0.306	0.078
Amphibians	11	22	0.336	0.082
Reptiles	9	21	0.231	0.047
Birds	4	19	0.145	0.042
Rodents	26	26	0.202	0.054
Large mammals[b]	4	40	0.233	0.037
Plants[c]	8	8	0.464	0.170

[a]Females are diploid, males haploid.

[b]Human, chimpanzee, pigtailed macaque, and southern elephant seal.

[c]Predominantly outcrossing species.

derive from being protected from herbivory and the disadvantage of being killed when frost disrupts the cell membrane, releasing toxins into the plant's tissues (Jones 1973).

2.5 Reduction in Variation

Inbreeding

Inbreeding (mating among close relatives) is certainly a reality in many social animals (Chapter 7). One severe form is self-fertilization or "selfing" in plants. The effects are usually deleterious, and, as is dramatically illustrated in Fig. 2.5, crossbreeding can reverse those effects.

The phenomenon of inbreeding depression has long been known. Viability and especially **fecundity** (or yield of crops) decline as populations become more inbred in the laboratory (Table 2.3). In human populations, the consequences include higher mortality, mental retardation, albinism, and other physical abnormalities (Stern 1973). Schemske (1983) noted that, in plants that breed by selfing or outcrossing, the fitness of

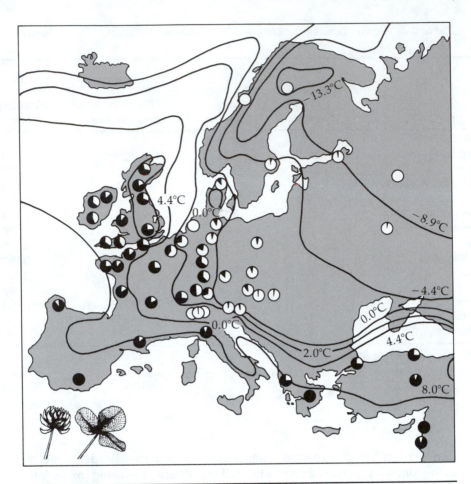

Figure 2.4 *Frequency of the cyanide-producing form in populations of white clover* (Trifolium repens), *represented by the black section of each circle.* The cyanogenic form is more common in warmer regions. Lines are January isotherms. (Redrawn from Daday 1954.)

offspring produced by self-fertilization was less than half that of offspring produced by outcrossing to an unrelated individual. The reason that inbreeding is so disadvantageous is that the frequency of homozygotes for recessive alleles is thought to increase. Recessive alleles are more likely to be deleterious for the simple reason that dominant deleterious traits are more quickly selected out of the population, leaving behind their recessive counterparts. The opposite, outbreeding, which increases the number of heterozygotes and conceals more of the recessive alleles, is termed *heterosis.*

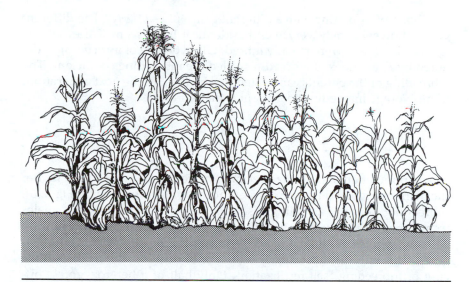

Figure 2.5 *Inbreeding depression and hybrid vigor.* The two corn plants at the left are of two inbred strains; the larger plant to their right is a hybrid of the two. All the plants to the right are successive self-fertilized generations from the hybrid. (Drawn after a photo by Jones 1924.)

The Effects of Small Population Sizes

The effects of inbreeding are more extreme in small populations. Suppose a population has N diploid breeding individuals in each generation; there are then $2N$ gene copies at each locus. If all matings are equally probable, the likelihood that, in generation t, a given gene will unite with a gene from a similar parent that is identical by descent is $1/2\,N$, and the

Table 2.3 *Inbreeding depression in rats. (After Lerner 1954.)*

Year	Nonproductive Matings (in Percent)	Average Litter Size	Mortality from Birth to Four Weeks (in Percent)
1887	0	7.50	3.9
1888	2.6	7.14	4.4
1889	5.6	7.71	5.0
1890	17.4	6.58	8.7
1891	50.0	4.58	36.4
1892	41.2	3.20	45.5

Note: The years 1887–1892 span about 30 generations of parent × offspring and sib matings.

likelihood of its uniting with a different gene is $1 - (1/2\ N)$. The different genes, however, may very well be identical by descent if the parents (generation $t - 1$) were already inbred. Often degree of inbreeding is denoted by F, an index of inbreeding called the inbreeding coefficient. The inbreeding of the parents can be denoted by F_{t-1}. In the next generation, t, the average inbreeding coefficient is therefore

$$F_t = \frac{1}{2}N + \left(1 - \frac{1}{2}N\right) F_{t-1}$$

so any population of finite size becomes more inbred in time. The higher the degree of inbreeding, the quicker the loss of heterozygosity (Fig. 2.6). Inbreeding does not actually cause a change in the frequency of alleles (remember the Hardy-Weinberg theorem), but it increases the proportion of homozygotes of all kinds and decreases the proportion of heterozygotes. This effect is most clearly seen if one considers the most extreme case of inbreeding, self-fertilization. Only half of a heterozygote's offspring are heterozygous, so in each generation the frequency of heterozygotes is halved.

The coefficient of inbreeding, F, is given by the frequency of heterozygotes relative to that expected under random mating (H_O)

$$F = \frac{2\ pq - H_F}{2pq}$$

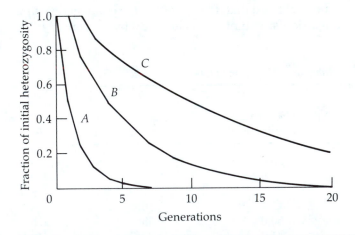

Figure 2.6 Decrease in heterozygosity due to inbreeding. Systems of mating are exclusive self-fertilization (curve *A*), sib mating (curve *B*), and double-first cousin mating (curve *C*). (Redrawn from Crow and Kimura 1970.)

or

$$F = \frac{H_O - H_F}{H_O}$$

where H_F is the observed proportion of heterozygotes and H_O is the theoretical proportion expected under the Hardy-Weinberg assumptions. If $(H_O - H_t)/H_O$ is substituted for F_t and $(H_O - H_{t-1})/H_O$ for F_{t-1} it can be shown that

$$H_t = \left(1 - \frac{1}{2}N\right)H_{t-1}$$

or

$$H_t = H_O\left(1 - \frac{1}{2}N\right)^t$$

In other words, the smaller the population (N), the faster heterozygosity declines (Fig. 2.7). This result has important consequences in the real world, where animal populations are constantly declining be-

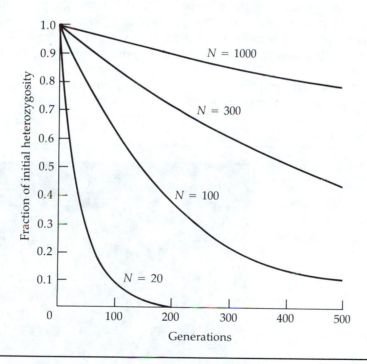

Figure 2.7 *Decrease in heterozygosity due to finite population size.* N equals population size. (Modified from Strickberger 1986.)

cause of shrinking habitats, and conservation science has become particularly concerned with the genetics of small populations. One species of concern is the California condor, shown in Photo 2.5. It remains to be seen whether the offspring of 26 captive individuals—the sole remnants of the species in 1986—can overcome the effects of inbreeding and form a fit and healthy population when released into the wild.

Genetic Drift

Even in large populations, there may be some random genetic change. In a population of N breeding individuals ($2N$ genes), the critical frequency of alleles A and A′ are p and $(1 - p)$ If $2N$ gene copies are chosen at random to mate in the next generation, the probability that exactly i of them will be A is given by the binomial distribution

$$\frac{2N(2N - 1) \ldots (2N - i + 1)}{i!} p^i (1 - p)^{2N - i}$$

The mean of the distribution is $\bar{x} = p$, and the variance $= V = p(1 - p)/2N$ (Futuyma 1986). That is, after one generation the average frequency of A among all the populations is still p, but the populations vary in gene frequency to the extent $p(1 - p)/2N$. Such random change in allele frequencies is termed **genetic drift.** It can occur even in large populations, but generally the smaller the populations, the greater the variation.

As genetic drift proceeds, variation among populations increases. For example, some colonies may drift from $p = 0.50$ to $p = 0.55$, with a

Photo 2.5   California condor, Gymnogyps californianus. *A total (captive) population of only 26 existed in 1986. (Photo by Glen Smart, U.S. Fish and Wildlife Service.)*

corresponding drift to 0.45 in others. Among these, some will drift even further in gene frequency, but this type of logic usually applies to neutral alleles (those that do not differ in their effect on survival or reproduction). Whether or not many alleles conform to this assumption is a subject of considerable controversy. Kimura (1983*a* and 1983*b*) and Nei (1983) hold that much molecular variation in allozymes, DNA, and proteins is neutral and that any divergence detected among species is likely to be the result of genetic drift.

At any point in time, the distribution of gene frequencies will be between 0 and 1. When $p = 1$, the A allele is said to be fixed; the population is monomorphic AA. When $p = 0$, the A' allele is fixed; the A allele has disappeared. Alleles cannot then reappear except by new mutation. Clearly the allele that is most common has the greatest chance of becoming fixed. Again, genetic drift has important implications for conservation of populations. The smaller the population concerned, the greater the likelihood of genetic drift, the fixation of alleles, and the loss of genetic variation.

Neighborhoods

Even in large populations the effective population size may actually be quite small because individuals only mate within a neighborhood. The number of individuals in this neighborhood is given by $4\pi s^2 D$, where D is the **density** of the population and s is the standard deviation of the distances between the birth sites of individuals and the birth sites of their offspring. By marking individual deer mice (*Peromyscus maniculatus*), Howard (1949) showed that at least 70 percent of males and 85 percent of females breed within 150 meters of their birthplaces. Even migratory species, such as birds, usually return to the vicinity of their birthplaces to breed. Barrowclough (1980) has also shown that, for noncolonial birds such as wrens and finches, local effective population sizes range from 175 to 7,700 individuals, and Levin (1981) has argued that, despite seed and pollen dispersal, effective size of local populations for plants is often in the dozens or hundreds.

Furthermore, even within a neighborhood, some individuals do not reproduce. If only half the individuals in a population breed, half the heterozygosity is lost per generation. In a population of 50, if only half the members breed, then the population has an effective size N_E of 25. In territorial species and those with harems, N_E is even lower, because the few males that reproduce contribute disproportionately to the subsequent generations and genetic drift is inflated. If a population consists of N_m breeding males and N_f breeding females, the effective population size is $N_E = 4N_m N_f / (N_m + N_f)$. In 20 groups of elephant seals, each with 1 male and 10 females, the effective population size is 73 rather than 220.

Bottlenecks

Effective population sizes can be further lowered if populations vary in size from generation to generation. The effective size N_E in this instance is estimated by the harmonic mean population size

$$\frac{1}{N} = \left(\frac{1}{t}\right) \sum_{i=1}^{t} \left(\frac{1}{N_i}\right)$$

Thus, if a population goes through five generations of size 100, 150, 25, 150, and 125 individuals, N_E is about 70, as opposed to the arithmetic mean of 110. N_E is more strongly affected by lower than by higher population sizes. Rarely in ecology have all the factors that impinge upon N_E been measured (dispersal distance, effective sex ratio, mating frequency, and so on). Where only one has been measured, variance in reproductive success, N_E in humans and *Drosophila* turns out to be less than 75 percent of the actual number of individuals (Crow and Kimura 1970). In many situations, N_E is likely to be an even lower percentage.

The northern elephant seal (*Mirounga angustirostris*) went through a severe bottleneck in the 1890's when its numbers were reduced to only about 20 by hunting. Remarkably, it recovered, and more than 30,000 are alive today. Because of the harem mating system (see Chapter 7), the effective population size must have been less than 20. Bonnell and Selander (1974) were able to demonstrate by electrophoresis that there is no genetic variation among today's northern elephant seals in a sample of 24 loci, whereas in the southern elephant seal (*M. leonina*) normal genetic variation exists. Southern elephant seals were never drastically reduced in numbers.

In many colonizing species a bottleneck occurs when a new habitat is colonized for the first time. This type of bottleneck is known as the **founder effect.** In dispersing species, for example, it is not unusual for only one or two members of a population to make it successfully to an island. A colony founded by a pair of diploid individuals can have at most four alleles at a locus, though there may be many more in the population from which they came. Bottlenecks and founder effects are, therefore, common occurrences for weeds and pests and should not always be regarded as universally deleterious.

The main conclusion is that, despite limitations by bottlenecks and neighborhoods, much variation exists in nature. The question of whether much of it is evolutionarily meaningful or is maintained by selection is open to debate (Lewontin 1974; Koehn, Zera, and Hall 1983). Probably the answer is that variation is neutral on prevailing genetic backgrounds in many environments but affects fitness on different genetic backgrounds in some environments. The cases in which it does are examples of natural selection operating in nature.

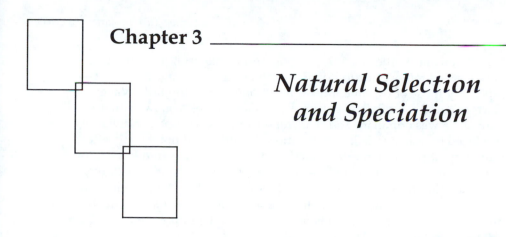

Chapter 3

Natural Selection and Speciation

Natural selection, the backbone of Darwin's theories on evolution and the mechanism that can explain how evolution occurs, is in essence the differential survival and reproduction of some individuals in a population and the death without issue (or with fewer issue) of others. Important though random changes and genetic drift are, a more active process, **natural selection,** is best invoked to explain many natural features of evolution in the real world. The feet of different types of birds are features that fit them to their ways of life and are best explained by natural selection. Never, however, is there a predestined form to which plants or animals are shaped as they evolve (Dawkins 1986); as the environment changes so usually do the organisms that live in it—no moral or ethical forces impinge on natural selection.

3.1 Natural Selection

Theoretically, natural selection can operate whenever different kinds of self-reproducing entities differ in survival or reproduction. Sometimes the genes themselves have been regarded as these entities (*selfish gene theory*, Dawkins 1976), but more commonly individual organisms are regarded as the units (see Sober and Lewontin [1982] for a critique of the selfish gene theory). Natural selection, then, occurs when genotypes differ in fitness.

Industrial Melanism: The Case of the Peppered Moth

One of the best-analyzed examples of natural selection of genotypes in operation is the change in color that has taken place in certain populations of the peppered moth, *Biston betularia,* in the industrial regions of Europe during the past 100 years. Originally, moths were uniformly pale grey or whitish in color; isolated dark-colored individuals were rare, less than 2 percent (Kettlewell 1973; Bishop and Cook 1980). Gradually, the dark-colored forms came to dominate the populations of certain areas—especially those of extreme industrialization such as the Ruhr Valley of Germany and the Midlands of England. Genetic tests showed that the dark allele was dominant and that crosses of dark with pale individuals produced typical Mendelian segregation. Pollution did not directly affect mutation rates, through caterpillars' feeding on soot-covered leaves, for example. Rather, it promoted the survival of dark morphs, which were normally quickly eliminated in nonindustrial areas by adverse selection; birds found them conspicuous. This phenomenon, an increase in the frequency of dark-colored mutants (carbonaria forms) in polluted areas, is known as industrial **melanism.**

The operation of natural selection on the peppered moth was elegantly illustrated by Professor H.B.D. Kettlewell of Oxford University. He showed that normal pale forms are cryptic when resting on lichen-covered trees, whereas dark forms are conspicuous. In industrialized areas, lichens are killed off, tree barks become darker, and the dark moths are the cryptic ones. Figure 3.1 illustrates the two forms of *Biston betularia*. Kettlewell suspected that birds were the selecting force, and he set out to prove it by releasing thousands of moths marked with a small spot of paint into urban and industrialized areas (Kettlewell 1955). In the nonindustrial area of Dorset he recaptured 14.6 percent of the pale morphs released but only 4.7 percent of the dark moths. In the industrial area of Birmingham, the situation was reversed; 13 percent of pale morphs but 27.5 percent of dark morphs were recaptured. Birds were clearly implicated in differential predation, eating more pale morphs in industrial habitats and more dark morphs in nonindustrial areas. As a test of his field observations, Kettlewell and companions set up blinds and watched birds voraciously gobble up moths placed on tree trunks. The action of natural selection in producing a small but highly significant

(a)   (b)

Figure 3.1 *Industrial melanism.* **(a)** Typical (light colored) form of *Biston betularia*. **(b)** Carbonaria (dark colored) form of *Biston betularia*. Drawn from photo by Kettlewell 1955.

step of evolution was seemingly demonstrated. However, the black form has not become fixed even in the most industrial of locales, so other factors may play roles in the maintenance of melanic frequencies (Lees 1981). Interestingly enough, the white form of the peppered moth is making a comeback in Britain. Sir Cyril Clarke has been trapping moths at his home on Merseyside, Liverpool, since 1960. Before about 1975, 90 percent of the moths were dark, but since then there has been a steep decline in carbonaria forms, and in 1984 only 60 percent of the moths caught were melanic (Clarke, Mani, and Wynne 1985). The mean concentration of sulphur dioxide pollution fell from about 300 μg m^{-3} in 1970 to less than 50 μg m^{-3} in 1975 and remained fairly constant for a number of years afterward. Although it appears that peppered moths may be a reasonable indicator species of environmental pollution (see Chapter 20), it is disconcerting that the numbers of dark morphs only decreased when pollutant levels had reached a low value. Clarke and friends cast doubts on the traditional peppered moth pollution story. They suggest that, despite the change in SO$_2$ levels, there was no change in the state of the lichens that were supposed to be hiding the moths. Further, in 25 years Clarke found only two moths resting on lichen-covered trees and walls in the daytime; their normal daytime hiding place may well be somewhere else.

The case of the peppered moth notwithstanding, the preponderance of the evidence still shows that many alleles or genotypes vary in fitness according to their environment. There is no automatic ''better'' genotype. Other cases of industrial melanism are known (Bishop and Cook 1981), and other examples of rapid evolutionary change have become apparent as, for example, more and more pests become resistant to insecticides. Even such a normally harmful allele as sickle-cell hemoglobin (which causes severe anemia) can be advantageous in areas where malaria is prevalent because heterozygous individuals, who carry the allele but are not severely affected by the anemia, are more resistant to malaria than are normal individuals. This mechanism for the maintenance of a polymorphism is known as *heterozygous advantage*. In the absence of malaria, the sickle-cell trait is quickly lost. In Norway rats, the allele for resistance to the pesticide warfarin lowers the animals' ability to synthesize vitamin K. Resistant varieties are thought to be at a 54 percent disadvantage to wild types in nature (Bishop 1981), but the allele is maintained in the population by the advantage it gives individuals that encounter warfarin.

Balanced Polymorphism

The unstable existence of two or more morphs in a population is called a transitional or directional **polymorphism**; one allele is replacing another in the population. In many cases, however, one allele does not completely replace the other, and the stable intermediate frequency is called a **balanced polymorphism**. An example can be seen in the land snail *Cepaea nemoralis*, in which six alleles affect the color of the shell, which can be several shades of brown, pink, and yellow. (See Fig. 3.2.) The relative abundances of the various forms differ, even between localities less than a mile apart, and fossils show that this polymorphism has persisted at least since the Pleiostocene epoch (Diver 1929). This European snail is common in a variety of habitats, woods, meadows, and hedgerows, and the maintenance of the polymorphism has again been shown to be due to bird-predation pressure, this time by the song thrush, *Turdus ericetorum* (Cain and Sheppard 1954*a*). Thrushes hunt by sight, and shell color plays the central role in the concealing coloration of these snails. The birds break open shells on suitable stones, or anvils, providing the experimenter with an ideal opportunity to compare the proportion of varieties in a colony with those taken from it by song thrushes. In beech woods, the leaf litter is red-brown, and the snails are brown and pink. In grassland habitats, red and brown forms are rare, and yellow forms predominate. The story is somewhat complicated by the fact that genotypes differ in their susceptibility to extreme temperatures (Jones,

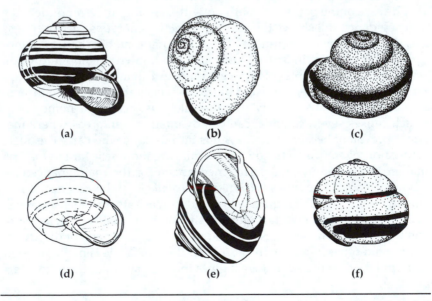

Figure 3.2   *The highly polymorphic land snail* Cepaea nemoralis. **(a)** *A yellow shell with five bands and a dark lip at the mouth of the shell.* **(b)** *A pink shell with no bands and a dark lip.* **(c)** *A brown shell with only the central band present.* **(d)** *A yellow shell with the bands present but unpigmented, making them translucent (the lip is also unpigmented).* **(e)** *A yellow shell with pigmented bands but an unpigmented lip.* **(f)** *A pink shell with the first two bands missing so that it has only the central and two lower bands present.*

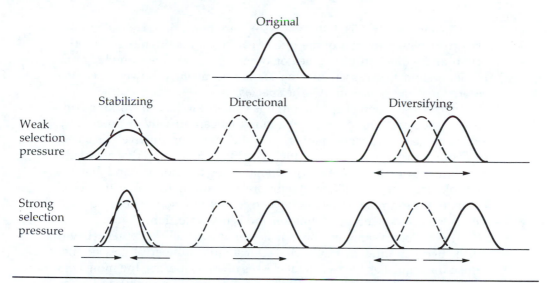

Figure 3.3 *Effects of stabilizing, directional, and diversifying selection upon variation for a quantitative character.*

Leith, and Rawlings 1977). In addition, the snails have a complicated pattern of thin black bands. This characteristic is again linked to habitat type, as unbanded snails are cryptic in uniform habitats and banded ones are cryptic in "rough" habitats (as are zebras) (Cain and Sheppard 1954*a*). Interestingly, an African land snail exhibits a similar variation in shell color and banding patterns (Owen 1966), so the phenomenon is likely to be of wide ecological occurrence.

The peppered-moth and land-snail examples show, fairly conclusively, the operation of natural selection on populations and the survival of the fittest individuals. They also show that natural selection can operate in more than one way; it can drive populations toward one type of morph or act to maintain more than one. Often it is difficult to tell whether variations on a theme in nature represent truly different species or merely different morphs of the same thing. In general, three types of natural selection are recognized: stabilizing, in which one morph is favored; directional, in which the population is driven to exhibit a different type of morph; and diversifying or disruptive, in which, as in the case of the land snail, more than one morph can be favored (Fig. 3.3). Presumably, when disruptive selection is strong enough, speciation results.

3.2 Speciation

Of all the smaller taxonomic groups specified, races, subspecies, biotypes, species, genera, and families, most are quite arbitrarily defined

(Wilson and Brown 1953; Futuyma 1986), and their value is questionable. Unfortunately, estimates of the rate of evolution are often based on the origin and extinction of families or other taxa, so some knowledge of higher systematics is essential. One taxonomic category, however, is considered real and nonarbitrary—the **species**.

In any discussion of speciation, it is valuable to have a working concept of species: ''groups of populations that can actually or potentially exchange genes with one another and that are reproductively isolated from other such groups'' (Mayr 1942, p. 120). Bear in mind that asexually reproducing organisms are also given names by taxonomists and that paleontologists recognize different temporal portions of the same lineage as distinct—for example, *Homo erectus* and *H. sapiens*—so the above quotation is but a good working definition and is not watertight.

The biological species concept is not universally accepted (Levin 1979). Even as fundamental a character as the number of cotyledons, which distinguishes the two great classes of angiosperms, the monocots and the dicots, varies intraspecifically in the shrub *Pittosporum*, and it has been possible to generate tricotyledonous strains of snapdragons by selection (Stebbins 1974). How much of a genetic difference is necessary to make a new species? Nobody knows. To paraphrase Mayr (1963), to try to determine the difference between species in terms of nucleotide pairs of DNA would be absurd, much as it would be to compare how different two books were on the basis of the number of letters they use. Species have many genes in common, much as books have words in common; it is the particular arrangement of each that is so critical.

The evolution of reproductive barriers is obviously a critical process in speciation, for two populations can only diverge if changes in gene frequencies in one are not immediately transmitted to the other by interbreeding. These barriers may be either extrinsic or intrinsic. Extrinsic barriers exist if the populations reproduce in different places (are allopatric) or at different seasons (are allochronic). Intrinsic barriers may consist of either premating isolating mechanisms (which prevent F_1 zygotes from forming) or postmating barriers such as hybrid inviability or sterility.

Allopatric Speciation

Most evolutionists consider **allopatric speciation** to be the most likely mechanism for the evolution of a species (Mayr 1942; 1963). Allopatric speciation involves separation of populations by a geographical barrier. For example, nonswimming populations separated by a river may gradually diverge because there is no gene flow between them (Fig. 8.18). Alternatively, the upthrusting of mountains often divides populations into many units, among which speciation then proceeds. In an area only 20 × 5 miles on Hawaii, 26 subspecies of land snail, *Achatinella mustelina*, have been recognized, each in a different valley separated from the others by mountain ridges. Some of the best-known instances of divergence among isolated populations are on islands, where a species that is rather homogeneous over its continental range may diverge spectacularly from the continental form in appearance, ecology, and behavior. Darwin's finches have speciated extensively within the Galapagos archipelago.

In many situations, the ranges of species that have evolved allopatrically come to overlap, and the species become sympatric once again. The isolating mechanisms between the species, however, have usually continued to evolve separately, and as a result the males and females of the different species are likely to have become incompatible. Some important isolating mechanisms are outlined in Table 3.1. Among birds and fish, species-specific male coloration seems to be an important isolating mechanism, whereas in insects it is often smell (moths), sound (grasshoppers), or even the correct flight path and flash patterns of lights (Lampyridae); in frogs and toads, chorusing is important.

Sympatric Speciation

The alternative to allopatric speciation is **sympatric speciation,** the appearance of new species in an area not geographically separated from other members of the population. Most models of sympatric speciation are highly controversial. Even if some members of a population began to inhabit slightly different parts of a population's range, gene flow between the two groups would probably still be sufficient to prevent speciation. Any slight morphological or behavioral changes in one group would be conveyed to the other. If a single mutation or chromosomal change were to confer complete reproductive isolation in one fell swoop, its bearer would be reproductively isolated. Given the unlikely scenario of the same mutation's occurring at the same time in another individual of the opposite sex, and given that these two individuals could find each other, a new species might form. Close inbreeding (self-fertilization or mating with sibs) may promote the likelihood of such events, and thus sympatric speciation has been proposed for some insect groups such as the Chalcidoidea, parasitic Hymenoptera in which many individuals develop from a single egg (a process called *polyembryony*) laid in a host (Askew 1968). Indeed, because insects are themselves so speciose, comprising an estimated 30 million species (Erwin 1982), it is sometimes difficult to imagine allopatric speciation in this order by means of 30 million geographic barriers, especially given the dispersal abilities of insects (Johnson 1969). In many groups of insects, closely related species are restricted to different host plants, and it has been argued that sympatric speciation has occurred there (Bush 1975a and 1975b; Wood and Guttman 1983). However, in the few cases that have been analyzed, host preference in insects appears to be controlled by many genes, a situation that would not be favorable to a quick, one-genetic-step method of sympatric

Table 3.1. *Summary of the most important isolating mechanisms that separate species of organisms.*

Mechanism	Mode of Action
Prezygotic	Fertilization and zygote formation are prevented.
Habitat	Populations live in the same regions but occupy different habitats; for example, different spadefoot toad species occupy different soil types (Wasserman 1957).
Seasonal or temporal	Populations exist in the same regions but are sexually mature at different times, for example, flowers that bloom in different months (Grant and Grant 1964) or fireflies that mate at different times of the night (Lloyd 1966).
Ethological (animals only)	Populations are isolated by different and incompatible behavior before mating, for example, courtship songs of birds or frogs and flash patterns of fireflies.
Mechanical	Crossfertilization is prevented or restricted by differences in structure of reproductive structures, for example, genitalia in animals and flowers in plants.
Postzygotic	Fertilization takes place, and hybrid zygotes are formed, but these are inviable or give rise to weak or sterile hybrids.
Hybrid inviability or weakness	Hybrids of the frogs *Rana pipiens* and *R. sylvatica* do not develop beyond the gastrula stage (Moore 1961).
Developmental hybrid sterility	Hybrids are sterile because gonads develop abnormally or meiosis breaks down before it is completed, for example, hybrids between fruit flies (Dobzhansky 1936).
Segregational hybrid sterility	Hybrids are sterile because of abnormal segregation to the gametes of whole chromosomes, chromosome segments, or combinations of genes; more common in plants (Grant 1981).
F_2 breakdown	F_1 hybrids are normal, vigorous, and fertile, but F_2 contains many weak or sterile individuals.

speciation (Futuyma and Peterson 1985). Furthermore, even when a species utilizes a new host plant to feed on, it often still interbreeds with individuals reared on other host plants. Thomas et al. (1987) showed that the butterfly *Euphydryas editha* has extended its range onto a new plant, *Plantago lanceolata*, within the last 100 years but that populations still interbreed with individuals reared on the old host, *Collinsia parviflora*.

How rapidly does speciation occur? The answer, of course, differs for different organisms. J.B.S. Haldane theorized that species of vertebrates might differ at a minimum of 1,000 loci and that at least 300,000 generations would be necessary for the formation of new species. Indeed a great many of the populations isolated for thousands of generations by the Pleistocene glaciations did not achieve full species status. American and Eurasian sycamore trees (*Platanus* sp.) have been isolated for at least 20 million years, yet still form fertile hybrids (Stebbins 1950). The selective forces on these two continents have obviously not been sufficient to cause reproductive isolation between these ecologically general species. However, several genera of mammals, for example polar bears (*Thalarctos*) and voles (*Microtus*) do appear to have originated relatively recently, in the Pleistocene (Stanley 1979). Many of the Hawaiian species of *Drosophila* flies have arisen in just a few thousand years, although their generation time is, of course, much shorter than that of mammals. Lake Nabugabo in Africa has been isolated from Lake Victoria for less than 4,000 years, yet it contains five endemic species of fish (Fryer and Iles 1972). Lake Victoria itself is only 500,000–750,000 years old but harbors about 170 species of cichlid. Again in Hawaii, at least five species of *Hedylepta* moth feed exclusively on bananas, which were only introduced by the Polynesians some 1,000 years ago.

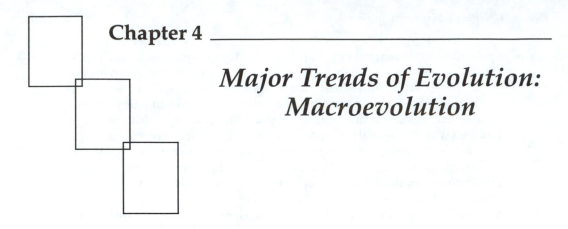

Chapter 4

Major Trends of Evolution: Macroevolution

T he natural progression of speciation can be seen as evolution. Single
lineages change direction and gradually become new species *(ana-
genesis)*, or some daughter lineages break off the phylogenetic tree and
both species persist *(cladogenesis)*. The group of individuals all derived
from one common ancestor is known as a **clade**. Reticulate evolution may
occur as a result of hybridization between related species or by the transfer of genetic material between taxa, but this process is not common. During the history of life on earth, millions of new species have arisen and
millions more have become extinct. Are there any patterns to these two
processes?

4.1 *Phylogenetics*

To understand evolution, one must be able to understand the phylogenetic relationships between species—the patterns of branching on a
cladogram (Fig. 4.1). In every lineage some characteristics have evolved
very little from the ancestral or primitive form. Other characteristics have
evolved a lot, and these are called *derived characteristics.* For example, the
large human brain is derived compared to the brains of other primates. In
contrast the pentadactyl limb is an ancestral state. The ancestral state can
be modified to different degrees in different lineages; the number of dig-
its varies from five in primates to four in sheep, two in bovines, and one
in horses. Some characteristics can of course be only recently derived, so
two species may look very different even though they have derived from
a common ancestor (for example, character u' in Fig. 4.1). The levels of
division in taxonomy—genera, families, orders, and so on—are called *categories*, and all members of a given category are referred to as a *taxon* (plural taxa). Felidae, the cats, and Papilionidae, the swallowtail butterflies,
are both family-level taxa. The classification of species into different taxa
by taxonomists is supposed to reflect the evolutionary history of the spe-

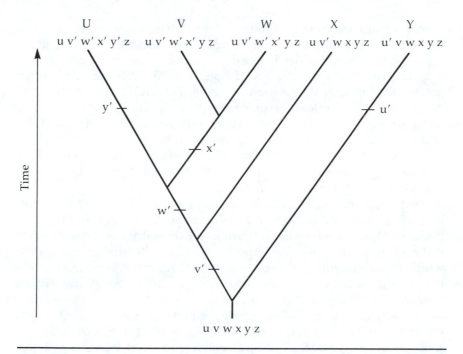

Figure 4.1　*A hypothetical phylogenetic tree, or cladogram, of five taxa U–Y derived from a common ancestor, with the evolution of six characters u–z indicated by transitions from the ancestral (plesiomorphic) to derived (apomorphic) states (indicated by primes). In this example, each character changes once at most, and taxa with more recent common ancestors share more derived character states than taxa with more remote common ancestors. Each taxon retains some ancestral character states.*

cies concerned; this branch of science is known as *systematics*. Species are not grouped together merely because they look alike; such a system would give rise to polyphyletic groups—that is, groups not derived from a single common ancestor. Some so-called "numerical taxonomists" or pheneticists still prefer to classify organisms by elaborate numerical methods, and they often arrive at phenograms that reflect species groupings based on overall morphological similarity rather than on evolutionary processes. The cladistic school of systematics, based on cladograms, seeks to draw up phylogenies based on relationships among species (Hennig 1979). If the rates of divergence of species were constant, one could infer a correct **phylogeny** from the overall difference among species. The rate of **phyletic evolution** varies through time, however, going in fast spurts at certain times and slowly at others, perhaps in direct relation to the environment. Convergent evolution may also occur, so the upshot is that, when phylogenies are inferred, variable rates of evolution and possible convergence must be taken into account; such data are often hard to obtain. **Convergent evolution,** the independent appearance of a character state in more than one lineage, is undoubtedly common in nature. For example, hole-nesting birds, whatever their phylogenetic affinities, lay white rather than speckled eggs. Rensch (1959) lists many other patterns of convergent evolution. Simberloff (1987a), however, has provided a statistical method for comparing phylogenetic cladograms, based on the degree of matching between the two. Although both drift and natural selection may cloud phylogeny (Simberloff 1988a), minor similarities between apparently dissimilar cladograms can be shown, by a matching methodology, to be so highly nonrandom as to implicate a common sequence of vicariant events.

Reversals are also common in evolution. Quite often a complex structure will degenerate and return to its original state. For example, some insect species that have lost their wings, such as lice and fleas, must be distinguished from primitive species such as silverfish (Thysanura), which never had wings. There is a general theory that, once lost, a complex structure can never be reformed. This principle is commonly known as *Dollo's law.* Exceptions do occur, however. Of the several molars once possessed by most cats (Felidae), only the first is retained, but in the lynx the second molar has reappeared (Kurtén 1963). The South American tree frog *Amphignathodon* has apparently re-evolved teeth in its lower jaw, though most other frogs have lacked teeth there since the Jurassic (Noble 1931).

4.2 The Fossil Record

Bearing in mind these caveats, one can make some generalities about the ancestry of species on the basis of the fossil record.

Methods of Fossilization

The term *fossil* is often used to mean any preserved remains of living things. Strictly speaking, there are a number of processes of ''fossilization.'' Examples of fossils preserved by some of the methods described below are depicted in Photo 4.1.

- **Imprisonment in ice:** Ice excludes bacterial and fungal action entirely and therefore preserves specimens nearly perfectly. Woolly rhinoceroses and mammoths from the Siberian ice even show what the animal was eating at the time of death.

- **Preservation in amber:** Amber is a very durable substance formed from the hardened resin of coniferous trees. It has allowed for the preservation of small specimens such as insects that became embedded in the original soft resin and then remained trapped for millions of years.

Photo 4.1   Some examples of fossilization. (a) Baby woolly mammoth dug out of frozen earth in Alaska. (b) A caddis fly preserved in amber. (c) The amphibian Buettneria per-fecta *preserved in rock. (d) Petrified wood at Blue Mesa Petrified Forest National Park. (Photo (a) Neg. No. 320497, by Thane L. Bierwert. Courtesy Department of Library Services, American Museum of Natural History. Photos (b) and (c) Copyright Smithsonian Institution. Photo (d) Dana C. Bryan.)*

■ **Preservation in oil and "tarry" substances:** The asphalt lakes of southern California have yielded teeth and bones in almost perfect condition, although softer animal parts have not fared so well. Skeletons of many wolves and saber-toothed tigers have been recovered from this area.

■ **Preservation in acid bogs:** The acid conditions in some bogs also prevent the decay of harder organic parts. In the peat bogs of northern Europe, woody plants and animal skeletons have been found, including "Irish elk" antlers and the almost perfect remains of "Pete Marsh," a 2,500-year-old Iron Age human dug up in Britain in 1984.

- **Preservation by dehydration:** Animal remains are well preserved in the extremely dry conditions of deserts, and large caches of bones have been found in caves.

- **Formation of impressions:** When fine-grained sediments are laid down over plants or animals and are gradually crushed to rock, the organismal remains decay and are lost, but an impression or cast remains. This method of fossilization has preserved articles as fine as leaf veins and jellyfish.

- **Formation of casts:** When the hollow mold of an impression is filled with mineral particles that become hardened under pressure, they form a cast. Many fossil shells of molluscs, insects, and other invertebrates are simply castings of molds formed by the empty shell.

- **Petrifaction:** "Real fossils" are those of organisms that have literally been turned into stone. Petrifaction occurs usually where there is an abundance of mineral-containing water. As the organic molecules are slowly washed away, calcium carbonate, silica, or pyrites are deposited in their place. Petrified wood is often preserved in great detail in this fashion, but petrified animals are not so common.

Dating Fossils

If the age of the rocks or sediments is known, the fossils they bear can be accurately dated. There are several methods of dating rocks:

- **Stratigraphy:** Stratigraphy is the analysis of the succession of rock strata that have been laid down, one after the other, in a more or less regular fashion. The simplest analyses are carried out where stream erosion has cut deep gashes in the Earth's surface or where strata have been tilted so that one can pass over the Earth's surface from older to progressively younger strata, as in the eastern foothills of the Rocky Mountains.

- **Biostratigraphy:** Biostratigraphy is a comparison of rock ages by the fossils they contain. Often, if rocks "on top" contain less advanced fossils than those "below," it is an indication of folding or contortion, perhaps accompanying mountain building.

These two techniques can be used to assess relative ages of rocks (which are older), but they cannot give absolute ages. They also cannot always reveal *how much* older one stratum is than another because the speed of deposition of rock varies dramatically with prevailing conditions. Volcanic eruptions may lay down several feet of rock in a few days, whereas limestones formed from fossilized protozoan shells in the sea take thousands of years for just one inch to form. During periods of elevation or drought, there is no deposition of this type of rock, and the geologist has no way of telling how long such an elevation may have lasted. Thus, although stratigraphy and biostratigraphy can give ''ballpark'' estimates of rock ages, the most accurate technique involves a third method.

■ **Radiometric dating:** Radiometric dating relies on the regular decay of atomic nuclei of certain elements, notably those from carbon-14 to ordinary carbon, from uranium to lead, from potassium to argon, and from rubidium to strontium. The parent element, say rubidium, eventually decays into another element, in this case strontium. Because the **half-life**—the time it takes for half of the element to decay—of each of these parent elements is known, rocks can be fairly accurately aged from measurements of the proportions of the parent elements and their decay products. The rate of decay is independent of pressure and temperature. Of course, this technique can only be used on igneous rocks; sedimentary rocks (those actually containing fossils) must be dated relative to the igneous formations that bracket them. Radiometric dating has made possible accurate estimates of the age of the Earth and of the major rock formations and fossil groups (Table 4.1, Fig. 4.2).

It is worth emphasizing that the fossil record is very incomplete, patchy, and biased. In most places where organisms die, their remains quickly decay. Fossils are best preserved under water in **anaerobic** conditions, so the fossil record is biased toward aquatic species, marine types, and periods in the Earth's history when shallow lakes rather than mountain chains covered the land. There are still no adequate fossils to show whether birds evolved from crocodiles, other reptiles, or mammals. Hard parts of organisms are preserved much more readily than are soft parts, so soft-bodied invertebrates are represented much less frequently than vertebrate skeletons. Despite these shortcomings, the fossil record has yielded a surprisingly great deal of information.

There are excellent fossil records showing how the small whippet-sized grazer *Hyracotherium* (or *Eohippus*), a five-toed animal, evolved into a one-toed grazer, *Equus*, in the Pleistocene. Such changes correspond to climatic changes during which, in North America, early Tertiary forests gradually gave way to Pliocene savannahs and prairies. Bigger teeth (and more of them) were needed to grind the harsher grasses, and stronger, faster bodies and bigger brains were needed to escape predators, because horses on open grassland were in full view of their enemies. Similar excellent fossil series exist to show the development of sea urchins, mollusc shells, and mammalian horns.

Table 4.1 *The geological time scale.*

Era	Period	Epoch	Millions of Years from Start to Present	Time on 24-Hour Clock	Major Events
Cenozoic	Quaternary	Recent (Holocene)	0.01		Repeated glaciations in northern hemisphere; extinctions of large mammals; evolution of *Homo*; rise of civilization.
		Pleistocene	2	11.59 P.M.	
	Tertiary	Pliocene	5	11:58 P.M.	Radiation of mammals and birds; flourishing of insects and angiosperms. Continents in approximately modern positions.
		Miocene	25		
		Oligocene	38		
		Eocene	55		Drying trend in mid-Tertiary.
		Paleocene	65	11:31 P.M.	
Mesozoic	Cretaceous		144	10:51 P.M.	Mass extinctions of marine and terrestrial life, including last dinosaurs. Angiosperms become dominant over gymnosperms. Continents well separated.
	Jurassic		213	10:18 P.M.	Dinosaurs abundant; first birds, archaic mammals, and angiosperms appear. Gymnosperms dominant. Continents drifting.
	Triassic		248	10:01 P.M.	Increase of reptiles, first dinosaurs; gymnosperms become dominant. Continents begin to drift apart. Mass extinction near end of period.
Paleozoic	Permian		286	9:43 P.M.	Continents aggregated into Pangaea; glaciations. Marine extinctions, including last trilobites. Reptiles radiate; amphibians decline.

Carboniferous (Pennsylvanian and Mississippian)	360	9:12 P.M.	Extensive forests of early vascular plants especially lycopsids, sphenopsids, ferns. Amphibians diverse; first reptiles. Radiation of early insect orders.
Devonian	408	8:45 P.M.	Fishes and trilobites diverge. First amphibians and insects. Mass extinction late in period.
Silurian	438	8:30 P.M.	Invasion of land by primitive tracheophytes, arthropods.
Ordovician	505	7:58 P.M.	Diversification of echinoderms, first agnathan vertebrates. Mass extinction at end of period.
Cambrian	570	7:30 P.M.	Sudden appearance of most marine invetebrate phyla, primitive algae.
Pre-Cambrian	1,000	4:00 P.M.	Trace fossils of marine algae, especially Cyanophyta. Origin of life in the dim past.
	3,000	12:00 A.M.	First life.

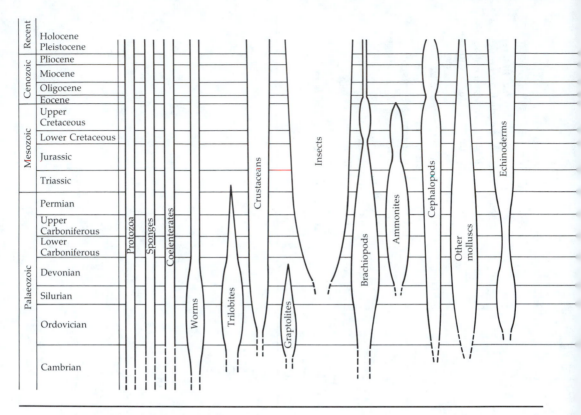

Figure 4.2 *The rise and fall of the main animal groups.* Species abundance is roughly indicated by the width of each group.

The Molecular Record

Another valuable source of information about the pathways of evolution is provided by proteins and nucleic acids. Each base pair in a DNA sequence is essentially a separate character in the code—hence there is a vast amount of potentially available information (Felsenstein 1982 and 1985). The most direct approach is to determine the nucleotide sequences of sections of chromosomes directly. Partial data on the sequence can be obtained from the use of restriction enzymes, which cleave DNA at particular sequences. Any segment of DNA, say mitochondrial DNA, can be cleaved into a series of fragments. If this process is carried out for each of the members of a cladogram under investigation, a comparison of the pattern of the fragments reveals the extent to which species differ in enzyme-specific nucleotide sequences. Other molecular techniques exist that test the dissimilarity of allozyme frequencies over a number of loci as a measure of the ''genetic distance'' among species (Avise and Aquadro 1982; Avise 1983; Buth 1984).

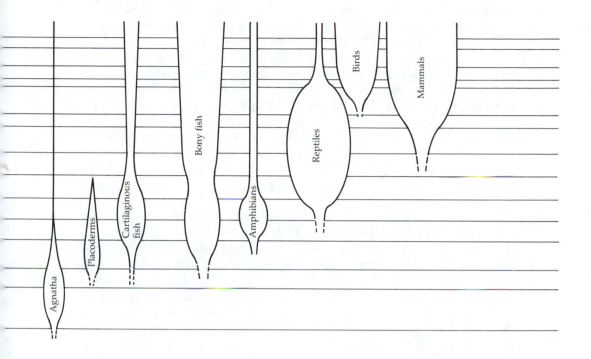

4.3 The History of Life on Earth

Early Life

Incredible as it may seem, the land mass, oceans, and atmosphere we know today can be traced to volcanic activity. The original form of Earth, formed by coalescence of material from the solar nebula 4.5 thousand million years ago, was largely a mixture of silicates, together with iron and sulphides. Water was not present in a free form but was bound to hydrated minerals such as mica in the Earth's crust. Estimates have been made that, at the present rate of activity, about 2.6 km³ of solid materials are contributed to the land mass each year by volcanic activity. Over 3–4 billion years, this amount is sufficient to account for the present continental land mass. What is more, 97 percent of volcanic gases consist of water, so water released from rocks condensed to form a major portion of the hydrosphere. Before life evolved, Earth had a reducing atmosphere, and only with the evolution of photosynthetic organisms about 3.2 billion

years ago did an oxidizing atmosphere begin to form. As terrestrial vegetation formed and decayed, organic soils began to form, dating from
about the Silurian (438 million years before present [B.P.]).

The essential step in the origin of life was the formation of replicating
DNA or DNA-like molecules possessing the properties now found in
genes (Dickerson, 1978). Miller's experiments (Miller and Urey 1953) and
Oparin's (1953) showed that molecules like amino acids and polypep-
tides could originate from an "organic soup" of chemicals such as hydrogen, water, ammonia, and methane with energy input from ultraviolet or
cosmic rays. The spontaneous appearance of nucleic acids has, however,
not yet been achieved, though 30 years of failure should not rule out the
origin of life on Earth by this process when 1,000 million years were available for its appearance.

For more than half the history of life on Earth, there were no recorded
living things. The earliest origins of life in the fossil record appeared
about 3.2 billion years B.P. at the beginning of the Precambrian era. Unicellular forms predominated until the sudden appearance during the
"Cambrian explosion" (600–700 million years B.P.) of multicellular forms.
In these early days, conditions were **anaerobic,** and fermentation provided most of the energy, but this process was inefficient and left most of
the carbon compounds untapped. **Aerobic** respiration releases much
more energy. The long period needed for the action of primitive eukaryotes to build up an oxygen layer through photosynthesis may explain the
2-billion-year gap between the origin of life and the appearance of multi-
celled animals, metazoans, which operate aerobically. The buildup of oxygen concommittently led to the demise of many of the early anaerobic
organisms—but not before they had their day; cyanobacteria (blue-green
"algae") and prokaryotes held sway for almost 2 billion years!

Multicellular Organisms

The stage of complex multicellular organisms occupies the last 700
million years of Earth's history. The appearance of the first eukaryotes
occurred about 900 million years ago and marked the appearance of chromosomes, meiosis, and sexual reproduction. These phenomena undoubtedly speeded up the process of evolution. In the 1970's and 1980's,
early "Ediacaran faunas" were recognized from the Precambrian era.
The organisms were soft-bodied and large. Without skeleton or shell for
muscle attachment, their movements would have been slow. They survived despite the absence of any protective shell because no predators
existed with sharp enough jaws to prey on them. Therefore, it has been
suggested that giant wormlike organisms existed in the distant past
where they could not survive today because of the diversity of predators
present (Southwood 1987).

The Appearance of Vertebrates

Coelenterates, arthropods, molluscs, and many other phyla were present in profusion by the Cambrian period. Famous among them are the now-extinct trilobites (Arthropoda), whose closest living relative is perhaps the horseshoe crab, *Limulus polyphemus* (shown in Photo 4.2) (Attenborough 1979). During the next 100 million years, these groups diversified widely and also suffered numerous extinctions, a process that carried on into the Ordovician period. During the Ordovician, the first chordates, jawless fish, were recorded. In the Silurian, the first jaws arrived with the "placoderm" fish, and it is likely that the two great fish classes, sharks and teleosts, also evolved at about the end of this time. Also in the Silurian, the first evidence of terrestrial life appeared: scorpions, millipedes, and simple vascular plants. In the Devonian, marine invertebrates, especially trilobites and corals, continued to diversify, and the first bony fishes appeared in the fossil record. The Devonian is sometimes known as the "age of fishes," and the sharks were very diverse. Devonian fish were often heavily armored to defend against predation, in contrast with modern fish, which emphasize speed. Amphibians appeared at the end of the Devonian, as did the first insects (collembolans), undoubtedly connected with the proliferation of bryophytes and gymnosperms, vascular plants.

In the Carboniferous period, insects radiated, though only cockroaches have remained relatively unchanged to the present day. The reptiles arose, and the amphibians radiated briefly. Huge, lumbering 5 m giants appeared, quite unlike the small forms present today.

Photo 4.2 *A living fossil. The horseshoe crab,* Limulus polyphemus, *has existed unchanged for hundreds of millions of years. It is the nearest surviving relative of the trilobites. Most live at considerable depths in the oceans off eastern North America and Southeast Asia, but every spring they appear along the coasts to mate and lay eggs. (Photo by P. Stiling.)*

Two hundred and eighty million years B.P., during the Permian period, the continents had aggregated into one central land mass—Pangaea. The extensive swamp forests of Pangaea in this period gave rise to today's rich coal beds. Reptiles and insects underwent extensive radiation, and metamorphic development appeared for the first time in insects (Neuroptera, Mecoptera, and Coleoptera). The amphibia suffered mass extinctions. Perhaps the most remarkable feature of this period, however, was the vast extinction of marine invertebrates, including the last of the trilobites and plankton, corals, and benthic invertebrates on a scale that commonly implies some worldwide catastrophe. Many plant species went **extinct** at the same time. At least 52 percent of the families of skeleton-bearing marine invertebrates became extinct—the greatest mass extinction the planet has yet known. This mass extinction can be explained by a species-area effect (see Chapter 18). Before Permian times there were two land masses, one lying near the South Pole and one at lower latitudes. When these two masses coalesced to form Pangaea, the area of continental shelf was much reduced, and the reduction may have precipitated the mass marine extinctions by leaving less area available to occupy (Simberloff 1974; Sepkoski 1976; Hallam 1984). The supercontinent must also have had extreme fluctuations in temperature, as large continents do today, which probably also caused many extinctions of terrestrial animals, though many fewer than occurred among the marine species.

During the Mesozoic era, beginning 230 million years ago, Pangaea started to split up into a southern continent, Gondwanaland, and a northern one, Laurasia. By the end of the era, Gondwanaland had formed South America, Africa, Australia, Antarctica, and India (which later drifted north), and Laurasia had split into Eurasia and North America. The land now called the Sahara Desert was probably located near the South Pole 450 million years ago and has since passed through every major climatic zone. Now still drifting northward at 1–2 cm yr^{-1}, the Sahara will move north 1° in the next 5–10 million years, and the climate and vegetation will change accordingly (Eckholm 1976).

Following the Permian extinctions, marine invertebrates began to diversify again, along with the reptilian fauna. By the Jurassic period, dinosaurs dominated the terrestrial vertebrate fossil records, and the first birds and mammals appeared. The whole of the Mesozoic, however, is generally known as the ''age of reptiles.'' Turtles and crocodilians had appeared, and giant dinosaurs stalked the earth. Two of the more famous species, *Triceratops* and *Tyrannosaurus*, are illustrated in Photo 4.3. Advanced insect orders such as Diptera and Hymenoptera were also evolving in conjunction with the first flowering plants, angiosperms. By the Cretaceous, however, dinosaurs had become extinct, as had many other animal groups, including, once again, much marine life, such as ammonites and planktonic Foraminifera. This extinction was, after the Per-

Photo 4.3 Dinosaurs. Artist's rendering of two large reptiles, Triceratops *and* Tyrannosaurus, *which lived 100 million years ago. (Photo Neg. No. 129205. Courtesy Department Library Services, American Museum of Natural History.)*

mian, the second greatest extinction in the history of life and was accompanied by a major drop in sea level. The explanation considered most likely now is that a severe change took place, especially a cooling of the climate. What brought this climatic change about is the subject of much debate, with meteorite collisions featuring prominently. On land, all vertebrates larger than about 25 kg seem to have gone extinct, but extinction was virtually undetectable in fish and plants. Snakes first appeared in the Cretaceous.

In the first two epochs of the Cenozoic era (from 65 million years B.P. to the present), most of the modern orders of birds and mammals arose, and the teleosts, angiosperms, and insects continued to diversify. By the middle of the Tertiary period, the world's forests were dominated by angiosperms. The continents arrived at their present positions early in the era but were connected and disconnected as the sea levels rose and fell. For example, during the Eocene and the Pleistocene, Central America formed a series of islands between North and South America. By the Miocene period, there were substantial numbers of vertebrate genera that persist today, most having first differentiated in the Paleocene. During this time existed *Baluchitherium*, an extinct rhinoceros. At 18 feet high at the shoulder, it is the largest land mammal known, weighing about 30 metric tons (modern elephants rarely weigh more than 10 tons). Photo 4.4 depicts *Brontotherium*, another huge rhinoceros, and other animals from the same period, all now extinct. The elephants did not appear until the Upper Eocene and evolved into a great diversity of forms, of which only two species survive today.

In the **Pleistocene,** about 2 million years ago, there were four Ice Ages, separated by warmer interglacial periods. During the Ice Ages, mammals adapted to cold conditions came southward—reindeer and arctic fox roamed in England, and musk-ox *(Ovibos)* occurred in the Southern United States. Conversely, in the interglacial periods, species spread northward from the tropics—the lion is known from northern England and the hippopotamus from the River Thames. During the **glacial periods**, some species that had moved south during the interglacial periods

Photo 4.4   Life as it may have appeared in the South Dakota-Nebraska region of the United States 35 million years ago. The large animal is a Brontotherium, *a huge type of rhinoceros. Also present are an ancestral tapir and a group of small three-toed horses,* Mesohippus *(left). (Photo © Copyright Smithsonian Institution, no. MNH-969.)*

became restricted to isolated pockets of cool habitat, especially mountain tops. The most important extinctions at this time were large mammals and ground birds, the so-called megafauna. For example, in North America, *Megatherium*, the giant 18-foot ground sloth, became extinct about 11,000 years ago. By 13,000 years ago, humans had crossed the Bering land bridge into the New World, and these extinctions could well represent the first of many extinctions caused by hunters (Martin and Klein 1984*b*). The fruits of neotropical trees today seem still to be adapted to dispersal by these extinct large mammals (Janzen and Martin 1982).

Patterns in the History of Life

Are there any trends in evolution? Some have been suggested. Increasing complexity of form is one, whereby, for example, ammonite shell coiling and structure became more complex. Another is that mammals have evolved toward larger body size. (This is known as *Cope's rule* but is not an invariable trend.) Likewise, brain size has increased, relative to body size, in many birds and mammals, but again this is not an invariable trend (Radinsky 1978). The most obvious feature of evolution is a lack of unidirectional change; adaptive radiation is the rule. Thus, although one may review the equid lineage from modern horse (*Equus*) back to dawn horse (*Hyracotherium*) and obvious trends in teeth, toe number, and body size become apparent, when the entire equid clade is considered, a complex branching pattern of lineages is evident, giving rise to various grazers and browsers of different shapes and sizes (Fig. 4.3). The reason, of course, is that no single force or plan directed the evolution of this lineage. As vegetational types and forms diversified to meet different environmental conditions, the members of the equine lineage diversified with them. In cases in which predator and prey shape one another's ad-

Hindfeet   Forefeet

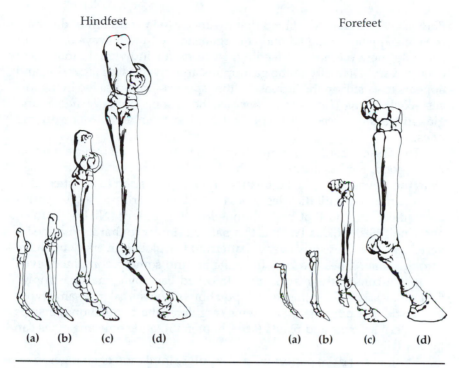

(a)   (b)   (c)   (d)   (a)   (b)   (c)   (d)

Figure 4.3 Trends in the evolution of horse feet. **(a)** Hyracotherium. **(b)** Mesohippus. **(c)** Merychippus. **(d)** Equus.

aptations, the changes may be more linear—the prey become faster as the fastest individuals escape to reproduce, the predators become faster as the fastest individuals catch more prey, and so on. **Mutualisms** (interactions in which species benefit one another) are another area in which evolution may seem to follow a linear path. Examples are given in Chapter 10.

Rates of Formation of New Species

What are the rates of formation of new species and the rates of extinction of old ones? What patterns issue from the fossil record? Marine invertebrates have left the best fossil records and have been the most intensely analyzed. Sepkoski (1984) documented a steady rise in the number of these families, which reached a plateau in the Ordovician, suffered a major extinction in the Permian, and have shown a steady increase in diversity ever since; at least 1,900 families of marine invertebrates are now recognized. Some of this increase has been attributed to *provincialization*, the appearance of differentiated regional biotas, but

Bambach (1983) has concluded that the increase in community diversity is a consequence of the addition of organisms with new ways of life, for example, new methods of feeding. Patterns for individual families, of course, vary. Diversity in the gastropods and bivalves is staggering and appears to be still on the increase. Other species—the so-called living fossils—coelacanths *(Latimeria chalumnae)*, horseshoe crabs *(Limulus)*, and ginkgo trees *(Ginkgo biloba)* represent the last members of once-diverse lines.

The fossil records for groups other than the marine invertebrates is not so good. For vascular plants, the diversity appears to have increased from the Devonian to the Permian, dropped, then risen to a plateau that was maintained until the Mesozoic. In the Upper Triassic, angiosperms diversified, a trend that has continued to the present (Niklas, Tiffney, and Knoll 1980; Niklas 1986). Is this pattern similar to that for marine invertebrates? Is there any overall pattern to the global diversity of all life through time? Does diversity fluctuate around some preordained level, or does it constantly increase until knocked back by some catastrophic natural disaster and extinction? Population biologists ask the same types of questions about what (if anything) regulates the population densities of modern animals and plants (see Chapter 15), and some analogies can be drawn.

Raup et al. (1973) performed a computer simulation of changes in diversity on the assumptions that a lineage could branch, go extinct, or remain unchanged in a given time period. Extinctions and speciation events were assumed to occur randomly and to occur, on average, with the same frequency. The results mirrored many real historical patterns of diversity, suggesting that the available fossil record is largely a result of random extinctions and speciation events through time. Some particular biological phenomena were not well predicted: the long-term survival of "living fossils" was one; another was that lineages often increase and diversify in the fossil record much more rapidly than they would at random.

Often the origin of new taxa is correlated with the extinction rate (Stanley 1979). The result, of course, is that diversity remains unchanged, leading some to believe that even historically distant communities were saturated with species and that new ones could succeed only in the place of old, extinct ones. Alternatively, it may be more likely that environmental changes affect both processes concurrently. Stanley (1975) assumed that, for newly arisen taxa, the number of species was able to increase in an exponential fashion following the equation $N_t = N_0 e^{rt}$ (see Chapter 9) because there were no competing species to usurp the existing "niche space." In this formula, r, the per-capita rate of increase, equals $O - E$, the rate of origin or speciation minus the rate of extinction. If the time t since origin is known from the fossil record, the initial number of species is assumed to be 1, E is calculated from the average life span of fossils, and N_t is the number of species existing, then O could be solved for. Stanley concluded that speciation rates of mammals were higher than those of bivalves (Table 4.2).

Sepkoski (1978, 1979, and 1984) has made the analogy between historical diversity and another population model, the logistic (see Chapter 9). This model assumes initial rapid population growth followed by a leveling off at an asymptote as resources for growth become limiting; the result is a **sigmoid curve**. The upper asymptote is depressed if other competing species are present to lower the level of available resources still further. For marine invertebrates in the entire Paleozoic, diversity fits a logistic model. This result implies that, as diversity increases, speciation rates decline and extinction rates rise, possibly because of competition between species. However, one must remember that in many cases a group has diversified only shortly after the demise of another lineage created a potential "empty niche." Thus the crocodilians invaded their present habitats only after the phytosaurs became extinct. The great decline of branchiopods at the end of the Permian was followed by an explosive radiation of clams. In fact, the evidence for this type of event outweighs that for competitive exclusion of one lineage by another (Raup 1984; Jablonski 1986a).

Table 4.2 *Estimated rates of speciation O, extinction E, and increase in diversity R. (After Stanley 1975.)*

Phylogenetic group	t (Million Years)	N (Species)	$\bar{r}$	r	E	O
Bivalvia				0.07	0.17	0.24
Veneridae	120	2400	0.06			
Tellinidae	120	2700	0.07			
Mammalia				0.20	0.50	0.70
Bovidae (cattle, antelopes)	23	115	0.21			
Cervidae (deer)	23	53	0.17			
Muridae (rats, mice)	23	844	0.29			
Cercopithecidae (Old World monkeys)	23	60	0.18			
Cebidae (New World monkeys	28	37	0.13			
Cricetidae (mice)	35	714	0.19			

Rates (Per Million Years) spans the $\bar{r}$, r, E, and O columns.

Human Evolution

That humans and apes are descended from common stock is undoubted. Only about 1.1 percent of the base pairs differ between human and chimpanzee genomes (King and Wilson 1975). Indeed, humans are now classified within the family Pongidae rather than as a separate family. The exact relationships among the great ape genera *Pongo* (the orangutan), *Pan* (the chimpanzees), *Gorilla* (the gorillas), and *Homo* (man) are still debated (Futuyma 1986).

Many anthropologists have discovered humanoid forms and argued forcibly for their priority over other forms as the likely ancestors of humans. As for so many other species, several such forms probably existed contemporaneously. Only one of these can actually have given rise to *Homo sapiens*—the others were merely other branches of the evolutionary bush—and debate rages over which one it was. The emotional nature of the subject means that arguments are more commonly infused with emotive pleas and, perhaps, more speculation than might normally be permitted in scientific debate.

The first undoubted hominids appeared 3.8–3.6 million years ago in Tanzania and Ethiopia. Fossil records of bipedal footprints and the bones of "Lucy," a bipedal *Australopithecus afarensis*, have been found. Lucy's species looked more like a lightweight gorilla than like modern man (Johanson and White 1979). The name comes from the Latin *austral* (southern) and the Greek *pithekos* (ape). The canine teeth were reduced, and in their place primitive wooden tools may have been used; no stone tools have been found.

About 2 million years ago, these African hominids diversified into a range of species, all but one genus of which, *Homo*, were to go extinct. The genus *Homo* is associated with a larger brain than the other groups with which it was contemporaneous. *Homo habilis* (handy man, first user of stone tools) gradually graded into *H. erectus* (upright man, a fire user who soon appeared in Asia by way of Arabia and survived there until 300,000 years ago), and then *H. sapiens* (wise man), which included the Neanderthals, who lived in the Neander Valley in Germany about 40,000 years ago. The different varieties of ancient *H. sapiens* discovered probably represent no more racial variation than is present between a pygmy and a tall Masai warrior today.

The major changes involved in the development of hominoids were development of bipedal gain, which left the hands free for tool making and use, a change in diet from frugivory to omnivory, and an increase in brain size. Dramatic though these physical changes were, in no way did they allow humans to shape their environment on the massive scale now observed (see Section Six). The ability to shape the environment is the result of cultural development—perhaps the only example of Lamarckian evolution, the evolution of acquired characteristics, at work. Language and, more importantly, a writing system could be passed on to future generations, enabling cultural evolution to proceed at an exceptionally high rate.

Historical Biogeography

The evidence for macroevolution comes not only from the fossil record but also from a study of the present-day faunas of different continental land masses. **Biogeography** is the branch of biology that deals with the geographic distribution of plants and animals. South America, Africa, and Australia all have similar climates, ranging from tropical to temperate, yet each is clearly characterized by its inhabitants. South America is inhabited by sloths, anteaters, armadillos, and monkeys with prehensile tails. Africa possesses a wide variety of antelopes, zebras, giraffes, lions, and baboons, the okapi, and the aardvark. Australia, which has no placental mammals except bats and the introduced mouse and dingo, is home to a variety of marsupials such as kangaroos and the peculiar egg-laying mammals or monotremes, the duck-billed platypus and the echidna. A plausible explanation is that each region supports the fauna best adapted to it, but introductions have proved this explanation incorrect; rabbits introduced into Australia proliferated rapidly. The best explanation is provided by macroevolution and **continental drift.**

The arrangement of the seas and land masses on Earth has changed enormously over time as a result of **plate tectonic** forces. The Earth consists of a molten mass overlain by a solid crust about 100 km thick. This crust is not a single continuous piece but is broken into a number of irregular pieces, called plates. As the molten material below rises along the cracks between the plates, it pushes them aside and cools to form new edges to the crack. The irregular, tumbled edges of these cracks are the midoceanic ridges (Fig. 4.4). As the plates are pushed aside, their opposite edges meet. Where they crumple or one edge is forced under the other, mountain chains are formed (Fig. 4.5). The continents and parts of them are carried along with the movement of their plates. The drift of these plates also has profound influences on climate and the geographical distribution of organisms. The theory of plate tectonics was first crystallized by a German meteorologist, Alfred A. Wegener, in 1910 (Wegener 1966) and has since been reinforced by a variety of evidence.

Since the separation of the continents by continental drift, the organisms on them have evolved independently of one another. The phenomenon called **convergent evolution**, a process in which similar conditions promote the evolution of similar appearance and lifestyle, has led to the emergence on each continent of grazers, anteaters, large predators, and so on, but the geographic separation has meant that these various types have arisen evolutionarily from different stocks (Gray and Boucot 1979; Bambach, Scotese, and Ziegler 1980; Hallam 1983).

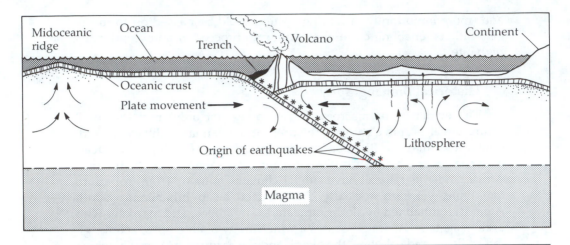

Figure 4.4 *Subduction.* The rigid crust of one tectonic plate is overridden by the edge of another, and a trench forms where they meet. Arrows in the more plastic layer show possible patterns of convective flow.

The breakup of a supercontinental land mass into constituent continents and the eventual reformation of a supercontinent is probably a cyclical event, with a distinct periodicity. The latest of these supercontinents, Pangaea, and its subsequent breakup into Laurasia and Gondwana (135 million years B.P.) and later into the present-day land masses are shown in Figure 4.6. This latest cycle is probably the most important in terms of influence on the present-day distribution of plants and animals. Throughout the breakup process, the climate varied tremendously. Eastern Pangaea was hot and humid, but there were glaciers in southern Pangaea in the Carboniferous. Sea levels rose several times, producing transgressions, the spread of epicontinental seas over the land, and then they fell again, producing regressions. Brown and Gibson (1983) give a detailed account of the movement of land masses during continental drift.

The most-cited example of a disjunct distribution caused by continental drift is probably the restricted distribution of monotremes and marsupials. These animals were once plentiful all over North America and Europe. They spread into the rest of the world, including South America and Australia, at the end of the Cretaceous when, although the continents were separated, land bridges existed between them (Fig. 4.7). Later, placental mammals evolved in North America and displaced the marsupials there entirely. However these "new" mammals could not invade Australia because the land bridge was by then broken. In the brief time that South America was connected to North America, many placental mammals crossed into the neotropics and flourished. A few mam-

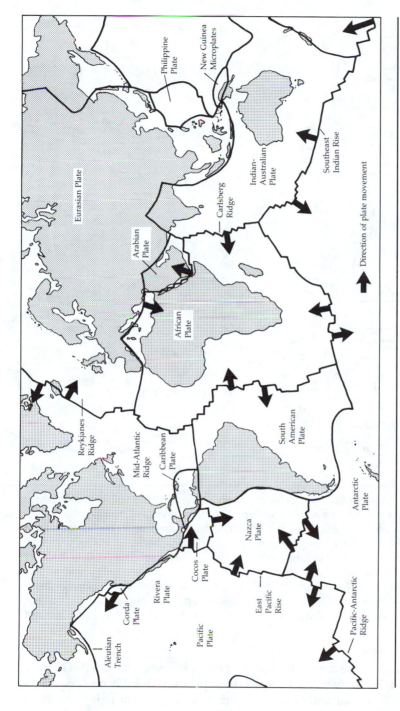

Figure 4.5 *Global map showing the location of the major oceanic and continental plates.* The oceanic trenches where plates are consumed and the oceanic ridges or rises along which seafloor spreading occurs are shown as thick lines. Arrows indicate the direction of spreading and, hence, the relative movement of each plate. (Redrawn from Uyeda 1978.)

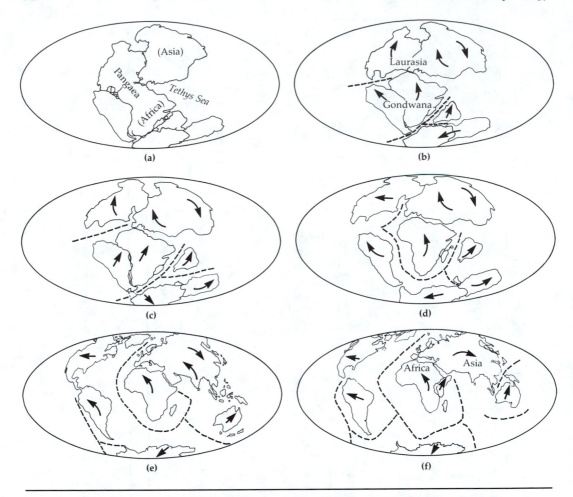

Figure 4.6 Continental drift. **(a)** *Supercontinent of Pangaea, 200 million years ago.* **(b)** *Pangaea breaking up into a northern supercontinent, Laurasia, and a southern supercontinent, Gondwana, 135 million years ago.* **(c)** *Continuing breakup, 100 million years ago.* **(d)** *Disappearance of dinosaurs and opening of the South Atlantic, 65 million years ago.* **(e)** *The Earth today.* **(f)** *The future, 50 million years from now.* The Atlantic widens, the Pacific shrinks, and, on the west coast of the United States, today's city of Los Angeles has drifted to the north of San Francisco. Arrows show general directions of plate movements. Broken lines show the approximate location of rifts. For simplicity, the plate boundaries, which are largely under the oceans, are not shown here.

mals, including the armadillo and the porcupine, were successful in trekking north, but only one marsupial, the opposum (see Photo 4.5), established itself permanently in North America. This distinct inequality of faunal exchange has led many to believe the placental mammals were competitors more often intrinsically superior to their marsupial ''cous-

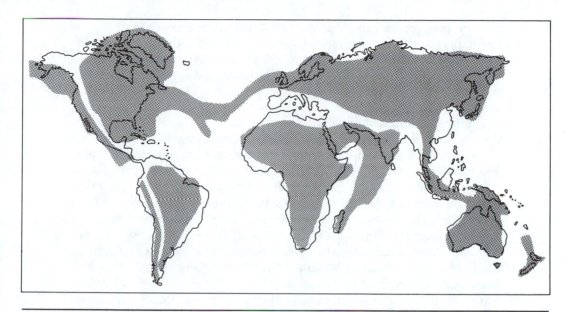

Figure 4.7 *Probable distribution of land during the Cretaceous period (shaded area) superimposed over a map of the modern world (solid outline).*

ins.'' Part of the reason for this inequality in the ''Great American Interchange'' was that North America had known many previous invasions. For South America, this was the first contact with many potential competitors and predators (Marshall 1988). Also, the total surface area of North America and Central America (24 million km^2) was greater than that of South America (18 million km^2). One could thus invoke the species-area effect (see Chapter 18) to explain why there was an average

Photo 4.5   The American opossum, Didelphis virginiana, *the only marsupial to cross the land bridge from South America and establish itself successfully in North America, a continent dominated by placental mammals. (Photo from U.S. Department of Agriculture.)*

of 60 percent greater generic diversity in the north and hence more potential dispersants. When this fact is taken into account and only true dispersants (those that did not evolve further) are considered, the interchange between North and South America is balanced—dispersants moved in proportion to the size of the faunal sources from which they came (Marshall 1988).

It must be stressed that plate tectonics can only explain the distribution of taxa that were already present when Pangaea, Laurasia, or Gondwanaland broke up. Thus the distribution of an essentially flightless bird family, the ratites—ostriches, emus, and rheas—in Africa, Australia, and South America is the result of continental drift. On the other hand, the Elephantidae and Camelidae have wide distributions, but both evolved after the breakup of the continents. Elephants evolved in Africa and subsequently dispersed on foot through Eurasia and across the Bering land bridge from Siberia to North America, where many are found as fossils. They subsequently became extinct everywhere except Africa and India. Camels evolved in North America and made the reverse trek across the Bering bridge into Eurasia; they also crossed into South America via the Central American isthmus. They have since become extinct everywhere except Asia, North Africa, and South America. Nevertheless, the breakup of the continents in general and of Gondwanaland in particular resulted in many disjunct distributions, particularly among southern continents such as the tips of South America, Africa, and Australia.

Modern Geographic Patterns

The result of continental drift and the evolution of different lineages on different land masses can be seen in the distribution patterns of today's flora and fauna. Although a few species, such as the barn owl, the painted lady butterfly, and various human-introduced pests like rats, are virtually cosmopolitan in distribution, most higher taxa are endemic (native or restricted) to a particular geographic region. Wallace (1878) was one of the earliest biogeographers to realize that many endemic taxa had more or less congruent distributions. For example, the distribution patterns of guinea pigs, anteaters, and many other groups are confined to Central and South America, from central Mexico southward. Although some of the distributions of the animals in this area are of course related to available habitat and extend north or south to slightly different degrees, the whole area was distinct enough for Wallace to proclaim it the "neotropical realm." Wallace went on to divide the world's biota into six major realms, or zoogeographic regions (Fig. 4.8), and these are still widely accepted today, though debate still continues about the exact locations of the boundary lines. Realms correspond largely to continents but more exactly to areas bounded by major barriers to dispersal like the

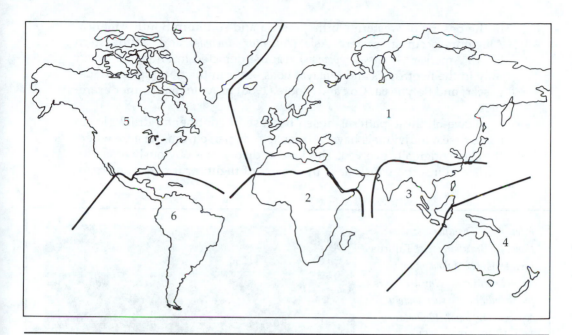

Figure 4.8 *The zoogeographic regions recognized by A.R. Wallace.* These regions are Palearctic (1), Ethiopian (2), Oriental (3), Australian (4), Nearctic (5), and Neotropical (6). Note that the borders (which are rather arbitrary) do not necessarily demarcate the continents.

Himalayas and the Sahara Desert. Very often, similar "ecological niches" in different biomes will be occupied by species with similar appearance and habits but from different taxonomic groups. For example, by convergent evolution the kangaroo rats of North American deserts, the jerboas of central Asia, and the hopping mice of Australia look similar and occupy similar hot, arid environments but they arise from different lineages (belonging to the families Heteromyidae, Dipodidae, and Muridae, respectively). Observations of this kind are strong arguments for the reality of evolution.

Many other geographical patterns are evident. Bromeliads (members of the Bromeliadaceae, the pineapple family) are virtually all neotropical, the Myrtaceae (*Eucalyptus* and its relatives) are restricted to Australia, and voles (Microtinae) and maples (Aceraceae) occur in the holarctic, the northern temperate realm of the New and Old Worlds. The Columbidae (the doves) are virtually cosmopolitan. Flight is not the only reason—the Todidae (small kingfisher-like birds) can fly but are restricted to the Caribbean islands, whereas skinks (Scincidae) are entirely terrestrial lizards but are cosmopolitan in most tropical and temperate areas. Many other distribution patterns cover smaller areas. Creosote bushes (*Larrea*) occur in the deserts of western North America and southern South America. Alligators occur in eastern Asia *(Alligator sinensis)* and southeastern North America *(A. mississippiensis)*. Tapirs, depicted in Photo 4.6, occur only in the neotropics and Malaya; boas occur in the neotropics, Madagascar, and Polynesia. Cox and Moore (1986) provide many other examples.

Have all these patterns arisen because of continental drift? No. In some cases, individuals have been able to disperse from the area where the clade originally evolved. This kind of dispersal is obviously easier for birds and insects, which have the power of flight, or for aquatic organ-

Photo 4.6   Disjunct distributions. (a) Brazilian tapir, Tapirus terrestris. *(b) Malayan tapir,* Tapirus indicus. *Why are tapirs found only in two such widely separated regions? (Photo (a) by Jerry Wolff, Mammal Slide Library, American Society of Mammalogists.) Photo (b) by Tracy S. Carter, Mammal Slide Library, American Society of Mammalogists)*

(a)

(b)

isms, which can drift with the tide (see Chapter 8). Bats, for example, are the only mammals native to Hawaii and New Zealand. Frogs and sala-manders are absent from most oceanic islands.

Other, more recent geological phenomena can explain the distribu-tions of some plants and animals. For example, during the glacial periods of the Pleistocene, lakes covered much of what is now desert in the U.S. Southwest. With the retreat of the ice, the lakes disappeared. Desert pupfishes *(Cyprinodon)* and other aquatic organisms were once widely dispersed throughout the Death Valley region but now occur only in iso-lated springs. Similarly, tapir fossils are found widely over the globe, and their present spotty distribution in South America and Malaya merely represents the relict populations of once-widespread groups.

4.4 *Extinctions: Causes and Patterns*

The extinctions of South American marsupials concurrent with the inva-sion of placental mammals from the north is just one of many purported causes of extinction. It is important to realize that extinction is the rule rather than the exception. Of the 500 million to 1 billion species that have existed during evolutionary time only perhaps 2–30 million are alive now. Humans may have contributed to the ultimate demise of some ani-mals, hunting large mammals to extinction in prehistoric times and, more recently, exterminating species such as the dodo and passenger pigeon. Others may well have fallen victim to disease. Within the last few de-cades, Dutch elm disease, *Cerastosomella ulmi,* has devastated English elms, and the American chestnut has been almost eliminated by the fun-gus *Endothia parasitica.* Smallpox and measles certainly facilitated the con-quest of South American Indians by the Spaniards. In all these cases, novel pathogens caused great havoc because the hosts were not adapted to combating them.

Ultimate Causes of Extinction

For most species in geologic time, and even for some in historical time, very little is known about the immediate causes of extinction (Simberloff 1986c). Predation, parasitism, and even competition can have severe impacts on populations of many species, but perhaps habitat alter-ation as a result of climatic change is the prime moving force in evolution over geologic time. Recently much extinction has been anthropogenous; indeed Soulé (1983) believes there are no modern documented examples of continental species extinguished by nonhuman agencies.

Over evolutionary time, Van Valen (1973) suggested that within most taxonomic groups the probability of the extinction of a genus or family is independent of the duration of its existence. Old lineages do not die out

more readily than younger ones. For example, among marine invertebrates, the average lifetime of a genus has been 11.1 million years and for
Carnivora 8 million years. Within these classifications, the figures vary
tremendously, from 78 million years for bivalves to 7.3 million years for
ammonites. Again, great care must be exercised in interpreting these records. From early to late Phanerozoic times, the number of species described per genus has generally increased. Under the influence of such a
trend, because extinction of a family requires extinction of all its species,
extinction rates of families will decline even if the probability of extinction
of species is constant (Flessa and Jablonski 1985). The same is true of genera. The assertion that the basal rates of extinction have speeded up at
certain times is a matter of some contention. The so-called periods of
''mass extinction'' at the ends of five geological eras have been argued to
be simply the quantitatively extreme cases in a basal array of extinction
rates (Quinn 1983). If each major extinction is different, however, catastrophic interruptions of the normal course of earthly events must be invoked. Perhaps most speculation has centered on the vanishing dinosaurs at the end of the Cretaceous. Alvarez et al. (1980) and others have
suggested that an asteroid or large meteorite collided with Earth at that
time and that the subsequent pall of dust darkened the skies, lowered the
temperatures, and killed off vegetation, both on land and in the sea. Evidence both for and against this argument has been accrued, and the debate rages on (see reviews by Jablonski 1984 and 1986b; Van Valen 1984).
The biggest smoking gun in the meteorite debate is the Manson crater in
Iowa, which was recently dated at 66 million years old, coincident with
the demise of the dinosaurs. This crater, now covered by a cornfield, is 22
miles wide and must have been caused by a giant asteroid, which crashed
into Earth at 43,000 mph with a force 10 times that of all the nuclear weapons on Earth combined. Ray Anderson, a geologist with the Iowa Department of Natural Resources, was quoted in a newspaper as saying, wryly,
that he didn't know whether that collision killed all the dinosaurs on
Earth but that it certainly wiped them out in Iowa (*Tallahassee Democrat*,
August 1, 1989). The literature abounds with theories concerning the extinction of the dinosaurs. Proposed explanations have included too cold a
climate, too hot a climate, primitive mammals that ate the dinosaurs'
eggs, overgrazing by herbivorous dinosaurs, and reversals of the Earth's
magnetic field that permitted lethal doses of cosmic radiation to reach the
Earth's surface. Raup and Sepkoski (1984) have suggested that mass extinctions occur regularly, with a periodicity of about 26 million years. If
they are right, then past adaptations of species provide little preadapta-
tion to extraordinary periodic conditions.

There is some evidence that the survivors of mass extinctions tended
to be the more ecologically and morphologically generalized species (Fig.
4.9). Specialization can hamper adaptation to changing conditions, and

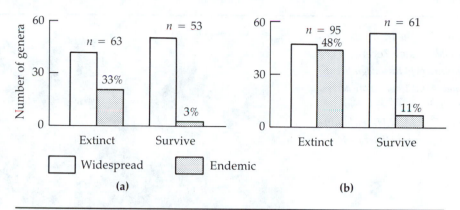

Figure 4.9 *The proportion of genera that survived the extinction event at the end of the Cretaceous period.* **(a)** *Bivalves.* **(b)** *Gastropods.* Among both bivalves and gastropods, the proportion that survived was greater for geographically widespread forms than for genera with narrow geographical distributions (endemics). The histograms plot the percentage survival and extinction of bivalves and gastropods of the Gulf and Atlantic coastal plain of North America, distinguishing those restricted to the region from those that were more widespread. (Based on Jablonski 1986*b*.)

overspecialization actually can work against a species. Although some modern workers disagree, the oyster *Gryphaea* once was argued to have developed such a coiled shell that the valves could no longer open and starvation ensued. As another example, the Irish elk, *Megaloceros giganteus*, depicted in Photo 4.7, stood 2 m at the shoulder and possessed a set of antlers measuring 4 m across and weighing over 20 kg. At one time, it was supposed that natural selection over many generations caused its antlers to become so huge, perhaps as a specialization for mating display or fighting among males, that the animal became slow and cumbersome and tended to tangle its antlers in trees, falling easy victim to predators. The elk, which lived across most of Europe and Asia (although most remains were found in Irish peat beds), went extinct about 11,000 years ago. A more likely explanation for its extinction is that it simply could not adapt to the tundra that developed following a glaciation or to the heavy forestation that followed a warm interglacial (Gould 1974). Habitat changes from climatic alterations have probably been the major cause of extinction in evolutionary time. Morphologically complex and specialized species tend to occur in more specialized environments and are thought to be more susceptible to local environmental changes. Generalists tend to have a greater breadth of geographic distribution, which appears to be important in enhancing survival.

Photo 4.7   The Irish ''elk'' (actually a deer), Megaloceros. *The huge antlers of this magnificent animal, eight feet or more across, were the biggest ever known. (Photo © F.M.N.H., Field Museum of Natural History, Neg. no. 63060.)*

Punctuated Equilibrium

Not all species that disappeared from the fossil record truly became extinct; some evolved into new organisms, sufficiently different to merit assignment of a new species name, a sort of taxonomic extinction. (This is the reason that in **palaeontology** many treatments of extinction deal with higher taxonomic categories, like genera and families.) Such gradual transitions, however, are rare. Fossils of new species usually appear suddenly, with few transitional types. This phenomenon has often been brushed off as simply an inadequacy of the fossil record. More recently, the theory of **punctuated equilibrium** has been introduced. It argues that the fossil record simply represents what has happened in nature (Gould and Eldredge 1977) and that new species *do* appear suddenly. Old species are not commonly transmuted evolutionarily into new ones; rather new types branch off very quickly from parental stock and are to be found contemporaneously with it. Any advantageous new mutation will become very quickly established in populations, often giving rise to new species. The parent stock may die out after the new species has become well established. This process is too quick for the slow process of fossilization to reflect accurately. Thus most evolution occurs not as ''ladders,'' in which one species slowly turns into another and in turn into another, but as ''bushes,'' of which only a few active tips survive in the long term. Evolution by natural selection will usually progress more rapidly than a coarse fossil record can document. Turner (1986) has written that population genetics predicts punctuational evolution. Long fossil series exist over time because species ''seek out'' preferred habitats, stay in them, and evolve little themselves. The tadpole shrimp, *Triops cancriformis,* a freshwater crustacean found in Africa, has existed unchanged for 180 million years. It is the oldest known living species on Earth (Stanley 1979). Dawkins (1986) has argued that the two camps, punctuated equilibrium and gradualism, are really not at all different, that the issue is a semantic one blown out of all proportion by the media. Even punctuated equilibrium, he argues, would be seen as a gradual process if the fossil record were only fine enough to show the details. Even what appear to be complex changes in morphology may arise from very simple changes in developmental rates. D'Arcy Thompson (1942) pioneered the study of shape and showed how quite simple mathematical transformations could turn one shape into another (Fig. 4.10). Similarly, Raup (1962; 1966) showed how all the complicated coils and patterns of shells could be generated by variation in only four basic parameters.

Biases in the fossil record can also affect estimates of speciation rates and of the number of species even if events have remained constant through time. Younger rocks have, for example, been subject to fewer destructive forces, such as erosion, than older ones. Therefore, the latest occurrences of a species are more likely to appear in the fossil record than

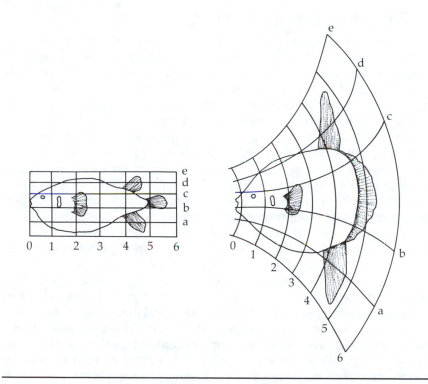

Figure 4.10   *D'Arcy Thompson's method of transformations.* If the coordinates at left are replaced by those at right, the shape of a puffer *(Diodon)* is transformed into that of an ocean sunfish *(Orthagoriscus* = *Mola).* (Redrawn from Thompson 1942.)

earlier ones. The geographic coverage of young rocks will also be greater. Also if an extant group is known from one fossil, its geological range is known to extend from then until now. A single occurrence in the fossil record of a nonextant species leads to an estimate of a very brief geological range. This phenomenon has been termed the ''pull of the recent'' (Raup 1979). Other biases abound.

The bottom line is that for many extinct lineages there is still no adequate explanation. The horse, *Equus*, had evolved entirely in North America, but 8–10,000 years ago it went extinct there. Yet, when introduced by the Spaniards in 1519, herds became established very quickly. Sometimes, after extinctions, the same morphological forms reappear in the fossil record, derived from new ancestors. More often, the chief impact of extinctions is the obliteration of forms of life whose like never reappears (Strathmann 1978). In summary, it is certainly hard to say why some species die and others persist. Out of the many genera of horse-like animals, why did *Equus* alone survive? Was it really structurally better adapted than the others, or was it merely lucky that its habitat persisted?

Proximate Causes of Extinction

Simberloff (1986c) has addressed the issue of **proximate causes** of extinction, the reasons behind the death of the last survivors of a relict population. He stresses that it would be nice to know the ultimate factors as well, the reasons why a population becomes small in the first place. However, in terms of conservation, the proximate reasons for extinction are valuable to managers who hope to guard against them when trying to conserve populations of rare plants and animals, especially in nature reserves. For example, introduced predators such as rats, cats, and mongooses have accounted for at least 112 of 258 recorded extinctions of birds on islands, or 43.4 percent (Brown 1989). Hunters account for 15.5 percent of these 258 extinctions, habitat destruction accounts for 20.1 percent, and disease accounts for 5.8 percent.

Four forces seem to be important in the final extinction of species— demographic stochasticity (see Chapter 9), genetic deterioration (Chapter 2), social dysfunction, and extrinsic forces, such as disease or weather. Simberloff (1986c) provides an interesting essay on the demise of the heath hen (Photo 4.8) in the northeastern United States in a seemingly safe refuge on Martha's Vineyard, an island in southeast Massachusetts. First, a fire spread throughout the reserve during a gale, then a hard winter followed, as did an unprecedented flight of predator goshawks. The population was reduced from over 2,000 to about 150, mostly males—many females were killed on their nests during the fire. The population was unable to recover from this skewed sex ratio, and the last bird

(b)

(a)

Photo 4.8   Extinction. (a) The heath hen went from a poulation size of 2,000 in 1915 to extinction in 1932 as a result of a series of unexpected setbacks. (b) The blackfooted ferret of Wyoming had a population size of only 31 in 1985 and would be extremely vulnerable to the kind of natural catastrophe that caused the extinction of the heath hen. Will it go the same way? (Photos by Luther Goldman, U.S. Fish and Wildlife Service.)

died in 1932. The story of the heath hen emphasizes how precarious is the survival of other species with even smaller population size. For example, in 1985, the black-footed ferret, illustrated in Photo 4.8, had a dangerously low population size of only 31 individuals. Similar detailed accounts of other recent extinctions are lacking (but see Pettersson 1985). Often extinction is brought about so quickly by ultimate causes that there is little time to study proximate causes (Gilpin 1987). It is worth bearing in mind then that disease could quickly wipe out whole herds of wild game in Tanzania. The best way to prevent this possibility is to establish more than one game park to safeguard a species.

Genetic deterioration is a definite threat to small populations and may arise from inbreeding or from **genetic drift**. Over the long term, the loss of genetic variation resulting from genetic drift occurs at a rate roughly proportional to $1/2\,N_E$, where N_E is the effective population size. Remember that effective population size is often much smaller than the censused population (Chapter 2). There is a dispute about just how small a population must be before genetic drift removes alleles in the face of

natural selection; estimates range around 500 (Simberloff 1988b). Inbreeding can cause more individuals to become homozygous, which is
likely to result in inbreeding depression, and population sizes of more
than 50 have been deemed necessary to prevent such phenomena
(Simberloff 1989). These figures, 50 and 500, have been thought of as
"magic numbers" in refuge design and have been raised to the status of
a rule in some management circles (Foose 1983; Lehmkuhl 1984;
Salwasser, Mealey, and Johnson 1984; Wilcox 1986). Soulé and others
(Soulé 1987), however, have shown just how incorrect this approach is,
despite the fact that Ralls and Ballou (1983) have shown how harmful inbreeding and genetic drift can be for primates, ungulates, and small
mammals in zoo environments. The **harem** mating structure of many
species (see Chapter 7) probably promotes inbreeding in these species in
the wild, and they seem not to be much affected by it (Bonnell and Selander 1974; Frankel and Soulé 1981). Templeton and Read (1983) demonstrated little inbreeding depression in a captive herd of Speke's gazelles.
Similarly, many self-fertilizing plants have little genetic variation (Clegg
and Brown 1983). Overall, however, nature seems to abhor inbreeding.
In a review of 12 species of birds and 15 species of mammals in the wild,
all except two show parent-offspring matings to be less than 6 percent of
the observed total (Ralls, Harvey, and Lyles 1986). The mechanism that
prevents inbreeding may be dispersal or specific kin-recognition mechanisms.

Soulé (1980) has suggested that, for conservation of wild, noninbred
animals, a per-generation rate of inbreeding of less than 1 percent (equivalent to $N_E = 50$) is necessary to prevent the fixation of deleterious alleles
by inbreeding. In practical terms, even if 500 is accepted as a minimum
viable population size, for some animals 500 individuals cover a lot of territory. Reed, Doerr, and Walters (1988) calculated that to maintain genetic variability a red-cockaded woodpecker population must contain 509
breeding pairs to be considered viable. It is likely that no existing population contains 509 breeding pairs. According to this criterion, the area required for a viable population would be at least 25,450 ha.

Small population sizes may also promote social dysfunction in many
ways. First, certain communal mating displays such as **lekking** (see
chapter section 7.3) make it easy for hunting to decimate a small population. Second, there may not be enough individuals to satiate predators
that depend on synchronous breeding to concentrate prey. Third, in very
small populations, there may not be enough individuals to stimulate synchronous breeding in cases where a number of females are needed to
stimulate ovarian development, for example in colonial species. Halliday
(1978) suggested that this factor was important in the final demise of the
colonially nesting passenger pigeon *(Ectopistes migratorius)*.

Various other compilations of traits that place species at higher risk of extinction have been provided, for example those proposed by Terborgh and Winter (1980), Frankel and Soulé (1981) Soulé (1983), and Diamond (1984), and an example is reproduced here (Table 4.3). Simberloff (1986c) has pointed out that there are always exceptions to the lists of Diamond, Frankel and Soulé, Soulé, and Terborgh and Winter—that is, species that have the "undesirable traits" but have not gone extinct. Often one must view a rare species as successful if it continues to persist without any danger of overexploitation of its resources.

Too often, however, it is difficult to study rare populations to see which of many factors is the most critical to their survival. Zimmerman and Bierregaard (1986) feel that theory will not be as valuable as an informed "naturalist's feel" in answering how to manage the survival of particular species. Rarity is a concept deserving of further attention; it is regarded in all four lists, together with "population size," as germane to the likelihood of extinction. Rabinowitz (1981) has shown that "rarity" itself may depend on three factors: geographic range, habitat breadth, and local population size. A species is often termed rare if it is found only in one area, regardless of its density there. A species that is widespread but at very low density can also be regarded as rare. Conservation by habitat management is much easier, and more likely to succeed, for species of the first type than for those of the latter. Schoener (1987) addresses the issue of rarity further.

Table 4.3 *Possible factors contributing to the extinction of local populations. (After Soulé 1983.)*

Rarity (low density)
Rarity (small, infrequent patches)
Limited dispersal ability
Inbreeding
Loss of heterozygosity
Founder effects
Hybridization
Successional loss of habitat
Environmental variation
Long-term environmental trends
Catastrophe
Extinction or reduction of mutualist populations
Competition
Predation
Disease
Hunting
Habitat disburbance
Habitat destruction

Finally, Wilson (1987) has addressed the question of extinction from the other viewpoint and has asked what causes ecological success, drawing largely from the literature on ants, an extremely "successful" group. He proposes that ecological persistence is based on four factors:

□ *Number of species in a monophyletic group:* The number of ant species is estimated to be 20,000; even if some of these go extinct, others will persist.

□ *Occupation of unusual adaptive zones:* Ants occupy unusual ecological "niches," living in rotting wood, farming fungal gardens, and herding homopterans (aphids). They have largely excluded other groups from these ways of life.

□ *Large effective population sizes that are stable over time:* Even if colonies shrink drastically and the queen is lost, when conditions improve other queens emerge, and the brood size is quickly increased.

□ *Breadth of geographic range:* Ants range from the Arctic Circle to the southern reaches of Tierra del Fuego.

4.5 Bioengineering

The advances made in the understanding of how genetic variation is constituted and maintained in populations and of "how evolution works" are truly staggering. Yet, even before we understand it fully, people are now attempting to experiment with genetic engineering. This is the nature of science. In this instance, however, the spectre of a plague of some deadly organism or disease inadvertently (or even deliberately) released upon the world is a possibility, however remote. Mary Shelley capitalized on fear of this kind of possibility in her novel *Frankenstein*.

What is the state of our abilities to make new organisms and new compounds using "designer genes"? The technology of genetic engineering is well explained by Cherfas (1982). It is based on restriction enzymes (of which even by 1980 over 200 were known [Roberts 1980]). Such enzymes may well have evolved to thwart viral infection and replication within cells. Almost any sequence of bases can be located and cut at will. Cohen (1978b) has outlined the next steps. Following breakage, DNA from different sources must be rejoined (to form so-called **recombinant DNA** or R-DNA). Next a suitable gene carrier is used to replicate both itself and the foreign DNA segment linked to it. Fourth, a means of introducing the composite DNA molecule (termed a **chimera**) into a functional cell is needed. Finally, a method of selection is needed to distinguish the clones of recipient cells (those into which the chimera has been successfully introduced) from the large population in which they have been cultured.

The applications of genetic engineering are many and varied. For example, 40 million people in the world suffer from diabetes, a disease that stems from lack of insulin. Injections of insulin help, and the chemical is usually collected from pigs and cows. Genetic engineering could provide (and has been shown capable of providing) an even more readily accessible and pure source. Similarly, growth hormones, reproductive hormones, and interferon (a defense against viruses) could be cultured in this way. Bioengineering companies like Biogen and Genentech have been formed to exploit this market. Because the Supreme Court, on June 16, 1980, allowed living things (specifically microorganisms) to be patented (Wade 1980), the speed with which genetically produced products and new organisms appear will undoubtedly increase. The first patent for a genetically engineered animal was granted by the U.S. Patent and Trademark Office to Harvard University for a mouse carrying an activated cancer-causing gene. (Bills that would halt animal patents have also been introduced [Booth 1988*a*].)

In the agricultural world alone, many benefits are apparent. Genes from cocoa trees have been incorporated into bacteria to produce cocoa extract, and genes from bacteria have been moved into plants for insect control (Vaeck et al. 1987). Indeed, two major areas of applied R-DNA projects involve microorganisms to improve plant growth, act as insecticides, or prevent frost damage and the engineering of plants themselves to ward off disease or insects (Gaertner and Kim 1988). However, even the best-intentioned genetic engineering can have unexpected results. Studies on the effects of bovine growth hormone (BGH) on pigs showed significant improvements in weight gain and feed efficiency and a marked reduction in fat but were offset by high incidence of gastric ulcers, arthritis, cardiomegaly, dermatitis, and renal disease (Pursel et al. 1989). It is clear that long-term multigenerational studies of such effects are needed. In plants an ironic feature is that some plant defenses involve synthesis of natural carcinogens, so plants that could be grown without pesticides could also be unsuitable as crops.

The potential disadvantages of bioengineering are many and varied. One is the potential for disaster when genetically engineered organisms are released into an environment, nature, that is completely novel to them. Nobody can really predict the effects. Some biologists claim that the risks differ little from those involved in classical plant and animal breeding programs; others state that the ability to move genetic material from one organism to another is unique and warrants concern. Some argue that one can look for analogies in the results of introductions of new and exotic animals to different parts of the world. Some such species have become pests or weeds, and others appear to have had no impact (Pimentel et al. 1989; see also Chapter 23). Others (Sharples 1983; Simberloff 1985) suggest that, because genetically engineered organisms are entirely unique, no useful parallels can be drawn in this way. The staunchest opponents suggest total abandonment of this type of technology (Rifkin 1983). It is true that the use of bioengineering for biological warfare is a distinct possibility.

There seem to be no easy generalizations as to which engineered organisms will succeed in nature, which will fail, and which could lead to ecological disaster. Each organism should probably be evaluated individually before release, with a focus on the ecology of the organism and its release environment (Simberloff 1985; National Academy of Sciences 1987; Simonsen and Levin 1988). Theory is most useful in suggesting a hierarchy of risks, raising the questions that have to be addressed in a case-by-case risk assessment (Simonsen and Levin 1988). One thing is certain, as ecologists have repeatedly stressed (Crawley 1988), releases of genetically engineered organisms, once made, should be regarded as irreversible. No realistic amount of funding could ensure subsequent eradication if something went wrong.

The first approved release of living genetically engineered bacteria took place on strawberry fields in California, April 24, 1987. At that time, Steven Lindow (1985) released "ice-minus" mutants of *Pseudomonas syringae* bacteria onto plants. Normal plant leaves support an array of bacteria that induce ice to form. Lindow's strain has been engineered to lack the ice-nucleating-protein gene; ice would not form so readily on plant leaves populated by this strain. Other releases included that of a genetically modified soil bacterium, *P. fluorescens,* in South Carolina on November 2, 1987, and the field testing of genetically altered tomato plants in Illinois begun June 2, 1987 (Gaertner and Kim 1988). Australian scientists have genetically engineered a new type of protein-rich alfalfa that boosts wool growth when fed to sheep (Ford 1988).

The ethical issues of the pro's and con's of genetic engineering and other social side effects are also varied. For example, bovine growth hormone (BGH) has the capacity to increase milk production by about 20 percent (Rauch 1987). However, use of BGH, at a time of milk surpluses, would probably decrease milk prices 10–15 percent and put an additional 25–30 percent of dairy farmers out of work. The use of microbes to produce synthetic cocoa, coffee, and tea extracts might damage these industries in developing nations (Buttel and Barker 1985).

Finally, it is worth bearing in mind that, despite a plethora of government regulations and the best intentions, some scientists will undoubtedly continue to introduce genetically engineered organisms into the environment anyway, to field test some aspect of their research (Crawford 1987; Holden 1987). In 1986 the Wistar Institute, a private research organization in Philadelphia, commissioned the Pan American Health Organization (PAHO) to conduct a trial with a new type of potential vaccine against rabies. Scientists from the institute had placed genes from the rabies virus into the vaccinia virus, a fairly harmless relative of the smallpox virus. Researchers at the institute had tested the vaccine in the laboratory but needed a field trial to show that the vaccine would work on cattle in the open. The institute chose Argentina as a field site and, instead of informing Argentine authorities, smuggled the vaccine into Argentina in a diplomatic pouch. Nobody would have been the wiser had a scientist not leaked the results to the Argentine Association of Research Scientists. Angry debate understandably followed, and the cattle were destroyed (Connor 1988).

Section Three

Behavioral Ecology

Why do some animals escalate fights over resources and others settle disputes more peacefully? Why do some species, like this leaf-footed bug from Trinidad, have such bizarre and apparently nonfunctional bodily adornments? These are some of the questions addressed by behavioral ecologists. (Top photo: Dana C Bryan. Bottom photo: P. Stiling.)

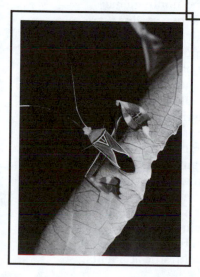

Niko Tinbergen, one of the founders of modern behavioral studies, or *ethology*, emphasized that there may be several different ways of answering the question why animals behave the way they do (Tinbergen 1963):

□ *Survival value:* to attract mates
□ *Causation:* because changes in day length trigger changes in hormone levels in the body
□ *Development:* offspring learn their songs from parents and neighbors
□ *Evolutionary history:* behaviors have devel-

oped or progressed into complex patterns from the simpler patterns of ancestors

It is important to distinguish between these different answers. All may be correct, but the ones influencing survival value are termed *ultimate*, whereas causal factors are referred to as *proximate*. This section deals mainly with the survival value and evolutionary significance of behavior. Students wishing for a more detailed treatment of how such behavior is achieved , how it is learned from parents, what part ''instinct'' plays, and how behavior is neurologically and physiologically achieved are referred to more detailed texts on animal behavior (for example, Alcock 1979).

Just like physical traits, changes in behavior may be caused by just one **gene.** Of course, a given behavior or physical trait may actually be coded for by many genes, but if one gene is altered (for example, the one for melanism in peppered moths), the whole behavior or physical appearance changes. One can make the analogy of baking a cake. A difference in one word of the recipe may change the whole taste of the cake, but that does not mean that the one word is responsible for the entire cake (Dawkins 1979). An excellent example of the effect of such a single-gene difference on behavior was demonstrated in W.C. Rothenbuhler's (1964) work on honeybees. Some strains of bees are hygienic; that is, they remove diseased larvae from the nest and discard them. This behavior involves two distinct maneuvers—the first is uncapping the wax cells, and the second is removing the larvae. Other strains are not hygienic and do not exhibit such behavior. Rothenbuhler demonstrated, by genetic crosses, that one gene (U) controlled cell uncapping and another gene (R) controlled larval removal. Hygienic strains were double recessives (uurr), and nonhygienic strains were double dominants (UURR). When the two strains were crossed, all the F1 hybrids were nonhygienic (UrRr). When the F1 hybrids were backcrossed with the pure hygienic strain (uurr), four different genotypes were produced: one-quarter of the offspring were hygienic (uurr), one-quarter were nonhygienic and showed neither behavior (UrRr), one-quarter uncapped the cells but failed to remove the larvae (uuRr), and one-quarter would remove the larvae but only if the cells were uncapped for them (Uurr). Genes for behavior actually act on the development of nervous systems and musculature, physical traits that evolve through natural selection as do any other.

It is important to realize that not every feature or behavior of a species is necessarily adaptive. Some differences between species may simply be alternative solutions to the same problem. Dawkins (1980) has illustrated this idea by reminding us that driving on the left side of the road in Britain and on the right side in the United States are both acceptable solutions to the problem of avoidance of traffic accidents.

Some differences between animals may be of a similar type. For fighting, sheep use horns, derived from bone, whereas deer use antlers, derived from skin (Modell 1969). One could of course argue, and many do, that there probably is a good reason for the difference, but it just has not been found yet. For example, antlers are dropped each year, but horns are retained, and the difference could be related to available food supply.

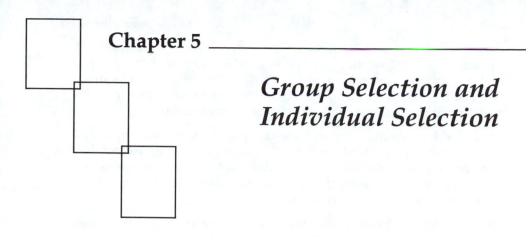

Chapter 5

Group Selection and Individual Selection

It used to be common to read statements such as, "Lions rarely fight to the death because if they did the survival of the species would be endangered," or, "Salmon migrate thousands of miles from the open ocean into a small stream, where they spawn and die, killing themselves with exhaustion to ensure survival of the species." Can behavior evolve for the good of the group or species? The main proponent of the idea that animals behave for the good of the group was V.C. Wynne-Edwards (1962). He suggested that if a population overexploited its food resources it would go extinct, so adaptations would evolve to ensure that each group of individuals purposefully controls its rate of consumption and its rate of breeding.

5.1 Drawbacks to Theories of Group Selection

Attractive though the idea of **group selection** is, it has several flaws:

- **Mutation:** Imagine a species of bird in which a pair lays only two eggs and there is no overexploitation of resources. Suppose the tendency to lay two eggs is inherited as a group-selection trait. Now consider a **mutant** that lays six eggs. If the population is not overexploiting its resource, there will be sufficient food for all the young to survive, and the six-egg genotype will become more common very rapidly. Gene frequencies in the population will change. This process would work for even larger brood sizes, and brood sizes would tend to increase until they became so large that the parents could not look after all their young, causing an increase in infant mortality. Thus the clutch size in nature evolves so as to maximize the number of surviving offspring. Field studies of great tits in Wytham Woods, England, for example, show a median clutch size of eight-to-nine eggs, above which adult birds cannot reliably supply sufficient food for all chicks to survive.

- **Immigration:** Even in a population in which all pairs laid two eggs and no mutations occurred to increase clutch size, "selfish" individuals that laid more could still migrate in from other areas. In nature, populations are rarely sufficiently isolated to prevent **immigration.**

- **Individual selection:** For group selection to work, some groups must die out faster than others. In practice, groups do not go extinct fast enough for group selection to be an important force. Individuals nearly always die more frequently than groups, so individual selection will be the more powerful evolutionary force.

- **Resource prediction:** For group selection to work, individuals must be able to assess and predict future food availability and population density within their own habitat. There is little evidence that they can.

Individual selfishness seems a more plausible result of natural selection. Group selection is probably a weak force and is only rarely very important (Maynard Smith 1976a). Any reduction in population sizes from **self-regulation** is likely to come from intraspecific competition, of a selfish nature, in which individuals are still striving to command as much of a resource as they can (see Chapter 11). Indeed we often see animals in nature acting in their own selfish interest. Male lions kill existing cubs when they take over a pride. The proximate cause may be the unfamiliar smell of the cubs. A similar effect, known as the Bruce effect, occurs in rodents, where the presence of a strange male prevents implantation of a fertilized egg or induces abortion in females (Bruce 1966). In the case of lions, the advantage of infanticide for the male lion is that, without their cubs, females come into the reproductive condition much faster, in 9 months as opposed to 25 if the cubs are spared, hastening the day when males can father their own offspring. A male's reproductive life in the pride is only two-to-three years before he in turn is supplanted by a younger, stronger male. Infanticide ensures the male will father more offspring, and the tendency spreads by natural selection (Bertram 1975). In human society, individuals also rarely act for the good of the group, as illustrated by the "tragedy of the commons" scenario (see Section Five). Where a common pasture is open to public grazing, farmers are likely to overload the land with cattle in an attempt to maximize their own returns. What usually happens is overgrazing, and the area becomes useless for anyone (Hardin 1968).

5.2 *Altruism*

Although natural selection favors individual rather than group selection, it is still common to see apparent cooperation. Animals of the same species groom one another, hunt communally, and give warning signals to each other in the presence of danger. How can this altruistic behavior be explained by natural selection?

We are not surprised to see a parent working hard caring for its young. All offspring have copies of their parents' genes, so parental care is genotypically selfish. Genes for **altruism** toward one's young will therefore become more numerous because offspring have copies of those same genes. The "selfish genes" themselves are increasing, by virtue of their effect on behavior and the copies of themselves in the bodies of other individuals. In meiosis, any given gene has a 50 percent chance of going into an egg or sperm. Thus each parent contributes 50 percent of its genes to its offspring. The probability that a parent and offspring will share a copy of a particular gene is a quantity, r, called the coefficient of relatedness. By similar reasoning, brothers or sisters are related by an amount $r = 0.5$, grandchildren to grandparents by 0.25, and cousins to each other by 0.125. It was W.D. Hamilton (1964) who realized the important implication of this relatedness for the evolution of altruism, though the idea was anticipated by two earlier giants in the field, Fisher (1930) and Haldane (1953). Just as gene replication can occur through parental care, so it can by care for siblings, cousins, and other relatives. (Perhaps we should talk about selfish genes, rather than selfish individuals [Dawkins 1976].) Selection for behavior that lowers an individual's own chances of survival and reproduction but raises that of a relative is known as **kin selection.** The conditions under which an altruistic act will spread by kin selection can be quantified as follows. If the donor sacrifices C offspring for which the recipient gains B offspring, then the gene causing the donor to act in this way will spread if

$$\frac{B}{C} > \frac{1}{r}$$

where r is the coefficient of relatedness of donor to recipient. Having made these calculations on the back of an envelope in a pub one evening, J.B.S. Haldane reputedly announced that he would lay down his life for two brothers or eight cousins (Krebs and Davies 1981). This illustration may sound a little farfetched, but in nature there are numerous examples of self-sacrifice.

Altruism Between Relatives

Many insect larvae, especially caterpillars, are weak, soft-bodied creatures. They rely on bad taste or poison to deter predators and advertise this condition by bright warning colors. For example, noxious *Datana* caterpillars, which feed on oak and other trees, have bright red and yellow stripes (Photo 5.1). Of course a predator has to kill and attempt to eat

Photo 5.1   Altruistic behavior. Datana *caterpillars exhibit a bright, striped, warning pattern to advertise their bad taste to predators. All the larvae in a group are likely to be the progeny of one egg mass from one adult female moth, so the death of the one caterpillar it takes to teach a predator to avoid the pattern benefits its close kin. (Photo by P. Stiling.)*

one of the caterpillars in order to learn to avoid similar individuals in the future. It is of no personal use to the unlucky caterpillar to be killed. However, warningly colored animals often aggregate in kin groups, so the death of one individual is most likely to benefit its relatives, such as siblings, and its genes will be preserved. Malcolm (1986) has documented the evolution of **aposematism** (warning coloration) in aphids by means of kin selection.

Another example of altruism again refers to lions. Lionesses tend to remain within the pride, whereas the males leave. As a result, lionesses within a pride are related, on average, by $r = 0.15$. Females all come into heat at the same time, one individual probably influencing the others' estrous cycles by means of pheromones. A similar phenomenon occurs in humans, where girls living in the same school dormitories may have synchronized menstrual cycles. In lionesses the result is the simultaneous birth of cubs, and females exhibit the apparently altruistic behavior of suckling other females' cubs. Because the females are related, the selfish-gene hypothesis accounts for this behavior. Similarly, when male lions depart, they may act in concert to take over a pride; each one's genes will be perpetuated both through his own offspring and through those sired by his brothers (Bertram 1976; Bygott, Bertram, and Hanby 1979).

Although an altruistic act toward two sisters or brothers may seem to be genetically equivalent to a similar act toward an offspring, there are sometimes other ecological or proximate factors that tip the scales in favor of the offspring. Young may be more valuable in terms of expected future reproduction, having a higher potential reproductive output than older siblings or parents. Progeny may also benefit more from a given amount of aid. One insect fed to a nestling contributes more to its survival than the same insect would to the survival of a healthy adult sibling. Young may thus be thought of as superbeneficiaries (West Eberhard 1975).

Castes

Perhaps the most extreme altruism is the evolution of sterile casts in the social insects, in which some females, known as workers, rarely reproduce themselves, but instead help others to raise offspring. The Amazonian termite pictured in Photo 5.2 is another example of a sterile caste dedicated to a specific function within the insect's social system.

The differentiation of one species into many different-sized castes in ants can be quite staggering (Fig. 5.1). Amazingly, all worker subcastes appear to have originated from allometric growth (Wilson 1985). The different forms, designated usually as minors, medias, and majors, differ from one another by size, and the size variation is accompanied by differences in body proportions. Thus majors (sometimes called soldiers) typically possess relatively larger heads than do the minors and medias. In ant species with simpler forms of differentiation, the relationship can be approximated by the equation $y = bx^a$, where x is one body dimension (such as pronotum width), y is another (say, head width), and a and b are fitted constants characteristic of the species (Fig. 5.2). Of the 263 living genera of ants known worldwide, 44 possess polymorphic species with caste systems. Furthermore, closely linked to allometric growth patterns is alloethism, a regular change in behavior patterns as a function of size. The explanation of the peculiar system of castes was thought to lie primarily in the particular genetics of most (at least hymenopteran) social insect reproduction (Hamilton 1967). Males develop from unfertilized

Photo 5.2   An Amazonian termite. Many species of insect (for example, ants, wasps, and termites) have sterile castes that do not reproduce themselves but are strictly confined to other functions, in this case defense. (Photo by P. Stiling.)

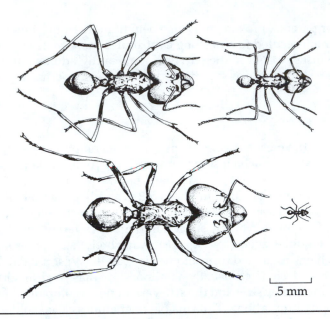

.5 mm

Figure 5.1 *The caste system of the genus* Atta. This caste system, here exemplified by *A. laevigata,* is among the most extreme found in ants, yet it is still based on single allometry curves combined with size variation. (Redrawn from Oster and Wilson 1978.)

eggs and are **haploid.** Male gametes are formed without meiosis, so every sperm is identical. Thus each daughter receives an identical set of genes from her father. Half of a female's genes come from her **diploid** mother, so the total relatedness of sisters is 0.75. Such a genetic system is called **haplodiploidy.** Thus, females are more related to their sisters than they would be to normal offspring. It is therefore advantageous to stay in the nest or hive and to try to produce new reproductive sisters.

The story is complicated a little bit when the interests of the queen are incorporated into the propensity for sterile castes. Queens are equally related to their sons and their daughters; $r = 0.5$ in each case. To maximize her reproductive potential, a queen should therefore produce as many sons as daughters, a 50:50 sex ratio (see Chapter 7). If she did, then sterile worker females would spend as much time rearing brothers (to which they are related only by 0.25) as sisters. The average relatedness of a female to her siblings would then be 0.5, and she would do equally well to breed on her own. From the workers' viewpoint, it is far better to have more sisters, and in this conflict with the queen they appear to have won, because in any colony there are more females than males, by a ratio of about 3:1 (Trivers and Hare 1976; but see also Alexander and Sherman 1977). Elegant though these types of explanations are, they do not provide the whole picture: Large social colonies exist in termites, too, but

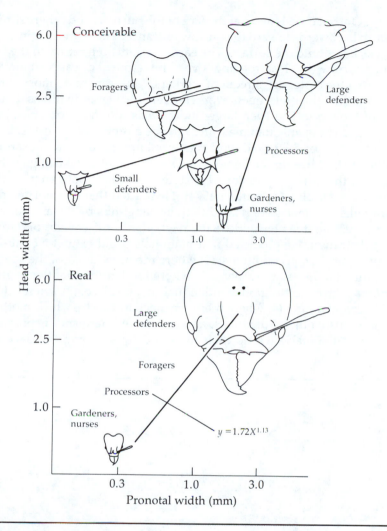

Figure 5.2 *A conceivable and an actual array of* Atta *castes.* A conceivable array of *Atta* castes is compared with the actual array in *A. sexdens,* in order to illustrate the principle that the evolution of ant castes is based on a single allometric curve and hence constrained in a fundamental manner. (Redrawn from Wilson 1980.)

theirs is not a haplodiploid system, and colony males have only half their genes in common, on average. Furthermore, naked mole rats, mammals living underground in South Africa, have a division of labor based on castes and cooperative broods, but the animals are, of course, **diploid** (Jarvis 1981). There is only one breeding female, the queen; she apparently suppresses reproduction in other females by producing a chemical in her urine that is passed around the colony by grooming after visits to a communal toilet. The other castes perform different types of work in burrowing. One rat chisels away at the face of the burrow, and others shuffle the earth backwards toward the burrow entrance. There another individual kicks it out onto the surface. The earth carriers then return to the face of the burrow by leapfrogging over the crouched earth removers (Fig. 5.3). One caste, the "frequent workers," appears to do most of this work; another, the "infrequent workers," are heavier individuals that do some of the work, and even larger individuals, the nonworkers, rarely work at all. These are often male and may be a reproductive caste. Even before the case of the naked mole rats was known, Richard Alexander, curator of the Museum of Zoology at the University of Michigan, had suggested that it is the particular life style of animals, not genetics, that promotes eusociality (Alexander 1974). He argued that in a normal diploid organism, females are related to their daughters by 0.5 and to their sisters by 0.5, so it matters little to them whether they rear sibs or daughters of their own. He predicted that mammals could exhibit a caste-like society under certain conditions: (1) where the individuals of the species are confined in nests or burrows, (2) where food is abundant enough to support a high concentration of individuals in one place, (3) where adults exhibit parental care, (4) where there are mechanisms by which mothers can manipulate other individuals, and (5) where "heroism" is possible, whereby individuals give up their lives and by so doing can save the

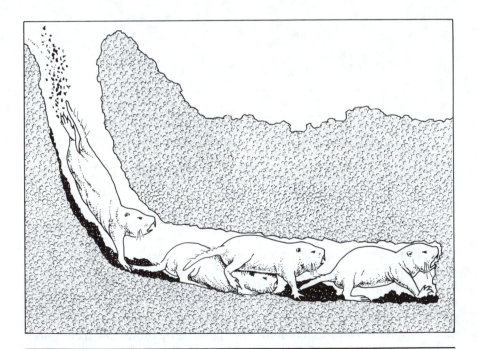

Figure 5.3 *Mole rat bucket-brigade.* (Based on Jarvis and Sale 1971.)

queen. These factors can immediately account for eusociality in the termites as well. In mole-rat colonies, where the burrows become as hard as cement—a heroic effort by a mole rat effectively stops a predator (commonly a snake) because predators cannot rip open the surrounding substrate. Self-sacrifice by a "worker" does translate into a genetic gain. In mole rats, the queen reigns supreme; all workers and nonworkers, whether male or female, develop teats during her pregnancy—testament to the power of her pheromonal cues. The superabundant food comes in the form of tubers of the plant *Pyrenacantha kaurabassana,* which weigh up to 50 kg and can provide food for a whole colony. Often the tubers remain only half-eaten and can regenerate in time. Why eusociality has developed in this, but not in other, systems of burrowing rodents such as prairie dogs is open to speculation. It could be that movement between mole-rat colonies is effectively zero, promoting intense inbreeding. Hamilton's genetic-relatedness theory seems even less appealing when one considers that the relatedness of workers in a hymenopteran colony is extremely close only if the colony is formed by a single queen who has mated once. When a queen mates twice and sperm mixes at random, the average relatedness between sisters is only 0.5.

Evidence has also come to light that, in a variety of hymenopterans, unrelated queens often initiate colonies together, dropping from mating swarms at the same time and engaging in pleiometrosis (cooperative nest digging and egg laying) (see references given by Strassman 1989). Some queens thus give up the opportunity to lay their own eggs, tending instead to the young of a cofoundress. In fire ants the number of cofoundresses commonly varies from two to five. How is this behavior explained? Females must not be able to predict which one will become the eventual egg layer when they begin a nest together. The benefits of nesting in a group must also be great. Part of the reason for cooperation apparently lies in the fact that many hymenopteran nests or colonies are clumped. Brood raiding by neighboring colonies is common (especially in ants), so attaining a large worker force quickly is critical to colony survival. In fire ants, especially, brood raiding is widespread, and in a study of newly initiated nests, W.R. Tschinkel (1990) documented the eventual merger of 80 nests into only two over the course of a month. Although this case was exceptional, brood raiding from four or five colonies over the space of a few hours is common, and the raids go back and forth from nest to nest until one colony eventually ends up with all the brood. Other advantages to large colony size may involve increased defense against predators and parasites and general lower adult mortality.

Not all acts of altruism result in such extremes as suicide and sterility. A common example of altruism is the raising of an alarm call by "sentries" in the presence of a predator. The alarm maker is drawing attention to itself and risking increased danger by its behavior. For some groups, like ground squirrels, *Spermophilus beldingi* (Sherman 1977), individuals near an alarm maker bolt down their burrows, and those close neighbors are most likely to be sisters or sisters' offspring; thus the altruistic act of alarm calling could be reasoned to be favored by kin selection (but sentries are often subordinate individuals who are driven to the edge of the group, where the risk is higher and they are forced to be alert for their own safety). Working out the exact costs and benefits in such a system of kin selection would, however, be a nightmare and has always been a stumbling block in behavioral ecology. To get a precise measure of natural selection, one must be able to calculate an individual's contribution to the gene pool. This might be the contribution of an individual plus 0.5 times its number of brothers and sisters, 0.125 times its number of cousins, and so on. This genetical octopus has been termed **inclusive fitness** by Hamilton (1964). More commonly, behavior or **adaptation** is recognized as beneficial if it shows a closely designed fit to some problem presented by the animal's environment (Williams 1966). At least for the alarm makers, some selfish motive is involved because if a predator fails to catch prey in a particular area it is less likely to return, and the number of potential attacks over the alarm maker's lifetime is likely to be reduced (Trivers 1971).

Altruism Between Unrelated Individuals

Not all acts of altruism are directed to close relatives. When a female olive baboon, *Papio anubis*, comes into estrus, a male forms a consort relationship with her, following her around and waiting for the opportunity to mate. Sometimes an unattached male enlists the help of another to engage the consort male in battle, while the solicitor attempts to mate with the female. On a later occasion the roles are reversed (Packer 1977). This is an example of reciprocal altruism (Trivers 1971). (Reciprocal altruism is common in human society where money is used to mediate its use.) Sometimes unrelated individuals will occupy the same territories as breeding individuals and help the parents raise offspring by foraging for additional food for the young. Although the helpers may be related to the parents, for example in the well-studied Florida scrub jay (Woolfenden 1975; Woolfenden and Fitzpatrick 1984), sometimes they are not, for example in mongooses (Rood 1978) and dunnocks, *Prunella modularis* (Houston and Davies 1985). By comparing nest or brood success of breeding pairs with helpers with those whose helpers were removed, Brown et al. (1978 and 1982) were able to show that helpers do significantly increase parents' fitness. Again, the motive seems to be a sort of reciprocal altruism. In these situations, not uncommon and occurring in more than 150 species of birds, the habitat is usually saturated with breeders, and helpers could probably not obtain a territory for themselves. What they do is help increase territory size of a breeding pair; they are later able to carve off a fragment of this territory themselves, or they can take over from the breeding pair after it dies. Brown (1982) has modelled the threshold at which owners should decide to share their territories (Fig. 5.4).

For many predators, reciprocal altruism in the form of social hunting allows bigger game to be caught. The benefits of a large kill outweigh the cost of having to share the meat (Caraco and Wolf 1975). Many acts of altruism between unrelated individuals, whether in defense, attack, or mating, are exhibited by individuals living in social groups.

Of course not all observed behavior can be explained by altruism.

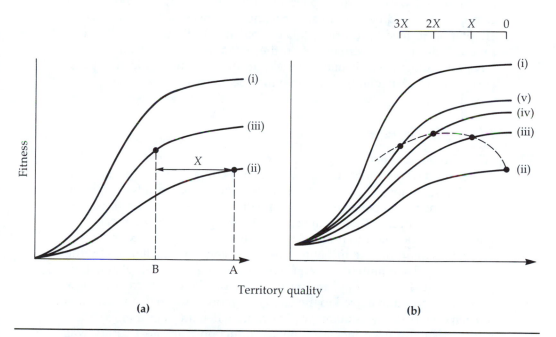

Figure 5.4 *Factors determining the feasibility of reciprocal altruism.* **(a)** *A model of sharing with one other individual.* Curve (i) is the maximum fitness an individual can obtain if there are no defense costs. Curve (ii) is the fitness an individual would obtain alone, including defense costs. Curve (iii) is the fitness if an individual shared its territory with another individual and thus only sustained half of the defense costs. The costs of sharing are represented as a decrease in territory quality. For example, with sharing costs of X, it would pay an owner with a territory of quality A to share it with a subordinate, because its fitness on a shared territory (B) would be increased. **(b)** *An extension of the model to include fitness curves from sharing with two others (curve iv) and three others (curve v).* Depletion costs from sharing are represented on the horizontal line above the graph. Sharing with one other individual causes a reduction in territory quality by an amount X, with two others 2X, and so on. The dots give the net fitness from sharing, and, in this example, an individual gains maximum fitness in a group of three. (Redrawn from Brown 1982).

Many adaptive explanations are in fact more the result of an author's in-genuity than natural consequences of the facts. Great care must be taken to avoid such errors; otherwise, the resultant theories will have no more substance than Kipling's *Just So Stories* (Gould and Lewontin 1979). Often some individuals are simply manipulated by others. Cuckoos, for exam-ple, have used the devoted parental care of adult birds to rear their para-sitic cuckoo young instead. Some authors have also suggested that in so-cial insects some females could actually do better, in terms of **inclusive fitness,** by reproducing themselves rather than being sterile, but they have been manipulated by their parents. Often more than one factor has to be invoked to explain any apparent pattern in animal behavior. As an example, it would be easy to say that the long neck of the giraffe is an adaptation to feeding on high foliage, but it could equally well aid in predator detection. There are often these confounding variables in be-havioral ecology, which make it tricky to decide which selective pres-sures cause a particular trait. Often it may be more than one.

5.3 *Living in Groups*

Why should gannets, penguins, and many other birds nest so close to-gether, and why should fish shoal? If dense congregations promote in-tense competition, there must be some high selective advantages of group living to compensate. In today's world, group living may be disas-trous in that animals that organize into dense aggregations can easily be harvested by people and reduced to very low numbers. For example, the control of vampire bats in Latin America was greatly facilitated by the bats' habit of communal roosting and grooming (Mitchell 1986). Bats groom for two hours a day, and pesticides applied topically to only a few bats were soon transferred to the others. A ratio of 15–16 dead vampires to each one treated was obtained. The annual benefit to farmers in Nica-ragua, resulting from increased milk production from host cattle, was U.S. $2,414,158 after vampire control. As annual costs were only $129,750, a favorable benefit-cost ratio of 18.6:1 was obtained.

Guppies, *Poecilia reticulata*, were first discovered in Trinidad in the 19th century by the Reverend P.L. Guppy. In their native habitat, gup-pies live in tighter groups when they are in streams in which predators are more common (Fig. 5.5), suggesting that being in a group helps an individual to avoid becoming a meal. Group living could reduce predator success in several different ways.

Increased Vigilance

For many predators, success depends on surprise; if a victim is alerted too soon during an attack, the predator's chance of success is low.

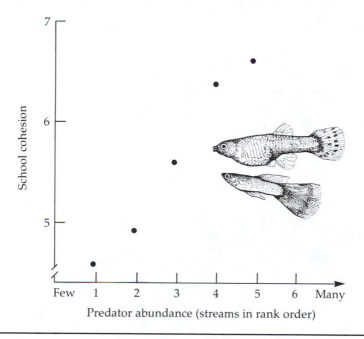

Figure 5.5   *Intraspecific variation in group size may be related to predators.* Guppies (*Poecilia reticulata*) from streams with many predators live in tighter schools than those from streams with few predators. Each dot is a different stream, and "cohesion" was measured by a count of the number of fish in grid squares on the bottom of the tank (Modified from Seghers 1974.)

Goshawks (*Accipiter gentilis*) are less successful in attacks on large flocks of pigeons (*Columba palumbus*) mainly because the birds in a large flock take to the air when the hawk is still some distance away. If each pigeon occasionally looks up to scan for a hawk, the bigger the flock, the more likely that one bird will spot the hawk early. Once one pigeon takes off, the rest follow (Fig. 5.6) (Kenward 1978). Of course, cheating is a possibility because some birds might never look up, relying on others to keep watch while they keep feeding.

Dilution Effect

Normally, predators take only one prey item per attack. An individual antelope in a herd of 100 has only a 1 in 100 chance of being attacked, whereas a single individual has a 1 in 1 chance. Of course large herds may well be attacked more, but a herd is hardly likely to attract 100 times more attacks than an individual. Associated with herds is a tendency to prefer the middle, because predators are likely to attack prey on the edge of the group. Part of the reason may be the difficulty of visually tracking large

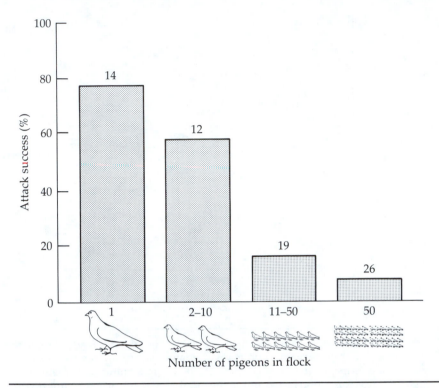

Figure 5.6 *The value of flocks of wood pigeons in detecting the approach of an avian predator, a goshawk.* The graph shows the dependence of the goshawk's attack success on the size of the flock. (Redrawn from Kenward 1978.)

numbers of prey. Throw three tennis balls to a friend, and the chances are that he or she will drop all three, whereas one ball can be tracked much more easily. A similar phenomenon may operate in predator-prey interaction, explaining why predators take peripheral individuals (Neill and Cullen 1974). It may also be physically difficult to get to the center of a group, with many herds tending to bunch close together when they are under attack (Hamilton 1971). Furthermore, large numbers of prey are able to defend themselves better than single individuals, which usually flee. Nesting black-headed gulls mob crows remorselessly and reduce the success of the crows at stealing gulls' eggs (Kruuk 1964). Natural predators are rare in the Hawaiian islands, and birds there seldom flock (Willis 1972).

Group Predation

Living in groups may confer advantages for the predator as well as the prey. Apart from predation, other advantages and disadvantages are

Table 5.1 *Examples of studies in which possible costs and benefits of group living other than those mentioned in the text have been measured. (After Krebs and Davies 1981.)*

Hypothesis	Example
Saving of energy by warm-blooded animals as a result of thermal advantage of being close together.	Pallid bats (*Antrozous pallidus*) roosting in groups use less energy than solitary neighbors.
Chance for small species to overcome competitive superiority of a large species by being in a group.	Groups of striped parrotfish (*Scarus croicus*) can feed successfully inside the territories of the competitively superior damsel fish (*Eupomacentrus flavifrons*).
Hydrodynamic advantage for fish swimming in a school. They save energy by positioning themselves to take advantage of vortices created by others in the group.	Measurements of distances and angles between individuals show that they are not correctly positioned to benefit according to the predictions of the theory.
Increased incidence of disease as a result of close proximity to others.	Number of ectoparasites in burrows of prairie dogs (*Cynomys* spp.) increase in larger colonies. Number of ectoparasitic bugs and fleas increases with size of cliff swallow (*Hirundo pyrrhonota*) colonies (Brown and Bomberger Brown 1986).
Risk of cuckoldry by neighbors.	In colonially nesting red-winged blackbirds (*Agelaius phoeniceus*), the mates of vasectomized males laid fertile eggs. They must have been fertilized by males other than their mates.
Risk of predation on young by cannibalistic neighbors.	In colonies of Belding's ground squirrel (*Spermophilus beldingi*), females with small territories are more likely to lose their young to cannibalistic neighbors than are females with large territories around their burrows.

listed in Table 5.1. Predatory groups often capture prey that are difficult for a single individual to overcome, either because the prey is too large (lions hunting adult buffalo) or because it it is too elusive (killer whales hunting dolphins). For species that feed on large ephemeral food clumps such as seeds or fruits, the limiting factor is often the location of a good site or tree. Once it has been found, there is usually plenty of food. It has been proposed, with good evidence from *Quelea* birds in Africa, that in large roosts, successful foragers may be followed by birds that had been previously unsuccessful. This "mutual parasitism" supposes poor foragers are in some way able to distinguish successful birds (Ward and Zahavi 1973). Brown (1988) has shown that cliff swallows (*Hirundo pyrrhonota*) in southwestern Nebraska nest in colonies that serve as information centers in which unsuccessful individuals locate and follow successful individu-

als to aerial insect food resources. Brown was able to factor out the confounding effects of increased ectoparasitism (by nest fumigation) on
larger colonies and colony location (by reducing certain colony sizes) to
show that increased nestling weight at larger colonies was due to more
successful foraging, attributable to more efficient transfer of information
among colony residents.

Apart from the avoidance of predation and the securing of food,
there are additional benefits and further drawbacks to group living
(Table 5.1). It is clear that the conflicting selective pressures operate on
group size to determine the eventual number of individuals in a herd or
flock. The operation of just two conflicting variables on the size of bird
flocks is illustrated in Fig. 5.7; the various effects of many different selective pressures on groups are sketched in Fig. 5.8.

Group Size

To explain why some animals live in groups of 50 and others in
groups of 10 is a little harder, but Clutton-Brock (1974) attempted just that
when comparing group sizes of red colobus monkeys (*Colobus badius*)
(group size 40) and black-and-white colobus (*Colobus geuereza*) (group size
11). The main conclusion was that group size in this case was set by patch
size of the food. The red colobus depends on the leaves of just one or two
trees, whereas the black-and-white colobus feeds on fruits and flowers of
a variety of species. Because the food of the black-and-white colobus is
predictable but scattered, small groups ensure that competition for food
is not severe. For the red colobus, food is rare but occurs in relatively
large clumps, whole trees, so once it is found large group size does not
result in competition. Clutton-Brock and Harvey (1977) went on to discuss why nocturnal primates are usually fruit or insect eaters, are small
and arboreal, live in small groups, and have small home ranges, whereas
diurnal terrestrial primates are large, live in large groups, and have large
home ranges. They suggested that nocturnal species feed from small,
narrow branches and therefore have small body size, are cryptic to avoid
predators, and thus are solitary and inconspicuous. At the other extreme,
diurnal terrestrial monkeys are inevitably conspicuous, so they use large
body size and group defense as protection against predators. These inter-
pretations are very plausible, but it is usually possible to erect other ways
of linking these types of behavior together. The real problem is to separate cause and effect.

Life in a group is not all roses however, as competition for food may
ensue. Hyenas and wild dogs all gobble food as fast as they can, as do
lions at a good-sized kill. When the kill is smaller, a definite pecking order
emerges, and size is a prime determinant—lionesses rob cubs of food,
and male lions rob lionesses and cubs, so cubs may starve in times of food
crises (Schaller 1972). Among wild dogs, however, the young feed before

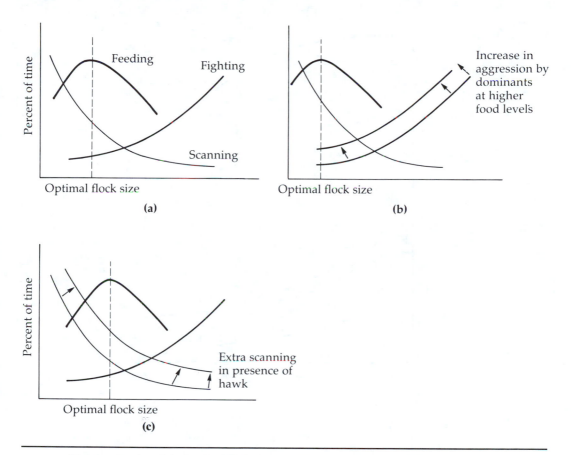

Figure 5.7 *A model of optimal flock size.* **(a)** *The trade-off between squabbling or fight-ing and scanning for predators.* As flock size increases, birds spend more time fighting and less time scanning. An intermediate flock size gives the maximum proportion of time feeding. **(b)** *The effect of an increase in resources on flock size.* When food is more plentiful, dominant birds can afford to spend more time attacking subordinates. The optimal flock size for the average bird therefore decreases. **(c)** *The effect of an increase in predation on flock size.* When predation risk is increased by the flight of a hawk over the flock, the scanning level should go up, and the optimal flock size is increased. (Modified from Krebs and Davies 1981.)

the adults. Why the difference? Usually wild-dog packs consist of highly related individuals, though only a few adult dogs actually reproduce. For this reason, the best way for a dog to raise its inclusive fitness is to help feed its relatives' pups. Lionesses, on the other hand, are less related to the cubs. On occasion, a male lion appears more tolerant to feeding cubs than a lioness, perhaps because he is much more likely to be related to them.

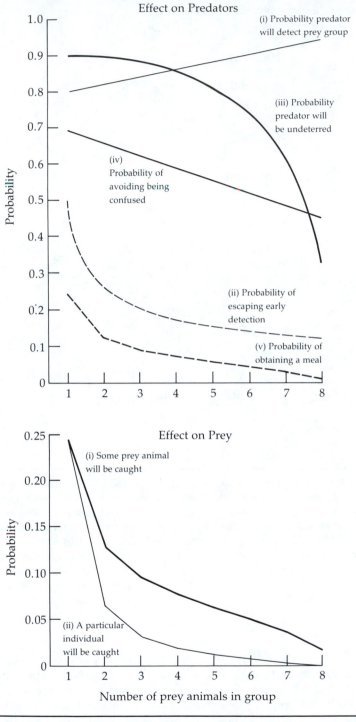

Figure 5.8 *The effects of prey group size on the probabilities of predator success and prey capture.*

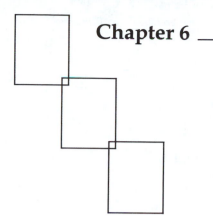

Chapter 6

Resources: Assessing, Obtaining, and Defending

In social animals, flock or herd size is a function of many opposing selective pressures. **Competition** between individuals for food or mates may present one of the strongest pressures. What dictates when competitors should fight and when they should give up and search for other resource areas? Solitary species also are faced with many conflicting pressures. When should an individual bother to defend a territory, and how does it assess habitats quickly when selecting a territory in the first place?

6.1 Resource Assessment

Imagine two habitats, one rich and one poor in resources. All individuals are free to go where they like, a situation referred to as "ideal free conditions" (Fretwell and Lucas 1970). As individuals arrive at the rich habitat, resources will gradually be depleted. Eventually a point will be reached where the next arrivals will do better by occupying the poor habitat, where, although resources are scarcer, competition is less (Fig. 6.1). A human social equivalent occurs at supermarkets, where customers distribute themselves among lines according to an ideal free distribution so that everyone has about the same waiting time. The express checkout lane is equivalent to a rich habitat, and the line of customers is usually longer there.

In an experiment using six sticklebacks, Milinski (1979) added *Daphnia* prey to one end of a tank at twice the rate of the other end. The sticklebacks distributed themselves with two at the "slow feed" end and four at the "fast" end, so that each animal was feeding at the highest rate. When the feeding regimes were reversed, the fish redistributed themselves accordingly.

For many animals, peaceful resource allocation does not occur. Indi-

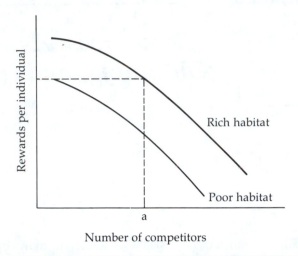

Figure 6.1 Resource sharing. There is no limit to the number of competitors who can exploit the resource. Every individual is free to choose where to go. The first arrivals will go to the rich habitat. Because of resource depletion, the more competitors, the lower the rewards per individual, so at point a the poor habitat will be equally attractive. Thereafter, the habitats should be filled so that the rewards per individual are the same in both.

viduals compete and fight over resources so that richer habitats always get filled up first before animals exploit the poorer resource (Fig. 6.2).

Many birds such as great tits and red grouse (*Lagopus lagopus*) are territorial and defend the richest territories (Krebs 1971; Watson 1967). After these are completely filled up, excluded birds occupy poor habitats, and when these too are filled, birds form floating groups where their chances of survival are low. Thus the strongest individuals are despots, who force others into low-quality areas. Although sticklebacks may exhibit the ideal free distribution, the chances are that one or two fish are better competitors than others and that the ideal free distribution comes about by the way the subordinates distribute themselves in relation to the despots.

In nature, despotism and resource sharing are not always mutually exclusive, but are combined in an overall behavioral strategy. In cottonwood aphids (*Pemphigus betae*), adult females settle on cottonwood (*Populus angustifolia*) leaf veins in the spring, sucking the sap and reproducing **parthenogenetically**. Eventually the expanding tissue forms a gall around the aphid. Large leaves provide the richest supplies of sap and allow formation of up to seven times the number of progeny produced by mothers on small leaves. As a result, large leaves are quickly colonized, often by more than one aphid per leaf. Of course, other colonizing females reduce the availability of sap, and less progeny are produced by each. In fact, the system fulfills the ideal free conditions such that, on one

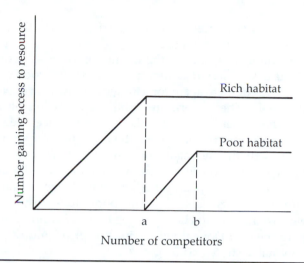

Figure 6.2 *Despotic behavior.* Competitors occupy the rich habitat first of all. At point a this becomes full and newcomers are forced to occupy the poor habitat. When this is also full (point b), further competitors are excluded from the resource altogether and become "floaters." (Redrawn from Brown 1969.)

large leaf with three aphids, the average number of progeny per female is equal to that on a medium leaf with two mothers or on a small leaf with a single mother. However, although the *average* success on different leaves is in accordance with an ideal free situation, not all competitors get equal rewards. The best place to be on a leaf is nearest the leaf petiole, in a position to tap the incoming sap first. Basally positioned mothers produce more progeny than more apical ones, so females compete for this optimal position in a despotic fashion, jostling for position; the bigger individuals usually win (Whitham 1978, 1979, and 1980).

Fighting and Game Strategy

In the aphid example, the females face the problem of whether to fight for a good leaf position or to retreat and look for a new leaf. How does an individual solve this dilemma? Why do some animals, like the *Pemphigus* aphids, fight and risk serious injury, whereas other animals settle disputes by display patterns?

John Maynard Smith and his associates (see Maynard Smith 1976b, 1979, and 1982) have tried to elucidate fighting in animals by considering contests in which there are different sorts of strategies. Imagine a game in which "Hawks" always fight to injure and kill their opponents (risking injury themselves) and "Doves" simply display and never engage in serious fights. These two strategies are chosen to represent the two possible extremes that can be seen in nature.

In this evolutionary game, let the winner of a contest score +50 and the loser 0. The cost of serious injury is −100 because the injured player may not be able to compete again for a long time; the cost of wasting time in a display is −10. These scores are arbitrary measures of fitness, and it will be assumed that Hawks and Doves reproduce their own kind faithfully in proportion to their scores. The next step is to draw up a two-by-two matrix with the average scores for the four possible types of encounter (Table 6.1). Consider what happens if all individuals in the population are Doves. Every contest is between a Dove and another Dove and the score, on average, is +15. In this population, any mutant Hawk could do very well, and the Hawk strategy would soon spread because when a Hawk meets a Dove it gets +50. It is clear that Dove is not an evolutionarily stable strategy (ESS).

However, the Hawk strategy would not spread to take over the entire population. In a population of all Hawks, the average score is −25, and any mutant Dove would do better because when a Dove meets a Hawk it gets 0 (which is not very good but still better than −25!). The Dove strategy would spread if the population consisted mainly of Hawks. Therefore Hawk is not an ESS either.

Nevertheless, a mixture of Hawks and Doves could be stable. The stable equilibrium is the point at which the average score for a Hawk is equal to the average score for a Dove. If the population moved away from the equilibrium, then one or the other strategy would be doing better and the population would not be stable. Each strategy does best when it is relatively rare, and the tendency in the evolutionary game will be for frequency-dependent selection to drive the frequencies of Hawk and Dove in the population so that they each enjoy the same success. For the values in Table 6.1, the stable mixture can be calculated as follows. If h is the proportion of Hawks in the population, the proportion of Doves is $(1 - h)$. The average score for a Hawk, H, is the score for each type of fight multiplied by the probability of meeting each type of contestant.

Table 6.1 *Average payoffs to the attacker in the game between Hawk and Dove.*

	Opponent	
Attacker	*Hawk*	*Dove*
Hawk	$\frac{1}{2}(50) + \frac{1}{2}(-100) = -25$	+50
Dove	0	$\frac{1}{2}(50 - 10) + \frac{1}{2}(-10) = +15$
Payoffs:	Winner +50 Injury −100	
	Loser 0 Display −10	

Therefore,

$$\overline{H} = -25h + 50(1 - h)$$

Similarly, for a Dove the average score will be

$$\overline{D} = 0h + 15(1 - h)$$

At the stable equilibrium (the ESS), $\overline{H}$ is equal to $\overline{D}$. Solving the two equations above by setting $\overline{H} = \overline{D}$ gives $h = 7/12$, and the proportion of Doves $(1 - h)$ must therefore be 5/12.

This ESS could be achieved in two distinct ways. The population could consist of individuals who played pure strategies. Each individual would be either a Hawk or a Dove, and the ESS would come about when 7/12 of the population were Hawks and 5/12 Doves. Alternatively, the population could consist of individuals who all adopted a mixed strategy, playing Hawk with probability 7/12 and Dove with probability 5/12, choosing at random which strategy to play in each contest. Either situation could produce stability in the game.

Now imagine another strategy in this game, "Bourgeois." An individual with this strategy plays Hawk if it is the owner of a territory and Dove if it is an intruder.

Keeping the same scores as before, imagine that a Bourgeois individual finds itself owner half the time and intruder half the time. (This assumption may be unrealistic in cases where territories are few and the "floaters" in the population far outnumber the owners.) The scores for the game involving three strategies are indicated in Table 6.2.

In this game, Bourgeois is an ESS. If all the members of a population are playing this strategy, no one ever engages in escalated fights because, when two individuals contest for a resource, one is owner and the other is intruder; the result is that the intruder always gives way. When everyone plays Bourgeois, the average score for a contest is +25. This strategy is stable against Doves, who would only get +7.5. For equal-sized male speckled wood butterflies (*Pararge aegeria*), Davies (1978) showed that territorial disputes were always won by the owner. If the original territory owner was removed and replaced by another male, the replacement won out in fresh disputes.

Despite the simplicity of these models, some broad but important conclusions can be drawn from real contests.

- **Fighting strategy is frequency dependent:** It depends on what other animals are doing. A Hawk strategy is good for an individual in a population of Doves, but not in one of Hawks.

Table 6.2   *Average payoffs to the attacker in the Hawk-Dove-Bourgeois game.*

		Opponent	
Attacker	*Hawk*	*Dove*	*Bourgeois*
Hawk	− 25	+ 50	+ 12.5
Dove	0	+ 15	+ 7.5
Bourgeois	− 12.5	+ 32.5	+ 25

Payoffs: Winner + 50   Injury − 100
                      Loser   0

- **The ESS is often a mixture of different strategy types, Hawk and Dove:** Such mixtures of displays and fighting are what is observed in nature.

- **The ESS is dependent on the values of scores in the game:** If the scores are changed, the stable mixture of Hawks and Doves will change. However, as long as the cost of injury exceeds the benefit of winning, probably a realistic assumption, then Hawk or Dove will never be an ESS. As a result, it would seem that, to predict realistically what happens in nature, we need to know the actual score values, not an easy thing to measure in the field. The source of one of the biggest positive scores is the opportunity to mate with a female, because failure to do so is equivalent to failure to pass on any genes—genetic death. Fights over mates are commonly severe, sometimes fatal, and Hawk-like strategies dominate. In musk-ox, 5–10 percent of the adult bulls may die each year from fights over females (Wilkinson and Shank 1977). As the positive score increases, the incidence of the Hawk strategy increases.

- **ESS is not necessarily the "best" strategy:** It merely is most resistant to invasion by alternative strategists.

In the real world, animals often construct their strategies according to the vigor of their opponents. It would be useless to adopt a Hawk strategy against a bigger opponent, even for a territory owner. Players using this strategy follow the rule, "If larger, behave like a hawk; if smaller, behave like a Dove; if equally matched, adopt the Bourgeois strategy." Hansen (1986) showed that pirating bald eagles, *Haliaeetus leucocephalus*, which stole food from conspecifics, assessed the size and hunger level of the food defenders and attacked those most likely to retreat. Age and territory may also enter into the equation. Though a young and an old animal may be equal in size, young animals may give up sooner in a fight because fighting incurs risks, and a young animal risks a larger proportion of its reproductive life; for an old male, each contest could represent his last chance to mate.

Ritual Displays

In many animals, fighting is ritualized into displays of fins or feathers or into sign waving and signal calling. These displays are then used as reliable signals of strength. In larger individuals the vocal cords are often bigger, and the pitch of the call is thus deeper. Many contests, for example between rutting stags (*Cervus elaphus*) and calling frogs, are often settled by vocal cacophonies (Davies and Halliday 1978; Clutton-Brock and Albon 1979; Clutton-Brock et al. 1979).

For animals whose body size increases gradually through life, young males may have weak voices and adopt a Dove strategy; as they get older, they become more Hawkish. Even the young males may get a few matings, however, by cheating. In bullfrogs, small "satellite" males sit silently in the proximity of larger calling males and try to sneak copulations with females who are attracted to the larger males' calls (Howard 1978). (See Fig. 6.3.) Young elephant seals sneak copulations by pretending to be females and joining harems (Le Boeuf 1974). In the case of the bullfrog, only 2 /73 matings in one study were by satellite males, but calling males do not have everything their own way—calling also attracts

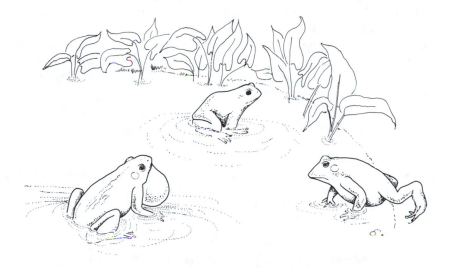

Figure 6.3 *Bullfrog mating strategies.* A male bullfrog (left) calls from his territory and attracts a female from the pond (right). The small male in the center is a satellite male in the larger male's territory; he sits silently and attempts to sneak a copulation with the female on her way to the caller.

predators. In the field cricket, *Gryllus integer,* male calls attract not only females but also parasitic tachinid flies (*Euphasiopteryx ochracea*) (Cade 1979; see also Fowler 1987). Louder calls attract more females and more parasites (Table 6.3) (see also Walker 1982). Sometimes parasites will lay their eggs on loudspeakers mimicking calling crickets! Cade (1979) found that, although 11 out of 14 calling males were parasitized by larvae of this fly, only 4 out of 29 satellite males were infected. In organisms with a potentially long life cycle, satellite males may actually father a good number of offspring as a result of their greater longevity than more active callers.

Rohwer and Rohwer (1978) tested, in an ingenious series of experiments using Harris sparrows (*Zonotricha quercula*), the reliability of signals as indicators of strength. In these birds, male breast plumage varies from pale to almost black. Darker birds are always able to displace pale birds from winter food supplies. Why shouldn't the pale individuals grow darker feathers and enjoy higher social status? The physiological mechanism is available, because in spring all males develop dark mating coloration. Rohwer and Rohwer attempted to create "cheats" by painting pale birds black. The "cheats" were simply attacked by other birds and failed to rise in status. Next, some pale birds were injected with testosterone. These birds became aggressive and fought more, but their opponents did not back down during disputes. Finally, when the two treatments—painting and testosterone—were combined, the cheaters won more fights over food, and their social status was elevated. In a reciprocal experiment, some dominant black birds were bleached white. They were attacked more by other birds, who tried to displace them, but after a lot of squabbling the bleached birds usually won. The conclusion was that plumage alone in birds, or signal strength in animals in general, is not an automatic passport to dominance; it must often be backed up by dominant behavior as well.

Table 6.3 *Results of an experiment in which the song of a male cricket,* Gryllus integer, *is broadcast from a loudspeaker, attracting not only females but also satellite males and a parasitic fly that can kill the cricket. (From Cade 1979.)*

	Attracted to Speaker		
Broadcast	*Females*	*Satellite Males*	*Parasitic Flies*
Silent	0	0	0
80 dB song	7	7	3
90 dB song	21	16	18

6.2 *Animal Communication*

Assessing an opponent's strength before a contest is obviously advantageous—the majority of individual decisions are based on what other individuals are doing. It is no use behaving like a Hawk in the presence of a bigger Hawk. It is no use behaving altruistically to unrelated individuals unless reciprocal altruism can be expected. To judge the likely behavior of others, animals must be able to communicate in some way.

The use of sensory channels by animals is very much dependent on the habitats in which they live. For example, color plays little role in the signals of nocturnal animals. Similarly, animals in dense forests often cannot see each other, so songs can be of prime importance in mapping out territories. Sound is a temporary signal, however; scent may last longer and is often used to mark the large territories of some mammals. The differences between these different sensory avenues and their functions in communication are outlined in Table 6.4.

Even between closely related genera, such as ants, communication signals may differ according to the ecological setting. In genera such as *Leptothorax*, which feed on immobile, easily subdued prey (like dead insects), scouting ants usually need only one additional worker to help bring the prey back to the nest. Rather than slowly laying down a scent trail, also energetically costly, the scout ant recruits a helper, which simply runs in tandem with the scout, with antennae touching the scout's abdomen. Fire ants (*Solenopsis*) attack larger, living prey, and many ants are needed to drag them back to the nest. In this case, a scent trail is laid by the scout, from a special abdominal gland. The scent excites other workers, which follow it to the prey. The marker pheromone is very volatile, and the trail effectively disappears in a few minutes, so there is no mass confusion of old trails. In South American leaf-cutting ants, the

Table 6.4 *Advantages of different sensory channels of communication. (From Alcock 1979.)*

Feature of Channel	Type of Signal			
	Chemical	Auditory	Visual	Tactile
Range	Long	Long	Medium	Short
Rate of change of signal	Slow	Fast	Fast	Fast
Ability to go past obstacles	Good	Good	Poor	Poor
Locatability	Variable	Medium	High	High
Energetic cost	Low	High	Low	Low

"prey," host leaves, are large and permanent. Trails are formed by constant wear and are permanent (Hölldobler 1977).

For animals living and calling near the ground, the earth itself can be a major cause of sound attenuation. This is the reason so many birds and insects use perches for singing—the sound travels farther. Furthermore, air turbulence is 14 times less at dawn and dusk than during the rest of the day (Henwood and Fabrick 1979), which explains the preference of most animals for calling at these times. Some insects have actually utilized the very herbage on which they feed as a medium of song transmission. Many leafhoppers and planthoppers vibrate their abdomens on leaves and create species-specific courtship songs, which are transmitted up to several meters away by adjacent and touching vegetation and are picked up by nearby conspecifics (Ichikawa and Ishii 1974; Claridge 1985).

Physical signals, such as preening movements in ducks during courtship, tooth baring in wolves in aggressive encounters, and urinating to mark territories, probably evolved as part of already existing incidental movements. Tooth baring is a natural prelude to biting in canids, and urination is a common nervous reaction on encountering a potential combatant. Though the study of the full repertoire of these behaviors is really the province of the ethologist, it is not difficult to see how they may have evolved (Table 6.5).

The pinnacle of communication in the insect world is the dance of the honeybee, elegantly studied by von Frisch (1967). Bees associate food locations with specific signals such as scents, colors, and geometric patterns provided by scout bees. The learning of specific signals can be rapid. The scouts constitute only a small percentage of "nonconformist"

Table 6.5 *Behavior patterns from which displays in birds, fish, and primates are thought to have evolved.*

Behavior or Response	Example of Display
Intention movement	Sky pointing in the gannet
Ambivalent behavior	Oblique-threat posture of black-headed gull
Protective response	Primate facial expressions
Autonomic response (such as sweating, urinating, rapid breathing)	Vocalizations (from rapid breathing); scent marking
Displacement activities	Preening in duck courtship
Redirected attack	Ground pecking in herring gulls

individuals that explore new potential resources. Army ants, *Eciton* sp. in the Americas and *Anomma* sp. in Africa, have even more efficient recruiting techniques. Thirty seconds after scouts returned to the main column, 50–100 ants were seen at a nearby bait (Chadab and Rettenmeyer 1975).

In the world of vertebrates a plethora of calls and signals exists for a variety of reasons. As antipredator behavior, alarm calls and signals have attracted a large amount of theory and only slightly less fact. The rump-patch signalling of white-tailed deer has been interpreted as a warning signal to other members of the herd. The problem with this theory is that the patch would be more visible on the flanks, like the markings of Thompson's gazelle. The rump patch is only visible to animals directly behind. Perhaps the strategy is to warn other members of the herd of an attack, thus reducing the likelihood of any predator success, but only after the warning individual has passed a few other members of the herd, putting some distance—and other deer—between it and the predator. Rump patches would seem to make the prey more visible to predators and thus easier to catch. Some enterprising ecologists have suggested that this type of signal, together with the deliberate prancing or stotting movements seen in some gazelles, constitutes a deliberate message, "Look at me and how healthy I am; I could easily outrun you." An un-healthy or injured animal would be unlikely to be as effective at stotting. Whether the alarm call or signal is directed at conspecifics or at the predator is uncertain in many cases (Harvey and Greenwood 1978). Imaginative ecologists can come up with many ways to explain a behavior. Caro (1986) reviewed 11 nonmutually exclusive hypotheses that have been proposed for the function of stotting (Table 6.6).

In mating and courtship, a vast number of calls and signals are used to impress mates and bring them together. Such showy maneuvers are bound to attract predators also (see also Table 6.3). Insectivorous bats have influenced the development of communication in neotropical katydids (Belwood and Morris 1987), causing males to sing shortened songs. It has also been suggested that species-specific tremulations through the host plant are used specifically because they are inaudible to bats. Many more subtle signals have evolved that can be used at night, in the absence of diurnal predators. Fireflies have developed species-specific light flashes (Lloyd 1966) (Fig. 6.4), and moths attract females by powerful chemical attractants called *pheromones*. Even the dance of the honeybee probably evolved so that information could be transferred in the darkness of the comb. Some signals may of course attract other males, potential competitors, who may be drawn to the action. Female elephant seals use this behavior of males to their advantage. When a male attempts to mate with a female, she screams loudly, attracting the attention of other would-be suitors, which fight over her. In this way, she is guaranteed a mating with the strongest male.

Table 6.6   *Summary of the hypotheses put forward to explain stotting. (After discussion by Caro 1986.)*

Number	Behavior Directed at	Hypothesis	Explanation
1	Predator	Pursuit invitation	Incites premature and therefore unsuccessful attack by predator
2	Predator	Predator detection	Indicates predator has been detected
3	Predator	Pursuit deterrence	Indicates to the predator that it has been detected and that its chances are therefore low of ambushing and catching the prey
4	Predator	Prey health	Indicates that the prey is so healthy that it could easily outrun the predator
5	Predator	Startle behavior	Causes predator to hesitate slightly
6	Predator	Confusion effect	Communal stotting confuses predators
7	Conspecifics	Social cohesion	Prey are better able to group together through visual signals
8	Conspecifics	Attracting mother's attention	Notifies mother that hidden fauns are changing their position
9	Nobody	Anti-ambush behavior	Allows prey to see over tall vegetation
10	Nobody	Play	No obvious benefits
11	Conspecifics	Warning	Altruistic warning to close kin

6.3 *Foraging Behavior and Optimality in Individuals*

The strategies of animals living in groups depend largely on what other members of the group are doing.

In some instances, decisions can be made purely on an individual basis, and single animals opt for or against one type of behavior that may maximize their benefits and minimize their costs. Such behavior is often employed in food gathering: whether to leave a resource patch and look for a new one or to stay a little longer and forage. Sometimes similar behavior is exhibited in searching for a mate. The analysis of these economic decisions is often performed in terms of optimality models, which study individual adaptations. These were first introduced to ecology by Robert MacArthur (MacArthur and Pianka 1966), a fertile mind in the field of ecology and an active researcher in population and community ecology until his early death in 1972. The logic for using optimality theory is that natural selection should produce animals that are maximally effi-

Figure 6.4 *Light signals of eight different species of male fireflies as they would appear in time-lapse photography.* Arrows show the direction of flight. Females normally respond to the pattern of their species and not to other patterns. (Some females mimic the signals of other species so that they can feed on the males they attract.) This illustration is diagrammatic; not all eight species occur in the same locality. (Redrawn from Lloyd 1966.)

cient at propagating their genes and also at performing all other functions that subserve this function in the end.

When shore crabs are given a choice of different-sized mussels, they prefer the size that gives them the highest rate of energy return (Fig. 6.5). Very large mussels take so long for the crab to crack open in its chelae that they are less profitable in terms of energy yield per unit breaking time than the preferred, intermediate-sized shells. Very small mussels are easy to crack open but contain so little flesh that they are hardly worth the trouble. Of course, the most profitable mussels may take a longer time to locate, so some less profitable sizes may be eaten, as they are encountered more frequently. The result is that the diet includes a range of sizes

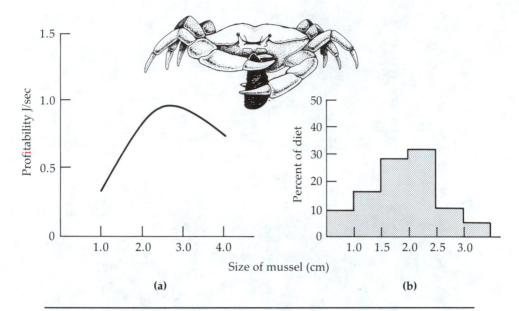

Profitability J/sec

Percent of diet

Size of mussel (cm)

(a)

(b)

Figure 6.5 *Shore crabs'* (Carcinus maenas) *preference for mussel size that gives the highest rate of energy return.* **(a)** *Calorie yield curve per second of time used by the crab in breaking open the shell.* **(b)** *Histogram of sizes eaten by crabs when they were offered a choice of equal numbers of each size in an aquarium.* (Redrawn from Elner and Hughes 1978.)

around the preferred, optimal size. John Krebs and associates (1977) tested this idea using caged great tits (*Parus major*). Various-sized mealworm prey were paraded past the birds on a conveyor belt. The mealworms were visible to the perching birds for only half a second. Once a mealworm was taken, the bird would miss the next worm on the belt because it would still be eating the first. An active choice was therefore involved. At low rates of encounter of large worms, birds took all sizes. As the encounter rate with large worms went up, the birds switched to taking just big worms.

Predators also hunt for food that is clumped or patchy in distribution, such as rotting tree stumps full of insects or bushes laden with berries. How long should a predator spend in each patch before leaving? The answer depends on how the quality of patches changes with time.

Behavioral ecologists have argued that patches change in quality most often as a result of predator activity, which depletes prey (Fig. 6.6). The predator's overall rate of food intake for the habitat is the average food intake per patch divided by average time in a patch plus travel time between patches. In order to maximize this quantity, the predator should choose to eat in each patch just long enough to make the slope of the line

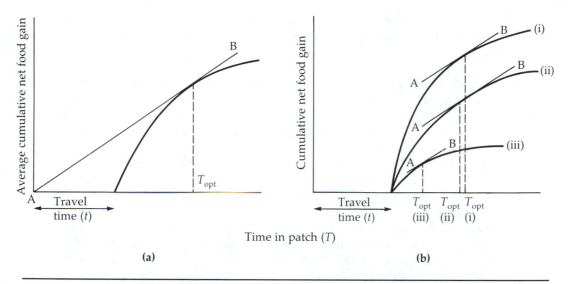

Figure 6.6 *A graphical solution of the optimal time in a patch with depletion of food.* **(a)** *Single-patch habitat.* The curve represents the average cumulative food gain *(f)* as a function of time in a patch *(T)*. The time spent travelling between patches *(t)* is also plotted on the *x* axis. The optimal predator should choose to stay in the patch just long enough to maximize the slope of the line AB (representing the average food intake per unit time for the habitat as a whole). In order to do so, the predator leaves a patch at the time T_{opt}. **(b)** *Multiple-patch habitat.* If individual patches in the habitat have different curves of *f(T)*, the predator should apply the same giving-up criterion to all of them.

AB, the net rate of food intake for the environment, as steep as possible. This line, of course, has to touch the curve of cumulative intake within a patch, and it does so where the slope of that curve is equal to *E / T* (the net rate of food intake). In experiments with blue tits foraging for mealworms, with bumblebees sipping nectar in desert flowers, and with parasitoids foraging for insect larvae to parasitize, it has been shown that foraging does occur according to this rule of optimality. That is, the predators stayed in each patch until their rate of intake (the marginal value) dropped to a level equal to the average intake for the habitat. A predator should not stay in a patch if it could do better by traveling to a new one.

This type of approach is known as the *theory of optimality,* and it predicts that all patches should eventually be reduced to the same marginal value and that the marginal value should equal the average rate of intake for the habitat. It is unusual in biology, in that the null hypothesis is often that selection has designed maximally efficient animals (Krebs and Davies 1981), rather than the more traditional null hypothesis that animals are behaving randomly rather than optimally. However, many physiological ecologists would argue that the result of natural selection is more often adequacy than perfection (Bennett 1987). There are many other features to this approach that have been severely criticized (Pierce and Ollason 1987).

■ Animals do not forage optimally, because they need an appropriate mix of nutrients for energy—fats, proteins, and so on. That food quality as well as quantity was important to foraging behavior was shown by Belovsky (1978). For moose, *Alces alces,* and for many herbivores in general, food selection is critical because many plants lack essential amino acids (see also Brodbeck and Strong 1987). Moose get energy mainly from browsing on deciduous forest trees. They also need to browse on aquatic vegetation, which is rich in sodium. Deciduous forest trees have low sodium levels, and aquatic vegetation is often a relatively poor energy source, so moose eat both. The intake of food types in diet can thus be represented by a plot of these two requirements against one another (Fig. 6.7). At least one other constraint is recognized in this system, the finite capacity of the moose's stomach, which limits daily intake. In the figure, the necessary daily intake of sodium is represented by a horizontal dot-and-dash line and the energy requirement by a solid line. Energy

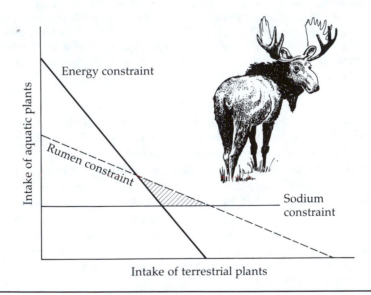

Figure 6.7 *The diet of moose constrained by the requirements for sodium and energy.*
The daily requirements are shown as the light and heavy solid lines respectively, and the moose has to eat a mixture of plants that lies in the space above these two lines. The third constraint is the size of the moose's rumen (broken line). Aquatic plants are bulkier than terrestrial ones, so fewer grams of these can be fitted into the rumen. The moose's diet was found to lie, as predicted, inside the region of the shaded triangle. (Modified from Belovsky 1978.)

can be obtained by different combinations of aquatic and terrestrial vegetation. Because the size of the moose's stomach is finite, it cannot dine on
aquatic vegetation alone because the necessary volume would not fit into
its stomach. The broken line represents the effect of this constraint. Thus
by simultaneous action of these three factors, the moose are constrained
to a fairly narrow diet (the shaded area in Fig. 6.7). Diet may explain
moose movement from patch to patch of vegetation.

Smallwood and Peters (1986) have shown that grey squirrels (*Sciurus
carolinensis*) also do not forage optimally. When feeding on acorns, they
do not forage in a manner that maximizes daily energy intake. The reason
is probably that squirrels respond to tannin concentrations in acorns and
spend less time eating acorns with elevated levels of tannins despite the
fact that some such acorns have added fat.

▪ Animals are designed not to maximize food intake but to minimize
risk of predation. They may only dart out to take food now and again. If
what animals do cannot be subdivided into independent activities, then
according to Lewontin (1978) we are left with the not very illuminating
result that a whole organism is adapted to its whole environment. The
animals that leave the most offspring are, by definition, the fittest.

▪ The environment may be changing so rapidly in ecological time that
any given species is not sufficiently in tune with its habitat to forage optimally.

▪ Optimal foraging presupposes that predators know the average rate
of intake for the habitat and also the instantaneous rate within each
patch. McNamara and Houston (1985) show that if an animal tried to
learn the optimum, as defined by a marginal value theorem, it would take
an infinite amount of time to converge upon the optimal solution. Some
authors have tried to circumvent these types of argument by introducing
the concept of stochasticity, such that patch quality, travel time, and
other variables vary in an unpredictable way, much as they would in nature. Suppose there are two types of habitat, one providing food at a constant rate and the other at a highly variable rate, but the mean rate is the
same for the two types. The constant habitat provides enough food to
meet the animal's normal energy requirements, whereas the variable
habitat has a 50 percent chance of providing more than enough food and
a 50 percent chance of not providing enough. The "normal" animal
chooses the safe, constant habitat. After several hours or days of starvation, the reverse behavior appears—the constant habitat cannot provide
enough food to make up for starvation, and the animal must take a
chance on the variable habitat; it becomes "risk-prone" (Stephens 1981).
This idea is sometimes called the *expected energy budget rule*—be risk-prone
if the daily energy budget is negative, be risk-averse if it is positive. Data
on foraging by yellow-eyed juncos (*Junco phaeonotus*) for seeds in different

environments were in accord with this idea (Caraco, Martindale, and Whitham 1980). It might be expected that about half of the risk-prone animals will die of starvation after a little while, but this is not usually the case. Why? Houston and McNamara (1982) explained that animals face a series of these sorts of choices throughout the day. If they start off risk-prone, about half will get enough food to ensure that they are risk-averse at the next decision, so by the end of the day far fewer risk-prone animals will be in a negative expected energy budget than expected.

▪ An animal may be able to survive and reproduce perfectly well under some conditions without behaving optimally. Its own "rules of thumb" are sufficient to ensure continued success (Simon 1956). There is an analogy here with parasites and predators, which do not always attack hosts in the best possible (density-dependent) way but opt to "spread their risk" by attacking a few hosts and moving on to new pastures. The risk here is that local catastrophes may destroy all hosts in any particular area but are unlikely to do so over a wide range of areas (see chapter section 15.2).

▪ Traits that are used at present for one purpose (feathers for flight) may have arisen for another (feathers for heat insulation). They may not appear to be an optimal solution to a problem.

Territorial Behavior

Apart from foraging optimally, many animals ranging from sea anemones to monkeys actively defend territories. There are economic costs to defending a territory as well as benefits, and territory owners should tend to optimize territory size according to the costs and benefits (Fig. 6.8). Territories may also be set up to reduce predation by conspecifics on eggs, to attract more females, or to spread out individuals to reduce density-dependent predation (Davies and Houston 1984). Such territories may be held for a season, a year, or the entire lifetime of the individual.

In studies of territorial defense in the golden-winged sunbird (*Nectarinia reichenowi*) in East Africa, the benefits of a territory could actually be measured as nectar content of flowers and the costs to the bird in terms of time-budget studies and laboratory analysis of the energy cost of different activities such as sitting, flying, and fighting. Gill and Wolf (1975) worked out that in defending a territory the bird actually saved 780 calories a day in reduced foraging activity (the flowers in its territory were higher in nectar) but spent 728 in additional defense, a net saving of 52 calories a day. Further refinements of this type of energetic and time-budget analysis may be used to predict optimal territory size (in the sunbird's case about 1,600 flowers).

The economics of territoriality were also investigated by Horn (1968).

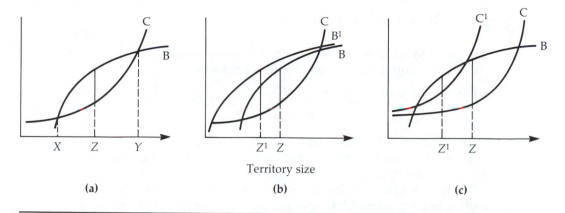

(a)

Territory size

(b)

(c)

Figure 6.8 *Hypothetical relationships between benefits B and costs C for various territory sizes.* **(a)** *Optimal cost-benefit curves.* Defense will be profitable (B ≥ C) only between points X and Y. Maximum net benefit (B - C) is at Z. **(b)** *Change with increasing benefits.* Increase in benefits (B curve shifts to B1) will decrease the thresholds at which defense is economical and decrease Z to Z^1. **(c)** *Change with increasing costs.* Increase in costs (C curve shifts to C^1) will likewise decrease the thresholds and decrease Z to Z^1. (Redrawn from Myers, Connors, and Pitelka 1981.)

Horn contrasted the average traveling distances involved to resources when species nested communally and when they held territories (Fig. 6.9). If food resources are distributed randomly and evenly, as in Fig. 6.9, then pairs of territorial animals use only the four resource points nearest to their nest or den. The average traveling distance to a resource point is then given by

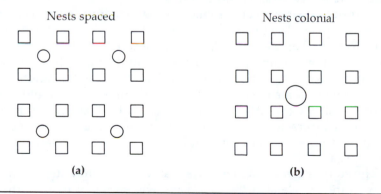

Figure 6.9 *Two diagrams of cost-benefit territoriality.* Nests are shown as circles, food as squares. (Redrawn from Horn 1968.)

$$\sqrt{(D/2)^2 + (D/2)^2} = 0.71D$$

where D is the distance between circles, resource points. If birds nest in one central colony, then the average travel distance is

$$\frac{4\sqrt{(D/2)^2 + (D/2)^2} + 8\sqrt{(3D/2)^2 + (D/2)^2} + 4\sqrt{(3D/2)^2 + (3D/2)^2}}{4 + 8 + 4} = 1.50D$$

For uniformly distributed resources, evidence from travel distance alone suggests birds should behave territorially. What if food resources are clumped? Travel distance from a colonial site still turns out to average $1.50D$, whereas from territorial nests it is $1.93D$. When food is clumped, colonial nesting should be expected. Despite the fact that this idea is based only on a single concept (travel distance), data from the field provide much validation of Horn's theory. For instance, most seabirds forage on transient schools of fish, clumped and unpredictable resources. Most seabirds nest colonially. Of course, small island or cliff size and predator avoidance could also be major contributors to such behavior.

It is worth noting that in most cases it is only conspecifics that are prevented from entering a territory. Where interspecific territoriality does occur it generally involves similar species (Cody 1974). There are probably two reasons: (1) the strongest competition for resources probably occurs between individuals that feed in the same manner or are likely to steal mates, and (2) it would be hard for a rattlesnake to exclude a hawk from an area even though both are feeding on the same rodents.

Mating Behavior and Optimality

The optimality approach used to analyze foraging decisions can equally well be applied to other kinds of behavior such as mating. Copulating dungflies serve as a good example.

A fresh cowpat in an English meadow is soon invaded by a swarm of yellow dungflies. The first to arrive are males seeking out females with whom to mate. Females arrive later and are soon "captured" by males. After copulation the females lay their eggs on the pat, where the larvae hatch out and grow. Males tend to copulate for about 36 minutes. Why this exact time? If males copulated for 100 minutes, they would fertilize all the female's eggs; in 40 minutes they can fertilize only 80 percent. However, a protracted mating leaves less time to find new females with whom to mate, so the flies cut their losses and cease copulation after 40 minutes (Parker 1978).

The story is complicated by rival males who, if they succeed in mating with the female after the first male, will supplant most of the first male's sperm, fathering 80 percent of the progeny themselves. Only 20 percent of the progeny will belong to the original male. Thus, after mating, males still sit on top of females, guarding them until they lay their eggs. Only then will males leave to find a new mate. The process of guarding and searching for a new mate takes an average of 156 minutes. The male cannot control this interval, but he can control his copulation time to maximize the number of progeny sired. His behavior according to the graphical model presented in Fig. 6.10 is nearly perfectly optimal; the theoretical copulation time is 41 minutes.

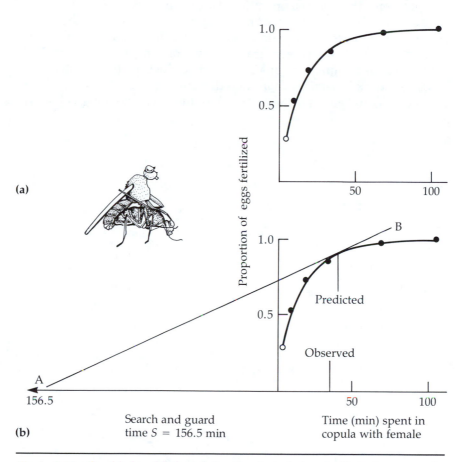

Figure 6.10 *Graphical model of copulation time for male dungfly (Scatophaga stercoriaria).* **(a)** *The proportion of eggs fertilized as a function of copulation time (results from sperm competition experiments).* **(b)** *The optimal copulation time.* The time that maximizes the proportion of eggs fertilized per minute, given the shape of the fertilization curve and the fact that it takes 156 min to search for and guard a female, is 41 min. The optimal time is found from the line AB. (Redrawn from Parker and Stuart 1976.)

Drawbacks to Optimal Foraging Theory

Studies of optimal foraging behavior are useful in that they allow us to predict the exploitation patterns of animals depleting patches. If the travel time between patches increases, optimal patch time increases, and the animal will spend longer searching (Fig. 6.11 top). If patch quality decreases, animals again spend longer searching (Fig. 6.11 bottom). Although these generalizations are useful in predicting search time for animals feeding on patchy resources, they take little account of other selection pressures that may be operating. House sparrows feed only on the edge of a freshly sown field, not on the rich resource of grain in the center of the field, because they would be more at risk from hawk predators in the center; they therefore feed near the cover of hedges at the edge (Barnard 1980). In the presence of kingfishers, stickleback fish (*Gasterosteus aculeatus*) attack low rather than high densities of prey. In this way the sticklebacks can better see the approach of the oncoming kingfishers. In

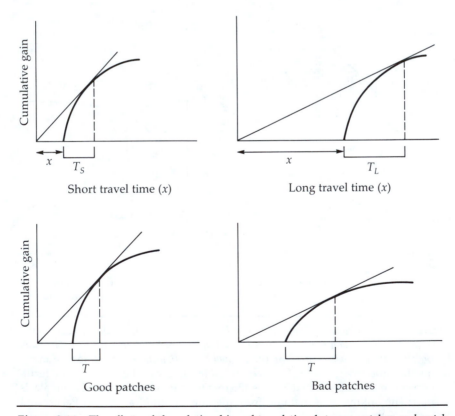

Figure 6.11   *The effects of the relationships of travel time between patches and patch quality on optimal foraging behavior.*

the absence of kingfishers, sticklebacks feed optimally on high-density prey (Milinski and Heller 1978). Nevertheless, optimality models may be useful in that they (1) generate quantitative predictions that often can be tested and (2) allow behavioral comparisons to be made between species through observations of whether each is following the same type of optimal behavior during feeding or copulation. Much further discussion of optimality, foraging theory, and maximization of feeding rates is provided by Stephens and Krebs (1986). These authors summarize many of the models and examine tests of models on optimal foraging, 125 studies in all. Most of the papers, 64 percent, test qualitative rather than quantitative predictions, taking perhaps the easier road. Seventy-one percent of the papers report agreement with models; only 13 percent clearly contradict predictions. Field tests showed as much support for models as did laboratory tests. Stephens and Krebs (1986) conclude that optimality models have now been thoroughly tested, usually fit the data, and, most importantly, lay the groundwork for future developments.

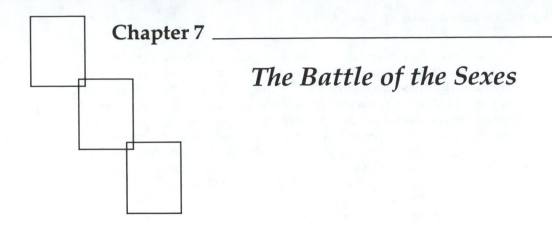

Chapter 7

The Battle of the Sexes

I f a single male can fertilize many females, why do many species have a 50:50 sex ratio? Why is it that males commonly fight over females but seldom vice versa? Because many species can reproduce asexually (or parthenogenetically), why have different sexes at all?

7.1 *The Maintenance of Sex Ratios*

Asexual Versus Sexual Reproduction

Part of the answer to the question "why have two sexes" is that the sexual process, through meiosis and fertilization, ensures genetic recombination, the creation in offspring of new combinations and permutations of the traits of the parents. The result is offspring with a much higher range of variability than those of asexual reproduction, which are genetically identical to their parents. Natural selection can work on this increased variability to cause much more rapid evolutionary adaptation to changing environmental circumstances. For example, if two favorable mutations a→A and b→B were to take place in different individuals, there would be no way, in an asexual population, that an AB individual could arise, except by the unlikely event of further mutations in either A or B. In a sexual population an AB individual could arise by recombination from a mating of the separate mutant individuals. Another advantage of sex is the so-called *Muller's ratchet* (Felsenstein 1974). Besides advantageous mutations, deleterious mutations occur in populations. In an asexual population, if the number of "perfect" individuals with no deleterious mutations is small, there is always a chance that, in some season, they will all fail to reproduce. Because back mutation is too infrequent to be important, the population would then have no "perfect" individuals—the ratchet has clicked around once. In a population with sexual reproduction, however, new "perfect" individuals can arise by recombination. It is often thought that sexual reproduction arose to correct genetic errors rather than to promote variability. Of course, these types of arguments are based on the good of the species and depend on group selection. In fact, a gene increasing fitness of the individual will spread throughout the population even if it reduces group fitness (Maynard Smith 1978).

In an attempt to explain the evolution of sex through advantage to the individual, Ghiselin (1974) has proposed the "tangled-bank" theory (Bell 1982), which suggests that the advantage of sexual reproduction results from the slight differences among offspring, which enable them to occupy microscopically different niches and to avoid competing as severely as they might. The theory is analogous to the one in economics that, in a saturated economy, it pays to diversify. By this logic, parthenogenetic species should be prevalent in unsaturated environments. An attractive new theory is that sex is an adaptation for the transfer of infectious (parasitic) genetic elements. These ideas are more fully discussed by Michod and Levin (1988).

Sex Ratio

Given the preponderance of sexual species, fighting over mates causes some of the most common, severe, and escalated contests in the animal kingdom. Next to finding food, the sex drive is the most important selective pressure. Yet it is nearly always males who do the fighting. Why don't females also fight for the best mate? The answer is that males produce many gametes, and one male could potentially fertilize the females of a whole population. Five milliliters of human semen contain enough sperm to fertilize twice the female population of the United States. Females produce very few gametes but they invest a lot in each relatively large egg. The consequence is that males have a much greater reproductive potential than females and are therefore under strong selective pressure to be good at seeking out and competing for females. By way of illustration, the maximum number of offspring per lifetime has been recorded at 200 in male elephant seals, as opposed to only 15 in a female. In humans, the record belongs to Moulay Ismail the Bloodthirsty, Emperor of Morocco, who fathered 888 children; the known maximum for a woman is 69 children from 27 pregnancies.

If one male can fertilize dozens of females, why are there no species with a sex ratio of, say, 1 male to 20 females? The sex ratio is nearly always about 1:1. The answer again lies with the selfish genes (Fisher 1930). If a population contained 20 females for every male, then a parent whose children were exclusively sons could expect to have 20 times the number of grandchildren produced by a parent with mostly daughters. Such constraints operate on the numbers of both male and female offspring, keeping the sex ratio at about 1:1. The most common exception is the situation in which it costs more to produce sons or daughters. Sup-

pose sons are twice as costly to produce as daughters, because they are
bigger and eat more. Sons are a bad investment, because each grandchild
is twice as costly to produce via a son as via a daughter. The sex ratio
would then swing toward a female bias until an ESS 2:1 females:males
was reached.

Local Mate Competition

One other exception to the 1:1 sex ratio is due to local mate competition, in which one male dominates in breeding and other males therefore
become superfluous because they are never likely to leave any offspring.
This phenomenon is most likely to occur in species with low powers of
dispersal in which brothers stay in the same place and inbreeding is frequent. In such situations, for example in parasitic Hymenoptera, whose
broods develop and mate in or around patches of host insect larvae, biased sex ratios are common (Werren 1980; Antolin 1989). Similar biased
sex ratios occur in Old World monkeys (Andelman 1986). Groups consisting of more than 10 females contain many breeding males, whereas
those consisting of five or fewer females contain one adult male.

In the viviparous mite *Acarophenox*, brood size is about 20, and each
brood contains only one son. The male mates with his sisters inside the
mother and dies before he is born (Hamilton 1967). Where local mate
competition is not so important or dispersal is more prevalent, more
males may be produced.

The effects of local mate competition on sex ratios can also be seen
where the effective patch size varies. That is, if the number of hosts, say
eggs in a clutch, were small, local mate competition would be increased.
A similar effect would result if the number of foundresses, female parasites attacking the patch, were large. In accord with these theories,
Waage (1982) reviewed the biologies of different species of Scelionidae,
parasites of insect eggs. In small patches, female-biased sex ratios were
produced. In large patches, presumably with little local mate competition, normal (1:1) sex ratios were produced, as they were for species attacking isolated eggs, for in these cases dispersal was very common. Both
Werren (1983) and Waage and Lane (1984) have also shown sex ratio to
vary with foundress number per host patch.

Tests of these theories of sex ratios are most commonly performed
with parasitic Hymenoptera, for it is only in these, and a few other
groups, that females can actively decide the sex ratio of broods by using
or not using stored sperm to fertilize eggs. Fertilized eggs develop into
females, unfertilized eggs into males. In most other groups, including all
vertebrates, sex is determined by sex chromosomes, and there is less evidence of any deviation from a 1:1 ratio.

Such effects are interesting not merely from a purely theoretical angle
but also from an applied point of view. Insect parasites are a prime line of

defense against insect pests, so-called biological control. Millions of parasitic wasps are bred in laboratories and released in the field, but many are reared under crowded conditions, and the proportions of males may be too high. In at least one situation the failure of an important parasite to control a pest was traced to a problem of sex allocation. The phenology of the ichneumonid *Tiphia popilliavora* in its native Japan is such that it attacks the large third-instar larvae of its host, the Japanese beetle. In the United States, where the beetle poses a big problem to lawn and turf, the wasp attacks mainly younger larvae, on which it lays mainly male eggs (Brunson 1939). Similar host-size effects have been implicated in the displacement of *Aphytis lignanensis* by *Aphytis melinus* in California (see also Chapter 11). The latter aphelinid parasite accepts smaller host-scale insects and lays a higher proportion of females on them than does *Aphytis lignanensis*, which it has displaced in most regions (Luck, Podoler, and Kfir 1982; Luck and Podoler 1985).

7.2 *Sexual Selection*

Mate Selection

Selection for the ability of males to acquire matings is very strong and is termed **sexual selection,** a term coined by Darwin (1871). Sexual selection works in two ways, by favoring competitive fighting (*intrasexual selection*) or by favoring traits that attract females (*intersexual selection*). Examples of traits favored by intrasexual selection are large horns in beetles, like those in Photo 7.1, and antlers in deer and their kin. A trait favored by intersexual selection is bright plumage in birds. The intensity of sexual

Photo 7.1   A neotropical Hercules beetle, Dynastes hercules. *Sexual selection operates on the males of this species to increase horn length. The horns may be used by males in fights over females. (Photo by P. Stiling.)*

selection depends on the degree of competition for males. This intensity in turn depends on two factors, the difference in parental effort between the sexes and the ratio of males to females available for mating. When parental effort is equal, for example in monogamous bird species in which both sexes feed the young, sexual selection is less intense than in species that have different levels of parental effort. If large numbers of both sexes come into breeding condition at the same time, sexual selection is again reduced because there is less chance for a few males to control access to a large number of females. Asynchronous breeding promotes sexual selection.

Some particularly bizarre examples of male competition for mates are given below.

- In dragonflies (*Calopteryx maculata*), males use special scoops on the penis to scoop out previously deposited sperm from their chosen mates before injecting their own (Waage 1979; Thompson and Dunbar 1988).

- In acanthocephalan worms, *Moniliformes dubius*, parasitic in rat intestines, the male cements up the female's genital opening after mating. Furthermore, the males apply the same cement to the genital regions of competing males (Abele and Gilchrist 1977).

- In the hemipteran insect *Xylocoris maculipennis*, the male simply injects his sperm through the body wall into the body cavity of the female, where the sperm swim around until they encounter the eggs. Males may also inject sperm into the bodies of rival males, where they swim to the testes and wait to be passed on to a female the next time the victim mates (Corayon 1974).

- In many fish, for example the blue-headed wrasse, *Thalassama bifasciatum*, of western Atlantic coral reefs, the largest males dominate nearly all the matings with females. Large size is correlated with age. As a result, many fish begin life as females and change sex to become males when big enough to command sufficient matings (Warner, Robertson, and Leigh 1975). In this way, potential reproductive output is maximized. Such a system of sex change is known as protogynous hermaphroditism. The reverse, protandrous hermaphroditism, or **protandry,** is much rarer but is exhibited by the clown fish *Amphiprion akallopisos*, which lives in close symbiosis with sea anemones on Indian Ocean coral reefs. There is only enough space on each anemone for only two fish, so they are forced to be monogamous. The species' reproductive ability is thus limited by the female's egg supply, and the pair does better if the larger individual is the female. These fish begin life as males and later change into females. If a female is removed, the male is joined by a smaller individual and changes sex (Fricke and Fricke 1977). Such sex changing is, of course, a physiologically complex event, and in more advanced animals, such as mammals, it is evidently too difficult.

Because females, in the great majority of species, are the chief providers of resources for the zygote and have a smaller number of potential offspring than do males, they must choose their mates very carefully. In many species males compete for and defend breeding territories that offer the richest supply of resources for potential offspring, and in these species, when a female chooses a mate, she is really choosing the territory. In choosing a good territory, she is presumably acquiring a vigorous mate. In other species, for example in the hanging fly *Hylobattacus apicalis*, food often limits a female's capacity to lay eggs, and males presenting large food packets are chosen. In turn, males presenting larger insect prey can copulate for longer and thus fertilize more eggs, while the female eats the food (Thornhill 1976). Similarly, male *Drosophila subobscura* provide females with a drop of regurgitated food during courtship, and this strategy increases their courtship success (Steele 1986). The story of how male hanging flies obtain their prey makes fascinating reading (Thornhill 1979), for males may hunt it themselves, steal it from other males, or adopt a female-like posture and try to dupe other males into giving it to them. Such "transvestite" behavior sometimes succeeds, but often the duped male will grab his food back once he discovers his mistake.

In a general sense, females are choosing mates with the best genes, males that increase the ability of their offspring to survive and reproduce. Partridge (1980) allowed groups of female fruit flies (*Drosophila*) to mate either freely or only with specific partners that she chose at random. The offspring of the "choice" and "no-choice" females were tested for competitive ability in bottles of a medium containing the larvae under examination and those of standard competitors, distinguished by a genetic marker. The offspring of the "choice" group fared slightly but consistently better than those from the "no-choice" group. Females can apparently choose the best types of mates. How they do so is unknown, although it may be that females preferentially select **genotypes** different from their own so as to confer heterozygous advantage on their offspring. This idea could explain why "opposites attract" in human society. However, the advantages that accrue to the offspring of this type of mating cannot be passed on to the next generation because heterozygotes do not breed true.

Being Attractive to Mates

How can we explain the attractiveness of the bizarre male adornments of some species, such as peacocks? In this case, the mechanism may be self-perpetuating—females that choose showy mates are more likely to produce attractive sons and thus to have more grandchildren. Why do males of all species not have huge showy tails? Such adornments have costs as well as benefits and often result in increased predation pressure. In the three-spined stickleback, some males have bright red throats and are preferred by females. Others have dull throats. Bright red throats, however, attract predatory trout, so weak individuals would profit little by cheating and exhibiting red throats (Moodie 1972). Marler and Moore (1988) provided further validation of this behavioral "balancing act" of conspicuousness and predation pressure using free-living male lizards, *Sceloporus jarrovi*. They noted that survivorship in normal animals was negatively correlated with body-weight index. Bigger males were more successful in male-male competition but usually could be expected to be eaten more quickly. Body size was linked to aggressiveness of behavior, but such behavior could also be induced by testosterone implants. Manipulated individuals were more aggressive and conspicuous than controls but were likely to survive for a shorter time. Marler and Moore stressed that natural testosterone levels had therefore to be maintained at an optimum between that promoting sexual selection and that increasing survivorship.

Zahavi (1975 and 1977) has argued that females choose males with the most bizarre adornments and therefore the greatest handicaps precisely because they *must* be stronger to survive in the face of such intense predatory pressure. "Look at me! See how strong I must be in order to survive with these heavy tail feathers." Recently, Hamilton and Zuk (1982) have suggested that, at least in birds, the males that develop showier plumage and more vigorous display actually are fitter—they have a lower load of blood parasites. Hamilton and Zuk found a highly significant negative correlation between blood parasite load and sexual vigor. However, some parasitologists disagree, saying that the data are far too weak to be used to support such a theory and that real infection levels are unlikely to have been accurately assayed by the type of samples Zuk and Hamilton took (Cox 1989). Møller (1988) showed that females select *for* male sexual tail ornaments in the monogamous swallow, *Hirundo rustica*. He examined mating frequency using normal birds, birds with their ornamental feathers clipped, and birds with experimentally elongated tails. Those with the elongated tails mated faster than controls. They were therefore often able to sire a second clutch of eggs, and their fitness was much increased.

7.3 *Polygyny*

It is most often the females that have to be very choosy in mate choice because it is they who must "stay home" and care for the offspring. Physiological and life-history constraints dictate that female mammals must care for the young, first with a long gestation period and then by giving milk. Because of these constraints, at least in mammals, males are able to desert, and most mammals have polygynous mating systems in which each male mates with several females but each female mates with only one male. In cases in which females are able to shed their young at an early stage and in which there is no parental care—for example in fish, amphibians, and, to some extent, birds—different strategies are adopted. As an example, it is the male sea horse, *Hippocampus*, who carries the fertilized eggs around in his brood patch; females desert to form reserves for more eggs. In most fish, the female deposits her eggs first, and then the male fertilizes them. Females are thus able to desert, leaving the male "holding the babies," sometimes literally in mouth breeders. In those fish with internal fertilization, however, it usually falls to the female to exhibit parental care. Lack (1968) concluded that over 90 percent of bird species are monogamous, a trait that may well engender empathy in many bird watchers.

Polygyny can be influenced by the spatial or temporal distribution of breeding females. In cases where all females are sexually receptive at the same time, there is little opportunity for a male to garner all of the resource for himself. Monogamous relationships are more common in these situations, and Knowlton (1979) has suggested that female synchrony has evolved specifically to enforce monogamy on males. Where female reproductive receptivity is spread out over weeks or months, there is much more opportunity for males to mate with more than one female. For example, females of the common British toad, *Bufo bufo*, all lay their eggs within a week, and males generally have time only to mate with one female. In contrast, bullfrog females, *Rana catesbeiana*, have a breeding season of several weeks, and males may mate with as many as six females in a season. Also, where some critical resource, say available breeding or nesting sites, is in short supply and is patchily distributed, not uniformly spread out over the habitat, there is great opportunity for certain males to dominate it and to breed with more than one visiting female. Male orange-rumped honeyguides, *Indicator xanthonotus*, defend bees' nests, and when a female comes to feed, the male mates with her, exchanging food for sex. The more nests he can defend, the more females he will attract (Cronin and Sherman 1977).

In the lark bunting (*Calamospiza melanocorys*), which mates in North American grasslands, males arrive first, compete for territories, and then display with song flights to attract females. The major source of nestling death in this species is overheating from overexposure to the sun. Prime territories are those with abundant shade, and some males with shaded territories attract two females, even though the second female can expect no help from the male in the process of rearing young. Males in some exposed territories remain bachelors for the season. Pleszczynska (1978) was able to predict with good success the status of males (bigamous, monogamous, or bachelor) on the basis of territory quality before the arrival of females. Furthermore, supplementing open areas with plastic strips to provide shade turned them from bad to good territories. Predation is a strong selective pressure that acts in a similar manner to force polygynous relationships on birds where females choose males with safe territories (Rubenstein and Wrangham 1986).

From the male's point of view, territory-based polygyny is advantageous; from the female's point of view, there are drawbacks. Although by choosing dominant males a female may be gaining access to good resources, she may also have to share these resources with other females. In the yellow-bellied marmot (*Marmota flaviventris*), males attract more females if they defend the best burrow sites. From the female's point of view, more females in a burrow means lower success per female (Fig. 7.1). Although it is best for a female to be with a monogamous male, it is best for the male to mate with two-to-three females. A compromise is often evident in which about two females are usually observed per territory (Downhower and Armitage 1971). Orians (1969*b*) formally modelled this type of optimization of number of females per male in a way reminiscent of the ideal free distribution used in foraging-optimization theory (Fig. 7.2).

Sometimes males simply defend females as a harem without bothering to command a conventional resource-based territory. This pattern is

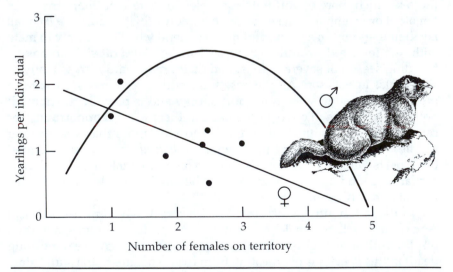

Figure 7.1 *Reproductive success in yellow-bellied marmots in relation to the number of females on a male's territory.* The success per female declines with number of females (solid dots). The success of the male, which is simply the success per female multiplied by the number of females, peaks at 2–3 females. (Modified from Downhower and Armitage 1971.)

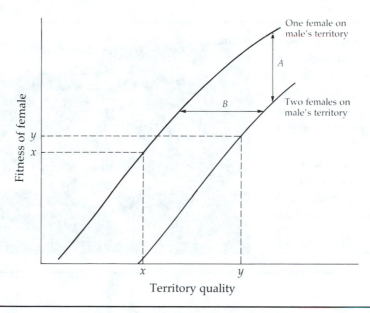

Figure 7.2 *The polygyny threshold model.* It is assumed that reproductive success of the female is correlated with environmental factors, such as the quality of the territory in which she breeds, and that females choose mates from the available males. In the model, a female suffers a decrease in fitness *A* by going to an already mated male from the fitness she could expect if she had a male all to herself. Despite this drop in fitness, provided the difference in quality between the territories is sufficient (*B* = the polygyny threshold), a female may expect greater reproductive success if she breeds with an already mated male. For example, a female who shared a male on territory *y* would do better than a female who had a male all to herself on territory *x*. (Redrawn from Orians 1969*b*.)

more common when females naturally occur in groups or herds, perhaps to avoid predation. Usually the largest and strongest males command all the matings, but being a harem master is usually so exhausting that males may only manage to remain at the top for a year or two. In the elephant seal, *Mirounga* sp., named for the enlarged proboscis of the male (Photo 7.2), males constantly lumber across the beach, squashing pups in the process. Because the offspring are likely to have been fathered by the previous year's dominant bulls, the havoc matters little to the present males (Cox and Le Boeuf 1977; McCann 1981). Sometimes, polygynous mating occurs where neither resources nor harems are defended. In some instances particularly in birds and mammals, males display in specific communal courting areas called **leks**. The females choose their prospective mates after the males have performed often-elaborate displays. A few males may perform the vast majority of the matings (see, for example, Wiley 1973). Perhaps the largest lek in the world is in Lake Malawi, Af-

Photo 7.2   Northern elephant seals, Mirounga angustirostris, *on Guadalupe Island, Baja California, Mexico. Note the enlarged proboscis of the male (raising his head), which gives the seals their name. Males control harems of females and will often fight each other over mates. (Photo by Dallas A. Sutton, Mammal Slide Library, American Society of Mammalogists.)*

rica, where as many as 50,000 male cichlid fish, one of which appears in Photo 7.3, may display on a sand bar 4 km long (McKaye 1983). Five main hypotheses have been proposed to explain leks (Bradbury and Gibson 1983). Males are thought to aggregate

- □ To decrease predation
- □ To increase the efficiency of attracting females
- □ To be close to ''hot spots'' through which the largest females pass

Photo 7.3   A male cichlid fish, Cyrtocara eucinostomus, *and its bower on a lek in Lake Malawi, southeastern Africa. (Photo by Kenneth R. McKaye.)*

    □ To make use of limited display sites
    □ To take advantage of female's preference for clumped males

Insufficient evidence is available to distinguish among these theories.

On the grounds that competition for mates is more intense between males of polygynous species than in monogamous ones and that large size increases an individual's fighting ability, it is not unrealistic to expect a relationship between sexual **dimorphism** and degree of polygyny, and such relationships have been observed in reptiles, amphibians, ungulates, carnivores, and primates (Clutton-Brock and Harvey 1984). Males are larger where they have the opportunity to monopolize more females. If this line of argument is taken one step further, it might be expected that the weaponry of males (horns, antlers, or teeth) might increase, relative to body size, in strongly polygynous species relative to monogamous ones. Such has been found to be the case in primates (Fig. 7.3). Further, competition occurs not only as fighting between males prior to copulation but also, in systems where females mate more than once, between the sperm of males inside the females. Volume of ejaculate is related to testis size, and it is also true in primates that males from species in which females copulate with more than one partner have larger testes relative to body size (Clutton-Brock, Guinness, and Albon 1982; Harvey and Harcourt 1982).

It is only in birds and mammals, however, that such large males are commonly found. In the great majority of lower vertebrates (frogs and snakes) and in invertebrates, it is the female that is the larger sex (Greenwood and Wheeler 1985). Male spiders are often much smaller than the female and have to use extreme caution when approaching a potential mate. In some deep-sea anglerfish, the male is merely a tiny fused appendage on the body of the female. In most cases, this situation is not surprising, given the number and volume of eggs females produce, together with the attendant food reserves. There are, of course, exceptions with large males, horned beetles and some crabs being prominent among them. The reasons for these differences in sexual dimorphism across the animal kingdom are not yet clear.

7.4 Polyandry

In most **polygamous** systems, those in which one individual mates with more than one individual of the opposite sex, the polygamous sex is the male, and such systems are termed *polygynous.* The opposite of polygyny, polyandry, in which the female is polygamous, is much rarer. Nevertheless, it is practiced by a few species of birds (Oring 1981). In Artic waders such as sanderling, *Calidris alba,* and Temminck's stint, *Calidris temminckii,* the males defend territories. The female lays one clutch of

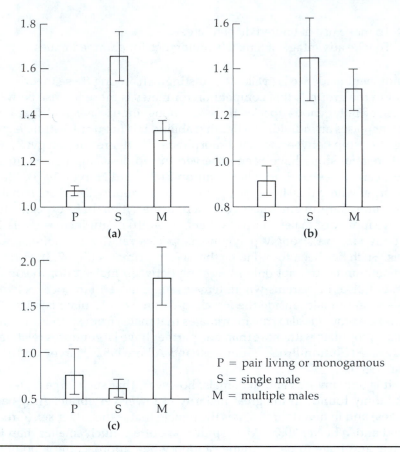

Figure 7.3 *Features of primate genera with different breeding systems.* **(a)** *Body size dimorphism (adult male divided by adult female weight).* **(b)** *Relative canine size (a measure of canine size dimorphism).* **(c)** *Relative testis size (a measure of testis size after body-size effects have been removed).* Bars indicate one standard error in each direction from the mean. (Based on data of Harvey, Kavanagh, and Clutton-Brock 1978 and Harcourt et al. 1981.)

eggs that the male incubates and another that she herself incubates. In the Artic tundra, the season is short but productive, providing a sudden but short-lived wealth of food. In the spotted sandpiper, *Actitis macularia,* the productivity of breeding grounds is so high that the female becomes rather like an egg factory, laying five clutches of 20 eggs in 40 days. Her reproductive success is limited by the number of males she can find to incubate the eggs, and females compete for males, defending territories where the males sit (Lank, Oring, and Maxson 1985). Many more examples of monogamous and polygynous patterns are discussed by Rubenstein and Wrangham (1986).

Section Four

Population Ecology

Do organisms, like this plant and its pollinator, commonly cooperate in nature for their mutual benefit, or do species more often compete for resources? Is predation so frequent in nature that some species, like this British beetle, Clytus arietis, *mimic the color patterns of dangerous wasps to avoid being eaten? These are some of the questions population ecologists strive to answer. Photos: P. Stiling.*

In this text, a population is loosely defined as a group of interbreeding organisms occupying the same general area at the same time. One can thus define, for example, a population of fish in a lake or the human population of New York City.

In this section, we are concerned with what limits species to their areas of distribution and what regulates their numbers. For example, some scientists argue that much of the Earth is covered with vegetation because the numbers of herbivores are kept low by natural enemies, predators and parasites. The population sizes of many herbivores in certain areas are fairly constant, but in others, for example in some forests and agricultural situations, outbreaks of pests such as gypsy moths or locusts occur. Naturally, there is great economic interest in determining the reasons for these population phenomena.

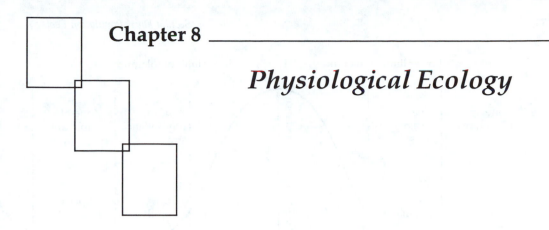

Chapter 8 _____

Physiological Ecology

T he local distribution patterns of most species are limited by certain physical or **abiotic factors** of the environment such as temperature, moisture, light, pH, soil quality, salinity, water current, and so on. Leibig's law of the minimum, coined by Justus Leibig in 1840, states that the distribution of species will be controlled by that environmental factor for which the organism has the narrowest range of adaptability or control. Often rather than just one limiting factor there are many factors, all interacting. Most environmental factors that affect an organism do so along a gradient, and the tolerance of the gradient varies from individual to individual and, or course, from species to species. Some species are tolerant of a wide range of environmental conditions (eurytopic), others of only a narrow range (stenotopic), but each functions best only over a limited part of the gradient, and this is termed a species' *optimal range* (Fig. 8.1). It must also be remembered that part of a preferred optimal range may already be occupied by a competitively superior species. In the field, species may not occupy their full ranges, as measured in the laboratory in terms of abiotic factors, because of competition with other organisms (Fig. 8.2, p. 155) (see also Chapter 11).

8.1 Abiotic Factors That Affect Species Distribution Patterns

Of all the abiotic factors limiting the distribution patterns of living organisms, temperature, especially tolerance to freezing, and moisture, specifically rainfall, are probably the most important. There are substantial temperature differentials over the Earth, a large proportion of which are due to variation in the incoming solar radiation. In higher latitudes, the sun's rays hit the Earth obliquely and are thus spread out over more of the Earth's surface than they are in the equatorial regions (Fig. 8.3, p. 156).

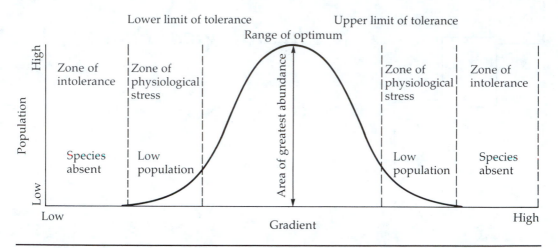

Figure 8.1 *Organismal distribution along a physical gradient.* (Modified from Cox, Healey, and Moore 1976.)

More heat is also dispersed in the higher latitudes because the sun's rays travel a greater distance through the atmosphere. The result is a much smaller (40 percent) total annual insolation in polar latitudes than in equatorial areas (Fig. 8.4, p. 156). In the summer, increased day length in high latitudes increases insolation, but shorter day length in winter decreases the daily total. The reason is that the Earth's axis of rotation is inclined at an angle of 23.5° (Fig. 8.5, p. 157); the Northern Hemisphere is treated to long summer days while the Southern Hemisphere has winter, and vice versa. At the summer solstice in the Northern Hemisphere (June 22), light falls perpendicularly on the tropic of Cancer; on December 22, it shines perpendicularly on the tropic of Capricorn. On March 22 and September 22 (the equinoxes), the sun's rays fall perpendicularly on the equator, and every place on Earth receives the same day length. These effects do not translate into a linear relationship between temperature at the surface and latitude—at the tropics both cloudiness and rain reduce mean temperature, and relatively cloud-free areas beyond this zone increase mean temperature relative to isolation (Fig. 8.6, p. 158).

Ocean currents also greatly affect the climates of land areas. Air-temperature differences between the poles create strong winds, and these are deflected by the Coriolis force to form the tradewinds (Figs. 8.7, p. 159, and 8.8, p. 160); these, together with the rotation of the Earth, create currents. The major currents act as ''pinwheels'' between continents, running clockwise in the ocean basins of the Northern Hemisphere and counterclockwise in those of the Southern Hemisphere. Thus, the Gulf Stream, equivalent in flow to 50 times all the world's major rivers combined, brings warm water from the Caribbean and the

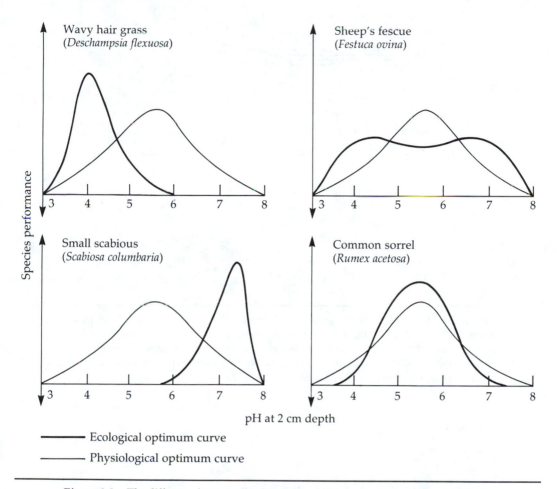

Figure 8.2 *The difference between the distributions of four plant species growing in the field (ecological optimum curve) and under noncompetitive conditions in controlled laboratory plots (physiological optimum curve).* (Redrawn from Collinson 1977.)

U.S. coasts to Europe, the climate of which is correspondingly moderated. The Humboldt current brings cool conditions almost to the equator along the western coast of South America (Fig. 8.9, p. 161). The climates of coastal regions may differ markedly from those of their climatic zones; many never experience frost, and fog is often evident.

The methods organisms use to cope with their physical environment are the realm of physiologists and physiological ecology. Townsend and Calow (1981) detail many examples of how plants and animals adapt to their environment and changes in it. For example, the growth form of a plant, often important in characterizing community types, is largely determined by its leaf structure. Leaves are of paramount importance in

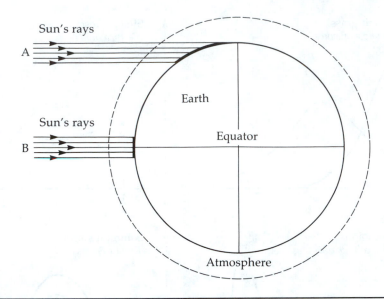

Figure 8.3 *Effect of the Earth's shape and atmosphere on incoming radiation.* In polar areas the sun's rays strike the Earth in an oblique manner (A) and deliver less energy than at tropical locations (B) for two reasons: (1) because the energy is spread over a larger surface in A and (2) because it passes through a thicker layer of absorbing, scattering, and reflecting atmosphere.

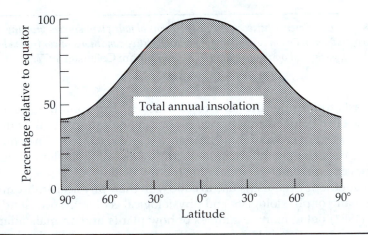

Figure 8.4 *Insolation at different latitudes during the year.* The amount of solar energy is expressed as a percentage of the amount at the equator.

Equinox

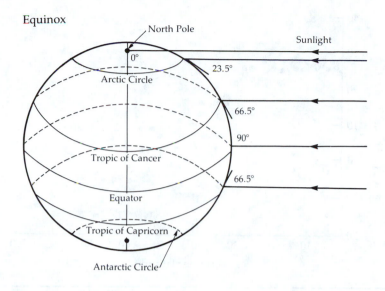

Summer solstice

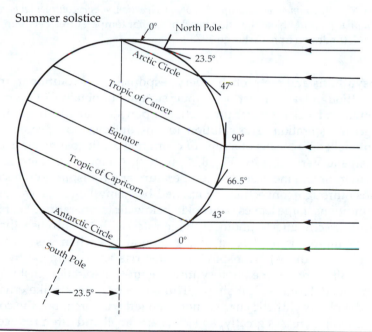

Figure 8.5 *Effects of Earth's inclined axis of rotation on amount of insolation.* The Earth's axis of rotation is inclined at an angle of 23.5°, which causes increasing seasonal variation in temperature and day length with increasing latitude.

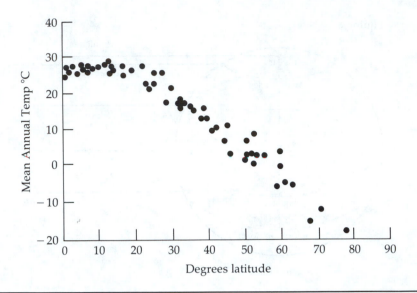

Figure 8.6 *Mean annual temperature (°C) of low-elevation, mesic, continental locations on latitudinal gradient.* Note the wide band of similar temperatures between 20°S and 20°N. (Redrawn from Terborgh 1973.)

photosynthesis and are probably very responsive to evolutionary pressure. Although leaves must open their stomata to obtain CO_2, a necessary building block for sugar production, open stomata mean water loss through transpiration. The solution to this optimization problem has been modelled by means of a type of economic profit-cost maximization (Givnish and Vermeij 1976) (Fig. 8.10, p. 162). An increase in leaf size in a hot environment increases leaf temperature and transpiration, because heat loss through convection is impeded by many large leaves. Under cool conditions, large leaves can cool to below air temperature. Leaf temperature directly affects photosynthetic rate (with the proviso that an upper asymptote is reached at high temperatures when gas exchange limits photosynthesis). Transpiration rates rise with temperature. The maximal difference in the photosynthetic gain-transpiration cost (profit-cost) model determines leaf shape. Thus, in a dry, desert-like environment, the photosynthetic curve is not changed with area, but the cost of transpiration increases greatly, so leaves are small and often reduced to spines.

Soil quality can also affect leaf size because photosynthetic rate is increased on good soils, and transpiration rate is unaffected (Fig. 8.10). The net result of these predictions is shown in Fig. 8.11: (p. 163) leaf shape converges to small and needle-like in both deserts and **taiga** forests; temperate forests have leaves of intermediate size; and warm, semi-shaded and wet tropical forests have the largest leaf sizes. Much variation exists

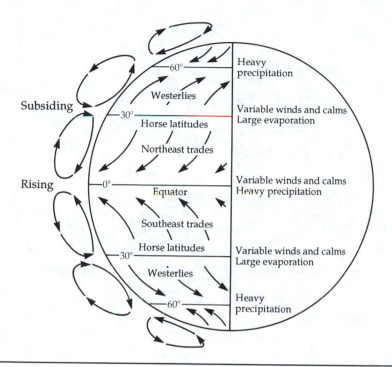

Figure 8.7 *Relationship between the vertical circulation of the atmosphere and horizontal wind patterns at the Earth's surface.* There are three convective cells in each hemisphere. As the winds move across the surface in response to rising and subsiding air masses, they are deflected by the Coriolis effect, producing the easterly tradewinds of the tropics and the westerlies of temperate latitudes. The Coriolis force is caused by the differential rates of movement of the Earth's surface at the equator and at points north and south of it. A point at the equator moves at 1,700 km h^{-1}, whereas at the same time a point north or south of the equator travels at a much lower speed. A southerly wind moving north from the equator continues to move at 1,700 km h^{-1} while the ground beneath it moves more slowly. Thus it appears to be deflected toward the right. This is the *Coriolis force.* It has the opposite effect in the Southern Hemisphere, where objects appear to be deflected to the left.

in leaf size and shape at a given locality, however, because many other pressures may be operating. For example, some leaves mimic feeding damage by insect herbivores in the hope that further herbivores will not eat this foliage. The rationale is that such herbivores avoid competition with the first supposed colonists or that they avoid predators and parasites that may be attracted to damaged leaves (Niemelä and Tuomi 1987).

King (1986) has analyzed the growth of tree form and height (using *Acer saccharum*, sugar maple) and its relation to susceptibility to wind damage. Trees should grow as rapidly as possible to escape shade condi-

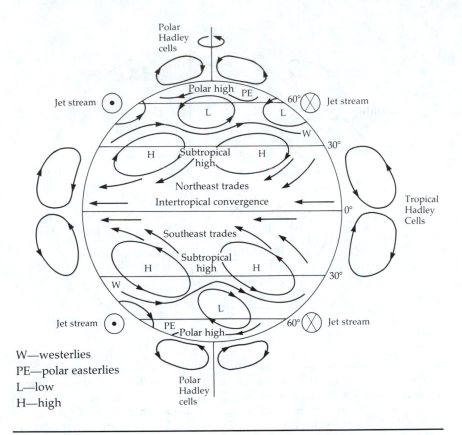

Figure 8.8   General air-circulation patterns, showing flows near the surface and cross sections of the principal upper troposphere circulation. Circled dot indicates jetstream coming toward viewer; circled X indicates jetstream going away from viewer.

tions. There is therefore a trade-off between trunk height and diameter—that needed to keep the tree upright. Support efficiency was analyzed in terms of the ratio of actual trunk diameter to the minimum required to keep the tree erect (the stability safety factor). A trade-off was expected between height-growth efficiency (maximized by a minimally designed trunk) and ability to resist storms. The lowest stability safety factors (1.8) were observed in saplings. The maxima were observed in mature canopy trees, which had trunks two-to-six times the minimum needed to keep trees erect in the absence of winds. However, large trees snapped more frequently in high winds than did the more supple saplings. Wind speed is higher in the canopy than closer to the ground, and large trunks are less flexible.

Wind can be a critically important ecological factor. Fifteen million trees in the south of England perished on October 16, 1987, in the wake

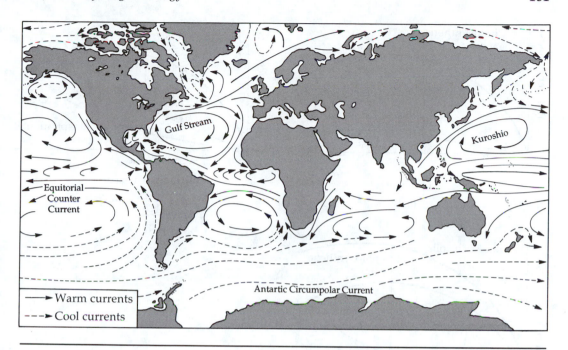

Figure 8.9 *Main patterns of circulation of the surface currents of the oceans.* In general the major circular gyres in each ocean move clockwise in the Northern Hemisphere and counterclockwise in the Southern Hemisphere. This pattern results in warm currents along the eastern coasts of continents and cold currents along the western coasts.

of a mighty storm (Kerr 1988). Records suggest that such winds had not hit the region for at least 300 years, so this may have been a rare event. Others suggest that such severe weather will become more common (Thompson 1988) because the world's climate is changing. Five of England's biggest freezes have come since 1978; the frequency of disastrous hurricanes in the South Pacific, especially over Fiji, has increased from 1 every 12 years to 1 every 7 years, and 6 storms were recorded between 1981 and 1985 alone. Woodley et al. (1981) present data to show that hurricanes can absolutely devastate coral reefs. Weather extremes can undoubtedly play a big part in the distributions of plants and animals.

Temperature resistance in plants, though poorly understood, is often critical to their distribution patterns. Under cold conditions, water must be moved outside cell walls or be bound up in such a chemical form that it cannot change to damaging ice, which would rupture the cellular machinery. Injury by frost is probably the single most important factor limiting plant distribution. As an example, the saguaro cactus can easily withstand frost for one night as long as it thaws in the day, but it will be

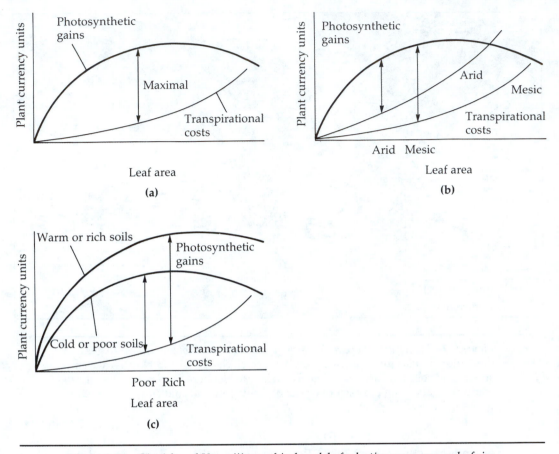

Figure 8.10 *Givnish and Vermeij's graphical model of selective pressures on leaf size.*
(a) *The cost and benefit curves indicating root costs associated with supplying water to balance transpiration and expected photosynthetic gain levels for each leaf size in a sunny location.* **(b)** *Cost curve altered by different habitats.* **(c)** *Benefit curve altered by temperature, humidity, wind, grazers, and nutrient levels.* Optimal leaf size for a given habitat is the point where the benefit curve most greatly exceeds the cost curve. (Redrawn from Givnish and Vermeij 1976.)

killed when temperatures remain below freezing for 36 hours. In Arizona the limit of the cactus' distribution corresponds to a line joining places where, on occasional days, it fails to thaw (Fig. 8.12, p. 164). For some plants, general coldness, not freezing, limits distribution. The northern boundary of the wild madder, *Rubia peregrina*, in Europe coincides with the January 4.5°C isotherm (Fig. 8.13, p. 165), and it has been suggested that this temperature is critical for the early growth phases of new shoots.

Frost injury has caused losses to agriculture of over $1 billion annu-

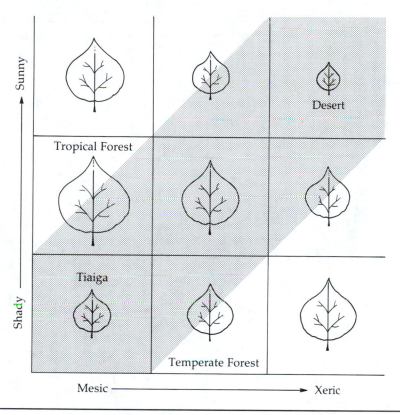

Figure 8.11 *Optimal leaf size predicted by the cost-benefit model of water use originated by Givnish and Vermeij.* The shaded area indicates the range of habitats likely to be encountered in nature, such as a vertical transect from rain-forest canopy (top right) to forest floor (lower left). (Modified from Givnish and Vermeij 1976.)

ally in the United States and has been considered an unavoidable result of subfreezing temperatures, but genetic engineering is changing the trend. Frost injury is precipitated by the ice-nucleation activity of just five species of bacteria that live on plant surfaces. Recently the DNA sequences conferring ice nucleation have been identified, isolated, and prevented from working in an engineered strain of one of them, *Pseudomonas syringae* (Lindow 1985). When such a strain is allowed to colonize plants, frost damage is greatly reduced, and plants can withstand approximately 5°C cooler temperatures before frost forms. The promise of this technique for the increase of agricultural yields and the alteration of normal plant distribution patterns is staggering (see Chapter 4).

Human actions that change temperature regimes can also have a drastic effect on ecosystems. In Denmark the brown weevil, *Hylobius*

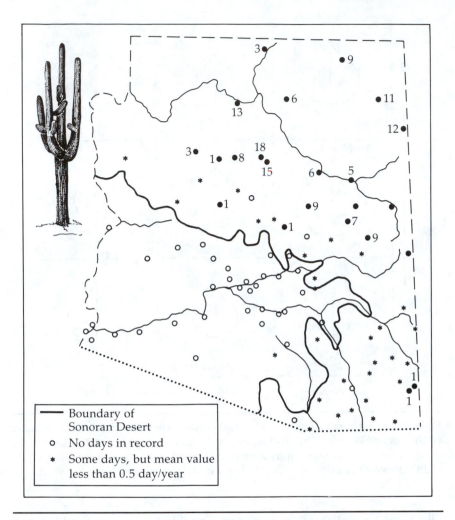

Figure 8.12 *Distribution of the saguaro cactus in Arizona.* There is a close correspondence between the northern and eastern edges of the cactus's range in the
Sonoran Desert and the line beyond which it occasionally fails to thaw during the
day. The numbers are mean numbers of days per year with no rise above freezing.
(Modified from Hastings and Turner 1965.)

abietes, normally takes three years to develop. When a forest is clearcut,
the sun warms the ground more than before, and the weevil matures in
two years. Thus **clearcutting** delivers more than one blow to forests by
increasing weevil damage to neighboring trees (Revelle and Revelle
1984).

In the real world, lethal temperatures, whether upper or lower, may
actually be very different from those recorded in the laboratory. For ex-

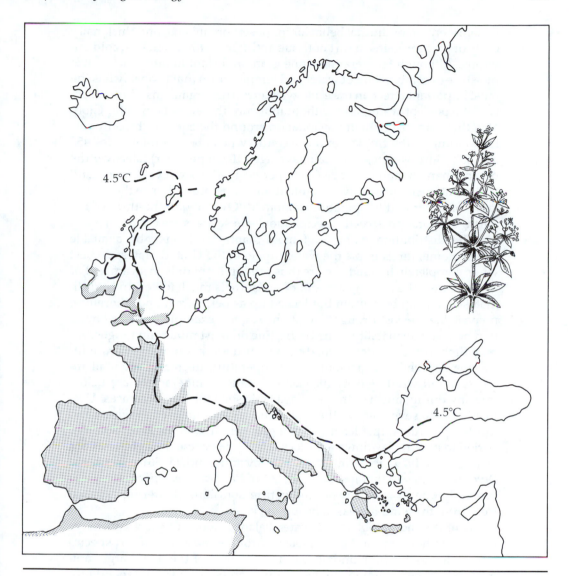

Figure 8.13 *The distribution of wild madder* (Rubia peregrina) *in Europe (shaded) and the location of the January isotherm for 4.5°C. (Modified from Cox, Healey, and Moore 1976.)*

ample, fish become torpid at high temperatures and are easy prey for predators. Populations may die from predation at certain temperatures long before they would die from actual temperature stress (Brett 1959).

The individual mechanisms by which animals circumvent heat stress under extreme physical conditions are really the concern of the physiologists, but most people are aware of a few: water evaporation (sweating), water retention or diurnal behavior for desert organisms, and thick hairy coats or glycoproteins, a sort of tissue anti-freeze, for animals in cold environs. Although temperature alone is an important limiting factor, it is moisture (a combination of water and temperature) that probably has the most important effect on the ecology of terrestrial organisms. Protoplasm is 85–90 percent water, and without moisture there can be no life. Globally, there is a belt of high precipitation around the equator, broadly corresponding to the tropics, and a secondary peak between latitudes 45° and 55°. Surprisingly, more rain (average 110 cm per year) falls over the oceans than on land (average 66 cm per year). Rates of evaporation and transpiration are primarily dependent on temperature, hence the importance of water and temperature combined. Over about one-third of the Earth, evaporation exceeds precipitation—these areas are the deserts.

The distribution patterns of most plants are limited by available water. Some, for example the water tupelo in the United States, do best when completely flooded and are thus predominant only in swamps. For many plants, the limiting amount of moisture is much lower. In cold climates water can be present but locked up as permafrost and, therefore, unavailable—a frost-drought situation. A good example of plant limitation by water availability is the **timberline** on most mountain ranges. Alpine timberlines are determined by winter desiccation or frost drought. As one proceeds up a mountain, temperature decreases, rainfall increases, and wind velocity increases. Because temperatures are below freezing during much of the year, available soil moisture decreases. Desiccation is so severe above the timberline, as a result of high winds and low temperatures, that leaves cannot grow enough in the short summer period to mature, lay down a thick cuticle, and become drought-resistant (Tranquillini 1979). Experiments by Hadley and Smith (1986) in southeast Wyoming showed that needle mortality of *Picea engelmannii* conifers was primarily due to winter wind and cuticle abrasion. Sheltering exposed trees from the wind increased needle survival.

Animals face problems of water balance, too, but most can move away from hot, dry, and intolerable environments. Many desert species are small and can hide underground in the heat of the day. Larger animals cannot be accommodated so easily, and because most depend ultimately on plants as food, their distributions are intrinsically linked to those of their food sources. The distributional boundary of the red kangaroo, *Macropus rufus*, in Australia coincides with the 400 mm rainfall contour, because the kangaroos are dependent on arid-zone grasses that are restricted to such low-rainfall areas (Fig. 8.14). In the wake of an extraordinary El Niño event (an irregular increase in water temperature in the eastern Pacific Ocean) in 1982–1983, the rainfall on Isla Genovesa, Galapagos, increased from its normal 100–150 mm during the rainy season to 2,400 mm from November 1982 through July 1983. Plants responded with

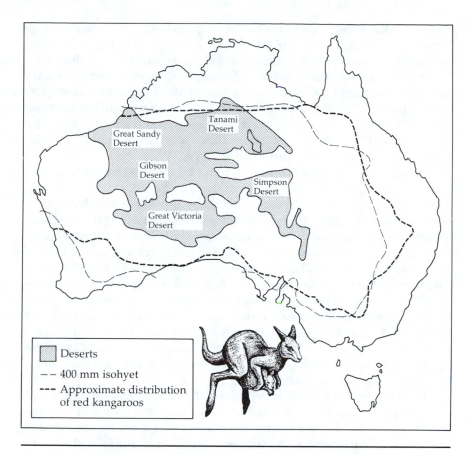

Figure 8.14 Distribution of the red kangaroo in the arid regions of Australia and the 400 mm (15 inch) rainfall line. Red kangaroos are relatively rare in the large desert areas shown. (Modified from Krebs 1985a.)

prodigious growth, and Darwin's finches bred up to eight times, rather than their normal maximum of three (Grant and Grant 1987).

Besides temperature and water, distribution patterns of species may be limited by other features, such as chemicals in the environment, salinity and oxygen content of water (a common factor for aquatic organisms), and even the frequency of fire.

Fire is a particularly interesting example. Before the arrival of Europeans in North America, fires started by lightning were a frequent and regular occurrence in some areas (Beaufait 1960), for example in the pine forests of what is now the southeastern United States. These fires, because they were so frequent, consumed leaf litter, dead twigs and branches, and undergrowth before they accumulated in great quantities. As a result, no single fire burned hot enough or long enough in one place to damage large trees—each one swept by quickly and at a relatively low temperature. The dominant plant species of these areas came to depend both directly and indirectly on frequent, low-intensity fires for their existence. The jack pine, *Pinus banksiana*, has serotinous cones, which remain sealed by resin until the heat of a fire melts them open and releases the seeds, and therefore depends directly on fire for its reproductive success. Much of the rest of the fire-adapted vegetation would be supplanted by other species if fires did not suppress those species periodically (Wade, Ewel, and Hotstetler 1980; Christensen 1981).

Current management practice, thought by many scientists to be unsound, attempts to maintain forests in their natural state by preventing forest fires completely, often with exactly the opposite result. First, trees like the jack pine simply stop reproducing in the absence of fire. Second, species like the longleaf pine and wiregrass that depend on fire to suppress their competitors are soon replaced by species characteristic of other communities. Finally, when a fire does occur, fuel has had a much longer period in which to accumulate on the forest floor and the result is an inferno—a fire that is so large and burns so hot that it consumes seeds, seedlings, and adult trees, native and competitor alike. Photo 8.1a shows a longleaf pine seedling; part b shows a small, rapidly moving "natural" fire of the sort that prevailed before current management practices were instituted; and part c shows a destructive fire of the sort that results when litter is allowed to accumulate for long periods.

The management practices arise from the mistaken assumption that most fires are caused by humans and that, before the arrival of Europeans, forests evolved in the virtual absence of fire. Lightning-caused fires are particularly frequent in the southeastern United States, but even in other areas many more fires are naturally caused than most people believe. For example, in the western half of the United States, nearly half of the yearly average of over 10,000 fires are thought to be started by lightning (Brown and Davis 1973).

Soil characteristics are also of prime importance to plant distributional patterns (Jenny 1980). Nitrogen availability is often crucial. In nitrogen-poor soils, such as bogs or poor sandy areas, only species that can supplement their nitrogen intake survive. Thus roots of alder (*Alnus glutinosa*) have nodules containing bacteria that fix atmospheric nitrogen, enabling the plant to grow in nitrogen-deficient soils. Similarly, bog myrtle, *Myrica gale*, has root nodules that are able to fix nitrogen, and bog myrtle occurs in Scotland in large areas of wet, acid, peat soils. In poor soils of the southeastern United States, some species (like the Venus flytrap and various pitcher plants) supplement their nitrogen intake by trapping insects, whose body fluids they can dissolve and then absorb.

Meteorological data not only determine the distribution patterns of individual tree species, but they also play an important part in species *richness*—the number of species in any given locale (see Chapter 17).

(a)

(b)

Photo 8.1 Adaptation to fire. (a) Longleaf pine seedling, show-
ing the dense cluster of long green needles that protects its grow-
ing tip from the low temperature, fast-moving fires that sup-
press its competitors. (b) A ''natural'' fire in a stand of longleaf
pine. (c) The results of unsound management practices—a de-
structive fire. Natural fires do not burn very hot because they
are frequent and not much litter accumulates before each burn.
When natural burning is stopped, litter accumulates repidly;
subsequent fires, of whatever origin, quickly get out of control,
not only killing the seedlings but leaping into the forest canopy
and devastating the mature trees. (Photos (a) by Dana C.
Bryan, Florida Park Service. Photo (b) from Florida Park Ser-
vice. Photo (c) from U.S. Department of Agriculture.)

(c)

Currie and Paquin (1987) showed that, of all the environmental data
available, evapotranspiration rates (strongly correlated with primary production and hence available energy) are the best predictors of tree species
richness in North America (Fig. 8.15).

By and large, it is the young stages of both plants and animals that are
most sensitive to environmental factors. Large trees are more likely to be
able to tolerate drought conditions because their roots can penetrate further into the soil and tap lower water levels. Adult animals are often more
mobile than juveniles and are better able to find shelter in times of stress.
In the long term, organisms evolve to survive in previously inhospitable
habitats. In less than 50 years, the grass *Agrostis tenuis* has evolved popu-

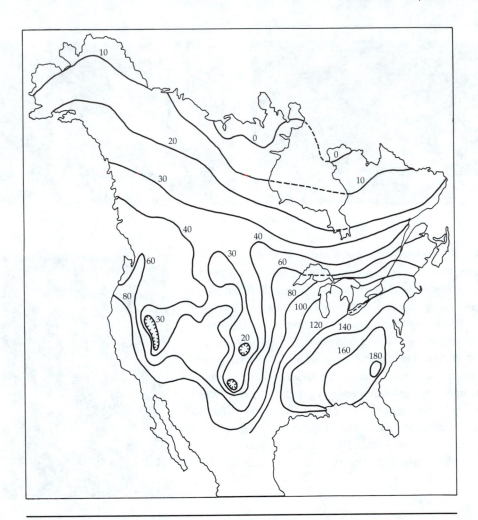

Figure 8.15 *Tree species richness in Canada and the United States.* Contours connect points with the same approximate number of species per quadrat. (Redrawn from Currie and Paquin 1987.)

lations that live on spoil tips in Great Britain, the areas of mine wastes often rich in noxious elements like lead, copper, or zinc (Antonovics, Bradshaw, and Turner 1971). In this case natural selection has favored the very few individuals tolerant to such areas, and these have prospered.

Of course, for every abiotic variable it must be asked whether the particular aspect that controls distribution patterns is the absolute maximum, the absolute minimum, a yearly average, or some combination of these. A good example of this type of approach is provided by Hocker's (1956) comparison of the distribution range of the loblolly pine (*Pinus taeda*) with meteorological data gathered from 207 weather stations in the southeastern United States, "home" of the loblolly. Hocker investigated (1) average monthly temperature, (2) average monthly range of temperature, (3) number of days per month of measurable rainfall, (4) number of days per month with rainfall over 13 mm, (5) average monthly precipitation, and (6) average length of frost-free periods. These data were collected from weather stations both inside and outside the natural range of loblolly pine. From the differences between these two groups, Hocker was able to map the climatic limits for the loblolly pine (Fig. 8.16) and to find good agreement between the observed limits of range and the theoretical limits set by meteorological data. The northern limit of this species is probably set by winter drought, when roots cannot take up sufficient water from cold soil to offset transpiration losses. However, loblolly pine

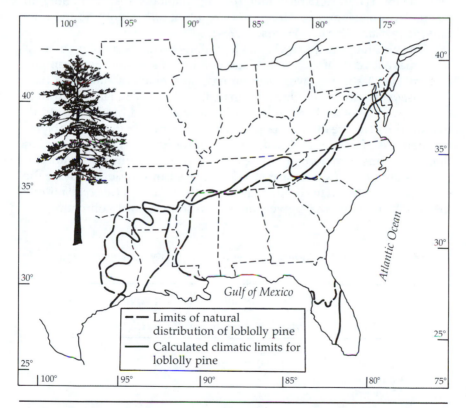

Figure 8.16 *Natural distribution limits and calculated climatic limits of loblolly pine* (Pinus taeda) *in the southeastern United States.* (Modified from Hocker 1956.)

has been artificially planted in Australia, New Zealand, Uruguay, Japan, South Africa, California, Tennessee, southern Illinois, southern Indiana, and central New Jersey (Parker 1950), areas well outside its natural range. Needles of trees in Ohio and southern Indiana were often killed by frost, and seedlings in Idaho were killed by low winter temperatures (Parker 1955), but pines near Stillwater, Oklahoma (370 km northwest of the natural range), have grown well for 25 years, producing many seedlings (Posey 1967). Taking the story one step further, Allen (1961) showed that seed from trees from different parts of the range responded differently when planted in one common environment in Virginia (Table 8.1).

All this information shows that local adaptation or **acclimation** can occur; genetic uniformity cannot be assumed to occur throughout the range of a species because much physiological adaptation will have taken place. Such variation in a plant can be **phenotypic** (environmental), in which case seeds transplanted from any locale will perform as if they were from resident individuals, or **genotypic,** in which case seedlings exhibit the form characteristic of their parents' habitat. Most commonly, in a wide array of examples, variation is a result of *both* genotypic and phenotypic variation (Heslop-Harrison 1964).

One general principle is clear; adaptations to one particular environment make it very difficult to live successfully in another. For example, Stearns (1980) studied two neighboring populations of mosquito fish, *Gambusia affinis,* in Texas. One lived in fresh water and had low reproductive output; the other lived in brackish water and had higher reproductive output. The freshwater population had osmoregulatory problems and had less ''free energy'' to devote to reproduction. This case implies that organisms commonly operate at close to maximum rates and efficiencies and that any increase in one process can only be made at the expense of another. This is often known as the *principle of allocation* (Rollo 1986), which states that, given limiting resources and competing de-

Table 8.1   *Response of loblolly pine seed from different parts of the range when planted in Virginia. (After Allen 1961.)*

	Results after Six Years in Virginia	
Seed Source	*Survival (in Percent)*	*Average Height (Meters)*
Virginia	90	2.4
Louisiana	88	2.3
Mississippi	81	2.3
Georgia	76	2.4
Florida	48	0.8

mands, natural selection leads to an allocation strategy that trades off various requirements so as to maximize fitness. The principle is implicit in many areas of ecology, for example in *r and K selection* (see Chapter 11). It has been challenged, however, by Rollo (1986) and others (see references in Rollo 1986). For example, the cockroaches *Periplaneta americana* and *P. brunnea* are sympatric and morphologically very similar, yet the former produces an ootheca only half the size of that of the latter. Rollo (1986) suggested that *P. americana* should compensate by producing the ootheca faster, yet laboratory mass budgets using controlled diets showed that the reverse was true. *Periplaneta brunnea* also grew faster than *P. americana*. These and other results suggest that the principle of resource allocation is not universal. One reason may be that some animals are operating below their potential and can increase many life processes (fecundity, growth rate, and so on) simultaneously when conditions are good. Perhaps the scheme for organismal design is based on avoiding failure (ensuring low-level persistence) rather than maximizing production. (In population terms, this is analogous to den Boer's spreading-the-risk theory; see Chapter 15.)

Finally, it is amazing how tolerant organisms can become in only a short evolutionary time. Six hundred years ago, all cotton plants were perennial shrubs confined to the frost-free tropics. Gradually, forms were selected to fruit early and produce a sizable crop in the first growing season. Early-fruiting varieties were suitable for cultivation in temperate regions, where they could produce cotton before winter. Cold winters and hot summers imposed an annual growth habit, and now all commercial cottons are obligate annuals that can be grown in cold-winter areas and semi-arid climates. All this—adaptation to previously lethal environments (Hutchinson 1965)—has been achieved in a maximum of 600 generations.

Because plants and animals adapt to certain environmental conditions, one can recognize characteristic assemblages of species for many of the major environmental settings on Earth. For example, tall evergreen trees and lush undergrowth grow rampantly in the tropics, whereas cacti are most prevalent in deserts. Each of the major types of floral and faunal assemblages is referred to as a **biome.** Thus, cooler, drier areas dominated by tall, deciduous trees form the temperate forest biome. A number of major biome types are recognized by ecologists, among which are the tropical rain forest, temperate **forest, desert, grassland, taiga, tundra,** tropical savannah, tropical scrub and seasonal forest, temperate rain forest, and chaparral. In aquatic situations there are coral reefs, the open ocean, estuaries, freshwater environments, and the intertidal biome. The meteorological conditions necessary for certain biomes and their physical appearances are discussed in Section Five, and their global distributions are outlined in Figs. 16.1 and 16.2.

8.2 *Dispersal*

Escape mechanisms such as migration and hibernation make it more difficult to apply limits of tolerance to situations in the real world. Houseflies can survive cold periods by coming inside homes for the winter. Explanations of these apparent inconsistencies may sometimes be provided by the behavior patterns or dispersive abilities of the organisms concerned.

It is well known that, on a large scale, geographic barriers can prevent species from colonizing potentially favorable areas. For millions of years the Atlantic Ocean provided an effective barrier over which the European starling, *Sturnus vulgaris,* a bird that evolved in the Old World, could not cross. In less than 100 years, since the deliberate introduction of over 100 birds to Central Park in New York City in 1890–1891, this species has colonized a huge area (Fig. 8.17). Similarly, the American chestnut, *Castanea dentata,* has been almost wiped out of its range in the eastern United States by the spread of chestnut blight, *Endothia parasitica,* probably introduced around New York City in 1900 from Asian nursery stock (Photo 8.2). The Mediterranean fruit fly (*Ceratitis capitata*), which is capable of devastating fruit crops in warm regions, especially Florida and California, and the ubiquitous tumbleweed (*Salsola iberica*) were both introduced from the Old World. Other examples abound (see Chapter 23). Furthermore, clearing of forest land and the creation of open scrubby habitats is thought to have been a major reason for the recent northward-range expansion of birds such as the cardinal (*Richmonderia cardinalis*) and the mockingbird (*Mimus polyglottos*). Opossums (*Didelphis virginiana*) have recently invaded southern Ontario, perhaps because of garbage dumps and trash cans, which provide feeding stations over the harsh winters.

What is less readily appreciated is that, even on a small scale, **dispersal,** or lack of it, can cause disjunct distribution patterns. (**Dispersal** is the process by which plants and animals arrive at the locations where they grow or live. Recall that the **distribution** of a species is the geographical area or areas where it lives, not the process by which it arrives there.) Off mainland Michigan, ruffed grouse, *Bonasa umbellus,* used to be found only on three islands in the Great Lakes, all within half a mile of the shore. Other islands further from the mainland remained uninhabited because birds were unwilling to fly more than 800 yards over water (Palmer 1962). Artificial stocking of more distant islands has since established successful populations (Moran and Palmer 1963). Heaney (1986) found that seawater channels no more than 15 km wide are major barriers for most mammals, and discounting one human introduction, he estimated that for nonvolant mammals such a barrier would be crossed only once every 250,000–500,000 years. For New World monkeys in South America, even major rivers are impenetrable barriers that individuals

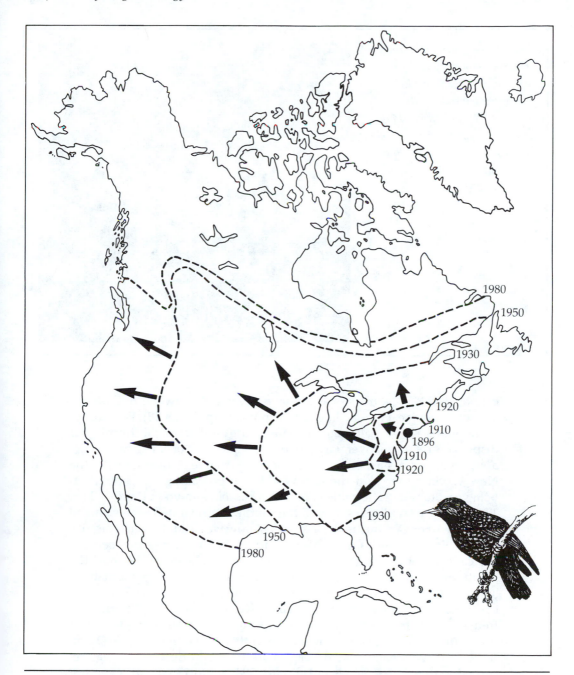

Figure 8.17 *Rapid expansion of the geographic range of the European starling* (Sturnus vulgaris) *following its successful introduction into North America.* (Modified from Brown and Gibson 1983.)

Photo 8.2 American chestnut. This species of tree was common in eastern United States deciduous forests but was essentially eliminated by an introduced fungal disease, chestnut blight. Only a few individuals remain. (Photo from National Archives, photo no. 95-G-250527.)

cannot cross. Thus, many species have developed between river drainages and provide a mosaic-like distributional map (Fig. 8.18). For strong-flying birds, large salt-water expanses are still a major hurdle for dispersal. Yet even over large expanses such as oceans, hurricane-assisted dispersal has, on occasion, permitted some colonizations of the New World by birds of the Old World, the prevailing wind direction being east to west. The cattle egret, *Bubulcus ibis,* shown in Photo 8.3 (p. 178), is now a familiar resident of the southeastern United States and all of South America. Only in the late 1800's, however, did this bird colonize eastern South America from Africa (Crosby 1972). During this century the cattle egret has colonized all the major Caribbean islands and the southern United States (Fig. 8.19, p. 179). From what was presumably a single colonization event, perhaps a small group of birds blown off course together, it has spread to occupy two continents. At the other extreme, in November 1987 a bald eagle, nicknamed ''Iolar,'' was found on the southern tip of Ireland. It had apparently crossed the Atlantic Ocean under its own steam (no bald eagles are native in Europe). It was returned to the United States in February 1988 (Pain 1988) and, even if it had been left to live out its life on the Irish coast, would presumably not have started a permanent colony, having no mate. We have no way of knowing, particularly for species less conspicuous than the bald eagle, how often such single individuals succeed in crossing major barriers to

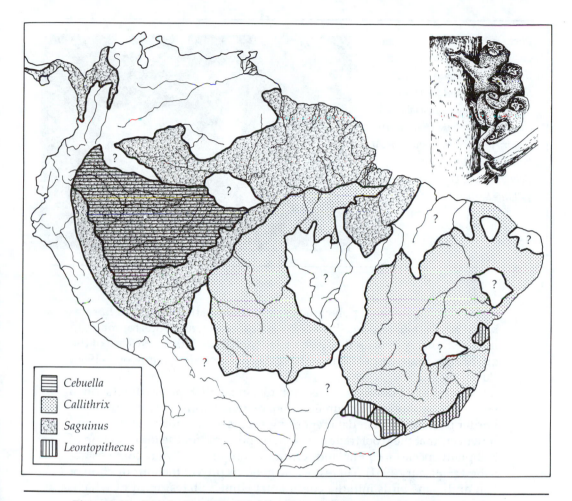

Figure 8.18 *Distribution of New World monkeys of the family Callithricidae.* Note that many of the taxa have adjacent but nonoverlapping geographic ranges that reflect the roles of major rivers in limiting their distributions. (Modified from Hershkovitz 1977.)

dispersal without lasting effect or what percentage of such colonists do establish permanent breeding populations.

In both birds and mammals, one sex usually disperses more than the other; thus close inbreeding is avoided. Mating of an individual with a close relative increases the chance that harmful recessive alleles will become homozygous in the offspring and cause low reproductive success. In birds, females usually disperse more than males. Males often help care for the brood, and male territories aid in persuading a female to mate. Often part of a territory may be inherited from the father. In mammals,

Photo 8.3   Cattle egrets, Bubulcus ibis, *an Old World species that colonized South America in the late 1800s and the United States in 1960, presumably by chance, after a long evolutionary history of inability to cross the Atlantic from Africa. (Photo by Bobby Brown, U.S. Fish and Wildlife Service.)*

males are usually more polygamous, and their mating system is harem-based rather than resource-based. Because it is advantageous for males to mate with many females (because males contribute little to the care of the offspring), male dispersal may have been favored (Greenwood 1980; Pusey 1987).

Commonly, however, it is not dispersal that places limits on a species' distribution. Documentation of failed introductions is much harder to come by than data regarding successful invaders. It is known, however, that the wool trade has been responsible for the introduction of 348 plant species into England. Of these, only four have become established (Salisbury 1961). There are also good records of the introduction of fish and game birds into the continental United States, most of which, between 70 and 90 percent, have failed. Exotic pest insects seem often to colonize the United States successfully, but again the ratio of successes to the number of introductions is probably low (Simberloff 1986b). Ninety-eight percent of the crop production of the United States, however, is based on plant species brought in from elsewhere (Schery 1972).

Plant seeds and small animals can often be transported by wind and would be found everywhere if dispersal were the only factor determining their patterns of distribution, as is quickly evident if one watches the recolonization of cleared ground. Helliwell (1974) showed that, 12 years after the completion of a major highway, the cuttings and embankments held nearly 400 species of plants, only 30 of which had originally been sown or planted. The largest natural experiments of this kind are provided by volcanic eruptions. On August 26, 1883, the small volcanic island of Krakatau in the East Indies was completely devastated and its biota destroyed by a massive volcanic explosion. Life on neighboring

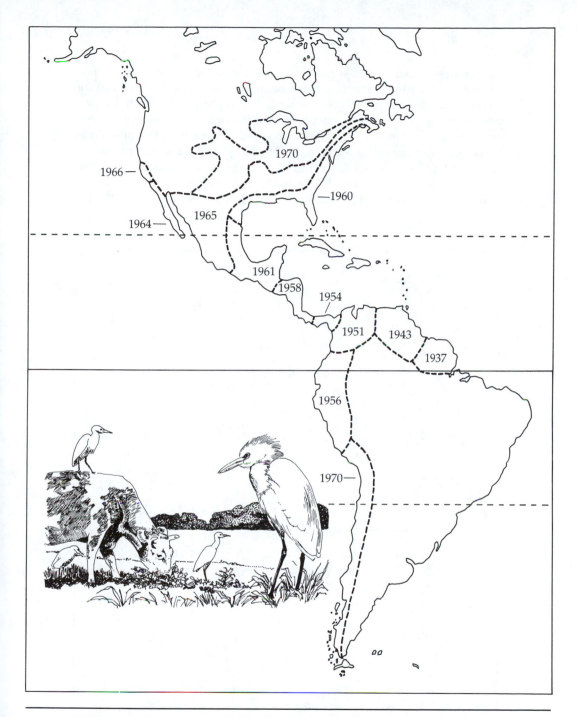

Figure 8.19 *Colonization of the New World by the cattle egret,* Bubulcus ibis.   The heron crossed the South Atlantic from Africa under its own power (possibly hurricane assisted), becoming established in northeastern South America by the late 1800's. From there it dispersed rapidly, and it is now one of the most widespread and abundant herons in the New World. (Modified from Smith 1974.)

Sertung and Rakata Kecil was obliterated, too. One part of Krakatau remained, and a new volcano, Anak Krakatau, formed in the center of the island group between 1927 and 1930 (Fig. 8.20). The nearest faunal source was an island 40 km away. After nine months only one species had been reestablished—a spider, probably ballooned in on silken threads. After three years the ground was covered with blue-green algae; eleven species of fern and 15 species of flowering plants were also found at this time. After ten years coconut trees began to grow, and after 25 years the island was thickly forested and contained at least 260 species of animals. Within

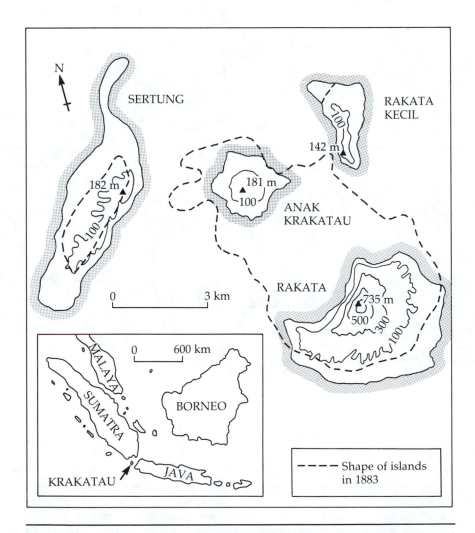

Figure 8.20 *The Krakatau Islands, Indonesia.* (Redrawn from Whittaker, Bush, and Richards 1989.)

50 years there were 47 species of vertebrates on Krakatau, of which 36 were birds and five were lizards; there were also three species of bats, a crocodile, a python, and one type of rat with two subspecies (Hesse, Allee, and Schmidt 1951). These figures have changed relatively little in the succeeding 50 years (Thornton 1987).

Dispersal for most organisms is a necessity. Many herbivores depending on fugitive weeds have to relocate to new host plants quickly as old weeds are replaced by other plants. Many species of insects therefore exhibit wing polymorphisms. That is, a flightless, short-winged (bra-chypterous) form exists in most habitats for most of the year, but a long-winged (macropterous) form is produced for dispersal (Harrison 1980). In the planthopper *Prokelisia marginata*, the production of long-winged forms is triggered by deterioration of the environment and by crowding of immatures (nymphs) and adults (Strong and Stiling 1983; Denno, Douglass, and Jacobs 1985 and 1986). Weeds themselves are great colo-nizers of temporarily vacant plots of land.

Of course, many species disperse seasonally or migrate (Pienkowski and Evans 1985). This phenomenon is extremely important in the man-agement of bird species that winter in tropical habitats undergoing exten-sive **deforestation.** Protection of summer feeding areas in temperate climes may have little effect if winter resources are the limiting factor. Fretwell (1972) showed that some populations are ultimately controlled by this one ''bottleneck'' season. Recognizing the problem, legislatures of different countries should band together to formulate treaties to pro-tect migratory bird species. As many as one-third of the birds that breed in the United States winter in South America, so habitat destruction in the tropics may have a severe impact on what is seen in North America. Similarly, if a population of salmon at sea is near K, the carrying capacity of the environment, then increasing hatching success in rivers might yield little benefit (Peterman 1984).

Very few species have abandoned dispersal altogether. Some of the few examples of an inability to disperse concern species on remote oce-anic island, such as the flightless cormorants of the Galapagos islands, where relaxed pressure from natural enemies has contributed to the re-duced need for flight. In steamer ducks, *Tachyeres*, and indeed many other aquatic species, flightlessness has been associated with extremely productive environments and a lessened need to migrate (Livezey and Humphrey 1986). Some species on ''ecological'' islands such as moun-tain tops have also become flightless. In the alpine zone of Mount Killimanjaro in Kenya there is a flightless crane fly (Carlquist 1974). Dis-tance between mountain tops is often large, and directional dispersal by small insects is difficult. Dispersal by strong winds is just as likely for wingless as for winged insects. In general, species occupying more per-manent habitats and big land masses have few adaptations for dispersal.

In other situations where an organism ought to exist but does not, dispersal is not the problem. In southern India, the mosquito *Anopheles culicifacies* is not found in rice fields after plants reach a height of 12 inches or more, even though their eggs, if transplanted to such fields, grow rapidly and successfully to adulthood. This species oviposits while flying and performing a hovering dance, never touching the water but remaining two-to-four inches above it. The mechanical obstruction provided by the rice plants interferes with this behavior and prevents oviposition (Russell and Rao 1942). Glass rods placed vertically in the fields have the same effect. Hence, some animals do not occupy their full potential range because of the requirements of a certain life stage, a phenomenon termed **habitat** selection. The wheatear, a common bird of open heaths in great Britain, nests in old rabbit burrows and is not found in newly forested heaths devoid of rabbits, despite an abundance of food there. It is excluded from otherwise suitable habitats by its nesting-site selection (Lack 1933). Other hole-nesting birds will nest in any type of forest (even where they would not normally occur) when provided with nest boxes (von Haartman 1956). Other criteria for habitat selection in birds include landscape and terrain; nest, song, feeding, and drinking sites; food; and other animals (Hilden 1965).

Chapter 9

Population Growth

W ithin their areas of distribution, plants and animals occur in varying densities. We recognize this pattern by saying an animal is
"rare" in one place and "common" in another. Schoener (1987) has
shown that, at least for Australian birds, a species we know as rare is in
fact more likely to be common in at least one part of its range and rare
everywhere else than to be rare throughout its range. Rarity, of course,
can be strongly influenced by human activities, especially habitat destruction, but it can also be produced naturally; some species are rare because they are **endemic** to limited areas, such as islands. For example,
more than 50 percent of the Canary Islands' plant species, which are
about 95 percent endemic, and about 66 percent of the 155 endemic
plants on Crete are considered **endangered** (Lucas and Synge 1978). Habitat destruction in these cases adds heavily to the perils of an already restricted range. Rarity can also be the result of the dynamics of food chains
(Chapter 19); top predators are normally less abundant than their prey,
and only about 1,400 panthers are thought to have occupied the whole
state of Florida before the arrival of Europeans (Cristoffer and Eisenberg
1985).

For more precision, and especially for management purposes, it is
desirable to quantify population density and more precisely to determine
what fractions of a population consist of juveniles and adults. Normally
density is calculated for a small area, and total abundance over an entire
habitat is estimated from these figures. Apart from pure visual counts of
organisms, sampling methods include the use of

- *Traps:* Live traps, snap traps, light traps (for night-flying insects), pit-
  fall traps (for crepuscular species), suction traps, pheromone traps
- *Fecal pellets:* For hare, mice, rabbits, and so on
- *Vocalization frequencies:* For birds or frogs
- *Pelt records:* Taken at trading stations for large mammals

□ *Catch per unit effort:* Especially useful in fisheries, where catch is often given per 100 trawling hours

□ *Percentage ground cover:* For plants

□ *Frequency of abundance along transects or in quadrants of known area:* For plants and sessile animals, which remain in place to be counted

□ *Feeding damage:* Useful for estimating the relative numbers of herbivorous insects

□ *Roadside spottings in a standard distance:* Often used in bird counts

Southwood (1977) provides an exhaustive review of these techniques and many more. From such data one can estimate not only the density of a population but also the relative frequency of juveniles and adults, larvae and nymphs, or other types of immatures. Great care must be taken in data collection; in particular, one must always consider whether the sampling regime is likely to bias results. For example, Mallet et al. (1987) showed how mark-recapture techniques strongly influenced butterfly behavior and hence population-size estimates. *Heliconius* butterflies avoid specific sites where they have been handled. Chase (personal communication) has documented how repeated visits to colonies of California gulls in Utah completely upset their behavior, and Spear (1988), in a fascinating article, showed that nesting gulls can recognize and distinguish different individual human investigators, mostly on the basis of facial features!

With data it is possible to construct life tables that show precisely how a population is age-structured. Life-table construction is termed *demography*. An accurate measure of age is essential here; the use of size as an indicator of age is tenuous at best and at worst leads to underestimates of juveniles and overharvesting. Basically there are two different types of life table: age-specific (**cohort** analysis) and time-specific (static).

9.1 Life Tables

Time-specific **life tables** provide a snapshot of a population's age structure at a given time. These tables were first developed by actuaries in the lifeinsurance business, who had a vested interest in knowing how long people could be expected to live. Time-specific life tables are useful in examining populations of long-lived animals, say herds of elephants, where following a cohort of individuals from birth to death would be impractical. An example of such a time-specific life table, prepared from a collection of skulls of known ages, for Dall mountain sheep (shown in Photo 9.1) living in Mount McKinley National Park, Alaska, is shown in Table 9.1.

The values given in the columns of life tables are symbolized by letters:

Photo 9.1   Dall Mountain sheep, Ovis dalli, in McKinley Park, Alaska. Collections of skulls, together with accurate age estimates of the skulls, permitted construction of an accurate life table for this species. (Photo by Adolph Murie, Mammal Slide Library, American Society of Mammalogists.)

x = age interval
n_x = number of survivors at beginning of age interval x
x' = age as percent deviation from mean length of life
d_x = number of organisms dying between the beginning of age interval x and the beginning of age interval $x + 1$

Table 9.1 Time-specific life table for Dall mountain sheep (Ovis d. Dalli) based on known age of death of 608 sheep dying before 1937 (both sexes combined). (Data from Murie 1944.)

x	n_x	x'	d_x	l_x	$q_x \times 1,000$	e_x
0–1	1000	− 100	199	1.000	199.0	7.0
1–2	801	− 85.8	12	0.801	15.0	7.7
2–3	789	− 71.6	13	0.789	16.5	6.8
3–4	776	− 57.5	12	0.776	15.5	5.9
4–5	764	− 43.3	30	0.764	39.3	5.0
5–6	734	− 29.1	46	0.734	62.6	4.2
6–7	688	− 14.9	48	0.688	69.8	3.4
7–8	640	− 0.8	69	0.640	107.8	2.6
8–9	571	+ 13.4	132	0.571	231.2	1.9
9–10	439	+ 27.6	187	0.439	426.0	1.3
10–11	252	+ 41.8	156	0.252	619.0	0.9
11–12	96	+ 56.0	90	0.096	937.0	0.6
12–13	6	+ 70.1	3	0.006	500.0	1.0
13–14	3	+ 84.3	3	0.003	1,000.0	0.5

Note: A small number of skulls without horns, but judged by their osteology to belong to sheep nine years old or older, have been apportioned pro rata among the older age classes. Mean length of life is 7.09 years.

l_x = proportion of organisms surviving to beginning of age interval x

q_x = rate of mortality between the beginning of age interval x and the beginning of age interval $x + 1$

e_x = mean expectation of life for organisms alive at beginning of age x (the goal of actuaries)

Age intervals can be set at any convenient length—a month, six months, a year, five years. An important point to remember about life tables is that each column is only a different way of expressing the same data. All (except x, of course) can be calculated from just one column. For example, d_x can be calculated from two adjacent items in the n_x column (n_x and $n_x + 1$):

$$n_{x + 1} = n_x - d_x$$

A numerical example using the first two n_x values from Table 9.1 is

$$n_1 = n_0 - d_0 = 1,000 - 199 = 801$$

Then, q_x can be calculated from the newly calculated d_x and n_x; thus,

$$q_x = \frac{d_x}{n_x}$$

A numerical example using the third row of Table 9.1 is

$$q_2 = \frac{d_2}{n_2} = \frac{13}{789} = 0.0165$$

and

$$q_2 \times 1,000 = 16.5$$

In turn, l_x is calculated from n_x and n_0 (the starting value of n_x):

$$l_x = \frac{n_x}{n_0}$$

A numerical example, from the fourth row of Table 9.1, is

$$l_3 = \frac{n_3}{n_0} = \frac{776}{1,000} = 0.776$$

In practice, the table should have an extra column to facilitate the calculation of e_x. For e_x one must first obtain the number of animals alive in each age interval, from x to $x + 1$. This number is known as L_x, where

$$L_x = \frac{n_x + n_{x + 1}}{2} = \frac{\text{number of individuals alive}}{\text{during age interval } x}$$

A numerical example, from the fourth and fifth rows of Table 9.1, is

$$L_4 = \frac{n_4 + n_5}{2} = \frac{776 + 764}{2} = 770$$

now

$$e_x = \frac{\sum L_x}{n_x}$$

that is, the sum of all the L_xs from age x to the last age, written as L_∞. So for the Dall mountain sheep,

$$e_{10} = \frac{174 + 51 + 4.5 + 1.5}{252} = 0.92$$

In Table 9.1, life expectancy is also shown by x', age as percentage deviation from mean length of life. Doing so permits comparisons between different species, as it would be meaningless to compare e_xs between animals that had different average age distributions. The n_x data are commonly plotted against time, age, or the x' data to give a survivorship curve as shown in Fig. 9.1. In survivorship curves, introduced to ecology by Raymond Pearl (see Pearl 1928), n_x values are plotted on a logarithmic scale because it is more valuable to examine rate of change than absolute numerical changes. Consider the following:

Population Size	No. Dying	Log No. Dying	Subsequent Population Size
1,000	500	3.00	500
500	250	2.70	250
250	125	2.40	125

Although the numbers lost are certainly different, the rate of change is identical. For the Dall mountain sheep, one might notice that the life expectancy actually increases during the first year of life. The reason is that wolf predation on juveniles is high; once young sheep have passed through this critical stage, their life expectancy is higher than that of new-

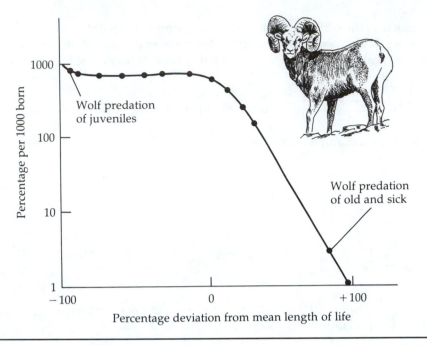

Figure 9.1   *Time-specific survivorship curve for the Dall mountain sheep* (Ovis dalli).
(Based on data from Murie 1944.)

borns. In some life tables an additional column, m_x, is included; this is
age-specific fertility. Such tables commonly only refer to females. The
product of each entry in the l_x and the corresponding entry in the m_x column is the total number of female offspring expected from that part of the
population. A **net reproductive rate,** R_0, is then often computed as

$$R_0 = l_x m_x$$

Clearly, values of R_0 greater than 1 indicate increasing populations, those
less than 1 decreasing populations, and those equal to 1 stationary populations.

   For organisms with short life spans, usually completed within a year,
or those with distinct breeding cycles, a snapshot, time-specific life table
may not give the correct picture. It will be severely biased toward the juvenile stage common at that moment. In these cases age-specific tables
are used, which follow one cohort or generation. Population censuses
must be conducted frequently, but only for a limited time (usually less
than a year). An age-specific life table for the spruce budworm is shown
in Table 9.2 and represented graphically in Fig. 9.2.

   Spruce budworms are larvae of tortricid moths. They excavate the

Table 9.2 *Age-specific life table for the 1952–1953 generation of the spruce budworm on plot G4 in the Green River watershed of New Brunswick. (After Morris and Miller 1954.)*

Age Interval x	Number Alive at Start of Age Interval n_x	Factor Responsible for Mortality	Number Dying during Age Interval d_x	Mortality Rate (in Percent) $100q_x$
Eggs	174	Parasites	3	2
		Predators	15	9
		Other	1	1
		Total	19	12
Instar 1	155	Dispersal	74.4	48
Hibernacula	80.6	Winter	13.7	17
Instar II	66.9	Dispersal	42.2	63
Instar III–VI	24.7	Parasites	8.92	36
		Disease	0.54	2
		Birds	3.39	14
		Other	10.57	43
		Total	23.42	95
Pupae	1.28	Parasites	0.10	8
		Predators	0.13	10
		Other	0.23	18
		Total	0.46	36
Moths	0.82		0.82	100
Total for generation (egg to adult)			173.18	99.53

Note: All numbers of animals expressed per 10 sq ft of branch surface.

terminal and lateral buds of conifers and are an economically important pest, especially in eastern Canada. Eggs are laid in August, and young caterpillars feed until the fall, when they pupate in hibernacula. In the spring the caterpillars resume feeding and pupate, and adult moths emerge in the early summer to lay new eggs, starting the cycle over again. Various sources of mortality operate at different stages in the life cycle, so, for example, eggs or larvae may be parasitized by small wasps, caterpillars may be eaten by birds or infected by fungi, and so on. In fact, one of the biggest decreases in density is caused by caterpillar dispersal— they balloon away on silken threads to try their luck on new trees. Because most do not land on a suitable tree, this is a big source of mortality.

As a result of both time-specific and age-specific demographic techniques, three general types of survivorship curves can be recognized (Fig. 9.3): Type III, in which a large fraction of the population is lost in the juvenile stages; Type II, in which there is an almost linear rate of loss; and Type I, in which most individuals are lost when they are older. Type I curves are often observed in higher organisms, especially vertebrates,

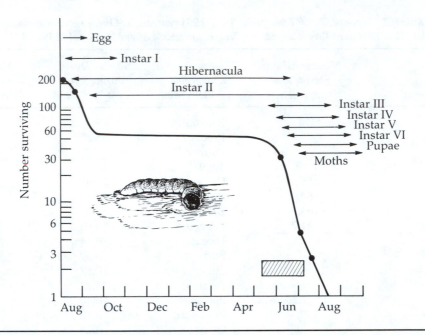

Figure 9.2 *Average survivorship curve for the spruce budworm during the 1945–1960 outbreak in New Brunswick.* This insect has one generation per year, and females can lay 200 eggs. Regular census points on the curve are indicated by dots, and the critical large larval stage in early summer is indicated by crosshatching. (Drawn after Morris 1963.)

that exhibit parental care and protect their young. Type III curves are often exhibited by lower organisms, especially invertebrates such as insects, many plants, especially weeds, and marine invertebrates. For example barnacles release millions of young into the sea but most drift off and are eaten by predators. Only a few survive and settle in the rocky intertidal (although, once there, they show excellent survivorship). Examples of a steady decline in survivorship, a Type II curve, are hydra and many birds.

Many other interesting divisions can be made on the basis of life-history characteristics. For example, some organisms, like salmon and plants like bamboo and yucca, are *semelparous.* They reproduce once only and die. This condition is in contrast to *iteroparous,* or repeated, reproduction. Even among iteroparous organisms strategies vary considerably. Weedy plants reproduce quickly and die, presumably before they are replaced by other colonists. Trees, however, often delay reproduction. The reasons for these sorts of differences in demographic processes are not always clear.

When a species can invade and successfully reproduce in a habitat,

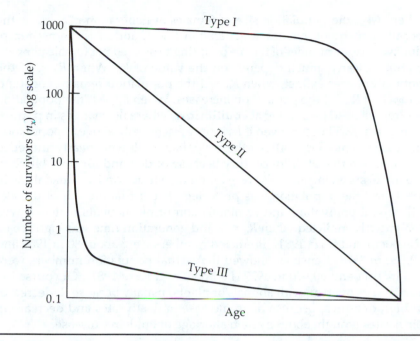

Figure 9.3 *Hypothetical survivorship curves.* Type I includes man, mammals, and higher animals, often with parental care of young. Type II includes some birds and some invertebrates such as *Hydra*. Type III includes some insects and many lower organisms with pelagic juvenile stages such as benthic invertebrates, molluscs, oysters, and fishes.

many factors contribute to its eventual population size. Such factors commonly include relationships with other organisms in the form of competition, predation, and herbivory. Before the effects of these selective pressures on population size can be examined, however, it must be known how populations grow naturally, in the absence of these pressures.

9.2 Deterministic Models

Discrete Generations

A population released into a favorable environment will begin to increase in numbers. Consider a **univoltine** insect with one feeding season and a lifespan of one year. If each female, on average, produces R_0 offspring that survive to the next year, then

$$N_{t+1} = R_0 N_t$$

where N_t is the population size of females at generation t, N_{t+1} is the population size of females at generation $t + 1$, and R_0 is the net reproductive rate, or number of female offspring produced per female per generation. Clearly, much depends on the value of R_0. When $R_0 < 1$, the population goes extinct, when $R_0 = 1$ the population remains constant, and when $R_0 > 1$ the population increases. When $R_0 = 1$, the population is often referred to as being at **equilibrium,** where no changes in population density will occur. Even if R_0 is only fractionally above 1, population increase is rapid (Fig. 9.4). Northern elephant seals were nearly hunted to extinction in the late 19th century because of demand for their blubber. About 20 surviving animals were found off Mexico on Isla Guadalupe in 1890, and the population was protected. The actual growth of the elephant-seal population and recolonization of old habitats matched well the growth predicted when $R_0 = 2$ and generation time was eight years (Le Boeuf and Kaza 1981). Predicted numbers were about 80 in 1906 and 40,960 in 1978. Censuses showed that actual population numbers were 125 in 1911 and 60,000 in 1977 (Le Boeuf and Kaza 1981). Of course the Earth is not overrun with animals or plants, usually because R_0 decreases at high densities, because of an increase in death rates and decrease in birth rates brought about by food shortage or epidemic disease.

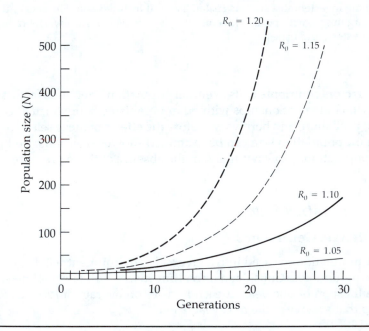

Figure 9.4 *Four examples of geometric population growth, discrete generations, constant reproductive rate.* $N_0 = 10$.

Overlapping Generations

For some species, reproduction occurs not seasonally but year round, and generations overlap. For such species the rate of increase is best described by

$$\text{rate of increase}\quad \frac{dN}{dt} = rN = (b - d)N$$

where N = population size, t = time, r = **per-capita rate of population growth,** b = instantaneous birth rate, and d = instantaneous death rate. r is related to the net reproductive rate R_0 by the equation

$$r \simeq \frac{(\ln R_0)}{T_c}$$

where T_c is the generation time (Southwood 1976). Again, the result is a curve of **geometric increase** similar to that in Fig. 9.4. As before, the multiplication rate is dependent on population size, density restrictions, and so on, so a more appropriate equation is

$$\frac{dN}{dt} = rN \left(\frac{K - N}{K} \right)$$

where K is the upper asymptote or maximal value of N, commonly referred to as the carrying capacity of the environment at the equilibrium level of the population. In essence this equation means:

rate of increase of population per unit time	=	rate of population growth per capita	×	population size times unused opportunity for population growth
	=	r	×	$N \left(\dfrac{K - N}{K} \right)$

When this type of growth is represented graphically, a **sigmoidal,** S-shaped, or so-called *logistic curve* results (Fig. 9.5). This equation was first described by Verhulst (1838) and was derived independently by Pearl and Reed (1920) to describe human population growth in the United States. Integrating the **logistic equation,** one can obtain N_t as

$$N_t = \frac{k}{(1 - e^{a - rt})}$$

where e = 2.71828 (base of natural logarithms) and a is an integration constant defining the position on the curve, relative to the origin. The

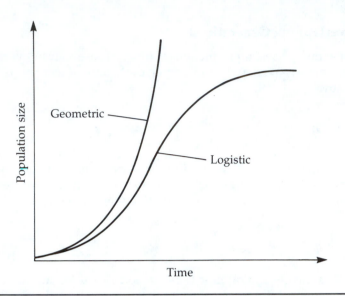

Figure 9.5   *Geometric growth in an unlimited environment and logistic growth in a limited environment.*

value of k can be estimated visually from the graph and a obtained from a plot of log $(k - N)/N$ against time. The Y intercept then gives a and the slope, r.

Logistic growth entails many important assumptions, of which the most important are

1. The population has a stable age distribution initially.
2. Density is measured in appropriate units.
3. The relation between density and rate of increase is linear.
4. The effect of density on rate of increase is instantaneous.
5. The environment is constant.
6. All individuals reproduce equally.
7. There is no immigration or emigration.

For many laboratory cultures of small organisms, such assumptions are easily met. Thus, early tests of these models using laboratory cultures of yeast or bacteria suggested they were valid (Gause 1934). For field populations of large animals, assumptions 3 and 4 are not so easily met, particularly as the effects of density on juveniles may take several months to be expressed through the resultant adult individuals.

How do these sorts of predictions hold up in the field? In 1911 reindeer were introduced onto two islands, St. Paul and St. George, in the

Bering Sea off Alaska. About 20 reindeer were introduced onto each island, both of which were completely undisturbed, having no predators and no hunting pressure (Fig. 9.6). The St. George population reached a low ceiling of 222 in 1922, then subsided to a herd of about 40. The St. Paul population grew enormously, to about 2,000 in 1938, but then crashed to eight animals in 1950. There appeared to be no ecological differences between the islands and no reason for the differences in population growth observed. Neither population had fit the pattern of logistic growth (Klein 1968). On the other hand, the rapid expansion of rabbits after their introduction into South Australia in 1859 seemed to confirm the logistic pattern. Sixteen years later, rabbits were reported on the west coast, having crossed an entire continent, over 1,100 miles, despite the efforts of Australians to stop them by means of huge, 1,000-mile-long fences (Fenner and Ratcliffe 1965). In general, however, in over 200 cases of mammalian introductions, most have failed to grow in an explosive or

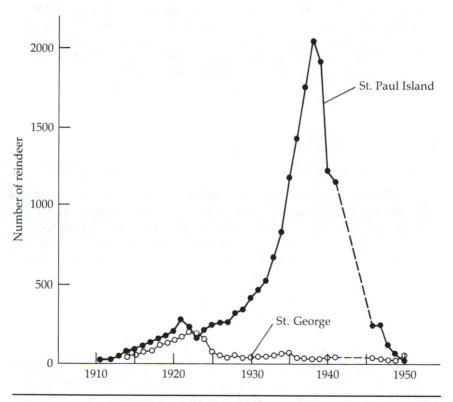

Figure 9.6 *Reindeer population growth on the Pribilof Islands, Bering Sea, from 1911, when they were introduced, until 1950. (Redrawn from Scheffer 1951.)*

geometric pattern of increase (De Vos, Manville, and Van Gilder 1956). No cases have yet been demonstrated in which the population of an organism with a complex life history came to a steady state at the upper asymptote or equilibrium level of the logistic curve (Krebs 1985*a*). Even Pearl and Reed's (1920) own prediction on human growth in the United States has not held true. On the basis of census data taken from 1790 to 1910, they projected an asymptote of 197 million to be reached around the year 2060. The census data for 1920 to 1940 fit the curve well, but since then the population has increased geometrically rather than logistically, resulting in projections of 260–350 million in the year 2025. To be fair to them, immigration rates have greatly increased and now contribute the same number of new people to the U.S. population as births. At least 500,000 immigrants enter the country each year legally, and probably one million illegally (over 1.5 million were caught along the U.S.-Mexican border in 1986 [Brewer 1988]).

It is obviously unrealistic to expect animals in real life to behave exactly like numbers in an equation. Part of the reason is the genetic variability between individuals that causes some females to produce more offspring, on average, than others, or some animals to be more resistant to climatic stress, predator pressure, and other factors. Given this variability, termed *stochasticity*, can we ever hope to predict population processes accurately or to incorporate such variability into a model?

9.3 *Stochastic Models*

Discrete Generations

Stochastic models of population growth are based largely on probability theory. Rather than exactly two offspring, one might assume that each female has a 0.5 probability of giving birth to two offspring, a 0.25 chance of producing three progeny, and a 0.25 chance of producing one.

For the discrete **deterministic model** discussed previously, if $R_0 = 2$ and $n = 5$, then

$$N_{t+1} = R_0 N_t$$
$$= 2 \times 5 = 10$$

For a stochastic model, a coin can be flipped to mimic the probability of the outcome, where tails/heads or heads/tails = two offspring, two tails = one offspring, and two heads = three offspring:

Parent	Outcome of Trial			
	1	*2*	*3*	*4*
1	2	3	3	2
2	3	1	1	1
3	3	1	2	2
4	1	1	3	3
5	3	1	1	1
Total population in next generation	12	7	10	9

Some of the outcomes are above the expected value of 10, and some are below. If this technique is continued, a frequency histogram can be constructed (Fig. 9.7).

Overlapping Generations

Stochastic models can also be developed for overlapping generations. Again, such a model is best explained by reference to the corresponding geometric equation, where

$$dN/dt = (b - d)N$$

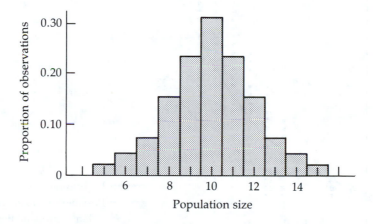

Figure 9.7 *Stochastic frequency distribution for size of a female population after one generation.* In this case, probability of having two female offspring = 0.5, probability of having three female offspring = 0.25, and probability of having one female offspring = 0.25.

and if $b = 0.5$, $d = 0$ (often true in new populations), and $N_0 = 5$, then

$$N_t = N_0 e^{rt} = 8.244$$

With a stochastic model, the probability of not reproducing in one time interval must be calculated as $e^{-b} = 0.6065$, and likewise the probability of reproducing once in one time interval is

$$1 - e^{-b} = 0.3935$$

Then for $N = 5$, the chance that no individuals will reproduce is $0.6065^5 = 0.082$. Similar calculations for other combinations eventually produce a frequency histogram, from which a population growth curve can be constructed (Fig. 9.8).

If death occurs in the population, there will be a chance that the population will become extinct. If birth rate is greater than death rate, then

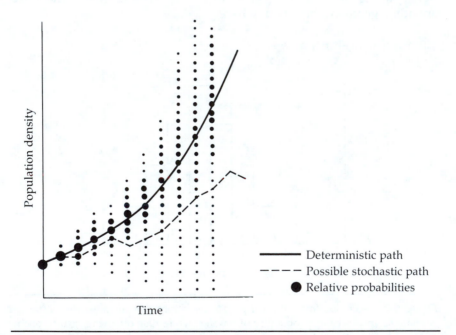

Figure 9.8   Stochastic model of geometric population growth for continuous, overlapping generations. (Redrawn from Krebs 1985a.)

$$\text{probability of extinction} = \left(\frac{d}{b}\right)^{N_0} \text{ as time} \to \infty$$

Thus, if $b = 0.75$, $d = 0.25$, and $N_0 = 5$, then

$$(d/b)^{N_0} = 0.0041$$

but if $b = 0.55$, $d = 0.45$, and $N_0 = 5$, then probability of extinction = 0.367. The larger the initial population size and the greater the value of $b - d$, the more **stable** the population. In reality $b - d$ is often zero, so $(d/b) = 1.0$ as time $\to \infty$. In other words, extinction is a certainty for a population given a long enough time span and is likely to occur more quickly for a small population. Fischer, Simon, and Vincent (1969) believe that probably 25 percent of the species of birds and mammals that have become extinct since 1600 may have died out naturally. Such stochastic effects are particularly important when the conservation of small populations of rare species is considered. For example, Schaffer and Samson (1985) have predicted that if N_e (effective population size) = 50 for grizzly bears, demographic stochasticity alone would cause extinction on average once every 114 years. A model of the spotted owl (*Strix occidentalis caurina*) suggests that demographic stochasticity is more likely than genetic factors to extinguish local subpopulations over the short term of decades (Simberloff 1986c).

Stochastic models introduce biological variation into population growth and are much more likely to represent what is happening in the field. The price paid is complicated mathematics as many new factors must be incorporated, such as the probability that a predator will kill a certain number of individuals or that there will be enough food available for herbivores. Stochastic models become more important as population sizes get smaller. If all populations were in the millions, one could throw away stochastic models—deterministic ones would do. These days, of course, populations of many mammals, except humans, tend toward the thousands rather than the millions.

Some closing generalizations can be made about population growth:

1. There is a strong correlation between size and generation time in organisms ranging from bacteria to whales and redwoods (Fig. 9.9).
2. Organisms with long generation times have lower per-capita rates of population growth (sometimes referred to as **innate capacity for increase**), r (Fig. 9.10).
3. Therefore, larger animals have lower rates of increase, r (Fig. 9.11). For any given size, warm-blooded animals, homeotherms, have a

higher rate of increase than heterotherms, which in turn are more fe-
cund than unicellulars.

Such generalizations are particularly important to conservation efforts because they underline how long it takes for populations of large
animals to rebound after ecological disasters or for large trees to reappear
after forest clearcutting. They may also point to reasons why larger species are more prone to extinction by people than are smaller species.

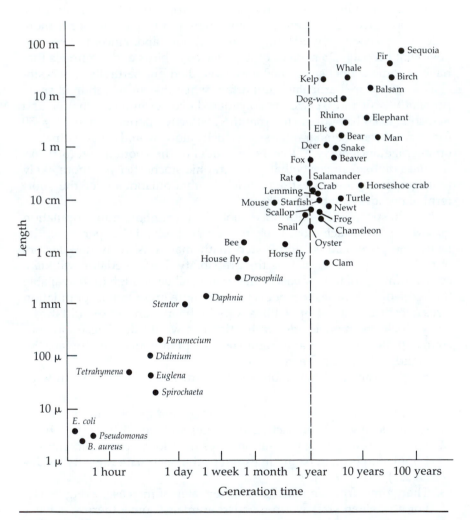

Figure 9.9   *The relationship of length and generation time, on a log-log scale, for a wide
variety of organisms.*   (Redrawn from Bonner 1965.)

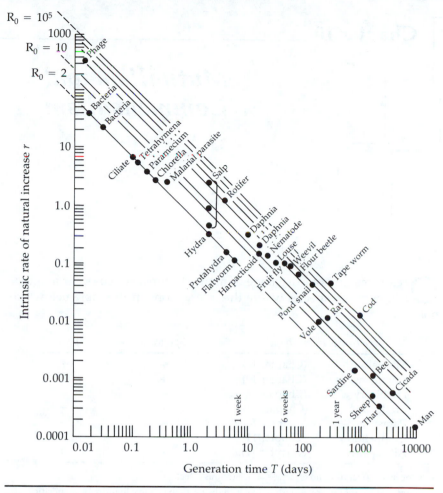

Figure 9.10 *The relationship of the intrinsic rate of natural increase,* **r**, *and generation time, with diagonal lines representing values of* R_0 *from 2 to* 10^5, *for a variety of organisms.* (Redrawn from Heron 1972.)

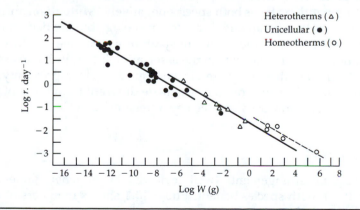

Figure 9.11 *The relationships of the intrinsic rate of natural increase to weight for various animals.* (Redrawn from Fenchel 1974.)

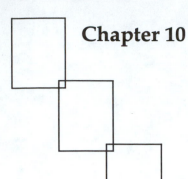

Chapter 10

Mutualism and Commensalism

Organisms do not exist alone in nature but instead co-occur in a matrix of many species where the interactions in the table below are possible.

Nature of Interaction		Species 1	Species 2
Mutualism*	(Chapter 10)	+	+
Commensalism*	(Chapter 10)	+	0
Herbivory	(Chapter 13)	+	−
Predation	(Chapter 12)	+	−
Parasitism*	(Chapter 14)	+	−
Competition	(Chapter 11)	−	−
Allelopathy	(Chapter 11)	−	0

+ = positive effect; 0 = no effect; − = deleterious effect.

*Examples of a symbiotic relationship, in which the participants live in intimate association with one another.

Herbivory, predation, and parasitism all have the same general effects, a positive effect on one population and a negative effect on the other. Competition affects both species negatively. **Mutualism** and **commensalism** are less commonly discussed in ecology but are tied together, with **parasitism,** under the banner of **symbiotic** relationships. In symbiotic relationships, the partners in the association live in intimate association with one another; they are always found in close proximity. The effects of mutualism and commensalism are different from those of parasitism, however, and are discussed separately here.

10.1 Mutualism

In mutualistic arrangements (well reviewed by Boucher, James, and Kebler 1982), both species benefit. Photo 10.1 shows members of some mutualistic relationships. In mutualistic pollination systems both plant

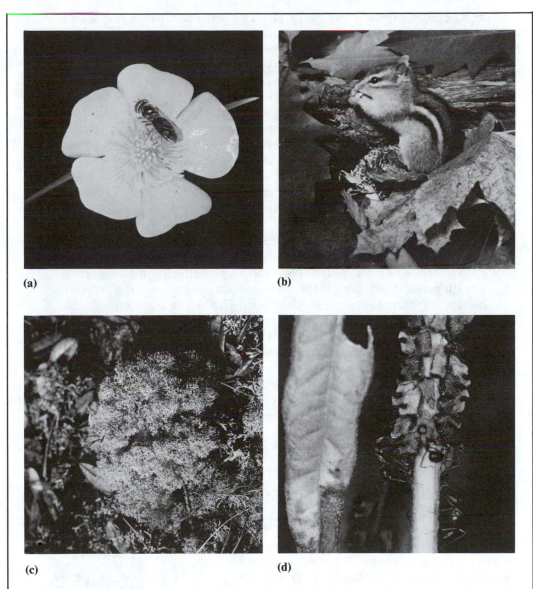

(a)

(b)

(c)

(d)

Photo 10.1   Some examples of mutualists. (a) Plant and pollinator. (b) Chipmunk, a species that eats seeds but also serves as a dispersal agent when its forgotten seed hoards germinate. (c) A lichen, Cladonia cristatella *(British soldiers), consisting of an outer layer of fungus, which provides mechanical support and a moist inner environment, and an inner layer of algae, which provides food to the fungus through photosynthesis. (d) Treehoppers,* Entylia carinata, *and the ants that tend and protect them in return for the honeydew they produce. (Photos (a), (c), and (d) by P. Stiling. Photo (b) from U.S. Department of Agriculture.)*

and pollinator (insect, bird, or bat) benefit, one usually by a nectar meal and the other by the transfer of pollen. From the plant's point of view, an animal pollinator can be viewed merely as a flying penis (Krebs and Davies 1981). In one extraordinary case, male euglossine bees visiting orchids in the tropics do not collect nectar or pollen but are rewarded instead with a variety of floral fragrances, which they modify to attract females (Williams 1983). The tightness of the relationship in some pollination systems is underlined by the phenomenon of *buzz pollination* (Erickson and Buchmann 1983). Certain flowers with anthers opening through pores at the top shed their pollen when subjected to vibrations emanating from the buzzing of bee wings. To ensure that some pollen falls on its target bee below, the pollen is negatively charged. As the pollen rains down, it is attracted electrostatically to the bees, which tend to have positive charges.

Both parties also benefit in mutualistic fashion from seed dispersal when fruits are eaten by frugivorous birds, bats, or mammals; the consumer receives a meal, and the plant receives an effective means of progeny **dispersal.** (According to the Krebs and Davies analogy, the disperser becomes a flying babysitter for the plant!) The lack of plant mobility has made many of them dependent on animals for pollination and seed dispersal. Temple (1977) has argued that the tree *Calvaria major* on the island of Mauritius has produced no seedlings for the past 300 years because its seeds do not germinate unless they have first passed through the digestive system of the now-extinct dodo, depicted in Photo 10.2. Corals are mutualists, too. The animal polyps contain unicellular algae. Some people even view the association of humans with domestic animals or crops as a mutualism.

Photo 10.2   The Mauritius dodo. This bird became extinct in 1681 as a result of the depredations of newly arrived Europeans and their domestic animals. Dodos were thought to be involved in a mutualistic arrangement with the tree Calvaria major, *because its seeds would only germinate after passing through the dodo's digestive system. (Photo Neg. No. 2A10281, Courtesy Department of Library Services, American Museum of Natural History. Photo by R.R. Logan.)*

Pollination studies are booming in modern ecology, partly because of the great diversity of tight, **coevolved,** and interesting systems to study and partly because money is available to study them. Over 90 crops in the United States alone are pollinated by insects. Comprehensive reviews of pollination biology are provided by Feinsinger (1983), Jones and Little (1983), and Real (1983). One of the finest studies in obligate mutualism involves figs and fig wasps as studied by Janzen (1979*a*). Over 900 species of *Ficus* exist, and virtually every one must be pollinated by its own species of agaonid wasp. The fig that we actually eat has an enclosed inflorescence containing many flowers. A female wasp enters through a small opening, pollinates the flowers, lays eggs in their ovaries, and then dies. The progeny develop in tiny galls and hatch inside the fig. Males hatch first, locate female wasps within their galls, and thrust their abdomens inside the galls to mate with them. The males then die, without ever having left the fig. Females collect pollen from the fig and then leave to search out new figs in which to lay their progeny. A similar mutualistic relationship occurs between yucca plants and yucca moths (Addicott 1986). Both mutualisms are highly coevolved. The distribution of each species of yucca or fig is controlled by the availability of its pollinator and vice versa. In the late 19th century, Smyrna figs were introduced into California, but they failed to produce fruit until the proper wasps were introduced to pollinate them. It is interesting that such highly coevolved systems arose, because on superficial examination the needs of the plants and those of the pollinators seem to conflict sharply. From the plant's perspective, an ideal pollinator would move quickly among individuals but retain a high fidelity to a plant species, thus ensuring that little pollen is wasted as it is inadvertently brushed onto the pistils of other plants. The plant should provide just enough nectar to attract a pollinator's visit. From the pollinator's perspective, it would probably be best to be a generalist and to obtain nectar and pollen from flowers in a small area, thus minimizing energy spent on flight between patches. This is obviously a problem in optimization theory (Chapter 6). One way in which the plants encourage the pollinator's species fidelity is by sequential flowering through the year of different plant species and by synchronous flowering within a species (Heinrich 1979). There are cases where both flower and pollinator try to cheat. In the bogs of Maine, the grass pink orchid (*Calopogon pulchellus*) produces no nectar, but it mimics the nectar-producing rose pogonia (*Pogonia ophioglossoides*) and is therefore still visited by bees. Bee orchids have even gone so far as to mimic female bees; males pick up and transfer pollen while trying to copulate with the flowers. So effective are the stimuli of flowers of the orchid genus *Ophrys* that male bees prefer to mate with them even in the presence of real female bees! Conversely, some *Bombus* species cheat by biting through the petals at the base of flowers and robbing the plants of their nectar without entering through the tunnel of the corolla and picking up pollen.

Finally, it is interesting to speculate about why ants, usually the most abundant insects in a given area, are so rarely involved in pollination. One reason might be that the subterranean nesting behavior of many ants exposes them to a wide range of pathogenic fungi and other dangerous microorganisms, to which they respond by producing large amounts of antibiotics. These antibiotics inhibit pollen function (Beattie et al. 1984; Beattie 1985). Peakall, Beattie, and James (1987) demonstrated that ants without metapleural glands, and therefore without these secretions, do successfully pollinate a certain orchid in Australia.

Mutualistic relations are highly prevalent in seed-dispersal systems of plants. In the tropics, some fruits are dispersed by birds that are strictly frugivorous. These fruits provide a balanced diet of proteins, fats, and vitamins (Proctor and Proctor 1978). In return for this juicy meal, birds unwittingly disperse the enclosed seeds, which pass unharmed through the digestive tract. Some plants, instead of producing highly nutritious fruits to attract an efficient disperser, simply produce abundant mediocre fruit in the hope that some of it will be eaten by generalists. Fruits taken by birds and mammals often have attractive colors—red, yellow, black, or blue; those dispersed by nocturnal bats are not brightly colored but instead give off a pungent odor to attract the bats. (In contrast, because birds do not have a keen sense of smell, fruits eaten by them are generally odorless.) A good introduction to the interesting literature on adaptations of plants and animals for seed dispersal is provided by Howe and Smallwood (1982) and Janzen (1983). In general, the relationships are not as obligately mutual as are plant-pollinator systems, because seed dispersal is performed by more generalist agents. Nevertheless, a wide array of adaptations exists; one has only to look at the impressive specialization of parrot beaks, strong and sharp to crack and peel fruits, to see that the mutualistic relationship between plant and seed disperser is strong in this case. Some bizarre strategies also exist; for example, in the floodplains of the Amazon, fruit- and seed-eating fish have evolved that disperse seeds (Goulding 1980). Microbes appear to be good at cheating the plant in this system, for they will readily attack the fruit without dispersing it. Janzen (1979c) has suggested that microbes deliberately cause fruit, and other resources like carcasses, to "rot," reserving it for themselves, by manufacturing ethanol and rendering the medium distasteful to vertebrate consumers. Animals eating such food are selected against because they become drunk and are easy victims for predators. As a countermeasure, some vertebrates have "learned" evolutionarily to tolerate these microorganisms and even to use them in their own guts to digest food. Alternatively, the enzyme alcohol dehydrogenase may exist to break down alcohol, though its action in many of us is insufficient to cope with social drinking.

Numerous other cases of mutualism can be gleaned from the literature, including the case of the ants on acacia thorns (Chapter 13) and the relationship of "cleaner" fish and "customer" fish on coral reefs (Ehrlich

1975), in which the cleaners nibble parasites and dead skin (which might otherwise cause disease) from their larger customers at specific cleaning stations. Such systems often leave their participants open to cheaters. Saber-toothed blennies of the genus *Aspidontus* bear a striking resemblance to the common cleaner wrasse, *Labroides dimidiatus.* Instead of performing a cleaning function, however, the blenny bites chunks out of the customers. Saber-tooths are protected from attack by their resemblance to the cleaners, though customers, in time, learn to avoid the cleaning stations that blennies frequent. In terrestrial systems one of the most commonly observed mutualisms is between ants and aphids. Aphids are fairly helpless creatures, easy prey to marauding ants. Yet in general ants tend to farm aphids like so many cattle. The aphids secrete honeydew, a sticky exudate rich in sugars, which the ants enjoy. In return, ants protect aphids from an array of predators, such as syrphid larvae, parasites like braconid wasps, and other competing insects, by vigorously attacking them (Fowler and MacGarvin 1985).

In some cases, the mutualistic relationship is so tight that neither participant could exist without the other and is called *obligatory mutualism.* Such is the case for many lichens, which are combinations of algae (which provide the photosynthate) and fungi. The ''lichenized'' fungi include within their bodies and near the surface a thin layer of algal cells, forming only 3–10 percent of the weight of the thallus body. Of the 70,000 or so species of fungi, 25 percent are lichenized. Lichenized forms occur in deserts, in alpine regions, and across a wide range of habitats. Non-lichenized fungi are usually restricted to being parasites of plants or animals or to being involved in decomposition. Many ruminants shelter symbiotic bacteria in their guts, which break down plant tissue to provide energy for their hosts; cellulose is otherwise indigestible for mammals (see Hungate 1975). Likewise, the roots of most higher plants (except the Cruciferae) are actually a mutualistic association of fungus and root tissue—the mycorrhizae. The fungi require soluble carbohydrates from their host as a carbon source, and they supply mineral resources, which they are able to extract efficiently from the soil, to the host (Harley and Smith 1983). The relationship between systemic fungal endophytes and vascular plants, usually thought to be a parasitic infection, has also been viewed as mutualism (Clay 1988). The fungi are thought to aid their hosts in defense against herbivory.

10.2 Commensalism

In commensal relationships one member derives benefit while the other is unaffected. Such is the case when sea anemones grow on hermit-crab shells. The crab is already well protected in its shell and gains nothing from the relationship, but the anemone gains continued access to new food sources. The same benefits accrue to members of a phoretic relationship (**phoresy**), in which the association involves the passive and more temporary transport of one organism by another, as in the transfer of flower-inhabiting mites from bloom to bloom in the nares of humming-birds (Colwell 1973). Some of the most numerous examples of commensalism are provided by plant mechanisms of seed dispersal. Many plants have essentially cheated their potential mutualistic seed-dispersal agents out of a meal by developing seeds with barbs or hooks to lodge in the animals' fur rather than their stomachs. In these cases, the plants receive free seed dispersal, and the animals receive nothing. This type of relationship is fairly common; most hikers have been plagued by "burrs" and "sticktights." Sometimes these barbed seeds can cause great discomfort to the animal in whose fur they become entangled. Fruits of the genus *Pisonia* (cabbage tree), which grows in the Pacific region, are so sticky that they cling to bird feathers. On some islands, birds and reptiles can become so entangled with *Pisonia* fruits that they die.

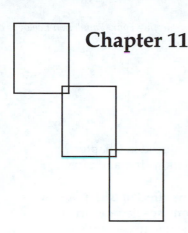

Chapter 11 _____

Competition and Coexistence

For many biologists, implicit in Darwin's theory of natural selection was a view of nature "red in tooth and claw" in which species scrambled to outcompete each other and leave the most offspring. How true a picture is this?

It is first worthwhile to be specific about what types of competitive event may occur in nature. **Competition** may be **intraspecific,** between individuals of the same species, or **interspecific,** between individuals of different species. Competition can also be characterized as scramble competition or contest competition. In scramble (**resource**) competition, organisms compete directly for the **limiting resource,** each obtaining as much as it can. Under severe stress, for example when fly maggots compete in a bottle of medium, few individuals can command enough of the resource to survive or reproduce. Such competition is most evident between invertebrates. In contest (interference) competition, individuals harm one another directly by physical force. Often this force is ritualized into threatening behavior associated with territories. In these cases strong individuals survive and take the best territories, and weaker ones perish or at best survive under suboptimal conditions. Such behavior is most common in vertebrates.

11.1 Mathematical Models

Mathematically, competitive interactions can be described by equations derived independently by Lotka (1925) in the United States and by Volterra (1926) in Italy, commonly called the Lotka-Volterra equations. For two species growing independently,

$$\frac{dN_1}{dt} = r_1 N_1 \left(\frac{K_1 - N_1}{K_1} \right)$$

and

$$\frac{dN_2}{dt} = r_2 N_2 \left(\frac{K_2 - N_2}{K_2} \right)$$

where r, N, and K represent the same variables introduced in Chapter 9, namely per-capita rate of population growth, population size, and carrying capacity.

Another term must be introduced to allow for the effect of each population on the other. In most cases, individuals of one species are larger than those of the other, and a conversion factor is needed to convert species 1 into units of species 2 such that

$$N_1 = \alpha N_2$$

where α is the conversion factor. Thus,

$$\frac{dN_1}{dt} = r_1 N_1 \left(\frac{K_1 - N_1 - \alpha N_2}{K_1} \right)$$

and

$$\frac{dN_2}{dt} = r_2 N_2 \left(\frac{K_2 - N_2 - \beta N_1}{K_2} \right)$$

where β is the conversion factor to convert N_2 into units of N_1.

Such relationships can be expressed graphically (Fig. 11.1). Population growth of N_1 continues to the carrying capacity of the environment K_1 in the absence of N_2. If there are K_1 / α individuals of N_2 present, no population growth of N_1 is possible. Between these two extremes are many combinations of N_2 and N_1 at which no further growth of N_1 is possible. These points fall on the diagonal $dN_1 / dt = r_1 = 0$, which is often called the *zero isocline*. Population growth of N_2 can be represented by a similar diagram. Combining the two figures and adding the arrows by vector addition illustrates what happens when the species co-occur (Fig. 11.2). Essentially there are four possible outcomes: species 1 goes extinct, species 2 goes extinct, either species 1 or species 2 goes extinct, depending on the initial densities, or the two species **coexist.**

11.2 *Laboratory Studies of Competition*

One must ask whether mathematical formulations represent real biological systems. One of the first and most important tests of these equations was performed in 1932 by a Russian microbiologist, who studied compe-

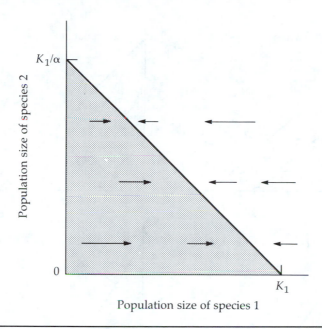

Figure 11.1 *Changes in population size of species 1 when competing with species 2.* Populations in the shaded area will increase in size and will come to equilibrium at some point on the diagonal line. Along the diagonal, $dN/dt = 0$.

tition between two species of yeast, *Saccharomyces* and *Schizosaccharomyces* (Gause 1932). Alone, both species grew according to the logistic curve; the asymptote reached was a function of ethyl alcohol concentration. Ethyl alcohol is a by-product of sugar breakdown under anaerobic conditions and can kill new yeast buds just after they separate from the mother cell. In cultures where the two yeasts grew together, population densities were lower than they were under single-species conditions. From these data, Gause was able to calculate that $\alpha = 3.15$ and $\beta = 0.44$; that is, 1 volume of *Saccharomyces* = 3.15 volumes of *Schizosaccharomyces*. Because alcohol is the limiting factor, Gause argued that he could determine α and β by measuring alcohol production of the two yeasts, which turned out to be 0.113 percent EtOH/cc yeast for *Saccharomyces* and 0.247 percent for *Schizosaccharomyces*. Thus $\alpha = 0.247/0.113 = 2.18$, and $\beta = 0.113/0.247 = 0.46$. The values of α and β obtained from the Lotka-Volterra equations were indeed in general agreement with those obtained independently by a physiological method.

In the late 1940's, Thomas Park and his students at the University of Chicago began a series of experiments examining competition between two flour beetles, *Tribolium confusum* and *Tribolium castaneum*. *Tribolium confusum* usually won, but in initial experiments the beetle cultures were

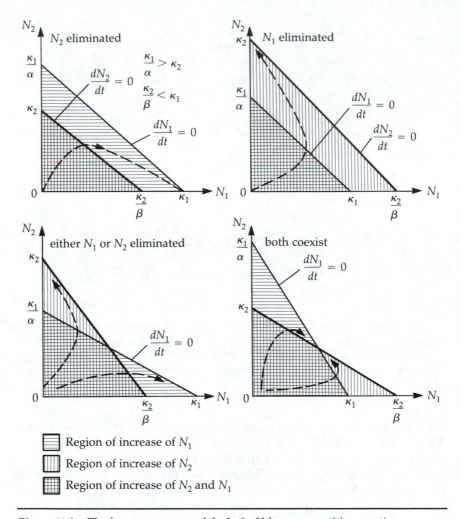

Figure 11.2 *The four consequences of the Lotka-Volterra competition equations.*

infested with a sporozoan parasite, *Adelina*, that killed some beetles, particularly individuals of *T. castaneum*. In these early experiments (Park 1948), *T. confusum* won in 66 out of 74 trials because it was more resistant to the parasite. Later, *Adelina* was removed, and *T. castaneum* won in 12 out of 18 trials. Most importantly, with or without the parasite, there was no absolute victor; some stochasticity was evident. Park then began to vary the abiotic environment and obtained the results shown in Table 11.1.

It was evident that competitive ability was greatly influenced by climate; each species was a better competitor in a different microclimate. However, single-species rearings in a given climate could not always be

Table 11.1 *Results of competition between the flour beetles* T. castaneum *and* T. confusum. *(After Park 1954.)*

Temper-ature °C	Relative Humidity (in Percent)	Climate	Single Species Numbers	Mixed Species (Wins in Percent)	
				T. confusum	T. castaneum
34	70	Hot-moist	confusum = castaneum	0	100
34	30	Hot-dry	confusum > castaneum	90	10
29	70	Temperate-moist	confusum < castaneum	14	86
29	30	Temperate-dry	confusum > castaneum	87	13
24	70	Cold-moist	confusum < castaneum	71	29
24	30	Cold-dry	confusum > castaneum	100	0

relied on to predict the outcome of mixed-species rearings (examine the entry for cold-moist climate). Later, it was found that the mechanism of competition was largely predation on eggs and pupae by larvae and adults. Park then varied the aggressive or cannibalistic tendencies of the beetles by selecting different strains; he obtained different results according to the strain of each beetle used (Table 11.2). Park had demonstrated

Table 11.2 *Results of competition experiments between the flour beetles* T. castaneum *and* T. confusum. *(After Park, Leslie, and Metz 1964.)*

T. castaneum Strain	T. confusum Strain	Number of castaneum Wins	Number of confusum Wins
CI	bI	10	0
	bII	10	0
	bIII	10	0
	bIV	10	0
CII	bI	1	8
	bII	0	10
	bIII	0	10
	bIV	4	6
CIII	bI	0	10
	bII	0	9
	bIII	0	10
	bIV	0	10
CIV	bI	9	1
	bII	9	0
	bIII	9	1
	bIV	8	2

Note: Cannibalistic tendency of strain goes from low to high as CI goes to CIV and bI goes to bIV.

a complete reversal of competitive outcome as a function of temperature, moisture, parasites, and genetic strains.

11.3 Competition in Nature

What of systems in nature where far more variability exists? One view holds that competition in nature is rare because by now, of all potential competitors, one has displaced the other. An alternative view holds that competition is a common enough force in nature to be a major factor influencing evolution. A third alternative is that predation and other factors hold populations below competitive levels. The question is important in applied situations, for example in biological-control campaigns, because it is vital to know whether releasing one natural enemy against a pest is likely to be more effective than the release of many, where competition between enemies might reduce their overall effectiveness. Circumstantial evidence suggests fewer natural enemies become established where many are released (Ehler and Hall 1982), yet sometimes this phenomenon does little to reduce overall effectiveness of control (Ehler 1979). Competitive effects between plants are also often thought to be of paramount importance in influencing crop yields, and many applied ecologists immediately assume all plants compete if resources are limiting (for example, Reynolds 1988). The fact is, relatively few watertight cases of active competitive displacement in nature have been documented in the literature, especially for groups of animals like insects (Lawton and Strong 1981; Strong, Lawton, and Southwood 1984). The most direct method of assessing the importance of competition is to remove individuals of one species and to measure the responses of the other species (Wise 1981). Often, however, such manipulations are difficult to make outside the laboratory. Three of the best examples of competition in nature involve barnacles, parasitic wasps, and chaparral shrubs.

Two barnacles, *Chthamalus stellatus* and *Balanus balanoides*, dominate the British coasts. Their distribution on the intertidal rock faces are often well defined (Fig. 11.3). Joseph Connell (1961) showed that *Chthamalus* could survive in the *Balanus* zone when *Balanus* was removed. In nature, *Balanus* grows faster on rocks of the middle intertidal zone, squeezing *Chthamalus* out. The limits of *Balanus* distribution are determined in the upper zone by desiccation and in the lower zone by predation and competition for space with algae. *Chthalamus* is more resistant to desiccation than *Balanus* and is normally found only high on the rock face. Thirty-four years later, in 1988, Connell repeated his competition experiments in the same area of the Scottish coast and, once more, observed strong evidence for competition (Connell, personal communication).

Three parasitic wasps of the genus *Aphytis* have been introduced into southern California to help control the red scale (*Aonidiella aurantii*), an insect pest of orange trees. *Aphytis chrysomphali* was introduced acciden-

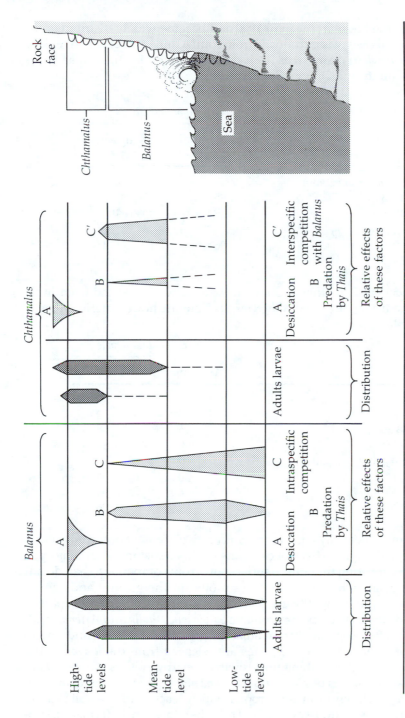

Figure 11.3 *Intertidal distribution of adults and newly settled larvae of Balanus balanoides and Chthamalus stellatus at Millport, Scotland, with a diagrammatic representation of the relative effects of the principal limiting factors. (Modified from Connell 1961.)*

tally from the Mediterranean in 1900 and became widely distributed. In 1948 *A. lignanensis* was introduced from south China and began to replace *A. chrysomphali* in many areas (Debach and Sundby 1963), such as Santa Barbara County and Orange County:

	Percent of Individuals	
	A. chrysomphali	A. lignanensis
Santa Barbara County		
1958	85	15
1959	0	100
Orange County		
1958	96	4
1959	7	93

In 1956–1957 another species, *A. melinus*, was imported from India and immediately displaced *A. lignanensis* from the hotter interior areas:

	Percent of Individuals	
	A. lignanensis	A. melinus
Coastal, Santa Barbara County		
1959	100	0
1960	95	5
1961	100	0
Interior, Santa Barbara County		
1959	50	50
1960	6	94
1961	4	96

The mechanism by which competitive displacement occurred there is not clear, although no two species pairs could coexist in the laboratory.

Plants are often thought to suffer more from competition than do animal populations because plants are rooted in the ground and cannot move to escape competitive effects. In southern California chaparral, grassland shrubs such as the aromatic *Salvia leucophylla* and *Artemisia californica* are often separated from adjacent grassland by bare sand one to two meters wide. Volatile terpenes are released from the leaves of the aromatic shrub; these inhibit the growth of nearby grasses (Muller 1966). Some plants, such as black walnut, *Juglans nigra*, produce similar chemicals (here, juglone) from their roots, which leach into the soil, killing neighboring roots (Massey 1925). This phenomenon is termed **allelopathy;** the action of penicillin among microorganisms is a classic case. In many cases such allelopathic chemicals are toxic to some competitors but not to others. Competitive interactions between plants are of paramount importance in *agroforestry*, a relatively new concept in which crops are grown under forest cover so that the land will yield both food and timber. Many species of tree, especially *Eucalyptus* in tropical regimes, are not suited to the practice of agroforestry because of their adverse effects on plants growing beneath them (Young 1988).

Other good but more anecdotal instances of competitive displacement in nature include the deliberate introductions of animals into areas for economic gain and the effects of habitat alteration. The introduction to Gatun Lake, Panama, of the cichlid fish *Cichla ocellaris* (a native of the Amazon) led to the elimination of six of the eight previously common fish species within five years (Zaret and Paine 1973; see also Payne 1987, Chapter 24). Also, the introduction of the game fish *Micropterus salmoides* (largemouth bass) and *Pomoxis nigromaculatus* (black crappie) led to the diminution of local fish and crab populations. Similarly, after construction of the Welland Canal linking the Atlantic Ocean with the Great Lakes, much of the native fish fauna was displaced by the alewife (*Alosa pseudoharengus*) through competition for food (Aron and Smith 1971). Introduced African dung beetles were successful in reducing the numbers of pest flies that competed for dung in Australia.

Of course active competition may not always lead to competitive displacement. In 1934 Gause argued that as a result of competition, two similar species scarcely ever occupy similar **niches,** but displace each other in such a manner that each takes possession of certain peculiar kinds of food and modes of life in which it has an advantage over its competitor. In more modern times this principle has been restated as the **competitive exclusion principle** by Hardin (1960), whose concise maxim is ''complete competitors cannot coexist.'' But can animals in close proximity compete partially yet still coexist? Two recent reviews suggest they can. Connell (1983) reviewed studies on active competition as reported in the literature. His review covered 215 species; competition was found in most of the studies, in over 50 percent of the species, and in 40 percent of 527 experiments involved. This result appears logical if one takes the following view. Imagine a resource set with, say, four species distributed along it. Then if only adjacent species competed, competitive effects would be expected in only three out of the six species pairs (50 percent). Of course the mathematics would be drastically different according to the number of species on the axis. For any given pair of adjacent species, however, competition would be expected, and indeed Connell found that, in studies of single pairs of species, competition was almost always reported (90 percent), whereas in studies involving more species the frequency was only 50 percent. In a parallel but independent review of 150 field experiments, Schoener (1983) reported competition in more than 90 percent of the studies and in 75 percent of the species studied. Both reviews may well, however, overrepresent the actual frequency of competition in nature because ''positive'' results demonstrating a phenomenon (here competition) may tend to be more readily accepted into the literature than ''negative'' results demonstrating patterns indistinguishable from randomness (Connell 1983). Some general patterns are evident nonetheless (Table 11.3). Folivorous insects (leaf feeders) and filter feeders (such as clams) showed less competition than plants, predators, scavengers, or grain feeders. Marine intertidal organisms tended to compete more than terrestrial ones and large organisms more than small ones. Some patterns like this might be expected; for example, given limited intertidal space, it would not seem odd to detect competition for space between sessile organisms. Seeds and grains also provide a limiting but very important nutrient-rich resource for desert granivores. Brown et al. (1986 and references therein) provide good evidence that all members of the grain-feeding guild from rodents and birds to ants compete for this resource.

Some taxa have themselves been analyzed in more detail. Strong, Lawton, and Southwood (1984) showed that only 17 out of 41 studies (41 percent) dealing with phytophagous insects showed competition, and even then sometimes it was demonstrated only for a very limited portion of the guild. Lawton (1984) has argued that there is much evidence of vacant **niche** space on the plants of the world. As evidence, he showed that bracken fern (*Pteridium aquilinum*) in Europe has a large array of chewing, sucking, mining, and galling insects in a wide range of habitats, but in the United States and especially in Papua New Guinea, whole guilds, for example gall formers, are missing (Fig. 11.4). With such vacant niches available, insects cannot be expected to compete so fiercely. The

Table 11.3 *Percentage of experimental studies showing interspecific competition. (After Connell 1983.)*

Subject of Study	Terrestrial Number of Experiments	Percent	Marine Number of Experiments	Percent	Freshwater Number of Experiments	Percent	Total Number of Experiments	Percent
Plants	205	30	31	68	2	50	238	35
Herbivores	45	20	13	69	0	—	58	31
Carnivores	36	11	5	60	3	67	44	20
Total	286	26	49	67	5	60	340	32
Invertebrates	57	16	37	32	0	—	94	22
Vertebrates	47	23	10	90	3	67	60	37

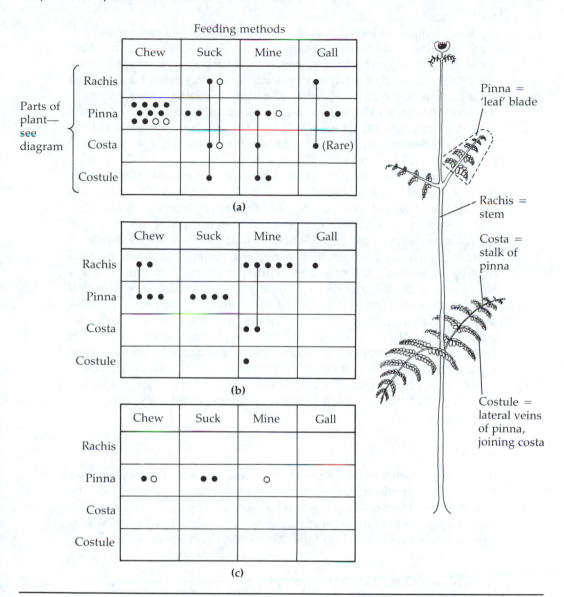

Figure 11.4 *Feeding sites and feeding methods of herbivorous insects attacking bracken* (Pteridium aquilinum) *on three continents.* **(a)** *Skipwith Common in northern England.* **(b)** *Hombron Bluff, a savannah woodland in Papua New Guinea.* **(c)** *Sierra Blanca in the Sacramento Mountains of New Mexico and Arizona.* Each bracken insect exploits the frond in a characteristic way. Chewers live externally and bite large pieces out of the plant; suckers puncture individual cells or the vascular system; miners live inside tissues; and gall-formers do likewise but induce galls. Feeding sites are indicated on the diagram of the bracken frond. Feeding sites of species exploiting more than one part of the frond are joined by lines. Closed circles represent open and woodland sites; open circles represent open sites only. (Drawn after Lawton 1984.)

fact that so many introduced insects, other animals, and plants have become established and thrive when introduced into new countries and novel habitats suggests that few niches in natural ecosystems are filled. The implication is that many empty ecological niches exist, but it is hard for a human observer to tell what constitute available niches independently of the species that occupy them. For example, could vampire bats later evolve in Africa to take advantage of the big game there? Could sea snakes, present now in the tropical parts of the Indian and Pacific oceans, evolve in the Atlantic region?

Lawton and McNeill (1979) also suggested that insects often "lie between the devil and the deep blue sea," that is, between a huge array of predators and parasites on the one hand and a deep blue sea of abundant but low-quality food on the other. As a result, they could scarcely become abundant enough to compete. Schoener himself had noted a dearth of competitive effects between **herbivores,** as compared to those between **carnivores,** plants, or detritivores. Thus carnivores and other groups with a lack of control from the next **trophic level** above could be reasoned to compete more. The foundations of this argument had been laid years earlier by Hairston, Smith, and Slobodkin (1960), who had essentially argued that the "Earth is green" and that the phytophagous insects that could eat this greenery must therefore be held in check by animals from the next trophic level up. Herbivores of course constitute a huge fraction of the Earth's biota—over 25 percent of the Earth's species are phytophagous insects alone—so such patterns must be taken seriously.

11.4 *Coexistence*

If species do compete in nature, the important question is perhaps not how much competition goes on but how similar can competing species be and still live together. This question has received more attention in ecology than any other single topic (Schoener 1974), but Lewin (1983) suggests it has led ecologists into futile works and blind alleys.

Theories Based on Morphology

In a seminal paper entitled "Homage to Santa Rosalia, or why are there so many kinds of animals?" G. Evelyn Hutchinson (1959) looked at size differences, particularly in feeding apparatus, between congeneric species when they were **sympatric** (occurring together) and **allopatric** (occurring alone) (Table 11.4). (The conceptual basis and explicit discussion of this approach had actually been laid out by Julian Huxley 17 years earlier in 1942 [Carothers 1986]).

Ratios between characters studied when species were sympatric

Table 11.4 *Size relationships, the ratio of the larger to the smaller dimension, between congeneric species when they are sympatric and allopatric. (From Hutchinson 1959.)*

Animals and Character	Species	Measurement (mm) When		Ratio When	
		Sympatric	Allopatric	Sympatric	Allopatric
Weasels	*Mustela nivalis*	39.3	42.9	1.28	1.07
(skull)	*M. erminea*	50.4	46.0		
Mice	*Apodemus sylvaticus*	24.8		1.09	
(skull)	*A. flavicollis*	27.0			
Nuthatches	*Sitta tephronota*	29.0	25.5	1.24	1.02
(culmen)	*S. neumayer*	23.5	26.0		
Darwin's finches	*Geospiza fortis*	12.0	10.5	1.43	1.13
(culmen)	*G. fuliginosa*	8.4	9.3		

ranged between 1.1 and 1.43, and Hutchinson tentatively argued that the mean value of 1.28 could be used as an indication of the amount of difference necessary to permit coexistence at the same trophic level but in different niches. This idea has come under heavy fire recently because, in a large series of animals, some ratios of 1.28 or greater would be expected to occur by chance alone (Simberloff and Boecklen 1981). Furthermore, size-ratio differences have too loosely been asserted to represent the ghost of competition past (Connell 1980). More importantly, little biological significance can be attached to ratios, particularly those of structures not used to gather food: ratios of 1.3 have been found to occur between members of sets of kitchen skillets, musical recorders, and children's bicycles (Horn and May 1977). Maiorana (1978) also found size ratios for a series of inanimate objects to have values near 1.3, and she argued that, in such cases, these values may simply reflect something about our perceptual abilities. However, there are some instances where the use of Hutchinson's ratios prove useful.

In the primeval forests of Canada, four indigenous parasitoids attacked the wood-boring siricid larva *Tremex columba*. Each had a different ovipositor length and laid an egg only when the ovipositor was fully extended. The ovipositor can be regarded as a food-provisioning apparatus for the larva; each species layed eggs in *Tremex* cocoons at a different depth in logs.

In the 1950's a fifth species, *Pleolophus basizonus*, was introduced into the area in an attempt to control another pest, the European sawfly. Its ovipositor length was intermediate between those of two of the existing species:

	Ratios
Mastrus aciculatus	1.05
Pleolophus basizonus	1.05
Pleolophus indistinctus	1.19
Endasys subclavatus	1.41
Gelis urbanus	

Even before the introduction of *P. basizonus*, the first three species were tightly packed but still maintained the minimum separation of about 1.1 noted by Hutchinson. However, when *P. basizonus* was introduced, strong competition ensued. *Pleolophus basizonus* or either of the two other species could have been displaced. In fact, *M. aciculatus* and *P. indistinctus* were forced out of the more favorable high-host-density sites (Price 1970) (Fig. 11.5).

Besides separation in size, species may also differ in their use of a particular set of resources, such as food or space. Consider three species normally distributed on a resource set (Fig. 11.6), where $K(x)$ is the re-

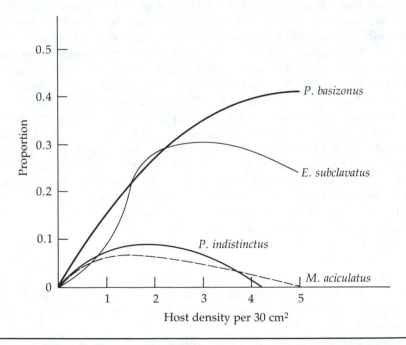

Figure 11.5 *The response of parasitoids to increasing host density illustrated by the change in the proportion of each species in the total parasitoid complex.* As the host density increases, competition between parasitoids becomes more severe. Note that as the introduced Pleolophus basizonus increases, the first species to be suppressed are those closest in ovipositor lengths. (Redrawn from Price 1970.)

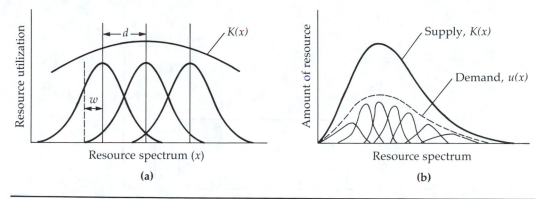

Figure 11.6 *Theoretical resource-utilization relationships.* **(a)** *The "simplest case" of three species with similar (and normal) resource-utilization curves.* d = distance apart of means, w = standard deviation of utilization, and d/w = resource separation ratio. (Modified from May and MacArthur 1972.) **(b)** *The more typical case with varying resource-utilization curves broadest in the region of fewer resources and less interspecific competition.* (Redrawn from Pianka 1976.)

source availability of x or carrying capacity, d is the distance between abundance maxima, and w represents one standard deviation, approximately 68% of the area on one side of the curve. It has been argued mathematically that, if $d/w < 1$, species cannot coexist, if d/w is < 3, there will be some interaction between species, and if $d/w > 3$, species coexist harmoniously (see Southwood 1978). The problem is that species are often not normally distributed.

Where the resource has a discontinuous distribution or occurs in distinct units, like leaves on a shrub, resource utilization can be illustrated graphically as in Fig. 11.7. The niche breadth of a species can then be quantified by Levins' (1968) formula:

$$\text{niche breadth} = \frac{1}{\sum\limits_{i=1}^{s} p_i^2 (S)}$$

where p_i = proportion of species found in the ith unit of a resource set of S units, such that $B_{max} = 1.0$ and $B_{min} = 1/S$. Proportional similarity between species is then given by

$$PS = \sum\limits_{i=1}^{n} p_{mi}$$

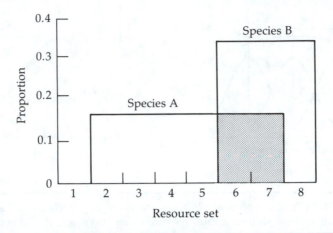

Figure 11.7 *Hypothetical distributions of a species, A, with a broad niche and a species, B, with a narrower niche, on a resource set subdivided into eight resource units. The species have the same proportional similarity (shaded zone), but species A overlaps B more than B overlaps A.*

where p_{mi} is the proportion of the less abundant species of the pair in the *i*th unit of a resource set with *n* units. Finally, the niche overlap of one species with another is represented by α_{ij} where

$$\alpha_{ij} = \sum_{h=1}^{n} p_{ih}\, p_{ij}\, (B_i)$$

where p_{ih} and p_{ij} are the proportion of each species in the *j*th unit of the resource set. Niche overlaps calculated in this way have been used as competition coefficients in classical Lotka-Volterra equations, although a safer method is to measure the effect of one species on another experimentally.

In accordance with Hutchinson's ideas, *PS* values of less than 0.70 have been taken to indicate possible coexistence, those greater than 0.70, competitive exclusion. However, species may differ not only along one resource axis, but along many, such as food, temperature, and moisture. For two resource axes, proportional similarity indices can be combined— proportional similarity values of 0.8 and 0.6 on two axes combine to give an overall *PS* of 0.48. Theoretically, coexistence would be permitted in cases where combined *PS* values $\leq 0.7 \times 0.7 \leq 0.49$ or less. Such analyses become more and more complex and less biologically meaningful as new axes are included.

Perhaps the most serious criticism of both *d/w* and *PS* treatment is that resource axes identified by the researcher as important may not accurately reflect limiting resources for organisms. Can sweep-net samples of insects on foliage reliably indicate food availability for birds? Furthermore, faced with the apparent contradiction that many ecologically similar species coexist with no apparent differences in biology, many researchers would argue that the correct niche dimensions had not yet been examined. This argument is similar to those proposed in Chapter 5 that all behavioral differences between species must be adaptive but that the reasons just have not been discovered yet.

It is worth noting that, in many situations, competing species have been found to differ hardly at all in morphology and yet still are found together. Two reasons have been proposed. First, in the presence of high levels of predation, competitively **dominant** species are likely to be selected by predators over less abundant prey. Thus good competitors will probably never be able to eliminate poor competitors totally if predation occurs; this is the idea of predator-mediated coexistence, and it will be addressed again in Chapter 17. Second, it is important to realize that many real populations in nature exist not in closed systems but in open areas where migration is possible and there is good connectance between populations. Caswell (1978) has theorized that, given good connectance between areas, immigration into an area of competitively inferior species, from areas where they do well, will be sufficient to maintain reasonable population sizes of both competitors for an indefinite time. Thus, if predators open up resources in an environment by killing members of the competitively dominant species, high connectance between populations means that competitively inferior species may first appear there by immigrating from other areas.

r and *K* Selection

One of the most useful concepts to come out of competition theory is the idea of the *r–K* **continuum** (MacArthur and Wilson 1967). Not all organisms are well suited to compete with others; some are better able to live in more hostile environments, often in a competitive vacuum. There is a continuum of reproductive strategies that encompass so-called *r*-selected species at one end and *K*-selected species at the other. *r*-selected species are fugitives with a high rate of per-capita population growth, *r*, but poor competitive ability. An example is a weed that quickly colonizes vacant habitats (such as barren land), passes through several generations, and then disappears, or more correctly is competitively excluded by individuals of more *K*-selected species. *K*-selected species compete well but tend to increase more slowly to the carrying capacity, *K*, of the environment. Some of the most important biological differences between these strategies are detailed in Table 11.5.

It is interesting to note that biological control, the control of pests by

Table 11.5   *Some of the correlates of r and K selection. (After Pianka 1970.)*

Parameter	r Selection	K Selection
Favorable climate	Variable and/or unpredictable; uncertain	Fairly constant and/or predictable; more certain
Mortality	Often catastrophic, nondirected, density-independent	More directed, density-dependent
Survivorship	Often type III	Usually types I or II
Population size	Variable in time, nonequilibrium; usually well below carrying capacity; communities saturated; no recolonization; portions thereof unsaturated; ecologic vacuums; recolonization each year	Fairly constant in time, equilibrium; at or near carrying capacity of the environment; communities saturated, no recolonization necessary
Intra- and interspecific competition	Variable, often lax	Usually keen
Selection favors	Rapid development; high reproductive rate; early reproduction; small body size; single reproduction	Slower development; great competitive ability; delayed reproduction; larger body size; repeated reproduction
Length of life	Short, usually less than one year	Longer, usually more than one year
Leads to	Productivity	Efficiency

natural enemies, usually insect parasitoids or predators, has proved more successful on pests that are closer to the *K* type than to the *r* type (Southwood 1977) (Table 11.6). Conway (1976) has proposed different control techniques for pests that lie toward either the *r* or the *K* end of the spectrum (see Table 11.7).

Table 11.6   *Cases of biological control of insects by imported natural enemies, grouped according to habitat characteristics on an r–K continuum.* (From Southwood 1977.)

Habitat Type	Cases Attempted		Cases Graded "Complete Control"	
	Number	Percent	Number	Percent
r end				
Cereals and forage crops	6	4	0	0
Vegetables	21	14	0	0
Sugarcane, cotton, pasture	19	12	2	8
Trees	107	70	23	92
K end				

Table 11.7 *Principal control techniques for different pest strategies. (After Conway 1976.)*

Technique	r Pests	Intermediate Pests	K Pests
Pesticides	Early wide-scale applications based on forecasting	Selective pesticides	Precisely targeted applications based on monitoring
Biological control	—	Introduction and/or enhancement of natural enemies	—
Cultural control	Timing, cultivation, sanitation, and rotation	—	Changes in agronomic practice, destruction of alternative hosts
Resistance	General, polygenic resistance	—	Specific monogenic resistance
Genetic control	—	—	Sterile mating technique

More recently, alternatives to the r and K continuum have been proposed. Gill (1974) suggested a three-way classification scheme with r, K, and α strategies, the last being characterized by high competitive ability. For plants, Grime (1977 and 1979) proposed the R, C, and S strategies, where R strategists (ruderals) are adapted to cope with habitat disturbance (especially man-made); C strategists (competitors) are adapted to live in supposed highly competitive environments like the tropics; and S strategists (tolerators) are adapted to cope with severe abiotic environmental parameters. Finally, Greenslade (1983) proposed the r, K, and A strategies, where A species are adapted to tolerate adverse environmental conditions. Useful though each of these schemes is, MacArthur and Wilson's original concept remains the rock on which each is based. Sometimes, however, the MacArthur-Wilson foundation seems a little shaky. Despite the apparently broad array of support for the r and K concept, from a wide variety of taxa, on closer examination the actual empirical evidence for this idea is difficult to assess. Different authors have used the terms r and K selection too loosely and in different senses (Parry 1981). It has thus become an ''omnibus'' term (Milne 1961), a term with such different intuitive definitions as to be ambiguous. For many people, r and K selection is what Hardin (1957) termed a *panchreston*—something that can explain almost anything.

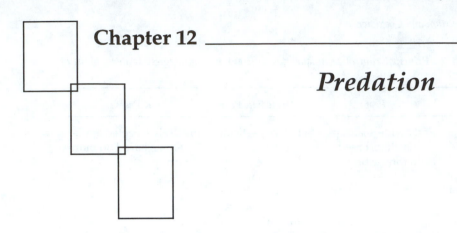

Chapter 12

Predation

A gain, if one envisages nature red in tooth and claw, predation looms large as a source of much of the bloodshed. Photo 12.1 illustrates how strong a selective force it can be. Several types of predation are recognized:

- *Herbivory*: Animals feeding on green plants
- *Carnivory*: Animals feeding on herbivores or other carnivores
- *Parasitism*: (a) Animals or plants feeding on other organisms without killing them and (b) parasitoids, usually insects, laying eggs on other insects, which are subsequently totally devoured by the developing parasitoid larvae
- *Cannibalism*: A special form of predation, predator and prey being of the same species

Although herbivory and parasitism could be viewed widely in the context of predation, each has special characteristics that tend to separate it out into a special subdivision of predation, and each subject will be discussed separately. Herbivory, for example, is often nonlethal predation on plants. Many leaves or parts of leaves of a plant can be eaten without serious damage to the ''prey''; in the animal world, on the other hand, predation generally means death for the prey. Parasitism, too, is a unique type of event, in which one individual prey is commonly utilized for the development of one predator. In the special case of insect herbivores on plants, there is great debate about whether such insects can be called plant *parasites*, because they seem to fit such a definition fairly well (Price 1980). Carnivory embodies the traditionally held view of predation, and cannibalism can be seen as a special case of it.

Photo 12.1   Testaments to the strength of predation as a selective force. (a) Katydid, Microcentrum retinerve, *mimicking foliage (exhibiting camouflage). (b) Three under-wing moths,* Catocala sp., *matching tree bark. (c) Underwing moth flashing brightly colored underwing to startle a predator that has spotted it. (d) Saddleback caterpillars,* Sibine stimulea, *exhibiting poisonous spines and the striking patterns by which simi-larly defended organisms often advertise their unpalatability. (e) Sumac-feeding beetles,* Blepharida rhois, *concealing themselves in their own excrement. (f) The same beetles with excrement removed. (Photos by P. Stiling.)*

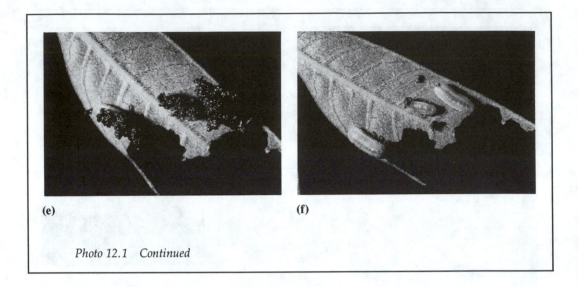

(e)   (f)

Photo 12.1   Continued

12.1 *Strategies to Avoid Being Eaten: Evidence for the Strength of Predation as a Selective Force*

In response to predation, many prey species have developed strategies to avoid being eaten. Edmunds (1974), Owen (1980), and Waldbauer (1988) have reviewed many case studies, which include the following.

■ **Aposematic, or warning, coloration:** Aposematic coloration advertises a distasteful nature. In the now-classic series of photographs shown in Photo 12.2, Lincoln Brower (1970) and coworkers (Brower et al. 1968) showed how inexperienced blue jays took monarch butterflies, suffered a violent vomiting reaction, and learned to associate forever afterward the striking orange-and-black barred appearance of a monarch with a noxious reaction. The caterpillar of this butterfly gleans the poison from its poisonous host plant, a milkweed. Many other species of animals, especially invertebrates, are also warningly colored, for example ladybird beetles. Caterpillars of many Lepidoptera are bright and conspicuous, too, because being noxious is their main line of defense—being soft-bodied, they would otherwise be very vulnerable to predators.

■ **Crypsis and catalepsis**: **Crypsis** and catalepsis are the development of a frozen posture with appendages retracted. This is another method of avoidance of detection by invertebrates. For example, many grasshoppers are green and blend in perfectly with the foliage on which they feed exhibiting **apatetic coloration.** Often, even leaf veins are mimicked on grasshopper wings. Stick insects mimic branches and twigs with their long slender bodies. In most cases, these animals stay perfectly still when

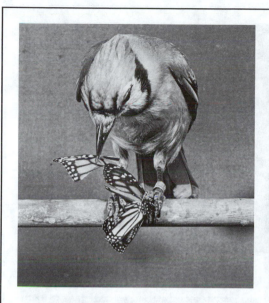 

Photo 12.2   Blue jay vomiting after eating a noxious monarch butterfly. Experienced blue jays subsequently avoid eating monarchs. (Photo by Lincoln P. Brower, University of Florida.)

threatened, because movement alerts a predator. Crypsis is prevalent in the vertebrate world, too. A zebra's stripes supposedly make it blend in with its grassy background, and the sargassum fish even adopts a grotesque body shape to mimic the sargassum weed in which it is found. Perhaps first prize should go the the chameleon, whose skin tones can be adjusted to match the background on which it is resting.

- **Mimicry:** Though many organisms may try to blend into their background, mimicking the foliage or background around them, some animals mimic other animals instead. For example, some hoverflies mimic wasps. Several types of mimicry can be defined. The three discussed here are illustrated in Photo 12.3.

  - *Mullerian:* The convergence of many unpalatable species to look the same, thus reinforcing the basic distasteful design, as for example with wasps and some butterflies (Muller 1879).
  - *Batesian* (after the English naturalist Henry Bates [1862], who first described it): Mimicry of an unpalatable species by a palatable one or of dangerous species (coral snakes) by innocuous ones (pseudocorals). Wickler (1968) has documented many cases in which flies, especially

Photo 12.3   Mimicry. (a) A south American hawk-moth cater-pillar, Pseudosphinz tetrio, *that may mimic a coral snake. (b) Mullerian mimicry by a British beetle,* Clytus arietis, *which, like many insect species, mimics the black and yellow banding pattern of wasps, normally avoided by predators. (c) Aggressive mimicry by a Florida bee-hunting fly,* Laphria *sp., a robber fly (Family: Asilidae) that mimics the bees on which it preys. (Photos by P. Stiling.)*

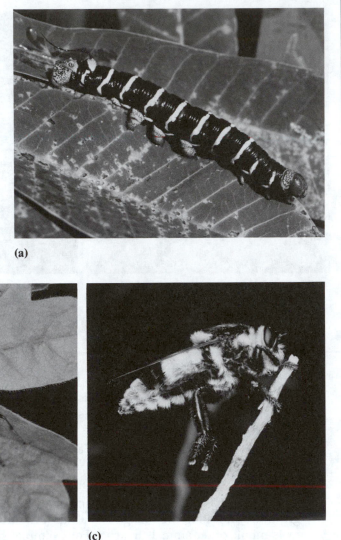

(a)

(b)　　　　　**(c)**

hoverflies of the family Syrphidae, are striped black and yellow to resemble stinging bees and wasps. Monarchs, aposematically colored themselves, are models for the mimic viceroy butterflies. Again, such phenomena are probably rampant in the butterfly world, but there are simply not the resources to study them all in detail. In at least one instance (Sternberg, Waldbauer, and Jeffords 1977), butterflies painted to mimic distasteful models were recaptured more frequently than those painted to mimic palatable forms, suggesting the great survival value of mimicry.

▫ *Aggressive*: In this case the body coloration permits individuals not so much to escape predation as to be better predators themselves. Many praying mantises mimic flowers so as to entrap insect prey. This strategy is shared by the crab spiders. Certain bottom-dwelling ocean fish also mimic the substrate so as to get in closer proximity to their prey. In extreme cases, such as anglerfish, parts of the body are modified to act as lures for prey. Some species also use light-emitting organs to attract prey. On land, aggressive mimicry is practiced by female fireflies of the genus *Photuris* (Lloyd 1975). Normally males respond to the species-specific light flashes of their females and move toward them. *Photuris* females mimic the flashing patterns of females of other species to lure the males of those species and eat them.

■ **Intimidation displays**: An example of intimidation display is a toad swallowing air to make itself appear larger. Frilled lizards extend their collars when intimidated to have the same effect.

■ **Polymorphisms**: **Polymorphism** is the occurrence together in the same population of two or more discrete forms of a species in proportions greater than can be maintained by recurrent **mutation** alone. Often this phenomenon takes the form of a color polymorphism; if a predator has a preference or **search image** for one color form, usually the commoner (Tinbergen 1960), then the prey can proliferate in the rarer form until this form itself becomes the more common (Cain and Sheppard 1954*b*) (so-called **apostatic selection,** Clark 1962). Stiling (1980) advocated just this type of choice of prey, by a visually searching **parasitoid,** to maintain the difference between two distinct color morphs, orange and black, in some leafhopper nymphs of the genus *Eupteryx*, though Stewart (1986*a*; 1986*b*) also implicated thermal **melanism** as an important agent of selection in some species. That is, black morphs occur in cooler climates because they heat up faster in the sun. Owen and Whiteley (1986) have pointed out that in many species the form of the polymorphism is such that *every* individual is slightly different from all others. This is true in brittlestars, butterflies, moths, echinoderms, and gastropods. They suggest that such a staggering variety of form thwarts predators' learning processes, and they suggest the term *reflexive selection* for this type of phenomenon.

■ **Phenological separation of prey from predator**: Fruit bats, normally nocturnal foragers, are active by day and at night on some small, species-poor Pacific islands such as Fiji. Wiens et al. (1986) suggest the fruit bats are constrained elsewhere to fly only at night by the presence of predatory diurnal eagles.

■ **Chemical defenses**: One of the classic defenses involves the bombardier beetle (*Bradinus crepitans*) as studied by Tom Eisner and coworkers (Eisner and Meinwald 1966; Eisner and Aneshansley 1982). These beetles possess a reservoir of hydroquinone and hydrogen peroxide in their abdomens. When threatened, they eject these chemicals into an explosion chamber, where they mix with a peroxidase enzyme. The resultant release of oxygen causes the whole mixture to be violently ejected as a spray that can be directed at the beetle's attackers. Many other arthropods have chemical defenses too, such as millipedes. This phenomenon is also found in vertebrates, as people who have had a close encounter with a skunk can testify.

■ **Masting**: Masting is the synchronous production of many progeny by all individuals in a population to satiate predators and allow some progeny to survive (Silvertown 1980). It is commonly documented in trees, which tend to have years of unusually high seed production. A similar phenomenon is exhibited by the emergence of 17-year and 13-year cicadas. It is worth noting in this context that both 13 and 17 are prime numbers, so no predator on a shorter multiannual cycle could repeatedly use this resource.

It is obvious that predation constitutes a great selective pressure on plant and animal populations. Despite the impressive array of defenses, predators still manage to survive by eating individuals of their chosen prey, often by circumventing the defenses in some way. The coevolution of defense and attack can be seen as an ongoing evolutionary arms race. According to Dawkins and Krebs (1979), the prey are always likely to be one step ahead. The reason is what they termed the *life-dinner principle*. In a race between a fox and a rabbit, the rabbit is usually faster because it is running for its life, whereas the fox is running "merely" for its dinner. A fox can still reproduce even if it does not catch the rabbit. The rabbit never reproduces again if it loses. This arms race has been argued to be run not only between predators and prey but also between parasite and host and between plant and herbivore. In the latter case, the race often proceeds by the production of toxins by the host and detoxifying mechanisms by the predator. Usually, however, at least for plants and herbivorous insects, phylogenetic relationships among herbivores do not correspond well to those of their host plants (Futuyma 1983; Mitter and Brooks 1983). The result is that at each defensive turn by a plant (or other host), it is as likely to run into a new set of enemies as it is to escape the old ones.

12.2 *Mathematical Models*

What effect has the predator on its prey population? The answer depends on many things, including prey and predator density and predator efficiency. Rosenzweig and MacArthur (1963) modelled predator-prey dynamics using a graphical method. First, they assumed that, in the absence of predators, the prey population increases exponentially according to the Lotka-Volterra formula

$$\frac{dN}{dt} = rN$$

The rate of removal of prey increases with an increase in encounter rate by predators dependent upon the number of predators (C) and the number of prey (N). The rate of removal is also dependent upon a "searching efficiency" or attack rate of predators, termed a'. The consumption of prey by predators is then $a'CN$, so

$$\frac{dn}{dt} = rN - a'CN$$

The rate of growth of the predators is given by

$$\frac{dC}{dt} = fa'CN - qC$$

where q is a mortality rate based on starvation in the absence of prey and $fa'\,CN$ is predator birth rate based on f, the predator's efficiency at turning food into offspring.

The properties of this model can be investigated by location of zero isoclines (regions of zero growth, neither positive nor negative). In the case of the prey, when

$$\frac{dN}{dt} = 0, \; rN = a'CN$$

or

$$C = \frac{r}{a'}$$

Because r and a' are constants, the prey zero isocline is a line for which C, the number of predators, is itself a constant (Fig. 12.1a). Along this line, prey neither increase nor decrease in abundance.

For the predators, when

$$\frac{dC}{dt} = 0, \; fa'\,CN = qC$$

or

$$N = \frac{q}{fa'}$$

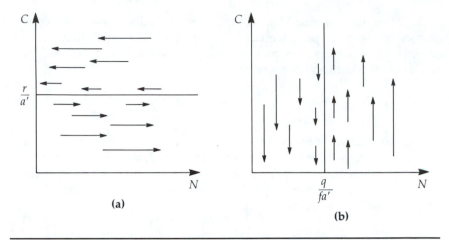

Figure 12.1 *A Lotka-Volterra type predator-prey model.* **(a)** *The prey zero isocline, with prey (N) increasing in abundance at lower predator densities (low C) and decreasing at higher predator densities.* **(b)** *The predator zero isocline, with predators increasing in abundance at higher prey densities and decreasing at lower prey densities.*

The predator zero isocline is also a straight line, one along which N, the number of prey, is constant (Fig. 12.1b). An assumption is that, if there are enough prey, a population of predators will increase and that, if there are not enough prey, they will starve. This is obviously a gross oversimplification. First, larger populations of predators require larger populations of prey to maintain them, so the zero isocline for predator growth should slant to the right (line B, Fig. 12.2). As predator density increases,

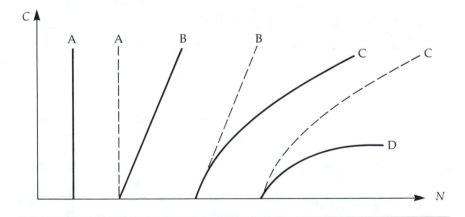

Figure 12.2 *Predator zero isoclines of increasing complexity, A to D.* A is the Lotka-Volterra isocline. B shows that more predators require more prey. C shows that the consumption rate is progressively reduced by mutual interference among predators. D shows that predators are limited by something other than their food.

so will mutual interference between predators, who will spend more time fighting one another and less time tackling prey. For zero growth, then, more realistically, there must be even more prey (line C, Fig. 12.2). Finally, at the highest densities of prey, it seems that the rate of growth of predators will be limited by something other than prey availability—say, social constraints on territory size or the availability of burrows or nest sites, so the zero isocline will appear as in line D in Fig. 12.2. Remember that at predator-prey combinations to the left of this line, predator numbers decrease, whereas to the right of it they increase.

The refinement of the prey zero isocline is dependent on two concepts. The first is that the recruitment rate of prey into the population is high when N is low but low when N is high and near the carrying capacity of the environment. The logic is that strong competition for resources at levels of N near K severely reduces recruitment rate; hence, the recruitment curve is semicircular (Fig. 12.3 top, p. 238). The rate of growth of the prey population is also dependent on the level of predation, and different predation rates are represented by the steepness of the dashed lines in Fig. 12.3 top. At each of the points where the consumption curve crosses the recruitment curve, the rate of population growth of the prey is zero, and these pairs of densities can be directly transferred to give the zero isocline for growth of the prey (Fig. 12.3 bottom).

The effect of different densities of predators can now be assessed by combination of the two isoclines (Fig. 12.4, p. 239). The highest levels of predation translate into large oscillations of predator and prey (i), sometimes so violent that one or both species goes extinct. Predators in this case are relatively efficient and abundant. Less efficient predators (ii) give rise to small and often damped oscillations in the system. At the lowest levels of predation, the system is highly stable, but only low levels of predators are supported (iii).

12.3 Field Data

How do the data on predator and prey abundances in the field compare with these types of graphical models? Predators of the Serengeti plains of eastern Africa (lions, cheetahs, leopards, wild dogs, and spotted hyenas) seem to have little impact on their large-mammal prey (Bertram 1979). Most of the prey taken are injured or senile and are likely to contribute little to future generations. In addition, most of the prey sources are migratory, and the predators, residents, are more likely to be limited in numbers by prey also resident in the dry season when migratory ungulates are elsewhere. Pimm (1979 and 1980) has argued that the importance of predation is dependent on whether the system is "donor-controlled" or "predator-controlled." In a donor-controlled system, prey supply is determined by factors other than predation, so removal of predators has no effect. Examples include consumers of fruit and seeds, con-

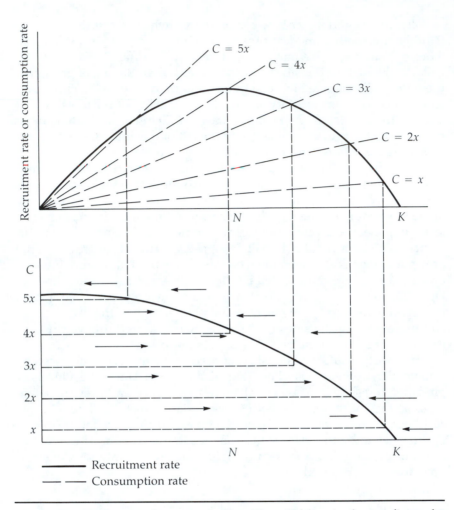

Figure 12.3   Refinement of prey zero isocline. The solid line in the top figure describes variation in prey recruitment rate with density. The dashed lines in the
upper figure describe the removal or consumption of prey by predators. There is
a family of dashed curves because the total rate of consumption depends on
predator density: increasingly steep dashed curves reflect these increasing densities. At the points where a consumption curve crosses the recruitment curve, the
net rate of prey increase is zero (consumption equals recruitment). Each of these
points is characterized by a prey density and a predator density, and these pairs
of densities therefore represent joint populations lying on the prey zero isocline
in the bottom figure. The arrows in the lower figure show the direction of change
in prey abundance. (Redrawn from Begon, Harper, and Townsend 1986.)

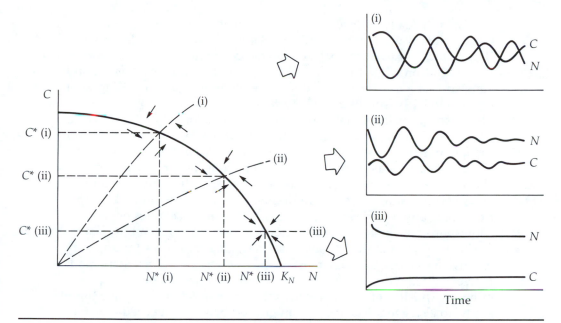

Figure 12.4 *A prey zero isocline with self-limitation, combined with predator zero iso-clines with increasing levels of self-limitation: (i), (ii), and (iii).* C* *is the equilibrium abundance of predators, and* N* *is the equilibrium abundance of prey. Combination (i) is least stable (most persistent oscillations) and generally has most predators and least prey: the predators are relatively efficient. Less efficient predators (ii) give rise to a lowered predator abundance, an increased prey abundance, and less persistent oscillations. Strong predator self-limitation (iii) can eliminate oscillations altogether, but* C* *is low and* N* *is close to* K_N*. (Redrawn from Begon, Harper, and Townsend 1986.)*

sumers of dead animals and plants, and intertidal communities in which space is limiting. In a predator-controlled system, the action of predator feeding eventually reduces the supply of prey and their reproductive ability. Removal of predators in a donor-controlled system is obviously likely to have little effect, whereas in a predator-controlled system such an action would probably result in large changes in abundance.

Many predator-prey systems appear stable; others appear to fluctuate dramatically. Determining which mechanism works in which situation is not easy. For example, in the Arctic there are two groups of primarily herbivorous rodents—the microtine varieties, lemmings and voles, and the ground squirrels. Ground squirrels exhibit the strongly self-limiting behavior of aggressive territorial defense of burrows (Batzli 1983). Their populations are remarkably consistent from year to year. On the other hand, the microtines are renowned for their dramatic population fluctuations. Thus, even in the same habitat, results for different species vary dramatically. One series of population fluctuations analyzed in great detail is that of the Canada lynx and snowshoe hare.

The Canada lynx (*Lynx canadensis*) eats snowshoe hares (*Lepus americanus*) and shows dramatic cyclic oscillation every nine-to-eleven years (Fig. 12.5). Charles Elton analyzed the records of furs traded by trappers to the Hudson's Bay Company in Canada over a 200-year period and showed that a cycle has existed for as long as records have been kept (Elton and Nicholson 1942). This cycle has been interpreted as an example of an intrinsically stable predator-prey relationship (Trostel et al. 1987), but Keith (1983) has argued that it is winter food shortage and not predation that precipitates hare decline. He showed that heavily grazed grasses produce shoots with high levels of toxins, making them unpalatable to hares. Such chemical protection remains in effect for two-to-three years, precipitating further hare decline. Predators, he argued, simply exacerbate population reduction. Thus, although lynx cycles depend on snowshoe hare numbers, hares fluctuate in response to their host plants. Subsequently, Smith et al. (1988) showed that, although food quality greatly affects hare **biomass,** most hares die of predation, not starvation. However, death due to predation is greatly exacerbated by poor quality of hares, which is of course greatly affected by food quality; thus there seems to be a good deal of common ground between the predation and starvation camps. No classical predator-prey oscillation has yet been proved to be driven only by prey and predator dynamics (Krebs 1985*a*), but the debate on the snowshoe hare continues. (Other predator-prey models based on the logistic equation, and modified to allow for the number of individuals eaten by predators, also predict population oscillation [Maynard Smith 1968; May 1976].)

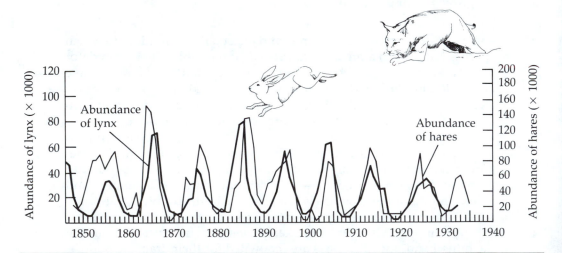

Figure 12.5 *The apparently coupled oscillations in abundance of the snowshoe hare* (Lepus americanus) *and Canada lynx* (Lynx canadensis) *as determined from numbers of pelts lodged with the Hudson's Bay Company.*

Some modification of the graphical model discussed so far is needed to reflect the instances in which prey populations have a refuge from predators. For example, many fossorial mammals escape from predators into underground burrows, and certain passerines exist in well-defined territories with abundant cover. Mammals outside their burrows or birds in suboptimal habitats are often exposed to predators, whereas those in their refuges are not. This situation can be represented graphically by a shift of the whole suite of consumption curves over to the right on the recruitment-consumption curve (Fig. 12.6 top). Such a shift leads to a

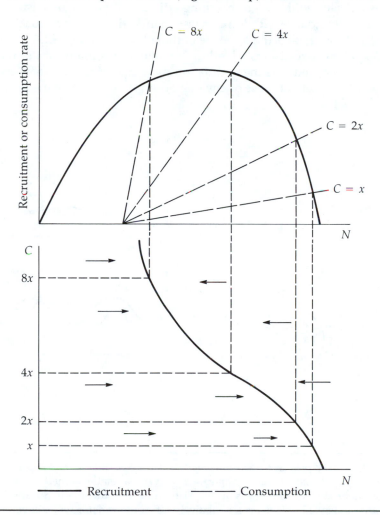

Figure 12.6 *Revised prey zero isocline allowing prey refuge from predators.* Here the total rate of consumption at low prey densities is zero, irrespective of predator abundance. The reason is a refuge that ensures complete safety of a small number of prey. This phenomenon leads to a vertical prey zero isocline at low densities. (Redrawn from Begon, Harper, and Townsend 1986.)

practically vertical prey zero isocline at low densities (Fig. 12.6 bottom). In other words, at low prey densities, prey can increase irrespective of predator densities—there are enough refuges for all individuals. At higher prey densities many prey exist outside the refuges and are available to predators. With respect to the relationship between predator and prey abundances, two different types of outcomes are possible (Fig. 12.7). First, the two isoclines for zero growth, that for predator and that for prey, can cross to the right of the prey isocline (situation ii), leading to oscillating abundances similar to those noted above. Second, if the predator is efficient, with a high searching efficiency, the outcome is a stable pattern of abundance with *N* much less than *K*, the carrying capacity (situation i). This result is the opposite of that obtained in Fig. 12.4.

As yet more realism is added to the model, it is important to realize that for many predators the population dynamics of the predator are actually independent of those of any one prey species because the predator

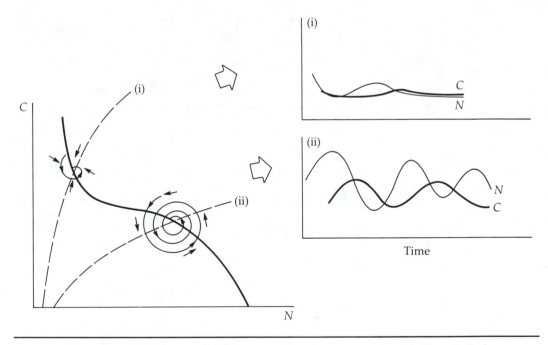

Figure 12.7 *The prey zero isocline appropriate when consumption rate is particularly low at low prey densities because of a prey refuge.* With a relatively inefficient predator, predator zero isocline (ii) is appropriate, and the outcome is not dissimilar from that of Fig. 12.4. However, a relatively efficient predator will still be able to maintain itself at low prey densities. Predator zero isocline (i) will therefore be appropriate, leading to a stable pattern of abundance in which prey density is well below the carrying capacity and predator density is relatively high. (Redrawn from Begon, Harper, and Townsend 1986.)

switches back and forth between various prey as they become abundant or rare. In such a situation the zero isocline for predators is a mere horizontal line (Fig. 12.8), which causes predators to regulate the prey at a low and stable level of abundance. Erlinge et al. (1984) give some support to this idea with a study on voles in Sweden. In southern Sweden, vole numbers are fairly low but stable. Their predators (foxes, weasels, owls) are able to switch to feed on other prey, notably rabbits and other rodents. In northern Sweden, such alternative prey are absent, and the field-vole population exhibits marked cycles.

Finally, if populations of prey become too small, even if a refuge exists, sometimes extinction can result. This is known as the *Allee effect* (Allee 1931). It may come about not because of predation but because of difficulties in finding mates or because of the need for social species to cooperate in order to obtain food. The implication is of a situation of no return when populations become too small, and it may explain why certain populations, for example populations of whales and fish, go extinct even though commercial usage could not actually catch the last specimen. One population result is that the consumption curve crosses the

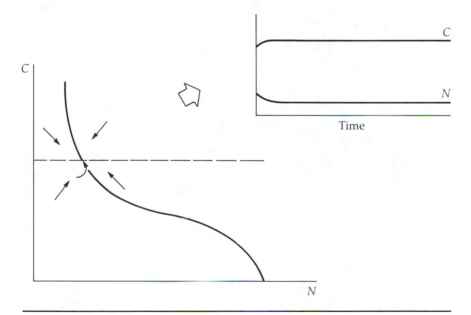

Figure 12.8 *The prey zero isocline when a predator exhibits switching behavior, that is, switches from one prey to another.* The predator's abundance may be independent of the density of any particular prey type, and the predator zero isocline may, therefore, be horizontal, that is, unchanging with prey density. This situation can lead to a stable pattern of abundance (inset) with prey density well below the carrying capacity.

recruitment curve twice (Fig. 12.9), implying more than one stable point of equilibrium. If a prey refuge is combined with an Allee effect, sometimes several equilibria are obtained (Fig. 12.10). This type of model predicts periodic population outbreaks, a situation not uncommon in nature,

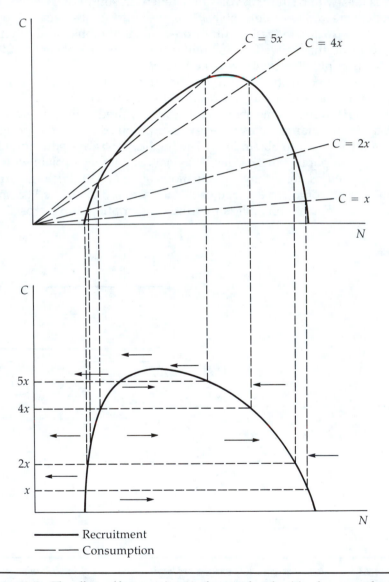

Figure 12.9 *The effects of low recruitment of prey at low densities, known as the Allee effect.* They may come about, for instance, because of difficulties in finding mates or because the prey need to cooperate with one another in order to obtain food. The consumption curves cross the recruitment curve twice, and the prey zero isocline has a hump.

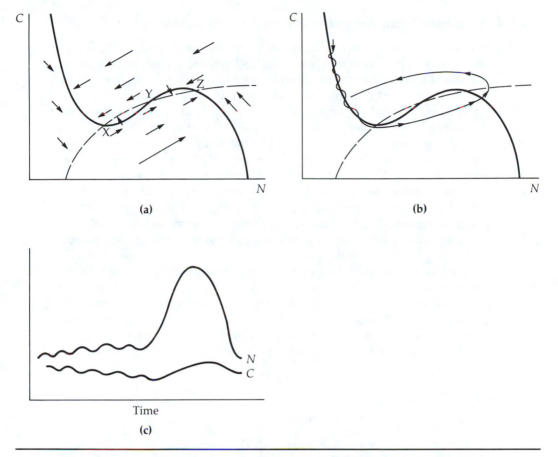

(a)

(b)

(c)

Figure 12.10 *A predator-prey zero isocline model with multiple equilibria.* **(a)** *The prey zero isocline with a vertical section at low densities and a hump.* The predator zero isocline can, therefore, cross it three times. Intersections X and Z are stable equilibria, but intersection Y is an unstable "breakpoint" from which the joint abundances move toward either intersection X or intersection Z. **(b)** *A feasible path that the joint abundances might take when subject to the forces shown in* (a). **(c)** *The same joint abundances plotted as numbers against time, showing that an intersection with the characteristics that do not change can lead to apparent "outbreak" in abundance.* (Redrawn from Begon, Harper, and Townsend 1986.)

especially outbreaks of pest insects in forests (Morris 1963; Varley, Gradwell, and Hassell 1973). The problem with this type of analysis and the prediction of multiple equilibria is that it is almost impossible to distinguish this type of subtle predator-prey cycling from population fluctuations resulting from environmental factors such as bad weather. The value of these models seems to decrease as their complexity increases.

12.4 *Evidence from Experiments and Introductions*

Perhaps the best way to find out whether predators determine the abundance of their prey is to remove predators from the system and to examine the response. Few such direct experiments have been done with barriers to prevent reimmigration of natural enemies, together with the appropriate control experiments. One of the best examples involves dingo predation on kangaroos in Australia (Caughley et al. 1980). The dingo, *Canis familiaris dingo,* is the largest naturally occurring carnivore in Australia and an important predator of imported sheep. Dingoes have been intensively hunted and poisoned in sheep country, southern and eastern Australia, and long fences (some up to 9,600 km) extend to prevent them from recolonizing areas, providing a classic experiment in predator control. The result has been a spectacular increase, 166-fold, of red kangaroos where the dingoes have been eliminated in New South Wales (Fig. 12.11), over their density in south Australia, where dingoes have not been molested. Emus (*Dromaius novaehollandiae*) are also over 20 times more abundant in dingo-free areas.

Another striking example of predation pressure has been provided by an inadvertant introduction by humans. Marine sea lampreys (*Petromyzon marinus*) live on the Atlantic coast of North America and migrate into fresh water to spawn. Adult lampreys feed by attaching themselves

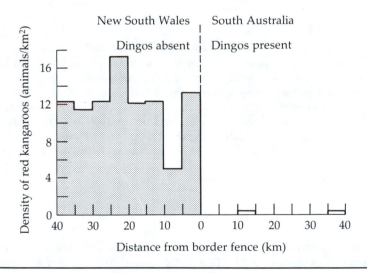

Figure 12.11 *Density of red kangaroos on a transect across the New South Wales–South Australia border in 1976.* The border is coincident with a dingo fence that prevents dingoes from moving from South Australia into the sheep country of New South Wales. (Redrawn from Caughley et al. 1980.)

to other fish, then rasping a hole and sucking out the body fluids. The passage of the lamprey to the upper Great Lakes was presumably blocked by Niagara Falls before the Welland Canal was built in 1829. The first sea lamprey was found in Lake Erie in 1921, in Lake Michigan in 1936, in Lake Huron in 1937, and in Lake Superior in 1945 (Applegate 1950). Lake trout catches decreased to virtually zero within about 20 years of lamprey invasion (Fig. 12.12, p. 248). Control efforts have been applied since 1956 to reduce the lamprey population, and attempts are being made to rebuild the Great Lakes fishery (Christie 1974).

In summary, we can conclude that in some but not all cases predators influence the abundance of their prey in the field. In the absence of controlled field experiments, a good analytical method for determining the importance of predators, or any other mortality factor, in the field is key-factor analysis, which is discussed in Chapter 15. Such information is important because biological control, the control of pests, relies heavily on the fact that introduced predators and parasites will have a substantial effect in the form of reduction of pest abundance.

Finally, it is worth noting that studies on the effects of predators are particularly timely now, given the desire of certain conservation groups to reintroduce large predators into certain areas. The U.S. Fish and Wildlife Service would like to reintroduce the wolf into Yellowstone National Park and to stabilize its numbers in Montana and Minnesota, the only states other than Alaska to possess viable populations. Cattle ranchers are fearful that wolves would decimate their herds. There is also considerable argument as to whether wolves constitute a large predation pressure on other prey such as caribou herds and should be controlled in times of declining caribou numbers (Bergerud 1980).

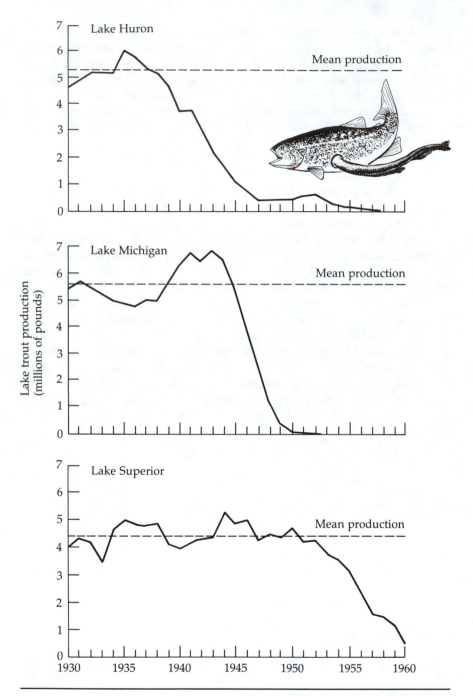

Figure 12.12 *Effect of sea lamprey introduction on the lake trout fishery of the upper Great Lakes.* Lampreys were first seen in Lake Huron and Lake Michigan in the 1930's and in Lake Superior in the 1940's. (Redrawn from Baldwin 1964.)

Chapter 13

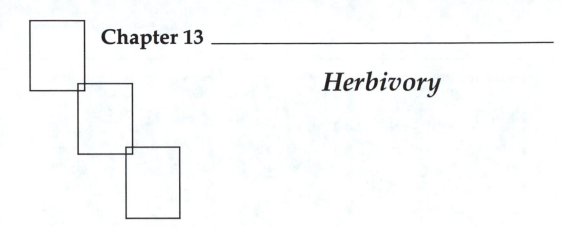

Herbivory

Plants appear to present a luscious green world of food to any organism versatile enough to attack and use it. Why could more of this food source not be exploited? After all, plants cannot even move to escape being eaten.

There are three possible reasons that more plant material is not eaten. First, natural enemies, predators and parasites, might keep **herbivores** below levels at which they could make full use of their resources. This idea is preferred by some ecologists, such as Lawton and Strong (1981). Second, herbivores may have evolved mechanisms of **self-regulation** to prevent the destruction of the host plant, perhaps ensuring food for future generations. Third, the plant world is not as helpless as it appears—the sea of green is in fact tinted with shades of noxious chemicals and armed with defensive spines and tough cuticles. Photo 13.1 illustrates some effects of herbivory and some defenses against it.

13.1 *Plant Defenses*

An array of unusual and powerful chemicals is present in plants, such as nicotine (an alkaloid) in tobacco, morphine and caffeine (also alkaloids), mustard oils, terpenoids (in peppermint and catnip), phenylpropanes (in cinnamon and cloves), and many others. A teaspoon of mustard should be enough to convince anyone of the potency of these chemicals. Such compounds are unusual—they are not part of the primary metabolic pathways of plants. They are therefore referred to as *secondary chemicals* and are thought to be synthesized specifically as a deterrent to herbivores.

Before this aspect of plant defense is discussed, it is important to note an alternative viewpoint—that secondary plant substances are merely the waste products of plant metabolism (Muller 1970). It must be remembered that excretion is a necessary part of metabolism even for plants, but

249

(a)   **(b)**

Photo 13.1   The effects of herbivory and defenses against it. (a) A large area of South American forest stripped bare by leaf-cutting ants. (b) An Amazonian palm tree with protective spines. As (a) shows, herbivores can devastate the plants on which they feed and put a strong selective pressure on plants to develop an array of chemical and mechanical defenses, as in (b). (Photos by P. Stiling.)

it must take a fundamentally different form (that is, storage) than it does in animals. Perhaps it is only coincidental that these waste metabolites have an effect on herbivores. It is argued that these chemicals are released into the environment in some situations to act as allelochemicals, which suppress the growth of potentially competing plant species (see Chapter 11).

Ehrlich and Raven (1964) were the first to crystallize the notion that secondary plant compounds evolved specifically to thwart herbivores. They proposed that most secondary compounds are produced only at a metabolic cost to the plant, not as energy-free by-products. This is the prevalent view among plant-insect ecologists today. Rhoades (1979) has formulated a general defense theory based on the idea that such plant compounds are costly to produce.

1. Higher herbivory levels lead to more defenses.
2. Higher costs of defense lead to fewer defenses.
3. More defenses are allocated to the most valuable tissues.
4. Environmental stress may lessen the availability of energy for defensive mechanisms.

5. Defense mechanisms are reduced when enemies are absent and increased when plants are attacked. This is known as the *theory of induced defense,* much in vogue in ecology in the 1980's. More is written about this theory below.

Types of Defensive Reactions

Defensive reactions in plants can be classified into several main types, of which the most commonly used are quantitative and qualitative.

Quantitative defenses are substances that gradually build up inside the herbivore as it eats and that prevent digestion of food. Examples are tannins and resins in leaves, which may occupy 60 percent of the dry weight of the leaf (Feeny 1976). In the case of tannin, the compound is not toxic in small doses, but it has cumulative effects. Tannins (the compounds in many leaves, like tea, that give water a brown color) act by binding with proteins in insect herbivore guts. The more leaf the herbivores ingest, the more difficult it is for them to digest it. Feeny (1970) was the first to document such a defense (by oaks against externally feeding caterpillars). Zucker (1983) proposed that there are two main classes of tannins, each with differing biological functions. *Hydrolyzable tannins* inactivate the digestive enzymes of herbivores, especially insects, whereas *condensed tannins* are attached to the cellulose and fiber-bound proteins of cell walls, thereby defending plants against microbial and fungal attack. Other authors have found that other insect herbivores are not much affected by quantitative defenses (Coley 1983; Karban and Ricklefs 1984; Faeth 1985; Mauffette and Oechel 1989), but the evidence that tannins affect vertebrate herbivory is stronger. Cooper and Owen-Smith (1985) showed that for browsing ruminants in Africa, kudus, impalas, and goats, palatability of 14 species of woody plants was clearly related to leaf contents of condensed tannins. The effect showed a distinct threshold; the browsers rejected all plants containing more than 5 percent condensed tannins. Cooper and Owen-Smith suggest that the reason is that ruminants depend on microbial fermentation of plant cell walls for part of their energy needs. For grey squirrels, acorns with higher concentrations of tannins are less preferred than acorns with lower concentrations (Smallwood and Peters 1986).

Qualitative defenses are, essentially, highly toxic substances, very small doses of which can kill herbivores. These compounds are present in leaves at low concentrations, like 1–2 percent of dry weight. Examples include alkaloids and cyanogenic compounds in leaves. Atropine, produced by deadly nightshade, *Atropa belladonna,* is a most potent poison. Of course, the plant must store many of these poisons in discrete glands or vacuoles or in latex or resin systems in order not to poison itself. Some compounds are stored as precursors and only become toxic when they are metabolized by the herbivore. For example, fluoroacetate found in certain Dichapetalaceae is metabolized by herbivores to fluorocitrate, a potent inhibitor of Krebs-cycle reactions (McKey 1979).

Good reviews of both these types of defenses, qualitative and quantitative, are given by Bernays (1981), Jacobson (1982), and Schoonhoven (1982). These two plant strategies are also correlated with plant "apparency" (Feeny 1976; Rhoades and Cates 1976). Apparent plants are long-lived and always apparent to the herbivores (for example, oak trees). Their defenses are thought to be mainly of the quantitative kind, effective against specialist herbivores with a long history of association with these *K*-selected plants. Unapparent plants are weeds, which are ephemeral and unavailable to herbivores for long periods. Their defenses are thought to be mainly qualitative, guarding against generalist enemies. Thus, trees nearly all contain digestibility-reducing compounds, and weeds contain the cheaper-to-make toxins. The value of these terms is of dubious value, however. An "unapparent" plant is not likely to be unapparent to the herbivores that specialize in finding it by chemical cues and eating it. Apparency probably best reflects the ability of human searchers to find plants. Nevertheless, the terms have spawned a great deal of research.

Cates and Orians (1975) set out to test the idea that early and late successional plants (see Chapter 18) might be differentially palatable to generalist herbivores. They found, using a slug as a test animal, that early successional plants were preferred. Is this result so surprising, though, given that the slug is an early successional animal, presumably predisposed to feeding on early successional vegetation? Quite the reverse was found to be true when late successional herbivores were used. For instance, gypsy moths, *Lymantria dispar*, prefer the leaves of trees and other late successional plants to those of weeds and other perennials (Bernays 1981). More recent evidence favors the traditional viewpoints. Coley (1988) studied growth, herbivory, and defenses in 41 common tree species in lowland Panamanian rain forest. Species with long-lived leaves had significantly higher concentrations of immobile defenses such as tannins and lignins. She also found trees with lower growth rates generally suffered higher herbivory.

Other defense mechanisms consist of the following:

- **Mechanical defenses**: Plant thorns and spines deter vertebrate herbivores, if not invertebrate ones. In Africa, in the presence of a large guild of vertebrate herbivores, much of the vegetation is thorny and spinose (Cooper and Owen-Smith 1986). Many neotropical plants are also armed in this fashion, for, although now absent, large browsers were abundant in this area until recently (Janzen and Martin 1982).

■ **Failure to attract**: Some plants may stop herbivory by failing to attract herbivores. They do so by lacking a certain chemical attractant that the herbivore uses as a cue.

■ **Reproductive inhibition**: Some plants, for example firs (*Abies* sp.), contain insect hormone derivatives that, if digested, prevent successful metamorphosis of insect juveniles into adults (Sláma 1969). In this way herbivory in the future is lessened by a decrease in the herbivore's reproductive output.

■ **Masting** (see also Chapter 12): The synchronous production of progeny, seeds, in some years satiates herbivores, permitting some seed to survive. Nilsson and Wästljung (1987) compared seed predation on beeches (*Fagus sylvatica*). In mast years, 3.1 percent of seeds were destroyed by a boring moth; in nonmast years, this figure was 38 percent. Vertebrate predation of seeds was 5.7 percent in mast years but 12 percent in normal years.

Some plants defend themselves against herbivores by enlisting the help of other animals. Such a relationship can, of course, be seen as **mutualism.** A very common example is that plants attract ants by providing sugary nectar secreted from extra-floral nectary glands (Barton 1986; Smiley 1986). African *Barteria* and neotropical *Cecropia* trees have hollow stems where ants maintain populations of scale insects. The ants are obligate occupiers of the trees; in return for food and shelter, they protect the trees from other herbivores and from encroaching vines, both of which they bite to death (Janzen 1979*b*). Schupp (1986) demonstrated the benefits to juvenile *Cecropia* by removing ants. The experimental plants suffered more damage from nocturnal herbivorous Coleoptera than unmanipulated controls. The *Acacias* of Central America have hollow thorns, which provide homes for extremely aggressive *Pseudomyrmex* ants, which also kill other herbivores and chop away encroaching vegetation (Janzen 1966). In numerical terms, the effects of ants can be quite substantial. Schemske (1980) found that seed production of *Costus woodsonii* in Panama was reduced 66 percent where ants were excluded. In England, Skinner and Whittaker (1981) used exclusion techniques to show that the wood ant, *Formica rufa*, was able to reduce herbivory from 8 percent of the leaves to 1 percent. Not all ant attendance is beneficial to trees, however, because many ants simply farm sap-sucking Homoptera and effectively protect their populations, a behavior that is of little use to the plant.

Of course, some defensive schemes backfire on the plants. Some chemicals that are toxic to generalist insects actually increase the growth rates of adapted specialist insects, which can circumvent the defense or actually put it to good use in their own metabolic pathways. Danaid (monarch) butterflies are attracted to milkweed plants that contain cardiac glycosides. These substances are vertebrate heart poisons, and cattle will not eat the milkweeds, but monarch butterflies can assimilate these poisons and use them in their own bodies as a defense against their own predators, advertising their distastefulness with bright colors (Brower 1969).

Induced Defenses

Plants do not necessarily keep their tissues permanently suffused with defensive and deadly chemicals. There is much evidence that chemicals are produced only as they are needed. The initiation of herbivore attack is usually sufficient to start the metabolic pathways of defense grinding. This defensive tactic is known as *induced defense* (Schultz and Baldwin 1982; Karban and Carey 1984; Edwards and Wratten 1985; Fowler and Lawton 1985; Faeth 1988). For example, Rhoades (1979) removed 50 percent of the leaves of ragwort, *Senecio jacobaea*, and detected a 45 percent increase in leaf alkaloids and N-oxidases in the undamaged leaves. Some of these effects can persist for long periods. Haukioja (1980) found that leaves of a birch tree were still poor-quality food for a moth three years after the tree had been damaged. This phenomenon may help to explain the severe oscillations of many forest insects. Sheep fertility was reduced six weeks after aphid attack on the alfalfa the sheep fed on had increased the production of its estrogen mimic, coumestrol (Schutt 1976).

Facultative defenses are not confined to chemical mechanisms. For example, the prickles on cattle-grazed *Rubus* plants were longer and sharper than those on ungrazed individuals nearby (Abrahamson 1975). Browsed *Acacia depranolobium* trees in Kenya have longer thorns than unbrowsed ones (Young 1987). Stinging nettles, *Urtica dioica*, exhibit increased density of stinging trichomes after herbivore damage (Pullin and Gilbert 1989). On holly trees, *Ilex aquifolium*, in Britain, lower leaves, subject to grazing, are heavily armed with spines; upper leaves, free from herbivory, are not (Crawley 1983). Interestingly enough, though the same phenomenon is apparent on American holly, *Ilex opaca*, Potter and Kimmerer (1988) regard it as an ontogenetic phenomenon rather than a defense against browsers. They suggest that poor nutritional quality of holly and high concentrations of saponins are more important as deterrents of vertebrate herbivores and that fibrous leaf edges deter invertebrate herbivores, which are unlikely to be affected by spines. Williams and Whitham (1986) have even argued that leaves infested with sessile insects, for example gall makers and leaf miners, are abscised prematurely as a defense against herbivory; the insects are killed as the leaf senesces on the forest floor. Stiling and Simberloff examined such a situation on oak trees in northern Florida. Although it is true that leaves infested with leaf miners do abscise earlier than noninfested leaves and that premature abscission kills miners inside the leaf, leaf abscission is not likely to be a complex, induced defense. It is more likely that premature abscission is a simple wound response on the part of the tree to rid itself of damaged leaves, possibly to avoid infection (Stiling and Simberloff 1989). For abscission to work as an induced defense in this situation, miners would have to show a high fidelity to a host tree; otherwise the tree would simply be reinfected by miners from neighboring trees. We have evidence that there is little fidelity in this system.

Though induced defenses undoubtedly occur in nature, their effectiveness as a deterrent is open to speculation and is probably less than that of permanent chemical defenses. Fowler and Lawton (1985) reviewed much of the work on induced defenses and concluded that as a result of such chemicals only small changes, generally less than 10 percent, occurred in such things as larval development time or pupal weights. Other studies have shown certain insects even benefit by feeding on previously damaged plants (Myers and Williams 1984 and 1987; Niemelä et al. 1984).

13.2 *The Field Frequency of Herbivory*

Plants do not have things all their own way in the plant-herbivore interaction. Herbivores can detoxify many poisons by four chemical pathways: oxidation, reduction, hydrolysis, and conjugation (Smith 1962). Oxidation occurs in mammals in the liver and in insects in the midgut. It is brought about by a group of enzymes known as mixed-function oxidases (MFOs). Conjugation, often the critical step in detoxification, involves the uniting of two harmful elements into one inactive and readily excreted product. Given that herbivores can circumvent plant defenses in certain situations, what is their effect on plant-population densities? On average, no more than 10 percent of net primary productivity seems to be taken by herbivores and about 90 percent by decomposers, in most natural systems (Crawley 1983). In a review of 93 cases of herbivory in terrestrial systems, an average of 7 percent was found to be consumed (Pimentel 1988). It must be remembered, of course, that such figures mask large and important variations. For example, the larch budmoth may take less than 2 percent of the net production of forest trees in some years but 100 percent in others.

Sometimes damage from insect feeding causes wilting or allows disease to cause loss of more plant **biomass.** Grasshoppers feeding on needlerush, *Juncus roemerianus,* in salt marshes feed in the middles of the tall, narrow leaves, so even though only the middle of the leaf is actually digested, the top half is cut off and added to the litter layer (Parsons and de la Cruz 1980). This author has observed that regions of *Quercus* *geminata* leaves distal to *Stilbosis* leaf mines often turn brown and senesce, even though the miners do not damage them directly.

Ultimately, the best way to estimate the effects of herbivory on plant populations is to remove the herbivores and examine subsequent growth and reproductive output. Some of the best evidence on the impact of herbivores on plants comes from the biological control of weeds. Following its importation from the Americas in 1839, the prickly pear cactus, *Opuntia stricta*, became a serious pest in Australia, occupying by 1925 over 240,000 km^2 of once-valuable rangeland (Fig. 13.1). After some initial imports of insects that failed to control the growth of the cactus, *Cactoblastis cactorum*, a moth, was introduced from South America in 1925 (Osmond and Monro 1981). By 1932, the original stands of prickly pear had collapsed under the onslaught of the moth larvae. Despite a small resurgence of prickly pear in 1932–1933, *Cactoblastis* has devastated prickly pear populations ever since, and the cactus is now confined to isolated areas. "Before" and "after" views appear in Photo 13.2.

There have been several other successes in the biological control of weeds, and these tend to overshadow the probably more numerous failures. Klamath weed (*Hypericum perforatum*), a pest of pasture land in California, was controlled by two French beetles (Huffaker and Kennett 1959). As illustrated in Photo 13.3, floating fern, *Salvinia molesta*, choked a lake in Australia and was controlled by the weevil *Cyrtobagus singularis*, introduced from Brazil, where the fern is native (Room et al. 1981). Alligatorweed was controlled in Florida's rivers by the so-called alligatorweed beetle, *Agasicles hygrophila*, from South America, and hopes are high that water hyacinth can be controlled biologically, too (Buckingham 1987). On the other side of the coin, large numbers of insects have been introduced to control *Lantana camara*, an introduced weed in Hawaii. Very few have had any impact on the growth of the plant, though its spread might have been slowed.

In successful cases of biological control of weeds, all the plants have been alien, and most were perennials. No native weeds and very few annual weeds have yet been controlled by insects. Annuals can produce huge amounts of seed even when heavily infested by insects. No seed-eating insect has yet controlled a weed plant. The weevil *Apion ulicis* was introduced into New Zealand to control gorse, *Ulex europaeus*, and has become one of the most abundant insects in New Zealand. Unfortunately, although the insect eats up to 95 percent of the gorse seed produced every year, there has been no appreciable effect on plant numbers (Miller 1970).

In a natural setting, removal of herbivores from their host plants has been done only rarely and with mixed results. Karban (1982) used this technique to demonstrate reduction in growth of apple trees due to insect infestation, and he detected a similar effect in scrub oak, *Quercus ilicifolia* (Karban 1980). The most serious effects, however, may be on reproductive output. Crawley (1985) removed herbivores from oaks in Britain by

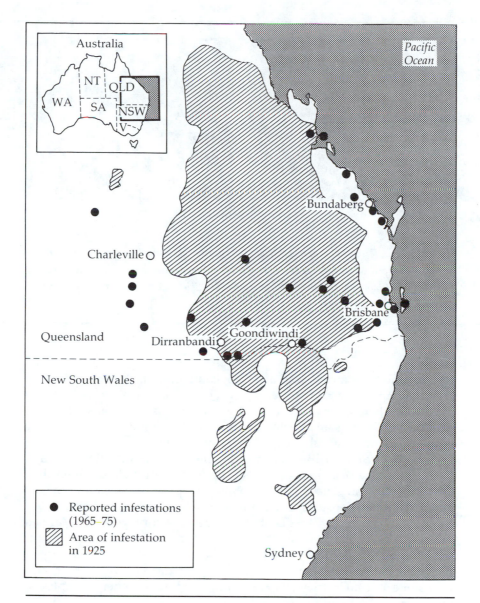

Figure 13.1 *Distribution of the prickly pear* (Opuntia) *in eastern Australia in 1925 at the peak of infestation and modern areas of local infestation.* (Redrawn from White 1981.)

spraying techniques. Though unsprayed trees lost only 8–12 percent of their leaf area, the sprayed trees consistently produced from 2.5 to 4.5 times the number of seeds produced by unsprayed plants. Waloff and Richards (1977) found almost three times more seed on broom bushes sprayed for insect control than on unsprayed ones. Once again, however,

(a)   **(b)**

*Photo 13.2   A terrestrial example of biological control. (a) Rangeland in Australia in-
fested with* Opuntia *cactus. (b) The same site after introduction of a cactus-eating moth,*
Cactoblastis. *(Photos courtesy of the Commonwealth Scientific and Industrial Research
Organization of Australia.)*

results tend to be mixed. Karban (1985) could find no effect of periodical
cicada nymphs on acorn production by *Q. ilicifolia* in New York State. In
instances where an exotic herbivore is introduced in the absence of its
enemies, the results are much more dramatic. Bermuda cedar, *Juniperus
bermudiana*, was virtually wiped out by an introduced scale insect,
Lepidosaphes newsteadi (Bennett and Hughes 1959; Cock 1985). Young
trees in forestry plantations can also suffer large mortalities when they
are girdled by introduced goats, rabbits, sheep, or squirrels. In times of
abnormally high densities, other animals can have these same effects. For
example, in 1958 only 24 percent of mature trees surveyed in an African
Terminalia glaucescens woodland were dead, but in 1967 after elephant
densities were boosted by immigration beyond the carrying capacity of
the region, almost 96 percent of the trees were dead (Laws, Parker, and
Johnstone 1975). Crawley (1983) has provided many other examples of
the impact of herbivory.

In most cases, herbivores cause subtle alterations of growth rates of
stems and roots rather than outright death of the plant. Flower, seed, and
fruit production can also be influenced, though it is likely that predators
of fallen seed and fruit are more important in a scheme of plant fitness. In
this respect, herbivores can be seen as successful **parasites** because they
do not kill their hosts—they merely reduce the growth rate. In an agricultural setting, of course, there are many estimates of losses of crops to herbivores. Damage is often severe enough not only to justify control but to

(a) **(b)**

Photo 13.3   An aquatic example of biological control. (a) Lake Kabufwe, Papua New Guinea, choked with the floating fern Salvinia in October 1983. (b) The same lake clear of the weed in November 1984, after release of the herbivorous weevil Cyrtobagus salviniae. (Photos by P.M. Room, courtesy of the Commonwealth Scientific and Industrial Research Organization of Australia.)

cause economic hardship to agriculturalists (May 1977; Pimentel et al. 1980; Barrons 1981). Even though plants are rarely killed outright, the effects of even minor damage on crop yields can be substantial. It is clear, however, that control measures are often initiated when populations of pests are so small that significant damage is unlikely to occur (Pimentel et al. 1980).

Beneficial Herbivory?

Some authors have argued that herbivory can be beneficial to plants (see McNaughton 1986; Crawley 1987; Owen and Weigert 1987). The rationale is that plants are stimulated to regrow after damage—they end up overcompensating, growing even more than they would have had they not been damaged. The result is often more seed production from more vegetative plant parts. Simberloff, Brown, and Lowrie (1978) noted that the action of isopod and other invertebrate root borers of mangroves tended to initiate new prop roots at the point of attack (Fig. 13.2). More prop roots meant greater stability of mangroves against wave and storm action, so root herbivory could in fact be beneficial. However, in a review of the 20 papers most commonly cited as evidence for beneficial herbivory, Belsky (1986) found fault with the logic, experimental design, or statistics of nearly all of them. Even newer papers that purport to support

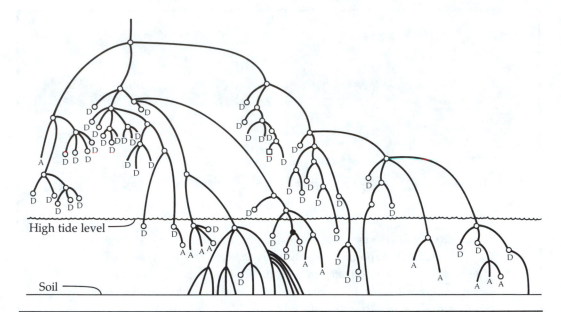

Figure 13.2 *Branching pattern for a single* Rhizophora mangle *root from Clam Key, Florida.* A = alive; D = dead; open square = bored by *Ecdytolopha* sp.; open circle (above water) = bored by unknown insect; open circle (below water) = bored by *Sphaeroma terebrans;* shaded circle = bored by *Teredo* sp. (Redrawn from Simberloff, Brown, and Lowrie 1978.)

the beneficial-herbivory theory are usually fraught with methodological or technical errors (Belsky 1987). In addition, Strauss (1988) has pointed out that very carefully designed experiments involving measurements of plant size before and after herbivory are needed because herbivores themselves naturally choose larger plants, which might be expected to show more growth than would stunted plants, even after herbivory. Because of the economic damage supposedly done by herbivores, especially insects, it is not trivial to assess whether or not defoliation is beneficial. Mattson and Addy (1975) have tried to model forest growth with and without insect herbivores. They examined two situations. In the first, aspen was defoliated by forest tent caterpillars. Forest tent caterpillars begin to infest a forest slowly, reach a peak in numbers, remain there for three or four years, and then subside. Stemwood production in the years of peak infestation is much reduced, but foliage production increases to compensate for insect defoliation. Within roughly 10 years after the infestation, the biomass production was identical to that of unaffected forests. In the short term, the caterpillars reduced wood production, but in the long term they had no major effect.

In a second example, balsam fir was defoliated by spruce budworms.

These larvae actually kill mature trees aged 55–60 years, but they leave young trees largely alone. The saplings grow quickly after their parents are killed, and a resurgence of the forest, from its juveniles or saplings, occurs. The end result is the same as in the previous case: in the short term there is a considerable effect on wood production, but there is not in the long term. Production rates in the young forest remain elevated above that of the mature forest for 15 years, because in a mature wood-land most trees have passed their rapid-growth phase. The role of forest-ers in this cycle is not at all clear (Holling 1978). Perhaps an effective ac-tion would be to harvest those trees in the center of budworm outbreaks because they would be killed in any case and because many larvae would be removed with them.

In herbivory on grasses, the placement of the meristem is an impor-tant question. In most species, the meristem is down low, safe from her-bivores and protected in the basal leaf sheaths. The few species that produce elongated vegetative shoots are very vulnerable to grazing. Be-cause growth is very slow from axillary buds, which must take over after the meristem is eaten, such plants tends to be outcompeted by species whose meristems survive. The inflorescences of most species, however, must extend upward into the air to ensure pollination of the flowers. Be-cause certain species die only after flowering, heavy herbivory or con-stant clipping by a lawnmower can, by preventing flowering, virtually ensure their immortality. Grasses in sports stadiums may effectively live forever.

The Effects of Plant Quality

There is much evidence that the herbivores themselves select the plants that are the most nutritionally adequate in terms of nitrogen con-tent of the tissue (Mattson 1980; Scriber and Slansky 1981; White 1984) or amino-acid concentration of the sap (Brodbeck and Strong 1987). Iason, Duck, and Clutton-Brock (1986) showed how red deer fed preferentially on grasses defecated upon by herring gulls (*Larus argentatus*). Where the number of gull droppings increased, so did the vegetation nitrogen con-tent. For birds, Watson, Moss, and Parr (1984) showed how food enrich-ment affects numbers and spacing behavior of red grouse. In some cases, however, such correlations between host-plant quality and herbivore density is absent, suggesting observed population patterns of herbivores are more dependent on other phenomena, such as predation or parasit-ism (Stiling, Brodbeck, and Strong 1982). There seems to be no easy way to predict when herbivore densities are controlled by the quality of their hosts. When food quality declines, many herbivores, especially verte-brates, respond simply by feeding at a higher rate or for a longer time. On the reverse side, deaths of herbivores due to depletion of food plants are witnessed very infrequently (Crawley 1983), perhaps because herbivores can leave an area of poor food availability, but they cannot easily escape the weather or an outbreak of disease. The relatively few examples of mass starvation due to overexploitation of plants come mainly from studies of insects that habitually undergo periodic outbreaks or from cycles of Arctic rodents. Also, vegetation often grows back after attack. A partly defoliated tree will live to photosynthesize another day, but an animal with a limb removed by a predator will rarely survive. This is a major difference between predation and herbivory. Therefore, a prudent herbivore will, like a careful gardener, exercise restraint in feeding, for then the resource will remain. Colonies of leaf-cutting ants, *Atta* sp., in South America are fairly permanent, yet food trees are rarely killed, so there is a source of food available close to the nest for many years (Rockwood 1975). That is not to say that food quality is not important. For example, Klein (1970) found that, in deer, food quality affects growth rate, dispersal, productivity of young, and sex ratio at maturity. Adult survival, however, was probably affected by other things.

Mathematical Models

Crawley (1983) has concluded that plants have a much more important impact on herbivores than herbivores have on the dynamics of plants. Mammals are especially prone to food limitation, though insect herbivores are probably more influenced by predation, parasitism, and disease. Plant death rates themselves, Crawley suggests, are largely determined by competition with other plants and by self-thinning. Nevertheless, Crawley has provided a series of models of plant-herbivore interactions. The most simple of these assumes that there is an upper limit, a carrying capacity, K, for a population of plants. The rate of change of a population of plants is given by

$$\frac{dV}{dt} = A - B$$

where A is the gains and B the losses and V is plant abundance. Similarly, for the herbivores

$$\frac{dN}{dt} = C - D$$

where C and D are gains and losses and N is herbivore numbers.

It is assumed that, in the absence of herbivores, plant populations increase exponentially such that gains, A, $= aV$ where a is the plant's intrinsic rate of increase. Losses for plants, B, $= bNV$ where b is the feeding rate of herbivores. For the herbivores, $C = cNV$ where c describes the numerical response of herbivores and $D = dN$ where d is herbivore death rate. The assumption is that, in the absence of plants, herbivores starve and their numbers decline exponentially. Assuming an upper carrying capacity, K, and following a basic Lotka-Volterra type of logistic model, then $A = aV(K - V)/K$. The other elements, B, C, and D, remain the same. Therefore,

$$\frac{dV}{dt} = \frac{aV(K - V)}{K} - bNV$$

and

$$\frac{dN}{dt} = cNV - dN$$

At equilibrium both dV/dt and dN/dt are zero; there is no population change. The equilibrium plant density V^* and equilibrium herbivore abundance N^* can then be solved for because $A = B$ and $C = D$. Therefore,

$$V^* = \frac{d}{c}$$

and

$$N^* = \frac{a(K - (d/c))}{bK}$$

The effect of each of the parameters a, b, c, and d is best assessed by graphical techniques. Each parameter can be made to vary while the others remain fixed. The effect of increasing the plant's intrinsic rate of growth, a, stabilizes and increases herbivore equilibrium density but has no effect on equilibrium plant abundance (Fig. 13.3, p. 264)—essentially, the faster the plants grow, the faster the herbivores eat them up. When b, the feeding rate of herbivores, is increased, there is a lower herbivore equilibrium, but the size of the equilibrium plant population is unchanged (Fig. 13.4, p. 265). The more efficient an herbivore is at food gathering, the rarer the herbivore is, because fewer herbivores can be supported by a given level of plant production. The numerical response of herbivores is described by c; the higher the value of c, the more herbivores can be borne per unit feeding. This is essentially the efficiency with which herbivores turn food into progeny. The higher this efficiency, the lower the equilibrium population of plants, because an efficient herbivore turns all production into herbivores (Fig. 13.5, p. 266). Perhaps more important is the effect of c on stability. When plant equilibrium density is low, much below K, the carrying capacity, the population is under lax control and tends to increase exponentially when herbivore numbers decrease. Thus the lower the equilibrium plant population, the lower the

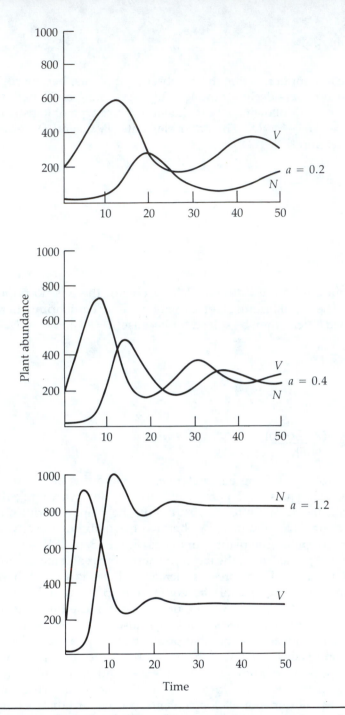

Figure 13.3 *The effect of the plant's intrinsic growth rate* a *on plant and herbivore abundance.* Increasing *a* from 0.2 to 1.2 increases stability and increases herbivore equilibrium density but has no effect on equilibrium plant abundance. Other parameters are *b* = 0.01, *c* = 0.001, *d* = 0.3, *K* = 1,000. As in all these types of models, the curve *V* represents plant abundance, and the curve *N* represents herbivore numbers (in arbitrary units). (Redrawn from Crawley 1983.)

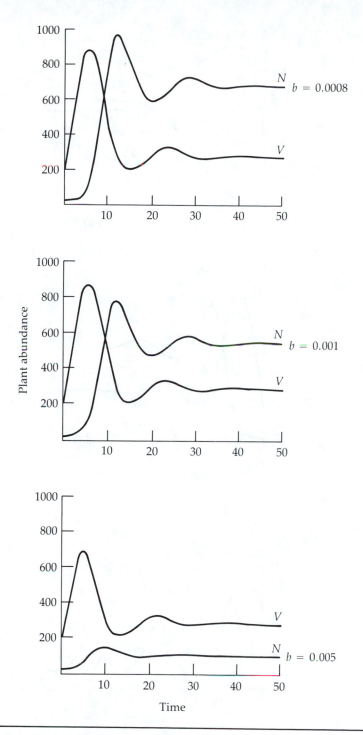

Figure 13.4 *Resource-limited plants—the effect of herbivore searching efficiency* b. Increasing b from 0.0008 to 0.005 reduces herbivore equilibrium density but has no effect on stability or on equilibrium plant abundance. Other parameters are *a* = 0.5, *c* = 0.001, *d* = 0.3, *K* = 1,000. (Redrawn from Crawley 1983.)

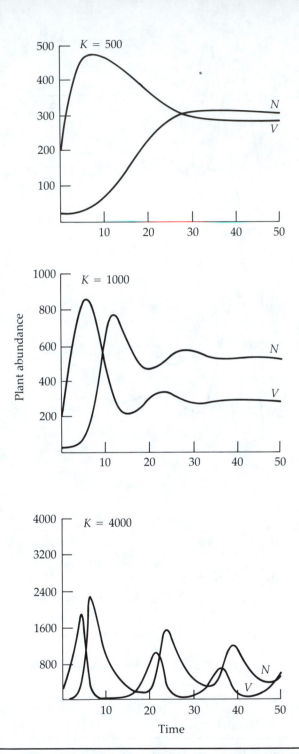

Figure 13.5 *The effect of herbivore growth efficiency* c. Increasing *c* from 0.0005 to 0.005 reduces equilibrium plant abundance and thereby reduces stability but increases equilibrium herbivore numbers. Other parameters are *a* = 0.5, *b* = 0.001, *d* = 0.3, *K* = 1,000. (Redrawn from Crawley 1983.)

stability of plant and herbivore numbers. The effects of herbivore death rate, d, are exactly opposite to those of c. Increasing herbivore death rate increases **stability** because it leads to an increase in equilibrium plant numbers. Finally, plant carrying capacity, K, is critical to the stability of these models. If K is high, then populations may be under lax control even where equilibrium plant numbers are quite high. An increase in K could lead to an increase in the number of herbivores, too. However, equilibrium plant density V^* is determined solely by c and d. Thus increasing K reduces the relative level of plant equilibrium populations, and control of plant numbers is lax. This effect explains why high values of K decrease the stability of a system (Fig. 13.6, p. 268). This is what Rosenzweig (1971) called the ''paradox of enrichment'' and may account for the unstable population cycles of herbivorous insects in the vast boreal forest of the Northern Hemisphere, where the carrying capacity for evergreen trees is enormous.

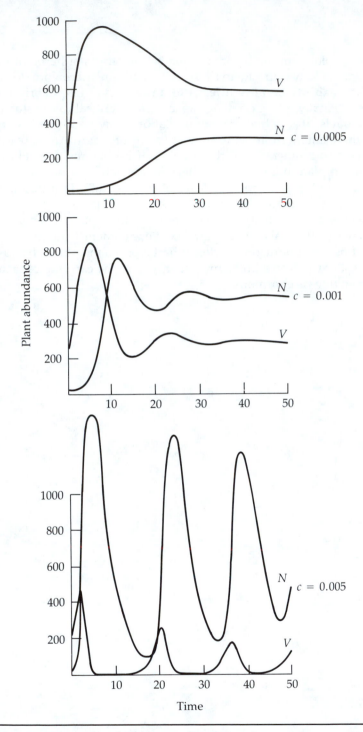

Figure 13.6 *The paradox of enrichment.* Increasing the carrying capacity of the environment for plants *K* from 500 to 4,000 increases herbivore equilibrium density but has no effect on plant equilibrium (note scale changes on *y* axis). The main effect of increasing *K* is to reduce stability. Other parameters are $a = 0.5$, $b = 0.001$, $c = 0.001$, $d = 0.3$. (Redrawn from Crawley 1983.)

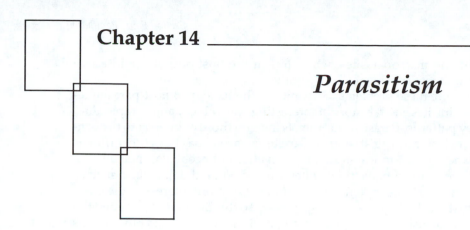

Chapter 14 _____

Parasitism

W hen one organism feeds off another but does not normally kill it outright, the predatory organism is termed a **parasite** and the prey a **host**. Some parasites remain attached to their hosts for most of their lives, like tapeworms, which remain inside the host's alimentary canal. Others, such as ticks and leeches, drop off after prolonged periods of feeding. Mosquitoes remain attached for relatively short periods. By this definition, many species of phytophagous insects are parasitic upon their "host" plants. Some studies treat them as such (Price 1980), but in this book such organisms are dealt with under the more conventional heading of herbivory. Still, there remain many problems of definition. Should organisms that feed off more than one individual, without killing them, be known as parasites or predators? For example, saber-tooth blennies (*Plagiotremus)* on the Great Barrier Reef dash out and bite chunks out of fish hosts/prey that swim by. Should the large ungulates of the Serengeti plains, wildebeest, zebra, and the like, be known as parasites? Although they feed off more than one individual host grass, the grass is not killed and will grow back later. Should we retain the term parasite for organisms that remain in intimate contact with their hosts? Mosquitoes develop as larvae in a nonparasitic manner in pools of water, and the adults only come into contact with "hosts" for short periods. Rhinoceroses live on top of their food supply for their entire lives. And what about parasites of insects? Many of these develop as internal parasites of caterpillars or other immature stages. In these cases, the host almost never survives, and the term **parasitoid** is used to refer to these "parasites," each of which uses only one host but invariably kills it. Even in this case, further gradation between parasitoid/parasite and predator is evident, as when an egg parasitoid hatches from a host egg and has to devour several more in the clutch before it is mature (Askew 1971). May and Anderson (1979) have tried to distinguish two types of parasites—microparasites, which multiply within their hosts, usually within the cells (bacteria and viruses), and macroparasites, which live in the host but release infective juvenile stages outside the host's body.

Despite these problems of definition, the biology of host-parasite relationships has a rich history of interesting, **coevolved**, and complex life-history patterns. Parasites on animals include those of interest to the conventional parasitologist—viruses, bacteria, protozoa, flatworms (flukes and tapeworms), thorny-headed worms (Acanthocephala), nematodes, and various arthropods (ticks, mites, and so on). Other parasites such as the parasitic Hymenoptera and Diptera are of more interest to the entomologist and biological-control specialist. To this list should be added the numbers of insects, nematodes, fungi, bacteria, viruses, and other plants parasitic on plants. By these definitions, over 70 percent of the British insect fauna have been regarded by Price (1980) as parasitic, a not inconsequential proportion and certainly bigger than the 4 percent that are predatory. Because about 75 percent of the known fauna on Earth consist of insects, then at least 50 percent of the animals on Earth might be considered parasitic. When the other large groups of parasites are considered—nematodes, flatworms, and so on—it is clear that parasitism is a very common way of life. A free-living organism that does not harbor several parasitic individuals of a number of species is a rarity. The frequency of human infection by parasites is staggering. There are 250 million cases of elephantiasis in the world, over 200 million of bilharzia, and the list goes on and on.

14.1 *Defenses Against Parasites*

The defensive reactions developed by hosts to resist parasites are as impressive as those to combat predation.

□ *Cellular defense reactions:* These reactions particularly are found in insect larvae as a defense against parasitoids, where eggs of the parasitoid are "encapsulated" or enclosed in a tough case rendering them inviable (Salt 1970).

□ *Immune responses in vertebrates* (Cox 1982): These responses are the vertebrate body's defense against the parasitic microbes that cause disease in humans and animals.

□ *Defensive displays or maneuvers:* These actions are intended to deter parasites or to carry organisms away from them. For example, gypsy moth pupae spin violently within their cocoons to deter pupal parasites (Rotheray and Barbosa 1984), and syrphid larvae often drop to the ground from the foliage they forage on to escape parasites (Rotheray 1981; 1986).

□ *Grooming and preening behavior:* This behavior is found in mammals and birds, respectively, to remove ectoparasites (Struhsaker 1967; Kethley and Johnston 1975).

14.2 *Models of Host-Parasite Interactions*

Circumstantial evidence such as that listed above suggests that parasites have at least as great an effect on the population dynamics of their hosts as do predators. Because the vast majority of parasites are actually parasitoids attacking insects, and because insect pests are such economically important organisms in agriculture, most population models of parasite-host relations have centered on parasitoid-host interactions. In many ways, these are similar to predator-prey relationships, because only a single host is killed.

Early models of parasitoid-host interactions centered on the **numerical response** of a population of parasites to host density. That is, as host density increases, more parasite eggs are laid per unit of search time. The first model was developed by the Australian entomologist Nicholson in the 1930's, who proposed that randomly searching parasitoids would be limited not by their own egg supply but by their ability to find hosts (Nicholson 1933; Nicholson and Bailey 1935). The average area one parasitoid searched in its lifetime was deemed to be constant and was termed the area of discovery a, as distinct from the area actually traversed. Thus, if 0.3 of the total habitat area is traversed by a parasitoid, the area of discovery may be only 0.254, because some ground may be covered twice. In other words, it may be expected that only 25.4 percent of the hosts will be parasitized instead of 30 percent. If this model is to be tested, a must be calculated, because measuring it is usually quite impractical. If N is the total number of hosts, P the number of searching parasitoids, and S the proportion of hosts not parasitized, then

$$a = \frac{1}{P} \log_e \frac{N}{S}$$

Once a is known, the proportion of hosts parasitized can be predicted from the number of searching parasites:

$$\log N_{(n+1)} = \log N_{(n)} \frac{aP_n}{2 \cdot 3} + \log F$$

where $N_{(n)}$ and $N_{(n+1)}$ represent successive host population, and F the host reproductive rate. The outcome of this model is usually to produce increasing oscillations in the populations of both species, and laboratory tests have often been in good agreement with the theory (Fig. 14.1) (see also Debach and Smith 1941). Nicholson was aware that increasing oscil-

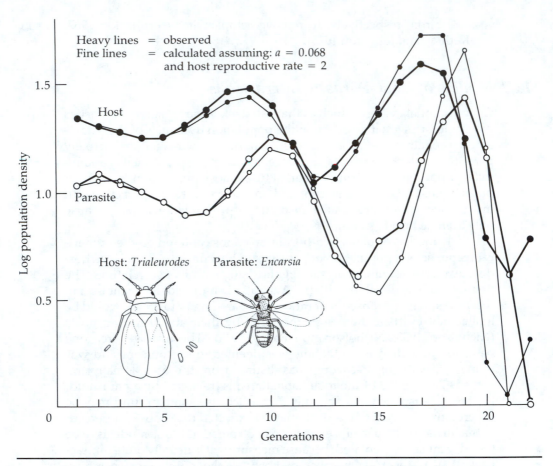

Figure 14.1 *Observed and calculated results of an interaction between* Encarsia, *a parasitoid, and its host, the greenhouse white-fly,* Trialeurodes. *Model calculated on basis of a constant area of discovery of 0.068 and a host reproductive rate of 2. (Redrawn from Varley, Gradwell, and Hassell 1973.)*

lations did not often occur under natural conditions, except for some species of alpine insects, pests of coniferous forests, which show regular oscillations of outbreak densities over a number of years (Klomp 1966). Nicholson envisaged a fragmentation of large host populations into small ones at high densities and thus a proliferation of subpopulations. Although this sort of interaction does occur in Australia between the moth *Cactoblastis cactorum* and its food supply, the prickly pear, once again such events have not been commonly documented in the field.

Included in the numerical response by parasites to host density is a response by individuals, called a **functional response**, in which the number of attacks per individual parasite changes with host density. C.S.

Holling (1959) demonstrated that, during a predator or parasite attack, "handling time" has a very important effect on the functional response. Handling time is the interval between the time at which a natural enemy first encounters a host or prey and the time at which search is resumed for a new one. It may be thought of as the time required to subdue prey and parasitize it, and it varies considerably between species from seconds to hours. Holling used blindfolded people searching for sandpaper discs on a table to mimic searching for parasites and then fitted an equation to the results. In this equation, commonly called the disc equation, the number of hosts encountered, N_a, is given by

$$N_a = \frac{Ta'N}{1 + a'T_h N} P$$

where T is total time available, T_h is handling time, a' is coefficient of attack, and N is initial number of available prey.

A typical functional response is shown in Fig. 14.2. In many cases,

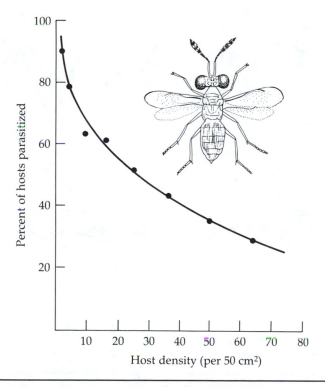

Figure 14.2 *The functional response of a single female of the chalcid parasite* Dahlbominus fuscipennis *searching for cocoons of the sawfly* Neodiprion sertifer *within cages of 50 cm² floor area, expressed as the percentage of hosts parasitized at different host densities.* (Redrawn from Varley, Gradwell, and Hassell 1973.)

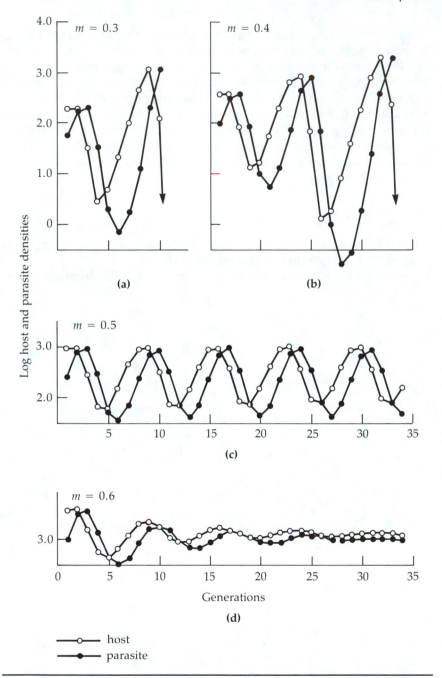

Figure 14.3   *Population models showing the increasing stability as the mutual interference constant* m *is increased from 0.3 in* (a) *to 0.6 in* (d).   (Redrawn from Hassell and Varley 1969.)

particularly when handling time is short, egg limitation of parasites or satiation of predators may become more important than actual handling time. Furthermore, some parasites change their behavior if other individuals of the same species are nearby or even leave the area, thus reducing their potential area of discovery.

The relationships between area of discovery and parasite density were investigated by M.P. Hassell and G.C. Varley (1969), who proposed

$$\log a = \log Q - m \log P$$

where Q is the "quest constant" or area of discovery when the parasite density is one, and m is the mutual interference constant. The outcome of some models based on quest theory are completely different from those of a Nicholsonian model, because they include *interference*.

The stability of the quest models increases with greater values of m as shown in Fig. 14.3. More importantly, there is a wide range of values for Q and m that allow the coexistence of two or more parasite species on a single species of host, a very common phenomenon in nature, yet one not accounted for in Nicholson's models.

Despite the added realism of newer models, there remain some serious flaws, the most important of which is the assumption of random search by predators and parasites. There is little evidence that random search is generally prevalent. Most parasitoids are attracted by the scent of their prey, or they remain in areas where they have previously been successful. In either case, the searching population will tend to aggregate in areas of high host density. Unfortunately, nonrandom search is difficult to incorporate into models, though field studies give the impression that it tends to increase the stability of the system.

14.3 *The Spread of Disease*

The dynamics of microparasites, the spread of disease, have been modelled by May (1981) and Anderson (1982). Instead of examining basic reproductive output, R_o, of a parasite, scientists more usually observe R_p, the average number of new cases of a disease that arise from each infected host. The reason is that in **epidemiology**, the study of the spread of disease, the number of infected hosts is the most important factor, not the number of parasites. The transmission threshold, which must be crossed if a disease is to spread, is therefore given by the condition $R_p = 1$. For a disease to spread, R_p must be greater than 1 and for a disease to die out it must be less than 1. The term R_p is influenced by

- N, the density of susceptibles in the population
- B, the transmission rate of the disease (a quantity correlated with frequency of host contact and infectiousness of the disease)
- f, the fraction of hosts that survive long enough to become infectious themselves
- L, the average period of time over which the infected host remains infectious

The value R_p is related to these factors by the equation

$$R_p = BNfL$$

Two generalizations can thus be made.

1. As L, the period of the host's life when it is infectious, increases, R_p increases. Some hosts remain infectious long after they are dead. This is especially true in plant parasites, which leave a residue of resting spores.
2. If diseases are highly infectious (have large Bs) or are unlikely to kill their hosts (have large fs), R_p increases. An efficient parasite therefore keeps its host alive.

By rearranging the above equation, we can obtain the critical threshold density N_T (where $R_p = 1$), the number of infected hosts needed to maintain the parasite population:

$$N_T = \frac{1}{BfL}$$

Now, if B, f, or L is large, N_T is small. Conversely, if B, f, or L is small, the disease can only persist in a large population of infected hosts. Cockburn (1971) has provided some interesting medical and anthropological evidence to back up these ideas, at least for humans. Measles, rubella, smallpox, mumps, cholera, and chicken pox, for example, probably did not exist in ancient times (Black 1975), because the hunter-gatherer populations were small, bands of 200–300 persons at most. These bands were too small to constitute reservoirs for the maintenance of infectious diseases of the types described. In a small population there should be no infections like measles, which spreads rapidly and immunizes a majority of the population in one epidemic. Instead, typhoid, amoebic dysentery, pinta, trachoma, or leprosy were probably the common afflictions, diseases for which the host remained infective for long periods of time. Malaria and schistosomiasis would still have been very prevalent because of the presence of outside vectors to serve as additional reservoirs. Paradoxically, civilization has increased the kinds and frequencies of diseases suffered by humans, by enlarging the source pools and by domestication of certain animals. Most modern diseases have arisen because of intimate association with animals and their viruses (Foster and Anderson 1979). Smallpox, for example, is very similar to the cowpox virus, measles belongs to the group containing dog distemper and cattle rinderpest, and human influenza viruses are closely related to those found in hogs. AIDS, of course, is similar to a virus found in monkeys in Africa.

14.4 *The Effects of Parasites on Host Populations*

Once again the best way to find out the effect of parasites on the population abundances of their hosts is to remove the parasites and to reexamine the system. Once again, this has rarely been done, even less frequently than predator removal, perhaps because of the small size of many parasites, which makes them difficult to exclude.

Evidence from the field, however, suggests that parasites do have a substantial impact on their hosts. In North America, chestnut blight has virtually eliminated chestnut trees. In Europe and North America, Dutch elm disease has devastated elms. In Italy, canker has had severe effects on cypress. The population dynamics of bighorn sheep in North America are dominated by a massive mortality resulting from infection by the lungworms *Protostrongylus stilesi* and *P. rushi*. This parasite predisposes the animals to pathogens causing pneumonia. A fetus can become infected through the mother's placenta, and mortality in lambs can be enormous (Hibler, Lange, and Metzger 1972). The lungworm-pneumonia complex is regarded as one of the most influential mortality factors in many sheep populations. Mortalities of 50–75 percent are reported (Uhazy, Holmes, and Stelfox 1973), and one might almost regard the lungworms as organizer species in early successional communities of western North America (see Section Five). Sometimes epidemics can be even more severe when the parasite is less specific. Rinderpest, caused by a virus, has at least 47 natural artiodactyl hosts (Scott 1970), most of which occur in Africa. The disease is usually fatal in buffalo, eland, kudu, and warthog and less fatal in bushpig, giraffe, and wildebeest. Other species, such as impala, gazelle, and hippopotamus, appear to suffer little. A major epidemic swept through Africa in 1896, leaving vast areas uninhabited by certain species, and even today distribution patterns reflect the impact (May 1983). For example, zebra were exterminated in an area of the Elizabeth National Park in Uganda, and they still had not recolonized the area by 1954 (Pearsall 1954). Because wildlife was eliminated, other parasites were affected, too. Tsetse flies became absent from large areas of Africa south of the Zambesi River (Stevenson-Hamilton 1957). Therefore, large areas became free from trypanosomiasis, sleeping sickness, a disease borne only by the tsetse fly. One parasite, rinderpest, thus had a severe impact on the pattern of life in an area. Furthermore, in the absence of tsetse flies, humans and cattle could move in, supplanting wildlife even further. Prins and Weyerhaeuser (1987) described two recent and major epidemics in wild mammals in a national park in Tanzania, east Africa. An anthrax outbreak lasted almost a year and killed more than 90 percent of the impala population, and a rinderpest outbreak of just a few weeks killed some 20 percent of the buffalo. The conclusion is that epidemics have a more severe impact on these populations than does predation. This is a disconcerting finding for conservation biology, for it means that certain populations on small reserves could be wiped out by disease unless recolonization is encouraged.

Another interesting situation exists in North America. The usual host of the meningeal worm *Parelaphostrongylus tenuis* is the white-tailed deer, *Odocoileus virginianus*, which is tolerant to the infection. All other cervids and the pronghorn antelope are, however, potential hosts, and in these species the worm causes severe neurological damage, even when very small numbers of the nematode are present in the brain. This differential pathogenicity of *P. tenuis* makes the white-tailed deer a potential competitor with other cervids because they cannot survive in the same area with white-tails. The deleterious effects of the parasite probably include direct mortality, increased predation, and reduced resistance to other disease. The activities of humans have altered the normal distribution pattern of the white-tailed deer. As northern forests were felled, the deer expanded their range from a stronghold in the eastern United States, eventually coming into contact with moose. In Maine and Nova Scotia white-tailed deer have replaced moose as the major cervid (Anderson 1972). They have also replaced mule deer and woodland caribou in some parts of their ranges, and reintroduction of caribou into regions now occupied by white-tailed deer is probably impossible (Anderson and Prestwood 1979). So, apart from direct mortality from parasites, competitive interactions between populations can be mediated by the action of parasites. This phenomenon is similar to that discussed in Chapter 11, in which Park (1948) compared competing populations of flour beetles with and without the parasite *Adelina triboli*.

Cornell (1974) has argued that distributional gaps between bird species, where apparently favorable habitat exists, are maintained by the capacity of vectors to travel between populations. The rationale is that each population has a pathogen to which it is adapted but the other species is not. This concept has led to the idea that populations might compete for parasite-free space. Because the same type of phenomenon might operate for predators, the idea of enemy-free space is more widely circulated. Crosby (1986) argues that the Old World diseases smallpox, measles, typhus, and chicken pox were so devastating to peoples of the New World that the Old World invaders, who had a limited measure of immunity, found subjugation of the people much easier.

Control of human parasites is, of course, of paramount importance to many people. Effective control usually involves a thorough knowledge of the life cycle of the parasite involved. Often two hosts are involved. This is true for malarial parasites *(Plasmodium)*. The life cycle of this protozoan alternates between two hosts, an *Anopheles* mosquito and a vertebrate. For four species of *Plasmodium,* the vertebrate is man (Fig. 14.4). People attempt to control this disease primarily by reduction of mosquito populations. The usual method is drainage of swamps and wet areas where the eggs are laid and the larvae develop. Sometimes, by alteration of agricultural practices to increase rice growing, further wet areas are provided, and mosquito populations increase (Desowitz 1981). Protozoan parasites are not unique in using an invertebrate and a vertebrate host to complete development. Certain blood flukes of the genus *Schistosoma* use snails and people as alternate hosts. In people, their attack causes a disease known as schistosomiasis or, in Africa, bilharzia.

14.5 *Parasites and Biological Control*

Not all parasites are detrimental to humans. Many are used as an effective line of defense against insect pests of crops, although only about 16 percent of classical biological-control attempts qualify as economic successes (Hall, Ehler, and Bisabri-Ershadi 1980), so we cannot abandon chemical control as yet. The theory of pest control by means of the release of parasitoids is so far not very advanced. Huffaker and Kennett (1969) have suggested five necessary attributes of a good agent of **biological control**:

- General adaptation to the environment and host
- High searching capacity
- High rate of increase relative to the host's
- General mobility adequate for dispersal
- Minimal lag effects in responding to changes in host numbers

Although these attributes seem necessary for a good control agent, they are clearly not sufficient. So far, the application of biocontrol agents has been carried out by a hit-or-miss technique rather than by a sound biological method. Some authors consider that this trial-and-error method probably makes best economic sense, given the high cost of research into the biology of natural enemies (van Lenteren 1980). Others have recommended new techniques, for example presenting novel parasite-host associations as the most likely avenue for control where hosts have not had the opportunity to evolve complex defenses against these parasites from foreign lands (Hokkanen and Pimentel 1984). Arguments still rage as to whether it is better to introduce one parasite at a

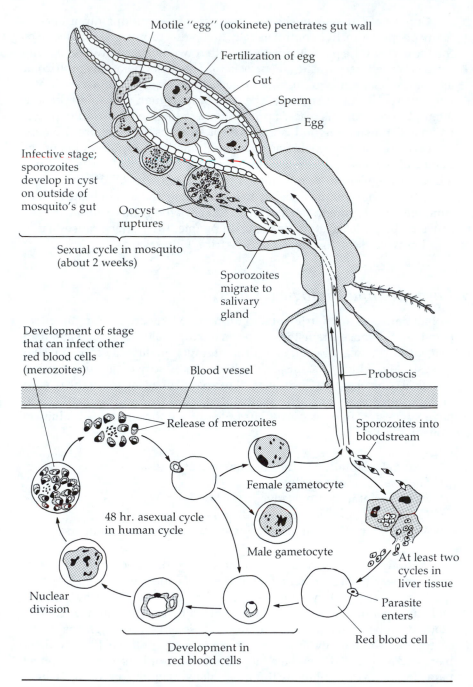

Motile "egg" (ookinete) penetrates gut wall

Fertilization of egg

Gut

Sperm

Egg

Infective stage; sporozoites develop in cyst on outside of mosquito's gut

Oocyst ruptures

Sexual cycle in mosquito (about 2 weeks)

Sporozoites migrate to salivary gland

Development of stage that can infect other red blood cells (merozoites)

Blood vessel

Proboscis

Release of merozoites

Sporozoites into bloodstream

Female gametocyte

48 hr. asexual cycle in human cycle

Male gametocyte

At least two cycles in liver tissue

Nuclear division

Parasite enters

Red blood cell

Development in red blood cells

Figure 14.4   *Diagram of the malaria* (Plasmodium falciparum) *life cycle.*

time or many. The problem is that, if more than one enemy is introduced, competition between parasites could ensue, lessening the overall level of control. Ehler and Hall (1982) provided evidence from a world review of 548 control projects to show that this phenomenon could be a problem; the more parasites released, the lower the rate of establishment, although this analysis was disputed by Keller (1984). It is probably fairest to say that in this, as in so many ecological situations, the jury is still out. Stiling (1990) reviewed the methods for determining success in biological control and determined that the result obtained was very much dependent on the method used. He also suggested that, of a broad array of biological characteristics, the factor of greatest importance was the climatic match between the control agent's locality of origin and the region where it was to be released. The result stresses the value of studies in physiological ecology (Chapter 8).

It is noteworthy that the shotgun introduction of parasites for biological control is a risky business for other, nontarget species and should be avoided. Howarth (1983), for example, lamented the reduction of native Hawaiian lepidopterans, partially due to wasp species introduced for biological control. He calls for a more narrowly focused release effort rather than a hit-or-miss campaign. Such a concern is even more important when release of insect enemies to control weeds is considered. In this case, stringent host-specificity tests are performed to ensure introduced insects will not turn to feed on valuable crops, even in times of starvation.

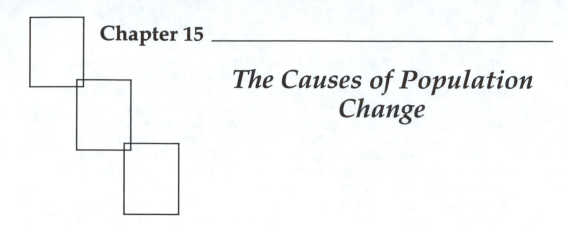

Chapter 15

The Causes of Population Change

If parasitism, predation, competition, and weather can all affect the population densities of living organisms, which effect is most important? Comparing the strengths of mortality factors is one of the most important pursuits in ecology, yet, sadly, few scientific papers are broad enough in scope to be able to do so. Instead, most papers are focused on a single issue.

15.1 Key-Factor Analysis

One of the best techniques for empirically comparing the importance of the effect of predators, parasites, or other factors on the size of field populations is key-factor analysis (Morris 1959). In this method, population density is expressed on a logarithmic scale and plotted against time, because population changes are more easily visible in log plots. The killing power, or k value, of each source of mortality is then given by log $N(t)$ − log $N_{(t + 1)}$, where $N_{(t)}$ is the density of the population before it is subjected to the mortality factor, say microsporidian disease, and $N_{(t + 1)}$ is the density afterwards. The k value that most closely mirrors overall generation mortality (K) is then termed the *key factor*. More precisely, individual sources of mortality or k values can be plotted on the *y* axis against total mortality, K, and the key factor is then the source of mortality with the biggest correlation coefficient, *r*, with K (Podoler and Rogers 1975). In oak winter moths in England (Fig. 15.1), a species probably subject to the most comprehensive key-factor analysis ever done (Varley, Gradwell, and Hassell 1973), the key factor is overwintering loss (Fig. 15.2). Young winter moth larvae emerge in the spring and feed on the newly developing foliage. To find fresh leaves, they often disperse by ballooning away on silken threads to new trees. This is obviously a chancy business, and many larvae are lost; hence the high k values. Overwinter loss is a key factor in many animal populations because of severe climatic stress. Key

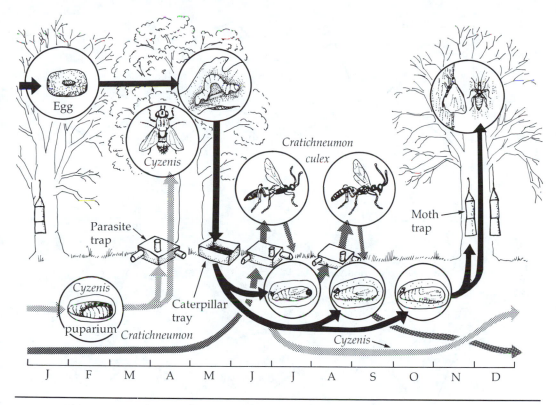

Figure 15.1 *Life cycle of the oak winter moth and sampling methods.* Adult female winter moths were counted in moth traps on the tree trunks; their larvae were counted in the caterpillar trays into which they fell when prepupal. Larvae of the parasite *Cyzenis* were counted by dissection of the fallen caterpillars, and adult *Cyzenis* and *Cratichneumon* were counted upon emergence from the soil into the parasite traps. (Redrawn from Varley 1971.)

factors can usually be detected only by analysis of many generations; in **univoltine** animals this analysis may take many years.

15.2 Density Dependence

Overwinter mortality in oak winter moths tends to disturb population densities away from the mean. Are there any compensatory mechanisms that tend to return populations to a mean value? The only way in which population densities might be regulated about a mean value is by some negative feedback process, commonly called a **density-dependent factor.** Density dependence can be determined from a plot of the k values of each source of mortality against the logarithm of the density of the life stage on which it acts. If a positive slope results and mortality increases with den-

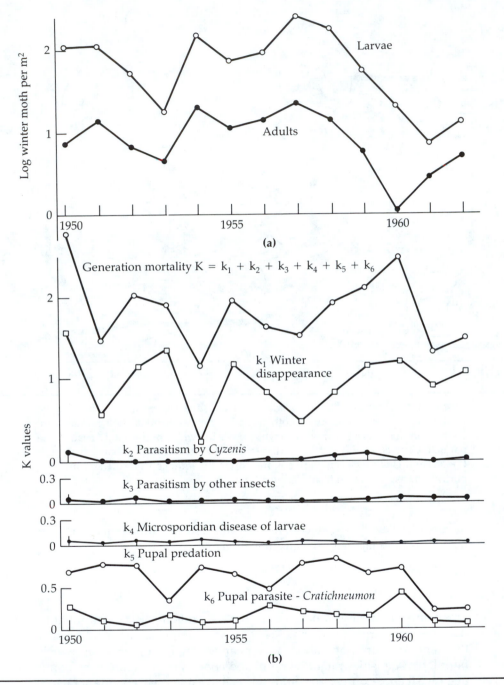

Figure 15.2 *Key-factor analysis of winter moth population changes.* **(a)** *Winter moth population changes expressed as generation curves for larvae and for adults.* **(b)** *Changes in the mortality, expressed as k values, showing that the biggest contribution to changing the generation mortality K comes from changes in* k_1, *winter disappearance.* (Redrawn from Varley, Gradwell, and Hassell 1973.)

sity, then the k factor is tending to affect less of sparse populations and more of dense ones and is clearly acting in a density-dependent manner. In the winter-moth example, pupal predation of overwintering moth pupae in the ground (by beetles) is the density-dependent factor (Fig. 15.3). In Canada the winter moth had been introduced in the 1930's, and parasites from Britain were introduced to control it, with great success it seemed. Surprisingly, in the similar environments of Britain and Canada, different factors seemed to be responsible for population control. In a brilliant reanalysis by Roland (1988), however, predation was again actually shown to be the density-dependent factor, though parasites helped because parasitized pupae lay longer in the soil and were available to predators for a longer period. Thus a single factor was actually shown to cause density dependence in one host in both countries.

Density dependence can also be detected where adverse effect, expressed as percent or raw numbers, is plotted against population density (Fig. 15.4). In such plots, factors that appear not to change with density, and thus do not contribute to population regulation, are termed **density independent.** Those sources of mortality that decrease with increasing population size are called *inversely density dependent.*

In some situations the effect of density-dependent factors on a population is delayed by one or two generations. In such cases, when percent mortality is plotted against host density and the points are joined in a time series, the **time lag** results in a counterclockwise spiral (Fig. 15.5). This phenomenon, termed *delayed density dependence,* is particularly common where cycles of host-parasitoid interactions occur over many generations.

Of course not all population densities are regulated by density-dependent predation or parasitism. The percentage of field studies of insects showing density-dependent parasitism is quite low, on the order of 25 percent (Stiling 1987; Walde and Murdoch 1988). In the absence of such factors, however, the question remains of what other regulatory processes operate. For many herbivores the answer may be plant quality. Bernays and Graham (1988) contend that in most cases generalist natural enemies are even more important than that. A group of nine articles in *Ecology* (1988, volume 4) that followed Bernays and Graham's article contested this point and stressed the importance of host quality. The density-dependent factor that is most important is certain to be different for populations of different species at different times, at least for insects. Karban (1989) examined the effects of different sources of mortality on three herbivores of the seaside daisy, *Erigeron glaucus.* For spittlebugs, *Philaneus spumarius,* damage caused by caterpillars increased rates of desiccation and mortality of nymphs. Thus competitive effects were strongest for this interaction. For the caterpillars, *Platyphilia williamsii,* predation by savannah sparrows, *Passerculus sandwichensis,* was the strongest effect, and protecting caterpillars from sparrows by enclosing colonies under

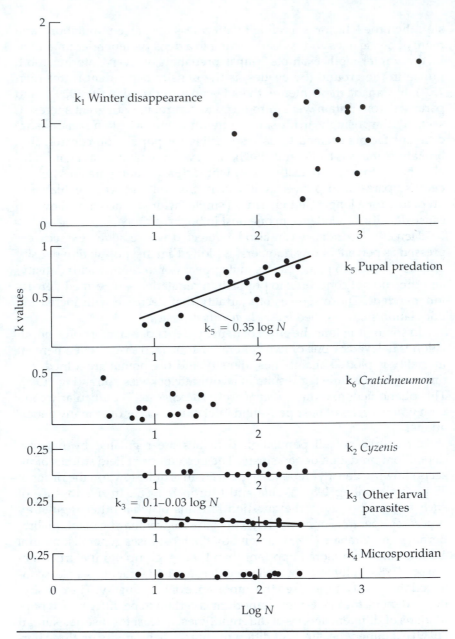

Figure 15.3 K *values for the different winter moth mortalities plotted against the population densities on which they acted.* k_1 and k_6 are density independent and vary quite a lot; k_2 and k_4 are density independent but are relatively constant; k_3 is weakly inversely density dependent; and k_5 is quite strongly density dependent. (Redrawn from Varley, Gradwell, and Hassell 1973.)

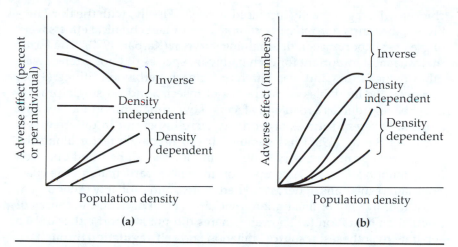

Figure 15.4 *Types of response to changes in population density.* **(a)** *Expressed as a percentage response to increasing population.* **(b)** *Expressed as the numbers of individuals affected with increasing populations.*

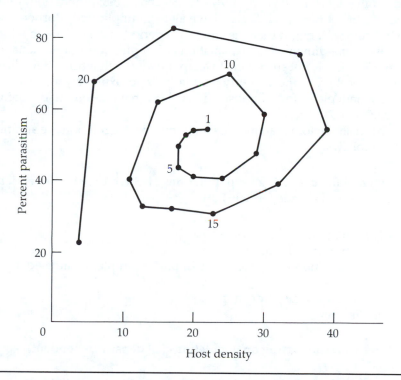

Figure 15.5 *An example of a delayed density-dependent relationship using the observed results from Fig. 14.1.* (Redrawn from Varley, Gradwell, and Hassell 1973.)

chicken-wire cages greatly increased survival. Finally, with the third herbivore, the thrips *Apterothrips secticornis*, host-plant chemical effects were of greatest importance in determining survival (Karban 1987). The factor that was most important for each herbivore species did not interact with other biotic factors and was different for each herbivore. Stiling (1988) reviewed the life tables of 58 species of insects, and no single process could be regarded as a regulatory factor of overriding importance in most studies. Different sources of mortality were important in different systems. As Strong (1988) has summed it up, no single factor along the gamut from plant chemistry to abiotic influences can be ruled out even for a minority of cases. A complex of influences participate in the coactions and regulations of herbs and herbivores, and probably other organisms, too. By way of illustration, Sinclair's (1986) data on the causes of population limitation for snowshoe hares did not support either of two separate hypotheses, **resource** limitation and **self-regulation** through social behavior. Results corroborated a multifactor hypothesis—that is, social behavior causes differential survival through weight loss only when food is limited. The population of Thompson's gazelles in the Serengeti National Park, Tanzania, declined by almost two-thirds over a 13-year period, from 660,000 in the early 1970's to 250,000 in 1985. Predation, interspecific competition, and disease have all been implicated (Borner et al. 1987). Ecology, it might seem, is entering a period of pluralism, in which simplistic, one-dimensional explanations of population phenomena are not sufficient. It is not surprising to learn, however, that this "new direction" in ecology appears fairly regularly, approximately once every decade, and pleas for pluralism in ecology have been made before (McIntosh 1987).

The major factors that impinge on population regulation are summarized as follows:

□ **External factors acting on populations** (1–3 are density dependent; 4–5 are density independent)

1. Predation, including parasitism
2. Competition for food
3. Competition for space, more important in plants and sedentary animals
4. Random stochastic change
5. Weather

□ **Internal factors acting on populations** (all density dependent)

1. Pathological effects in response to crowding, shock disease, adrenopituitary exhaustion

2. Dispersal polymorphism, switched on in crowded conditions (locusts)
3. Evolution of social checks on population size, possibly through territorial behavior

External factors acting on populations, such as predation, parasitism, and competition for food or space, have largely been discussed already. Random stochastic change can have a big influence, as evidenced by the stochastic mathematical treatment of population growth outlined in Chapter 9. The effects of the weather are profound and certainly influence population distribution patterns. That weather also severely affects population abundances of plants and animals within their range is underappreciated but certainly true (Chapter 8), as Andrewartha and Birch (1954) have argued specifically for populations of tiny thrips insects infecting flowerheads in Australia and for other plants and animals in general. They suggest temperature and rainfall have the greatest effect on population numbers. For example, climatic variation has often been implicated in the outbreaks of forest insect pests (Martinat 1987), and severe winters nearly always take a heavy toll of bird populations.

The internal factors acting on populations are many, and the cause of self-regulation of populations was championed by Wynne-Edwards (1962), who later softened his position (1977) (see also Chapter 5). Conscious self-regulation of population, however, smacks of group selection, a concept shown in Chapter 5 to be highly dubious. Individual selection is much more likely to occur, as an individual is unlikely to stop feeding or reproducing for the good of the species. Plants and animals have no crystal ball with which they can see into the future; each strives to leave as many offspring as possible. Nevertheless, through the effects of intraspecific competition, several mechanisms exist by which populations can apparently regulate themselves. These include

□ Maintenance of strict territory sizes in habitats, which, in birds, is what first stimulated Wynne-Edwards' ideas
□ Pathological effects like shock disease, which can kill individuals already under strain and which is especially prevalent in rodents kept at high densities (Christian 1971)
□ The existence of dispersal **polymorphisms,** "switched" on in times of crowding, especially prevalent in insects such as aphids and locusts (Chapter 8)

Spreading the Risk

Related to the dispersal polymorphism mentioned above is the fact that plants and animals may disperse widely anyway, even in the absence of crowding. Such behavior is analogous to not putting all one's eggs in one basket. Thus a searching parasite may not always oviposit in a density-dependent fashion, laying more eggs in dense concentrations of caterpillars; she may save some of her eggs to oviposit elsewhere, even in seemingly suboptimal places. The evolutionary reason may be that some local catastrophe could occur in the best area, wiping out all her progeny. This phenomenon has become known as *spreading the risk* (den Boer 1968 and 1981) and, despite the attractiveness of neat, intuitively correct, mathematical models of density dependence, spreading the risk may provide the best explanation for the apparently random and chaotic patterns that are so often observed in the field. In some ways, arguments against density dependence follow those arrayed against optimality theory (see Chapter 6); plants and animals are unlikely to behave optimally or in a density-dependent fashion in the face of a huge array of conflicting pressures. Their life-history patterns are adaptations to survival, not maximization of fitness. Emigration and immigration are often thought to cancel each other out in terms of population effects, and **dispersal** is thus rarely studied. Studies that have carefully looked at dispersal dispel this myth. In a long-term study of great tits in Oxford, England, a bird with a supposed "low level of dispersal," 57 percent of the breeding birds were immigrants rather than residents (Greenwood, Harvey, and Perrins 1978). In a review of dispersal in mice and voles, there were many studies where more than 50 percent of the population emigrated each week or where the population was made up of more than 50 percent recent immigrants (Gaines and McClenaghan 1980).

Spatial Phenomena

Analysis of population change is not only restricted to examination of a long series of population data through time. Similar problems exist on a spatial scale as well as on a temporal one. For example, variations of population densities of herbivorous insects commonly exist from host plant to host plant in a great many different systems. Africa's abundant large herbivores are very heterogeneously distributed, both geographically and regionally. Within a region, some localities contain dense animal concentrations, although areas nearby may be virtually unoccupied (McNaughton 1988). Stiling and colleagues have worked on differences in densities of a leaf-mining moth, *Stilbosis quadricustatella*, among different individuals of its host tree, sand live oak, *Quercus geminata*. Levels of infestation range from less than 1 percent of the leaves mined on some trees to over 70 percent on others. The advantages of studying spatial phenomena in this system are many. The main ones are that individual leaf miners are restricted to feeding in the leaf tissue between the surfaces of just one leaf and that leaf mines themselves leave excellent records of the fates of the miners. As Photo 15.1 shows, larvae that emerge successfully cut crescent-shaped holes in the lower surface of the mine and drop to the

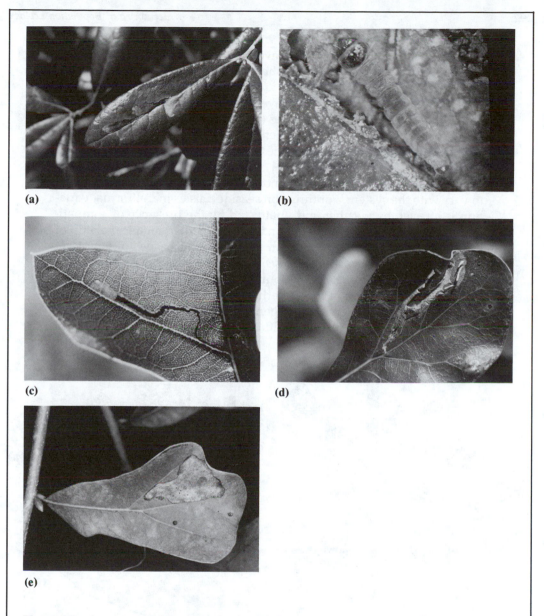

(a)

(b)

(c)

(d)

(e)

Photo 15.1   *A system useful for the study of population density control. (a) Four unharmed* Stilbo-sis quadricustatella *leaf mines on a* Quercus geminata *leaf. (b) A developing* Stilbosis *larva inside a leaf. (c) A* Stigmella *leaf mine from which the mature larva has escaped through a semi-circular exit hole. (d) A* Tischeria *mine on water oak ripped open by ants. (e) A* Cameraria *mine on water oak with five small holes in its upper surface, indicating that parasites have emerged from this mine. Because the state of the mine serves as a record of the fate of the miner, researchers on this system can determine the relative importance of the various factors that influence density without actually observing every instance of disease or predation. (Photos (a), (b), (d), and (e) by P. Stiling. Photo (c) by Daniel Simberloff, Florida State University.)*

forest floor to pupate; parasitoids leave circular holes in the upper surface of the leaf; and predators rip open the mine in dramatic fashion. In addition, leaves can be ground up and analyzed for quality by means of assays for amino acids and nitrogen levels. Levels of parasitism, predation, and amino acids differ between trees (Fig. 15.6), but none correlates negatively with miner densities between trees. Variation in leaf abscission rates are the likely cause of why some trees are preferred over others (Stiling, Brodbeck, and Simberloff, 1991). Just as with temporal studies, the reasons for population variation among different spatial areas are likely to vary with the system. For thrips on seaside daisies in California, variation is affected by weather and unaffected by the presence of competitors, but it is due mainly to variation in the particular genetic makeup of the plants (Karban 1987), though in what particular aspect it is not yet clear. Also, from the work on *Stilbosis*, no clearcut patterns yet emerge. In a similar fashion, Strauss (1989) found it difficult to detect reasons for plant-to-plant variation in insect herbivores on sumac clones in Minnesota, although she observed both genotype and environmental effects. These results illustrate just how difficult it is to pinpoint the causes of

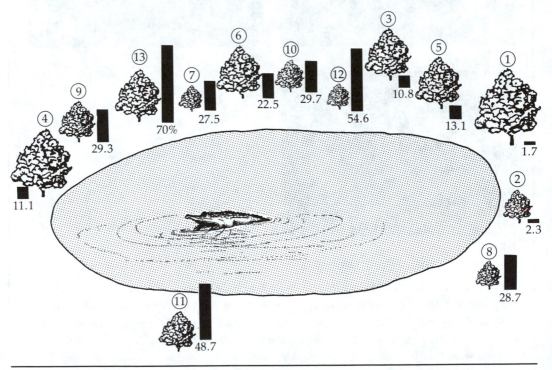

Figure 15.6   *Tree-to-tree* (Quercus geminata, *sand live oak) variation and percentages of leaves mined by larvae of a moth,* Stilbosis quadricustatella, *around a lake in northern Florida.* Numbers are code numbers for individual trees, and percentages and bars reflect percentage and intensity of mining on leaves.

population change even on a spatial scale, a point well known to most agricultural practitioners trying to combat hordes of insect pests.

Schoener (1986) has attempted to provide some generalities about sources of mortality by ranking communities along certain environmental and biotic axes. For example, he has listed six environmental factors: severity of physical stress, trophic position in food chain, resource input (from closed, a lake, to open), spatial fragmentation (fragmented to continuous), long-term climatic variation (high to low), and partitionability of resources (low, bare rock, to high, a complex leaf or prey of different sizes). To characterize organisms, Schoener has developed another set of axes: body size (small to large), recruitment, generation time (short to long), individual motility (sessile to mobile), and number of life stages (low to high, for example, holometabolous insects). He has proposed that, at some time in the future, it will be possible to describe communities in terms of such axes, with the result that, for "similia-communities" (his definition), say short-lived ephemeral marine algae, similar processes would shape the community. In this case, the algae are less affected by competition than by predation (herbivory). The final analysis of such a technique is a long way into the future and is perhaps more in the realm of community ecology (see Section Five) than of population ecology.

Section Five

Community Ecology

Why do certain climatic regimes produce similar communities of plants and animals all over the world? Why are there many more different types of species in tropical forests than in temperate forests? Community ecology is the study of questions like these. Photos: P. Stiling.

Why are there many more different types of species in tropical forests than in temperate forests? Why are there generally a few common species and lots of rare species in a given area? Why do we see weeds and herbs in an old field gradually being replaced by shrubs and trees? How are populations of different species linked together to form a food web, and how do energy and nutrients flow between members of that web?

Community ecology deals with populations not in isolation but together as a whole in the natural areas where they interact. Communities can occur on a wide range of scales and can be nested—the tropical forest community encompasses the community living in the water-filled recesses of bromeliads, which in turn encompasses the microfaunal communities of cellulose-digesting insects' guts.

Once an **association** between species has been recognized and a community identified, the usual comparison is conducted by means of analysis of species diversity, an index of the number of species present, and their abundance. We can also describe the basic type of community present, determine the **trophic** structure (who eats whom), and determine the relative **biomass** of individual components. The basic comparison, however, is still that of species diversity.

The understanding of communities can become particularly important in modern agriculture, where emphasis today is on the integration of pasture, tree crops, and livestock (so-called *agroforestry*), especially in tropical regions. Problems associated with particular facets of the community may develop; for example, in tropical Asia cattle can cause damage to young trees, and more importantly, their dung can serve as a breeding place for rhinoceros beetles, one of the major pests of coconut (Reynolds 1988). Adjusting the species mix for optimum yields is a complex problem (Young 1988).

Some ecologists view community ecology as a poor excuse for a science, arguing that communities are no more than the sum of their individual components. They claim that a grassland prairie community is simply a collection of populations of several species of grasses, each of which has the same environmental requirements. Others view the integration of populations as unique communities with special properties, in much the same way that salt has unique characteristics (taste, for example) and does not simply combine the attributes of sodium and chlorine. Such unforeseen characteristics of a community are often termed *emergent properties.* This concept may be important in applied situations, because human interventions may alter not only the population biology of certain species but also these higher-order interactions. In this view some members of a community are thought to change their habitat subtly and fractionally, enabling other species to exist. For example, in a log community, the first invaders attack and weaken the wood slightly, facilitating the entry of other organisms, though both can be found at the same time and can be thought of as part of the same community.

Chapter 16

The Main Types of Communities

The concept of the **community** is valid over a wide range of scales—we can refer to the community of a rotting log, the aquatic community of a lake, or the desert community. One of the strongest arguments for the existence of communities is that similar environmental conditions produce similar growth forms in plants. Desert plants all over the world have evolved a series of morphological features like small leaves (see Chapter 8); these adaptations can be found in different plant families in deserts all over the world. Such large-scale communities are referred to as **biomes,** and the major ones are tropical **forest,** temperate forest, **grassland, desert, tundra,** coral reef, coastal shelf, and open ocean. The meteorological conditions that produce different biome types, together with a rudimentary map of their global distribution patterns, are shown in Fig. 16.1 and 16.2. Each biome type shows general features characteristic of the community. Thus a tropical forest community may show a distinct vertical structure, whereas a single population of one plant species would not. Most communities do, in fact, show a vertical structuring. Leaves are very efficient at intercepting light—little passes through them. In low light, leaves are arranged in a virtual monolayer, and not enough light passes through to permit other plants to grow underneath. In high-intensity light, more light is reflected from leaf to leaf, and more penetrates between the leaves, so some leaves grow beneath the upper canopy to utilize this light. This phenomenon has been rigorously approached by Horn (1971), who measured the approximate numbers of layers of leaves in an oak-hickory forest in New Jersey and found numerical agreement with these ideas. In canopy trees, the mean number of layers of leaves was 2.7, in understory trees 1.4, in shrubs 1.1, and in ground cover 1.0. Light, it must be remembered, is, in times of abundant water and nutrients, a **limiting resource,** and one would expect leaves to be optimally placed to intercept it. This limitation often means that the successful plants are those that grow a little taller than their neighbors. Height of leaves in a canopy is usually greater where more herbaceous

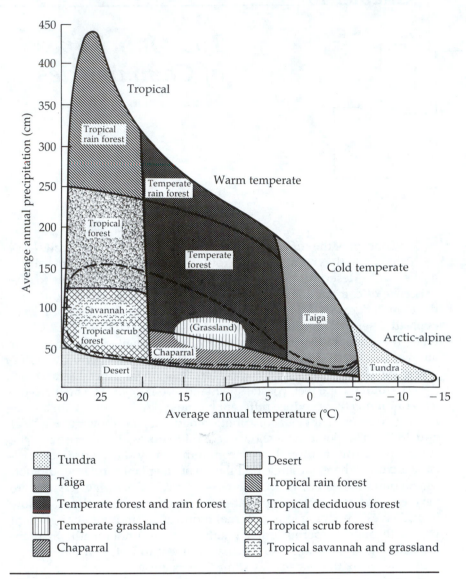

Figure 16.1 *Temperature and precipitation conditions that give rise to the world's major biomes. Dashed line represents the area where soil type and fire frequency determine whether woodlands or grasslands occur in the area. (Modified from Whittaker 1975.)*

cover exists (Fig. 16.3, p. 301). Plants on barren ground suffer no competition for light and remain short. There are trade-offs, of course, in growing tall, because more energy must be devoted to stem support (Givnish 1982). This relationship is illustrated in Fig. 16.4 (p. 302). Aquatic communities also show distinct vertical differentiation. For example, lakes and

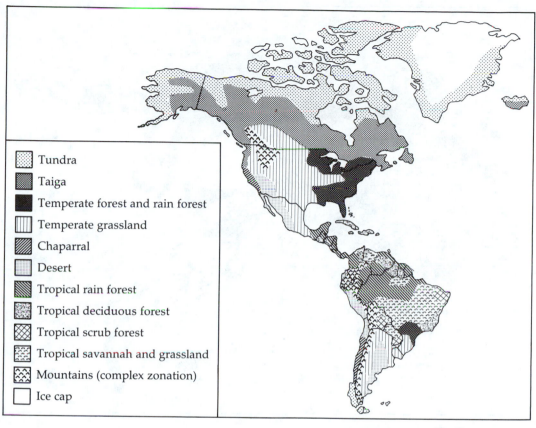

(Continues on p. 300)

Figure 16.2 *Distribution of the major biomes throughout the world.* (Modified from Odum 1971.)

oceans can be divided into **epilimnion, thermocline,** and **hypolimnion** on the basis of temperature or into **euphotic** and noneuphotic zones on the basis of light availability (see chapter sections 16.8 and 16.9).

As the complexity of the vegetational structure of a community increases, so does the animal diversity, in terms of numbers of species. Erwin (1982 and 1983) has argued that the richest diversity of insects on Earth exists in the canopies of tropical rain forests. Some animals may indeed migrate between different layers of the canopy. Vertical migration is a common phenomenon in communities of **zooplankton,** where individuals may migrate over 100 m from the bottom in daylight to the surface at night. Although there are many possible reasons for this phenomenon, the chief one is probably the daytime avoidance of visually oriented surface-feeding predators such as fish and whales (Iwasa 1982).

Communities can also differ in their seasonality. It is often assumed

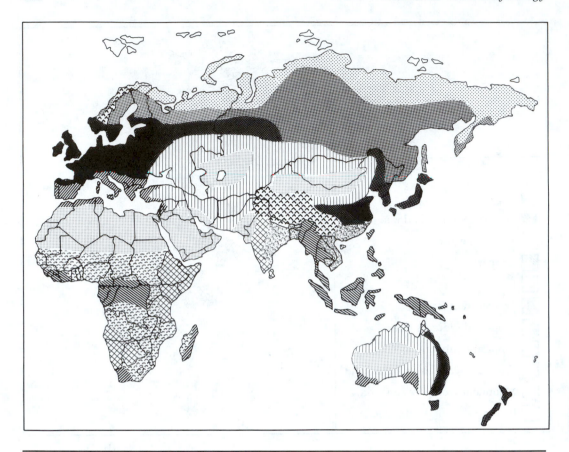

Figure 16.2   Continued

that tropical communities are aseasonal and that members are able to reproduce year-round. Although it is true that there are no hot and cold
spells in tropical environments, seasonality is often based on wet and dry
periods. Even in the Amazon, there are distinct wet and dry seasons; indeed, few forests exist in the world on which approximately the same
amount of rain falls in every month of the year. The main effect of seasonality in the tropics is that leaves are generally shed quickly and replaced
in the dry season (of course some plants produce leaves at a constant rate
all year). Thus, although some leaves are available year-round for generalist foliage feeders, leaf supply still changes seasonally. Young succulent
leaves appear at specific times, and their appearance is reflected by a distinct seasonality of many monophagous tropical insects (Wolda 1983;
Wolda and Broadhead 1985). Wolda (1986, p. 93) has stated, ''All available information suggests that tropical animals do not differ from temper-

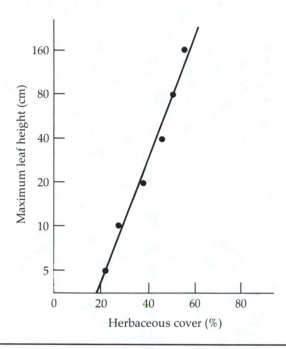

Figure 16.3 *Evolution of height in forest herbs: relationship of average leaf height to the amount of herbaceous cover in a transect from dry oak woods to floodplain forest in Virginia.* (Redrawn from Givnish 1982.)

ate zone species in terms of temporal stability." Flowering is also seasonal for individual species, although each species may flower at a different time in the year, enabling pollinators, hummingbirds and insects, to exist year-round. Tropical forests are never a blaze of color, as temperate forests are in the spring, and the paucity of blooms can be quite a shock to the first-time visitor to the tropics.

16.1 *Tropical Forest*

The most species-rich and complex communities on Earth are tropical rain **forests,** an example of which is shown in Photo 16.1 (p. 303). Studying these jungles is quite difficult, as is eloquently expressed by Sanderson (1945 p. 124):

> The principal difficulty encountered in studying a jungle is that you can never see it. This may sound absurd. An average jungle is about 100 feet tall, and is shaped like a much-flattened and inverted saucer, so that its edges slope down all around to meet the earth. When you enter it you become lost. You can't see the wood for the trees, and you often can't

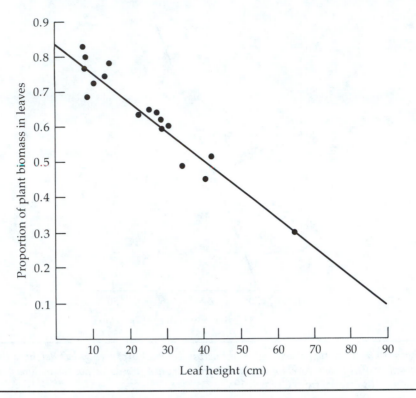

Figure 16.4 *Allocation of biomass to leaves or to support tissue in 18 species of forest herbs.* (Redrawn from Givnish 1982.)

see the trees for the creepers, lianas, and epiphytic green plants that grow all over them. If you fly over a jungle in an airplane you see even less, for nothing but a gently undulating, bumpy green mat is unfolded beneath you. If you climb a tree, you still do not gain any real conception of the jungle generally, for ants, leaves, hummingbirds and a riot of tangled vegetation obscure the view and bother you.

These forests are generally found in equatorial regions where annual rainfall exceeds 240 cm a year and the average temperature is more than 17°C. Thus, neither water nor temperature is a limiting factor. Surprisingly, soils in such areas can be fairly poor yet still support a luxuriant vegetation. Much of the ''goodness'' is **leached** out by heavy rainfall. There is no rich humus layer as there is in temperate systems; fallen leaves are quickly broken down and nutrients returned to the vegetation, where most of the mineral reserves are locked up. Consequently, cleared tropical forest land does not support agricultural practices well.

Tropical forests cover much of northern South America, Central

Photo 16.1   Tropical rain forest. (a) Rain forest in Trinidad. (b) High-elevation rain forest, termed cloud forest because it is often bathed in clouds. (c) Bromeliads (members of the pineapple family). (d) Parrots. Tropical forests typically support a large variety of epiphytes, like those in (c), and specialized feeders (d) that depend on a year-round supply of fruit. (Photos by P. Stiling.)

America, western and central equatorial Africa, and some of Madagascar, southeast Asia, and various islands in the Indian and Pacific oceans. The total amounts to a land area of about 3,000 million ha, 23 percent of the world total (Bunting 1988). The human population in these areas in 1984 was about 1,000 million, around 21 percent of the world total. The diversity in tropical forests is staggering, often reaching more than 50 tree species per ha; indeed the record for most tree species in an area alternates back and forth between southeast Asia and South America as different areas are censused. Gentry (1988) has recently recorded 283 tree species in one hectare of Peruvian rain forest. Sixty-three percent of the species in a 1 ha plot were represented by a single tree, and there were only twice as many individuals as species. Rain-forest trees are often smooth barked and have large oval leaves narrowing to "drip-tips" at the apex so that rainwater drains quickly before **photosynthetic** efficiency is impaired. Many trees have shallow roots with large buttresses for support (Warren et al. 1988). The tallest trees reach heights of 60 m or more and emerge above the tops of lower trees, which interdigitate to form a closed canopy. Little light penetrates this canopy, and the understory is often sparse. Tropical rain forests are also characterized by **epiphytes,** air plants that live perched on trees and are not rooted in the ground. Bromeliads are common epiphytes in New World forests. Climbing vines or lianas are also common.

Animal life in the tropical rain forests is also diverse; insects, reptiles, amphibians, and birds are well represented. As many butterflies can be found in a single rain forest as occur in the entire United States—500–600 species. Tropical rain forests are the great reservoirs of diversity on the planet; as many as half the species of plants and animals on Earth live in them. The island of Trinidad, off northwest Venezuela, is only a few hundred square miles in area yet can boast the same number of butterflies as the United States, largely as a result of the presence of tropical rain forest. Bright protective coloration and mimicry are rampant. Large mammals, however, are not common, though monkeys may be important herbivores. Though the genealogy of the major species in rain forests is different in different parts of the world, many species converge to a similar body form because they are adapted to a similar lifestyle (Fig. 16.5).

16.2 *Temperate Forest*

Temperate forest is the type of forest with which many people in the United States are most familiar. It occurs in regions where temperature falls below freezing each winter, but not usually below −12°C, and annual rainfall is between 75 and 200 cm. Large tracts of such habitat are evident in the eastern United States, east Asia, and western Europe. Commonly, leaves are shed in the fall and reappear in the spring, though there are exceptions. In the Southern Hemisphere, evergreen *Eucalyptus* forests occur in Australia, and large stands of southern beech, *Nothofagus*, occur in southern South America, New Zealand, and Australia. Species diversity is much lower than in the tropics; one or two tree species are dominant in a given locality—for example, oaks, hickories, or

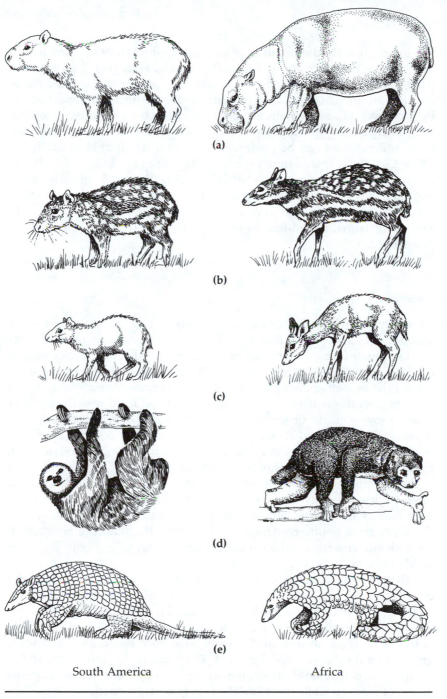

South America   Africa

Figure 16.5 *Convergence in mammals of tropical rain forests in South America and Africa.* **(a)** *Capybara and pigmy hippopotamus.* **(b)** *Paca and African chevrotain.* **(c)** *Agouti and royal antelope.* **(d)** *Three-toed sloth and Bosman's potto.* **(e)** *Giant armadillo and pangolin.* The first three South American animals are rodents, and the first three from Africa are ungulates, but convergence in shape and body size is still strong. (Modified from Ehrlich and Roughgarden 1987.)

maples are usually dominant in the eastern United States. Many herbaceous plants flower in spring before the trees leaf out and block the light
(Heinrich 1976) though, even in the summer, the forest is usually not as
dense as in tropical situations, so there is often abundant ground cover.
Epiphytes and lianas are absent. Soils are richer because the annual leaf
drop cannot be as quickly decomposed. With careful agricultural practices, soil richness can be preserved, and as a result agriculture flour-
ishes. Like the plants, animals are well adapted to the vagaries of the climate, and many mammals hibernate during the cold months. Birds
migrate, and insects enter **diapause,** a condition of dormancy passed
usually as a pupa (though sometimes as an egg, larva, or adult instead).
The reptile fauna, dependent on solar radiation for heat, is somewhat impoverished.

16.3 *Deserts*

Deserts are biomes suffering from water deficit. They are found generally
around latitudes of 30°N and 30°S, between latitudes of tropical forests
and temperate forests or grasslands. Over one-third of the Earth's surface
is occupied by these hot, dry regions. One reason for the locations of deserts is the movement of winds in the Earth's atmosphere. Warm, moist
air rises over the equator; as it cools, rain is produced, and the air rises
again, moving north or south of the equator. This rising, raining, and
movement continue to about 30° latitude, where the now dry and relatively cold air begins to sink groundward. It warms by compression and
produces a downward flow of warm dry air at about 30° north and south
of the equator. Such patterns of circulation, in which the drier air then
filters back to the equator, are known as *tropical Hadley cells.* They produce
deserts at characteristic latitudes, including the Sahara of north Africa,
the Kalahari of southern Africa, the Atacama of Chile, the Sonoran of
New Mexico and the southwest United States (shown in Photo 16.2), the
Gobi of central Asia, and the Simpson of Australia.

Deserts are characterized by two main conditions, lack of water and
usually high daytime temperatures. Lacking cloud cover, they quickly ra-
diate their heat at night and become cold. The degree of aridity is reflected in the ground cover. In true deserts, plants cover 10 percent or less
of the soil surface; in semi-arid deserts, like thorn woodlands and some
grasslands, they cover 10–33 percent. Only rarely do deserts consist of
completely lifeless sand dunes, but such places do exist: in the Atacama
desert of western Chile no rainfall has ever been recorded.

Three forms of plant life are adapted to deserts: (1) the annuals,
which circumvent drought by growing only when there is rain; (2) succu-
lents, such as the saguaro cactus (*Carnegiea gigantea*), and other barrel
cacti of the southwestern deserts, which store water; and (3) the desert

Photo 16.2 Sonoran Desert, Arizona. The prominent plants include the tall, columnar saguaro cactus, Canegiea gigantea; the green spray-like ocotillo, Fouquieria splendens; and smaller cholla cacti, Opuntia sp. (Photo by P. Stiling.)

shrubs, such as the spray-like ocotillo (*Fouquieria splendens*), which have short trunks, numerous branches, and small thick leaves that may be shed in prolonged dry periods. As a strategy against water-seeking herbivores, many plants have spines or an aromatic smell indicative of chemical defenses, although the physical structure of the desert plants—their few leaves and sharp spines—is probably also linked to water conservation and heat load. The above-ground parts of desert plants are more widely spaced than those of their forest counterparts because their roots are longer and occupy greater areas to ensure maximum water-gathering potential (Fig. 16.6). Typical perennial plants of the desert are the succulent and thorny cacti of the Western Hemisphere and succulent thorny members of the milkweed (Asclepiadaceae) and spurge (Euphorbiaceae) families in African deserts. In North America the creosote bush (*Larrea*) is widespread over the southwestern hot desert, and sage brush (*Artemisia*) is more common in the cooler deserts of the Great Basin.

Like the plants, desert animals have also evolved many ways of conserving water, like dry excretion (uric acid and guanin), heavy wax "waterproofing" in insects, and generally crepuscular habits and burrow 0pliving in the day.

Irrigated deserts can, because of the large amount of sunlight, be extremely productive for agriculture, though large volumes of water must flow through the system or detrimental salts may accumulate in the soil because of the rapid evaporation rate. Desert civilizations that harnessed the flow of such rivers as the Tigris, Euphrates, Indus, and Nile dominated early human history. Unlike tropical forests, deserts seem to be expanding under human influence because overgrazing, faulty irrigation, and the removal of what little hardwood exists all speed up desertifica-

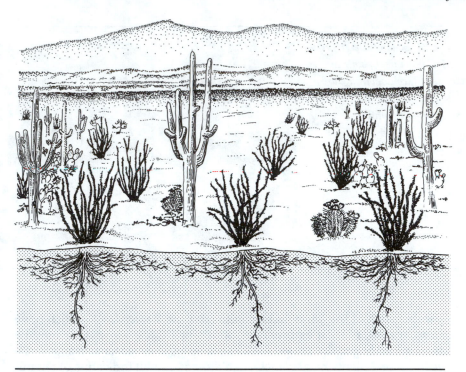

Figure 16.6 *Illustration of the regular spacing that allows desert plants to maximize their water uptake after rains.*

tion. The Sahel region, a narrow low-rainfall band south of the Sahara whose name is derived from the Arabic word for border, is an extreme case in point. The acacia tree, ubiquitous in many arid zones and useful as firewood and forage, was common around the Sudan capital, Khartoum, as recently as 1955; by 1972 the nearest trees were 90 km south of the city.

16.4 *Grassland*

Grasslands occur in the range between desert and temperate forest in which the rainfall, between 25 and 70 cm, is too low to support a forest but higher than necessary to support desert life forms. Alternatively, in the view of some ecologists (Bragg and Hurlbert 1976; Kucera 1981), the extensive grasslands of central North America, Russia, and parts of Africa are zones between forest and desert in which fire and grazing animals have worked together to prevent the spread of trees. Often forest removed from nearby areas is replaced by grassland. From east to west in

North America and from north to south in Asia, grasslands show differentiation along moisture gradients. In Illinois, with about 80 cm annual rainfall, tall prairie grasses about 2 m high such as big bluestem (*Andropogon*) and switchgrass (*Panicum*) dominate, whereas along the eastern base of the Rockies, 1,300 km to the west, where rainfall is only 40 cm, shortgrass prairies exist, rarely exceeding 0.5 m in height and consisting of buffalo grass (*Buchloe*) and blue grama (*Bouteloua*). Similar gradients occur in South Africa (the veldt) and in Argentina and Uruguay (the pampas). Nowadays, little of the original prairie remains. Prairie soil is the richest in the world, having 12 times the humus found in a typical forest soil. Historically and where the grasslands remain, large mammals are the most prominent members of the fauna; examples are bison (buffalo) and pronghorn antelope in North America, wild horses in Eurasia, large kangaroos in Australia, and a diversity of antelopes, zebras, and rhinoceroses in Africa, as well as their associated predators (lions, leopards, cheetahs, hyenas, and coyotes).

16.5 Taiga

North of the temperate-zone forests and grasslands lies the biome of coniferous forests, known commonly by its Russian name, **taiga.** Most of the trees are evergreens or conifers with tough, narrow leaves, needles that may persist three-to-five years. Spruces (*Picea*), firs (*Abies* and *Pseudotsuga*), and pines (*Pinus*) dominate, but some deciduous species such as aspens, alders, larches, and willows may occur in disturbed areas or along water courses. Many of the conifers have conical shapes to reduce bough breakage from heavy loads of snow. As in tropical forests, the understory is thin because of the dense year-round canopies; soils are also poor and acidic because of the slow decay of fallen needles. Snakes are rare, and few amphibians exist. Insects are strongly periodic but may often reach outbreak proportions, on the dense wood and foliage available, in times of climatic relaxation. Mammals such as bears, lynxes, moose, beavers, and squirrels are heavily furred. The taiga is famous for cyclic population patterns, of which the abundances of hares and lynxes are a well-known example. These patterns have given rise to the notion that diversity and stability are linked in ecological systems (Chapter 17). In the Southern Hemisphere, little land area occurs at latitudes at which one would expect extensive taiga to exist.

16.6 Tundra

The **tundra** is the last major biome and exists only in the Northern Hemisphere, north of the taiga. It is treeless, as precipitation is generally less

Photo 16.3   Typical tundra vegetation with erupted permafrost, Denali National Park, Alaska. (Photo by Dana C. Bryan.)

than 25 cm per year. What little water does fall is locked away for a large part of the year as **permafrost** (see Photo 16.3). Annual temperatures are less than −5°C, and even in the long summer days the ground thaws to less than one meter in depth. Vegetation occurs in the form of fragile, slow-growing lichens, mosses, sedges, and a few dwarf trees such as willow, which grow close to the ground. In some places vegetation cannot exist, and desert conditions prevail because so little moisture falls. Like the taiga animals, those of the tundra, such as the lemming and the predators that feed on it, are often cyclic. Agriculture is virtually impossible, and the tundra biome is relatively unexploited. Soils there often exhibit a

Photo 16.4   Tropical savannah, Ngorongoro Crater, Tanzania, Africa. (Photo by Dana C. Bryan.)

Photo 16.5   Temperate rain forest, Olympic National Park, Washington. In this habitat epiphytes such as mosses and ferns abound. The most extensive temperate rain forests exist along the coasts of northwestern United States, western Canada, and southern Alaska, where the extensive Tongass rain forest is located. (Photo by P. Stiling.)

polygonal or hummocked appearance that results from alternate freezing and thawing of the ground.

Of course, not all communities fit neatly into these six major biome types. As with most things ecological, there exist characteristic regions where one biome type grades into another. For example, in east Africa, tropical seasonal forests grade into grassland in areas known as thorn scrubs or tropical savannahs (Photo 16.4), where big game is plentiful. Tropical seasonal forests themselves are apparent where rainfall is heavy (between 125 and 250 cm a year) but occurs in a distinct wet season, as in India or Vietnam. In such monsoon forests, leaves may be shed in the dry season. Other distinct but smaller biome types include temperate rain forests like those in Washington state and Oregon (Photo 16.5). In these forests precipitation is high but temperatures are low, and chaparral, a Mediterranean scrub habitat adapted for fire, is common along the coastlines of southern Europe, California, South Africa, and southwest Australia.

16.7 *Mountain Communities*

Mountain ranges must be treated still differently. Biome type relies predominantly on climate, and on mountains temperature decreases with increasing altitude. This decrease is a result of a process known as *adiabatic cooling.* Increasing elevation means decrease in air pressure. When wind is blown across the Earth's surface and up over mountains, it expands because of the reduced pressure; as it expands it cools, at a rate of about 10°C for every 1,000 m, as long as no water vapor or cloud formation occurs. (Adiabatic cooling is also the principle behind the function of a refrigerator—freon gas cools as it expands coming out of the compressor.) Higher elevations are also cooler because the less dense air allows a higher rate of heat loss by radiation back through the atmosphere. A vertical ascent of 600 m is roughly equivalent to a trek north of 1,000 km. Precipitation changes with altitude, too, generally increasing in desert elevations but decreasing on the leeward side of slopes, which are in a rain shadow. Approaching clouds have usually dumped all their moisture on the windward side. Thus, biome type may change from temperate forest through taiga and into tundra on an elevation gradient in the Rocky Mountains, as is shown in Photo 16.6, and even from tropical forest to tundra on the highest peaks of the Andes in tropical South America.

Within aquatic environments, biome types can also be recognized, such as rivers, freshwater lakes, and, within saltwater oceans, the intertidal rocky shore, sandy shores, the neritic zone (encompassing shallow waters over continental shelves), coral reefs, seagrass beds, and the pelagic zone or open ocean. In each of these, the physical "climate" is different, varying in such parameters as salinity, oxygen content, current strength, and availability of light.

Photo 16.6   Mountain biome, Colorado. Habitat type varies with elevation. The tree line, above which trees cannot grow, is often distinct. (Photo from U.S. Department of Agriculture.)

16.8 Marine Communities

Marine environments are among the most extensive and uniform on Earth. The vast expanses of continually moving ocean provide a buffer against temperature extremes, and their aquatic communities are much less exposed to them. Sufficient sunlight penetrates to permit **photosynthesis** only in the upper 2 percent of the ocean volume, so **autotrophs,** mainly photosynthetic plants, are confined to a narrow surface zone. Within these depths in warm, tropical water, coral reefs form. Corals are animals with symbiotic algae that live within their tissue, transferring photosynthate to the coral animals, which depend on them. Most corals are found at depths of less than 50 m. Their calcareous skeletons build up to produce the massive coral reefs we know today, far larger than the Egyptian pyramids or any known human edifice. Coral reefs and the rocky intertidal zone are unusual habitats in that space is often the limiting factor for animals. Waves and currents constantly bring a new supply of food to the environment. Food, therefore, cannot easily be depleted by the foraging activities of animals, unlike that in most terrestrial systems, where green plants can be entirely consumed. As a result, **sessile** animals compete for space and position and actually form the dominant structure elements of communities. Coral reefs are among the most species-rich habitats on the planet; some 6,000–8,000 species of fish are associated with them, a figure that constitutes probably 30–40 percent of all fishes (Ehrlich 1975). In the open ocean, the **pelagic** zone, the primary producers are **phytoplankton,** fed upon by a whole array of **zooplankton** from protozoa, worms, jellyfish, and krill to the minute larvae of many benthic or bottom-feeding marine organisms such as crabs and the offspring of intertidal barnacles and oysters. Distinguished by their ability to swim against the current are the **nekton:** squid, fish, turtles, and marine mammals. Unlike terrestrial environments, marine habitats are devoid of insects, save for some water striders, hemipterous bugs of the genus *Halobates.* At the bottom of the open oceans live yet other communities adapted to live under high pressure and to feed on the corpses and feces that rain down from above. Recently other communities have been discovered that exist near the openings of deep-ocean volcanic vents (black smokers) in midocean ridges (Ballard 1977). The primary producers there are giant worms, which are nourished by symbiotic chemosynthetic bacteria that produce ATP by oxidizing sulphides and reducing carbon dioxide to organic compounds. In other areas, animals that harbor chemoautotrophic bacteria seem to be the most common. Hessler, Lonsdale, and Hawkins (1988) discuss the common occurrence of a "hairy snail" in the vents near the Philippines, which contains bacteria in its gills that oxidize sulphur to produce energy, and Smith (1985) discusses the dense beds of mussels, *Bathymodiolus thermophilus,* that can be found along the Galapagos rift. Water in these areas is often 20°C warmer than in surrounding areas.

16.9 Freshwater Communities

Freshwater habitats are traditionally divided into standing-water **lentic** habitats (from the Latin *lenis*, calm—lakes, ponds, and swamps) and running-water **lotic** habitats (*lotus*, washed—rivers and streams). Natural lakes are most common in regions that have been subject to geological change within the past 20,000 years, such as the glaciated regions of northern Europe and North America. They are also common in regions of recent uplift from the sea, such as Florida, and in regions subject to volcanic activity. Volcanic lakes formed either in extinct craters or in valleys dammed by volcanic action are among the most beautiful in the world. Geographically ancient areas such as the Appalachian Mountains of the eastern United States contain few natural lakes.

The ecology of lentic habitats is largely governed by the unusual properties of water. First, water is at its least dense when frozen; ice floats. From a fish's point of view this property is advantageous because a frozen surface insulates the rest of the lake from freezing. If ice sank, all temperate lakes would freeze solid in winter, and no fish would exist in lakes outside the tropics. Water is at its densest at 4°C. Thus, as long as no water in the lake is colder than 4°C, the warmest water is at the surface, and temperature declines with depth, though not in a linear fashion. Normally several layers are present (Fig. 16.7). There is an upper layer, called the **epilimnion,** that is warmed by the sun and mixed well by the wind. Below lies the **hypolimnion,** a cool layer too far below the surface to be warmed or mixed. The transition zone between the two is

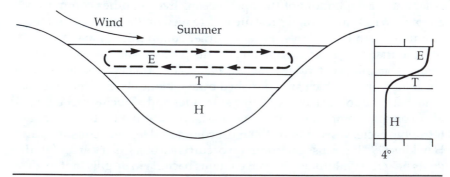

Figure 16.7 *Cross section of lake stratification and profile of temperature with depth.* E = epilimnion; T = thermocline; H = hypolimnion. Temperature scale on the right starts at 0°C and increases to the right.

known as the **thermocline.** There are other divisions within a lake based on light availability. The upper layer, where light penetrates, is the **autotrophic** or **euphotic** zone. Below, in darkness, is the profundal zone, where **heterotrophs** live, depending on the rain of material from above for subsistence. The depth of the euphotic zone depends on light availability and water clarity. The level at which photosynthate production equals energy used up by respiration is the lower limit of the euphotic zone and is known as the *compensation level* or compensation point. In the summer, in temperate lakes, the compensation level is usually above the thermocline. Oxygen-producing green plants cannot live in the hypolimnion, which becomes oxygen-depleted, a phenomenon known as *summer stagnation.*

The degree of summer stagnation in temperate lakes is partly determined by the degree of "productivity" of a lake. The least productive lakes are termed **oligotrophic.** Such lakes generally have low nutrient contents, largely as a result of their underlying substrate and young geologic age. Young lakes have not had a chance to accumulate as many dissolved nutrients as have older ones. Oligotrophic lakes are relatively clear, and their compensation levels may lie below the thermocline. In this situation, photosynthesis can take place in the hypolimnion, adding oxygen. Low nutrient concentrations keep the algae and rooted plants in the epilimnion sparse, and little debris rains down upon the inhabitants of the hypolimnion. As a result, oligotrophic lakes are clear and often contain desirable fish such as trout. Even though few nutrients are present in oligotrophic lakes, eventually they do begin to accumulate; sediments are deposited and both algae and rooted vegetation begin to bloom. Organic matter accumulates on the lake bottom, respiration of bottom dwellers increases, the water becomes more turbid, and the oxygen levels of the water go down. Fish such as trout are excluded by bass and sunfish, which thrive in warm water and at low oxygen concentrations. This process of aging and degradation is natural and is termed **eutrophication;** its end result is a eutrophic lake. Eutrophication, however, can be greatly speeded up by human influences, which increase nutrient concentrations by the input of sewage and fertilizers from agricultural runoff. A measure of eutrophication is given by the dissolved oxygen concentration or **biochemical oxygen demand (BOD).** The BOD is the difference between the production of oxygen by plants and the amount of oxygen needed for the respiration of the organisms in the water. It is normally measured in the laboratory as the number of milligrams of oxygen consumed per liter of water in five days at 20°C.

The stratified nature of temperate-zone lakes in summer does not last all year (Fig. 16.8). In the fall, the upper layers cool and sink, carrying oxygen to the bottom of the lake. The lake is thoroughly mixed by this action and by frontal storms, and the thermocline disappears. In winter the surface usually freezes, and no turnover of water occurs; once again a

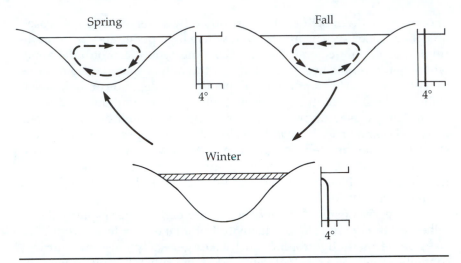

Figure 16.8   *Annual cycle of a lake. See Fig. 16.7 for further details.*

gradient is set up. Then, as spring returns, the ice melts and sinks, producing another mixing, the spring overturn. In contrast to temperate lakes, tropical lakes are often *isothermal* (that is, all at one temperature) or at most exhibit only a weak temperature gradient from top to bottom. Little mixing occurs, and deep lakes are generally unproductive, with oxygen-poor, fishless lower depths, as the builders of tropical dams learn (to their chagrin). Worse still, most water from dams is drawn off from the base, meaning that the streams below dams are much less oxygen-rich than those above them. Shallow lakes do not show thermal stratification in any region.

Abiotic phenomena other than nutrient content and dissolved oxygen content may be important to the lake communities also. Water **pH** is particularly relevant, as fish enthusiasts know. For example, some *Poecilia* species, such as live-bearing mollies, breed only in alkaline (high pH) waters, whereas *Hyphessobrycon*, neon tetras, breed only in low pH. Changes in the pH of lake water are more frequent these days because of the impact of **acid rain** (Chapter 22). Fish have been exterminated in over 300 lakes in the Adirondack Mountains of the northeastern United States by acid rain.

Lotic or running-water habitats generally have a fauna and flora completely different from those of lentic waters. Plants and animals are adapted so as to remain in place despite an often strong current. Nutrient inputs and phytoplankton blooms do not occur because each would be quickly washed away. Current also mixes water thoroughly, providing a well-aerated regime. Animals of lotic systems are therefore not well adapted for low-oxygen environments and are particularly susceptible to high-BOD (oxygen–reducing) pollutants.

When the freshwater environment is considered as a whole, the algae are the most important producers, and the aquatic spermatophytes rank second. Among the animal consumers, the bulk of the biomass is due to aquatic insects, Crustacea, and fish. Other orders rank much lower in importance, though in specific instances any one may loom larger in the "economy" of the system.

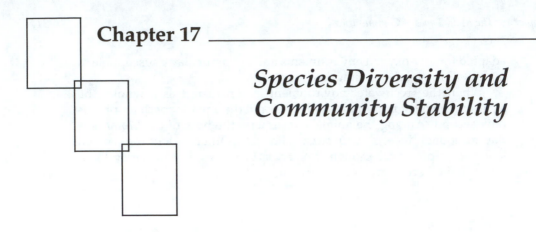

Chapter 17

Species Diversity and Community Stability

A. R. Wallace (1878) was perhaps the first to point out that animal life was generally more varied and abundant on an area-for-area basis in the tropics than in more temperate latitudes. The same is true for plant life. A naturalist may have an intuitive feel for diversity when observing and comparing two different habitats, but to express this feeling on paper, various indices of species diversity are often used. The simplest measure of diversity is to count the number of species; the result is termed the *species richness*. Diversity of communities can be compared this way, but it is often difficult to identify a true resident species of the community, as opposed to a transient. Although resident and migratory birds, for example, may be easy to classify, it is a little harder to decide whether a bottomland tree on a ridge top represents an accidental occurrence, a planting, or a true resident species (Krebs 1985a). Furthermore, in a sample of 100 individuals from a habitat, the species richness of a community of two species, each of population size 50, would equal that of a community in which the population sizes of the same two species were 1 and 99. In actuality, the first community must be considered more diverse; one would be much more likely to encounter both species there than in the second community. To incorporate both species richness and abundance, ecologists have turned to other measures.

17.1 *Logarithmic and Lognormal Series*

A common characteristic of communities is that they contain a few species that are common and many more that are rare. From personal experience in rearing parasites of leaf miners, this author knows that a few species make up the bulk of the parasites and that rare species continue to "trickle in" with continued rearings. Leaf miners are notorious for their large parasite complexes (Askew and Shaw 1974), and one is never sure that all the attacking parasites have been found. Light-trap catches of moths provide good data with which to examine this phenomenon (Fig. 17.1); indeed some tropical entomologists have based entire careers on these comparisons. Such data can make possible comparisons between temperate and tropical light-trap catches of moths, mist-netting catches for birds, or tree counts per hectare. It was Fisher, Corbet, and Williams (1943) who first attempted mathematical descriptions of these comparisons. They concluded that the curve was best fitted by the logarithmic series

$$\alpha x, \frac{\alpha x^2}{2}, \frac{\alpha x^3}{3}, \frac{\alpha x^4}{4} \cdots$$

where αx = number of species in the total catch represented by one individual, $\alpha x^2 / 2$ = number of species in the total catch represented by two individuals, and so on, where

$$\sum_{x}^{x^n} \alpha x, \frac{\alpha x^2}{2}, \frac{\alpha x^3}{3}, \ldots, \frac{\alpha x^n}{n} = S, \text{ the number of species in a sample}$$

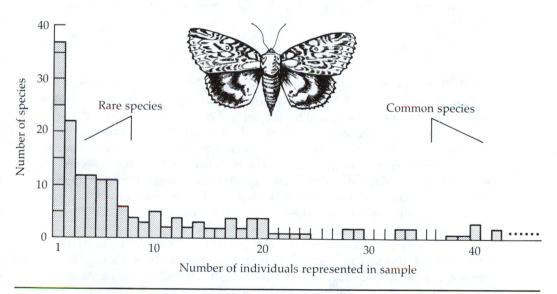

Figure 17.1 *Relative abundance of Lepidoptera (butterflies and moths) captured in a light trap in Rothamsted, England, in 1935. Not all of the abundant species are shown. There were 37 species represented in the catch by only a single specimen (rare species); one very common species was represented by 1,799 individuals in the catch. A total of 6,814 individuals were caught, representing 197 species. Six common species comprised 50 percent of the total catch. (Modified from Williams 1964 .)*

Another relationship giving S is

$$S = \alpha \log_e \left(\frac{1 + N}{\alpha} \right)$$

where N = number of individuals in the sample and α = index of diversity. Comparisons of alphas between communities allow comparisons of species diversity to be made. **Diversity,** α, is low when the number of species is low and high when many species are present, and it is independent of sample size—an important attribute. The type of curve illustrated in Fig. 17.1, a logarithmic series, implies that the greatest number of species has minimal abundance and that the number of species represented by a single specimen is always maximal. This is not always the case (Fig. 17.2). In samples of breeding birds from Quaker Run Valley, New York, the greatest number of bird species are represented by 10 breeding pairs. For comparative purposes, Preston (1948) suggested expressing a geometric scale of size classes on the x axis. Each class or grouping could be based on a different scale (Table 17.1, p. 322) using a constant multiplier. After this type of conversion, many sets of abundance data take the form of bell-shaped curves (Fig. 17.3, p. 322). Because the x axis is a geometric or logarithmic scale, this distribution is called *lognormal*; it is described by the formula

$$S_R = S_O e^{-(aR)^x}$$

where S_R = number of species occurring in the Rth octave (a section of the $\log_2$ scale of Table 17.1) to the left or right of the modal class, S_O = number of species in the modal octave (largest class), a = a constant describing the spread of the distribution, and e = 2.71828 (a constant).

The **lognormal distribution** arises in communities where the total numbers of species are large and the relative abundance of these species is thought to be determined by many factors operating independently (May 1975). It is thus the expected statistical distribution for many biological communities. By contrast, the logarithmic series arises with communities of relatively few species and is thought to occur where a single environmental factor is of dominating importance. That the lognormal distribution is due to many factors operating independently has also been argued mathematically (Ehrlich and Roughgarden 1987). If a random-number generator (or a phone book) is used to generate numbers from 1 to 1,000, the frequency distribution would be a flat line (a uniform distribution): each number would have an equal probability of occurrence. Now, if pairs of these numbers are taken and their averages, $\bar{x}$, are taken, the distribution of $\bar{x}$s will be more normally distributed than the original numbers, the $\bar{x}$s themselves. In fact, the distribution of $\bar{x}$s will look like a triangle with a peak at 500. If $\bar{x}$s are generated from triplets of numbers,

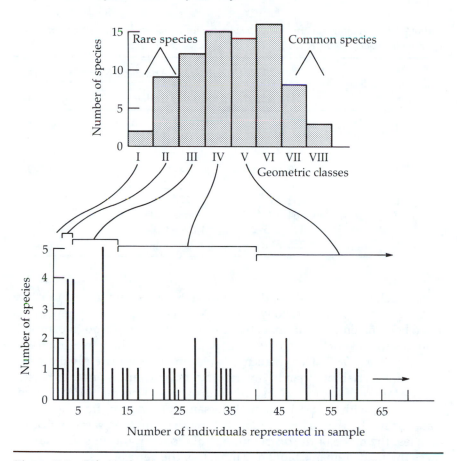

Figure 17.2 *Relative abundance of nesting bird species in Quaker Run Valley, New York.* The lower figure shows the distribution on an arithmetic scale, and the upper figure shows the same data on a geometric scale with × 3 size groupings (1, 2–4, 5–13, 14–40, 41–121, and so on). (Redrawn from Williams 1964.)

the distribution of $\bar{x}$ would appear even more normally distributed. As n, the number of xs averaged to form $\bar{x}$, increases toward infinity, the distribution of x tends toward an exact normal distribution. For practical purposes, the distribution is statistically indistinguishable from the normal distribution where $n \geq 10$. This phenomenon is described by the *central limit theorem* of statistics, which states that a normal distribution results for a quantity whose values are determined by the product of many small random effects; thus lognormal series tend to indicate randomness in nature.

The lognormal distribution appears in other contexts, apart from frequency of occurrence. The distribution of body sizes among the species of a community may be approximately lognormal (Fig. 17.4, p. 323). Some-

Table 17.1   *Groupings of arithmetic scale units into geometric scale units for three types of geometric scales. (From Krebs 1985.)*

Geometric Scale Number	Scale Used to Group Arithmetic Numbers		
	× 2 (*log$_2$ scale*)	× 3 (*log$_3$ scale*)	× 10 (*log$_{10}$ scale*)
1	1	1	1–9
2	2–3	2–4	10–99
3	4–7	5–13	100–999
4	8–15	14–40	1,000–9,999
5	16–31	41–121	10,000–99,999
6	32–63	122–364	100,000–999,999
7	64–127	365–1,093	—
8	128–255	1,094–3,280	—
9	256–511	3,281–9,841	—

times such graphs are skewed to left or right, as is true for mammalian (Fig. 17.5, p. 324) or bird body sizes in Michigan, and such phenomena have been taken to mean that character displacement through competition has spread out the distribution of body sizes (Kirchner in Ehrlich and Roughgarden 1987). This character displacement is apparently more important in vertebrate than in invertebrate taxa. There is another reason for the failure of lognormal distributions to appear normally distributed. It has to do with the problem of undersampling. Again, from light-trap data for moths (Fig. 17.6, p. 325), it is evident that, as sample size increases, the mode of the distribution shifts to the right. In small sample sizes the mode remains hidden to the left of the vertical axis.

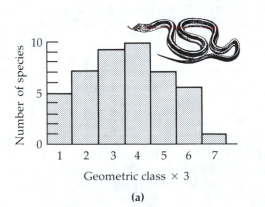

(a)

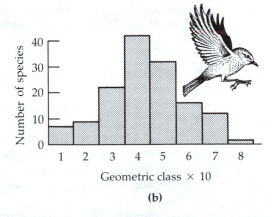

(b)

Figure 17.3   *Log-normal distribution patterns in two communities.* **(a)** *Panamanian snakes.* **(b)** *British birds.* (Modified from Williams 1964.)

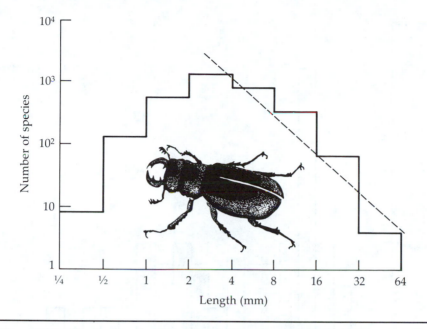

Figure 17.4 *The number of species of British Coleoptera, classified according to length.* (Redrawn from May 1978.)

17.2 *Diversity Indices*

The final approach to the study of community diversity involves use of diversity indices. The most common index of species diversity is the Shannon-Wiener diversity index H' (sometimes mislabelled the Shannon-Weaver index, a different entity), where

$$H' = -\sum_{i=1}^{S} p_i \log_e p_i$$

and p_i = proportion of the ith species in the total sample of S species. A community is less diverse in which species are unevenly abundant:

Number of Species in Community	Relative Abundance of Species			
	Species 1	Species 2	Species 3	H'
2 species	90	10		0.33
2 species	50	50		0.69
3 species	80	10	10	0.70
3 species	33.3	33.3	33.3	1.10

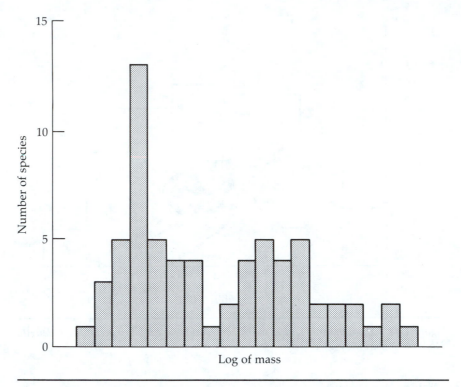

Figure 17.5 *Distribution of size for the mammals of Michigan.* (Redrawn from Ehrlich and Roughgarden 1987.)

The maximum possible diversity for a given number of species present, H' max, can be calculated on the assumption that all species are equally abundant. Then a measure of evenness of the actual community is determined as follows:

$$J', \text{evenness} = \frac{H'}{H' \max}$$

where J' ranges from 0 to 1. Many ecologists hoped that such shortcuts to community analysis would obviate the need for much more detailed work on population ecology of different species, but the sad fact is, of course, that the species present in a community can change substantially without the change's being reflected in the diversity index, if one species were simply replaced by another. Götmark, Åhlund, and Eriksson (1986) tested the value of species **diversity** (and four other conservation indices) as a measure of applicability to the conservation of birds in southwest Sweden. Census sites were first evaluated and ranked by the authors,

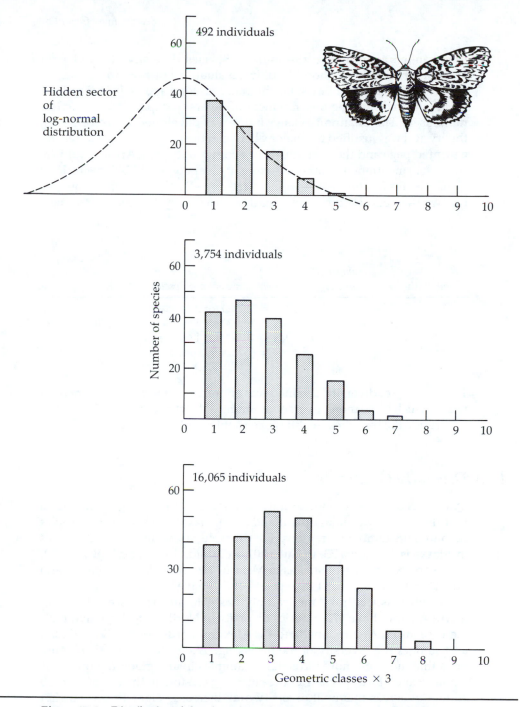

Figure 17.6 *Distribution of abundance in moths captured by light traps in England.* In the top figure, the true distribution of abundance is hidden behind the veil of the *y* axis. As sampling becomes more intense, the distribution pattern moves to the right to reveal the distribution of rarer species. (Modified from Williams 1964.)

then ranked according to the indices. Species diversity showed poor agreement with the authors' intuitive evaluations of how to rank areas for the conservation of birds. Furthermore, as long ago as 1971, Hurlbert (1971) showed that the measurement of diversity depended on the formula used. The Shannon-Wiener index was developed for information theory and was justified by analogy between the proportions of letters on a printed page and the individuals in a community. MacArthur and Wilson (1967) mentioned another measure of diversity, $\Delta_3 = 1/\Sigma p_i^2$. Hurlbert compared the Shannon-Wiener index $(H' = -\Sigma p_i \log_e p_i)$ to this one using two theoretical communities, each with a total of 100,000 individuals:

Community	Number of Species	Abundance of Each Species	Measure of Diversity H'	Measure of Diversity Δ_3
A	6	$N = 18,000$ for 2 species $N = 16,000$ for 4 species	0.78	5.98
B	91	$N = 40,000$ for 1 species $N = 667$ for 90 species	2.70	5.00

The measure of diversity obtained was entirely dependent on the method used. Which is the more diverse community? It is hard to say. No one has yet argued the biological realities of either formula.

17.3 *Diversity Gradients*

Questions of diversity indices notwithstanding, it has long been known that the diversity of many plant and animal taxa is higher in the tropics than in more temperate regions. For example, the number of ant species in Alaska is 7, in Iowa 73, in Cuba 101, in Trinidad 134, and in Brazil 222. There are 293 species of snakes in Mexico, 126 in the United States, and only 22 in Canada. Over 1,000 species of fish have been found in the Amazon, whereas Central America only has 456, and the Great Lakes of North America have 172. Species diversity of North American mammals increases from Arctic Canada to the Mexican border (Figs. 17.7, 17.8a), and so does the diversity of the birds (Fig. 17.8b). It should also be noticed that there are more birds than mammals in any given region of the United States. Bird species richness increases 12-fold in the 60° of latitude shown, whereas mammal diversity only increases eight times. Further examples of latitudinal gradients in species diversity are given in Table 17.2 (p. 330). Sometimes a reverse pattern is found, such as those for the sandpipers, family Scoloparidae, whose diversity increases toward the Arctic regions (Cook 1969), and for aphids, whose diversity is highest in temperate realms. Diversity of trees in North America is also not well

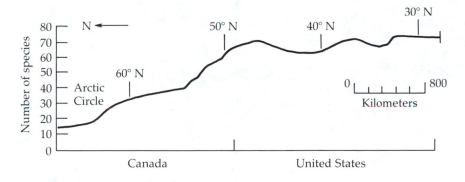

Figure 17.7 *Species densities of North American mammals from the Arctic to the Mexican border along the 100th meridian.* (Redrawn from Simpson 1964.)

linked to latitudinal gradients (Currie and Paquin 1987). Many theories for such temperate-to-tropical progressions have been advanced, and although it is difficult to "prove" any one of them, some of the more feasible are listed below:

- **Time theory:** Proponents of this theory argue that communities diversify with time and that temperate regions have younger communities than tropical ones because of recent glaciations and severe climatic disruption. According to this view, species that could possibly live in temperate regions have not migrated back from the unglaciated areas into which the Ice Ages drove them, or resident species have not yet evolved new forms to exploit vacant **niches.** Lake Baikal in the Soviet Union is an ancient unglaciated temperate lake and contains a very diverse fauna; for example there are 580 species of benthic invertebrates (Kozhov 1963). A comparable lake in glaciated north Canada, Great Slave Lake, contains only four species in the same zone (Sanders 1968). In a useful analogy, tropical and temperate habitats can be compared to equal-sized libraries. The numbers of species, books in each library, is dependent on different things. In the tropics, the library is full, and size of books and available shelf space dictate the number of volumes held. In a temperate situation, there is plenty of available shelf space, and the number of books depends on their rate of purchase by the library and the length of time since the library opened.

   Another test of the time hypothesis has been provided by areas recently colonized by plants and animals, where species richness can be assessed over time. Since the end of the last Ice Age, trees have recolonized Britain, and in the last 2,000 years, humans have introduced trees as well. Southwood (1961) was the first to examine the number of insect colonists associated with each tree, and he found good correlations of insect diversity with length of tree tenure in Britain. Strong (1974*a*; 1974*b*),

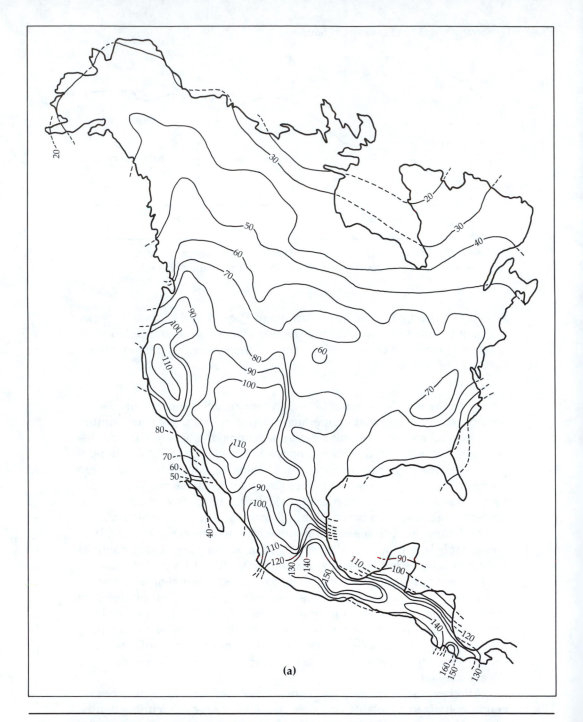

Figure 17.8 *Geographic variation in species diversity of North America.* **(a)** *Mammals.* **(b)** *Land birds.* Note the pronounced latitudinal gradients in both groups and the high diversity in the southwestern United States and northern Mexico, a region of great topographic relief and habitat diversity. [(a) Redrawn from Simpson 1964; (b) Redrawn from Cook 1969.]

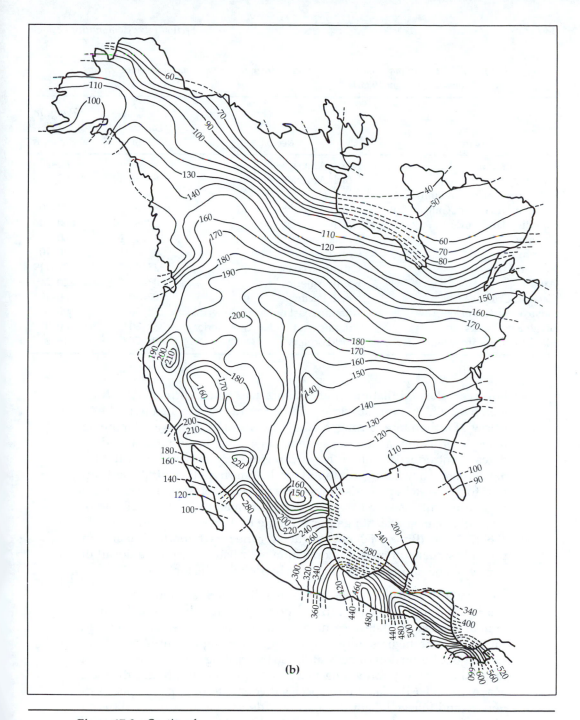

Figure 17.8 *Continued*

Table 17.2 *Latitudinal gradients in species diversity in various taxonomic groups.*
(From Brown and Gibson 1983.)

Taxon	Region	Latitudinal Range	Range in Species Richness
Land mammals	North America	8°–66°N	160–20
Bats (Chiroptera)	North America	8°–66°N	80–1
Quadrupedal land mammals (all orders except Chiroptera)	North America	8°–66°N	80–20
Breeding land birds	North America	8°–66°N	600–50
Reptiles	United States	30°–45°N	60–10
Amphibians	United States	30°–45°N	40–10
Marine fishes	California coast	32°–42°N	229–119
Ants	South America	20°–55°S	220–2
Calanid Crustacea	North Pacific	0°–80°N	80–10
Gastropod mollusks	Atlantic coast of North America	25°–50°N	300–35
Bivalve mollusks	Atlantic coast of North America	25°–50°N	200–30
Planktonic Foraminifera	World oceans	0°–70°	16–2

however, showed that insect abundance was better correlated with the area over which a tree species could be found (see the area hypothesis, below). Furthermore, Strong, McCoy, and Rey (1977) provided more detailed information on another system, sugarcane, its pest loads, and the dates of introduction into at least 75 regions of the world over the last 3,000 years. They found no support for the time hypothesis but good support for the area hypothesis, though there have been several criticisms of this argument, usually based on the poor quality of data available on areas of sugarcane plantings (for example by Kuris, Blaustein, and Alio 1980). Others (Birks 1980) have argued for the time hypothesis using improved estimates of time of colonization of British trees from radiocarbon dating of pollen profiles, but the time hypothesis still looks weak.

■ **Spatial-heterogeneity theory:** Generally, there are more plant species in the tropics, which in turn support higher numbers of herbivorous animal species and hence carnivores. Diversity of vegetation increases numbers of herbivores in two ways, by increasing the numbers of monophagous herbivores directly and also by creating a more diverse architectural complexity, providing more niches to occupy. MacArthur and MacArthur (1961) related bird species diversity to both plant species diversity and foliage-height diversity. (Birds usually visit many different plants for food, so their diversity is often linked to that of their plants. In contrast, insects often spend their entire lives on one type of plant.) Of course, the spatial-heterogeneity theory does not address the reason for the higher numbers of plant species themselves. Presumably it is due to microhabitat or topographical diversity, although this hypothesis remains untested and rather circular.

■ **Competition theory:** It has been argued that, in temperate climates, natural selection is controlled mainly by harsh physical extremes and that species are generally *r* selected. In the more constant tropical temperatures, species are thought to become more *K* selected, to compete more keenly, and to interact more. This keen competition would reduce species niche breadth, allowing more species to pack along the resource axes (Dobzhansky 1950). Again, no critical evaluation of this theory has been performed, and niche parameters have not been measured for a sufficient variety of species groups to determine how niche breadths are affected by tropical-to-polar gradients.

■ **Predation theory:** This theory, proposed by Paine (1966), runs contrary to the competition hypothesis and argues that there are more predators and parasites in the tropics and that these hold populations of their prey down to such low levels that more resources remain and competition is reduced, allowing more species to coexist. The increased diversity in turn promotes more predators. Paine provided evidence to support this theory from studies on the intertidal communities of the U.S. Northwest coast, where the food web was fairly constant and the starfish *Pisaster* was the top predator:

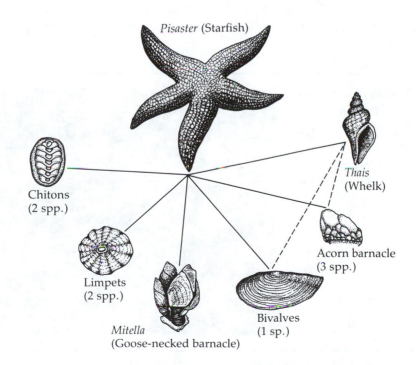

Pisaster (Starfish)

Thais
(Whelk)

Chitons
(2 spp.)

Acorn barnacle
(3 spp.)

Limpets
(2 spp.)

Bivalves
(1 sp.)

Mitella
(Goose-necked barnacle)

Thais preyed on bivalves and acorn barnacles. *Pisaster* preyed on those groups, on *Thais*, and on chitons, limpets, and *Mitella* as well. After removal of *Pisaster* from a section of the shore, diversity decreased from 15 to 8 species. A bivalve, *Mytilus*, increased, crowding out other species. In unmanipulated sections of shore, *Pisaster* tended to remove *Mytilus* and other species, preventing any one species from monopolizing space. Removal of any other single species from the system would not affect species diversity so drastically as *Pisaster* removal. For this reason, species like *Pisaster* are often termed *keystone species*, by analogy with the keystone that holds all the other stones in an arch in place. Top predators commonly specialize on the most abundant prey, developing a **search image** for it (see Chapter 12). Such a phenomenon, wherein predation allows the **coexistence** of more prey species, was noted even by Darwin (1859), who observed more grass species coexisting in areas grazed by sheep or rabbits than in ungrazed areas. Otters act as keystone species in the kelp beds of the Pacific coast (Estes, Jameson, and Rhode 1982), and elephants have been argued to function as keystone species on the African savannahs (Krebs 1985*a*). Keystone species are not always predators. In the southeastern United States, gopher tortoises (Photo 17.1) have been regarded as keystone species because their burrows provide homes for an array of mice, possums, frogs, snakes, and insects. Without tortoises' burrows, many of these creatures would be unable to survive in the sandhill areas where they are found (Sunquist 1988).

Coexistence of competitors, when promoted by predators, is sometimes termed *exploiter-mediated coexistence*. Of course, for such a system to

Photo 17.1 Adult gopher tortoise and its burrow. This tortoise is an example of a keystone species because its burrows provide homes for a variety of snakes, mice, and insects, as well as the tortoise itself. (Photo © Copyright Florida Game and Fresh Water Fish Commission.)

explain tropical diversity, the predation would have to be intense on the majority of species at all **trophic levels,** and few data are available as yet to test such an idea. In some systems, however, tropical rainstorms can have an analogous effect by creating gaps, essentially bare areas, which are invaded by rapidly colonizing but competitively inferior species (Paine and Levin 1981).

▪ **Climatic stability:** Temperate organisms are, of necessity, well adapted to harsh physical conditions and to climatic change. The tropics are climatically much more stable, and organisms there are less well adapted to cope with small temperature and moisture fluctuations. Because in fact there are more minor fluctuations in the tropics than major ones in temperate regions, the climatic-stability hypothesis argues, more species will specialize and adapt to each of the many small resultant niches in the tropics. In support of this argument are the relatively low numbers of organisms in harsh deserts and in extreme environments like hot springs or the Great Salt Lake of Utah. The organisms that do survive are often common, and competition has thus been deemed more important in temperate areas than in tropical ones. (Note that this argument is contrary to that outlined in the competition theory above.) As further evidence, there are, surprisingly, more species of benthic invertebrates in deep-sea environments, where conditions are relatively stable, than in shallow waters where physical factors are much more variable (Sanders 1968).

▪ **Productivity theory:** This theory has been termed the *species-energy hypothesis* by Wright (1983), who first proposed its use. The theory proposes that greater production results in greater diversity; that is, a broader base to the energy pyramid permits more species in that pyramid. A common modification is that there is "room" for obligate fruit-eating birds (like parrots) or raptorial reptile eaters in the tropics, but not in temperate regions (Orians 1969*a*). Fruits appear year-round in the tropics, but a parrot would starve in a temperate winter. It is also argued that a longer growing season not only increases productivity but also allows component species to partition the environment temporally as well as spatially, thereby permitting the coexistence of more species. Thus for species with annual life cycles, such as insects, some species could feed on leaves early in the year and others later. Recently, Currie and Paquin (1987) have shown that species richness of trees in North America is best predicted by evapotranspiration rate (see Fig. 8.15). Realized annual evapotranspiration is correlated with primary production and is therefore a measure of available energy. Turner, Gratehouse, and Carey (1987) demonstrated a correlation between the diversity of British butterflies, exothermal species, and sunshine and temperature during the months they were on the wing, again suggesting a relationship between energy and species diversity. Finally, a simple prediction from this theory is that the number of resident species in seasonal habitats should change according to the seasons. Turner, Lennon, and Lawrenson (1988) have shown that this is true for British birds. The number of birds present in Britain in the winter is less than that in summer, and this pattern is consistent with the amounts of energy present.

- **Area theory:** This idea is based on the notion that in larger areas the chances of isolation between populations increase, with corresponding increases in the chances of speciation (Terborgh 1973). It has also frequently been shown that larger areas support more species (see chapter section 18.2). Thus, large areas of climatic similarity will have greater species diversity. On a worldwide scale, there is a symmetry of climates between polar regions and temperate areas, but only in the tropics do we see the symmetrically opposite climates adjacent, creating one large area. In a slight variation of this idea, Darlington (1959) argued that most dominant species evolved in the largest areas (which he also argued were the equatorial zones) and diffused out, creating a species-diversity gradient. Neither theory, however, seems able to explain why, if diversity is linked to area, there should not be more species in the vast contiguous land mass of Asia.

- **Animal-pollinators theory:** In the tropics and other humid parts of the world, winds are less frequent and of lower intensity than in temperate regions. This effect is accentuated by dense vegetative cover. Therefore, most plants are pollinated by animals: insects, birds, and bats. Even some grasses that are typically wind-pollinated throughout most of the world are probably pollinated by insects in the tropics. Usually, and particularly in the case of bee pollinators, associations build up between plants and specific pollinators, increasing the reproductive isolation between plant populations with a consequent increase in speciation rates. Coevolution of plants and pollinators then ensures high animal-pollinator speciation as well.

Of course these eight theories are not mutually exclusive and can be combined in many permutations. Doubtless, new theories will be advanced. New empirical data suggest that for some groups, such as aphids and ichneumonids, highest species diversity is not in the tropics but in the temperate regions (Janzen 1981; Dixon et al. 1987), though for most vertebrate taxa and many invertebrates the trend of higher diversity in the topics still holds. Why the large concern over species diversity? There are at least two major reasons. First, species diversity is the basis for many fields of scientific research and education (U.S. Congress Office of Technology Assessment, 1987). A variety of species provides a variety of raw materials for biomedical research and drug synthesis. Desert pupfishes, found only in the U.S. Southwest, tolerate salinity twice that of sea water and are valuable models for research on human kidney disease. The armadillo is one of only two nonhuman species known to contract leprosy. These animals now serve as research models with which to find cures for human diseases (see Myers 1983; Chapter 20). Second, most ecological research is based on trying to find order in seeming chaos, to discern patterns and create holistic theories from a mass of biological data. One of the hallowed tenets of ecology is that diversity begets stability. More diverse communities are supposedly more stable and less subject to change. Stable communities are perhaps easier for humans to manage. Addressing this problem requires examination of the concept of **stability.**

17.4 *Stability*

Stability can be thought of in two ways: (1) the ability of a community to resist change following a disturbance, sometimes termed *resistance*, or (2) the ability of the community to bounce back to its original configuration after a perturbation, sometimes called *resilience* or elasticity. These ideas are perhaps best explained by reference to Fig. 17.9, in which the community is represented as a black ball on a range of topographic conditions. In Fig. 17.9a, the community is locally and globally stable and will always return to the same point of equilibrium. In Fig. 17.9b, the community is locally stable at X but will move to a new configuration, Y or Z, if perturbed severely enough. Such is often the result when exotic species are introduced. In Fig. 17.9c, the community exhibits **global stability** but not local stability, easily moving between X and Y. Local stability can easily be upset by some ecological disaster, an oil spill or other pollution event. In Fig. 17.9d, the community is again locally stable but may go extinct if the disturbance is too great, leaving open the possibility of colonization by other species. Some systems seem to be particularly vulnerable to perturbation; lakes, for example, act as natural traps and sumps for runoff. Rivers, on the other hand, can cleanse themselves more rapidly and can be said to be elastic, with a great ability to recover after damage (Fig. 17.9a). Deserts are essentially resistant to change in the first place. Estuaries can be said to be poorly resistant but elastic; the rapid and two-way flow of fresh and salt water essentially cleanses them very quickly. At the risk of being considered pedantic, Connell and Sousa (1983) suggested that there are no true multiple stable states for communities in nature. Thus, if a community were perturbed enough to move from condition X in Fig. 17.9b to condition Y or Z, it would in reality constitute a somewhat different community with slightly different biological components arranged to reflect slightly different physical variables. There is no evidence that a natural community will exist in two or more states in nature.

That diversity causes stability is suggested by several lines of circumstantial evidence (Elton 1958).

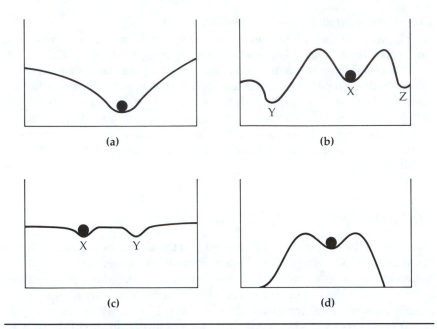

Figure 17.9 Local and global stability concepts where the community is represented as a
black ball on a series of topographic surfaces. **(a)** Community that is both locally and
globally stable. **(b)** Community that is locally stable at x but that, if perturbed enough, will
reach a new configuration at either Y or Z. **(c)** Community that is globally stable but locally
unstable, in that a slight change will shift it from X to Y. **(d)** Community that is stable but
in which, because there are no alternative states, a large disturbance will result in wholesale extinctions and colonizations by different species.

1. Laboratory experiments by Gause (1934) confirmed the difficulty of
   achieving numerical stability in simple systems.
2. Small, faunistically simple islands are much more vulnerable to invading species than are continents. Most natural species on remote
   oceanic islands have been selected for high dispersal ability, not competitive dominance.
3. Outbreaks of pests are often found on cultivated land or land disturbed by humans, both of which are areas with few naturally occurring species.
4. Tropical rain forests do not have insect outbreaks like those common
   in temperate forests.
5. Pesticides have caused pest outbreaks by the elimination of predators
   and parasites from the insect community of crop plants.
6. In a review of 40 food webs, the complexity of food webs in stable
   communities has been found to be greater than the complexity of
   food webs in fluctuating environments (Briand 1983).

Such evidence suggests that stability should increase as the number of links in food webs increases; simple mathematical analysis of food webs bears out this prediction (MacArthur 1955). If diversity is equated with stability, then stability equals $-\Sigma p_i \log_e p_i$. In a food web with four links, one predator and four prey (Fig. 17.10a), each link carries 0.25 of the total energy in the food web, and stability $= -(4 \times 0.25 \times \log 0.25) = \log 4 = 1.38$. Adding another predator that eats all the prey doubles the number of routes to 8, and stability $= \log 8 = 2.08$ (Fig. 17.10c). It is also worth noting that a given stability can be achieved by way of a large number of species, each with a fairly restricted diet (Fig. 17.10d), or by way of a smaller number of species, each eating a wide variety of other species. Maximum stability for m species results when there are m trophic levels with one species each, and each species eats all species below (Fig. 17.10e). Restricted diets lower stability in general, but in practice specialization may be essential for the most efficient exploitation of prey (in terms of caloric investment and return). A compromise is evident between these two strategies in different ecosytems. In Arctic regions, where few species coexist, stability may be hard to achieve. Foxes in

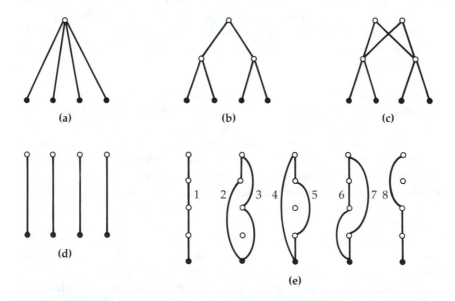

Figure 17.10 *Feeding links in a community between, for example, herbivores (•) and predators (○) where equal amounts of energy pass up each link.* Arrangements such as (a), (b), and (d) lead to equal stability in the system, but stability in (c) is higher. In a five-species system, maximum stability is achieved as in (e), where all species can feed at all trophic levels below, providing eight routes up which energy can flow. Thus the system in (e) with five species is as stable as the system in (c) with eight species.

these regions have a very broad diet, but it does not fully compensate for the small number of species, and populations fluctuate considerably. In the tropics, where there is a large number of species, stability can be achieved with fairly restricted diets, and species can specialize in their feeding habits, feeding on only one or two trophic levels.

More recently, however, considerable doubt has arisen as to the linkage between diversity and stability. Murdoch (1975) indicates that there is no convincing field evidence that diverse natural communities are generally more stable than simple ones.

1. The fluctuations of microtine rodents (lemmings, voles, and so on) are as violent in relatively complex temperate ecosystems as they are in simple Arctic environments.
2. Goodman (1975) has argued that the stability of tropical ecosystems is a myth and that there are reports of cases in which insects nearly completely defoliated Brazil-nut trees and monkeys succumbed in large numbers to epidemics.
3. Rain forests seem particularly susceptible to man-made perturbations (May 1979).
4. Agricultural systems may suffer from outbreaks not because of their simple nature but because their individual components are often bioengineered and have no coevolutionary history whatever, in complete contrast to the long associations evident in forest biomes (Murdoch 1975).

Furthermore, May (1973) has argued that increasing complexity actually reduces stability in hypothetical models. When trophic links are assembled at random, the more diverse communities are more unstable than the simple communities. May cautioned ecologists that, if diversity causes stability in nature, it does not do so as a direct consequence of the mathematics of the situation. Zaret (1982) has also criticized the diversity-stability notion by means of studies on fish-community diversity, which showed that diverse communities are more easily perturbed by introduced predators.

One of the few experimental tests of the diversity-stability hypothesis was performed in the laboratory with microorganisms (Hairston et al. 1968). *Paramecium*, when cultured with bacterial prey, showed less tendency to go extinct with only one species of prey present (extinction rate = 32 percent of cultures) than with two or three prey species available (extinction rates = 61 percent and 70 percent respectively). When the number of *Paramecium* species was increased to three, the results depended on which particular species was added to which other two. It was evident that species were not simply interchangeable numerical units. Finally, when a third trophic level was added, in the form of the predator protozoa *Didinium* and *Woodruffia*, there was a further decrease in stability because *Paramecium* were usually forced to extinction regardless of how many species of *Paramecium* were present or whether one or two predator species were present. In these simple systems, diversity did not automatically lead to stability. Most important, species cannot always be regarded as simply interchangeable. This point casts doubt on mathematical models of stability, all of which treat species as simple equivalents.

What then is the conclusion about how communities are organized around such concepts as stability and diversity? The older, more conventional view can be termed the *equilibrium* hypothesis, in which local population sizes fluctuate little from equilibrium values, which are thought to be determined by predation, parasitism, and competition. In this view, communities are stable and disturbances are damped out; species diversity is determined by the diversity of available niches. In the more modern nonequilibrium hypothesis, species composition is viewed as constantly changing and never in balance. Stability is elusive, and persistence (the length of time for which a community exists) and resilience are the most apt measures of community behavior. The most probable mechanism for the nonequilibrium hypothesis is the intermediate-disturbance hypothesis of Connell (1978) (Fig. 17.11), in which highest diversity is maintained at intermediate levels of disturbance. The rationale is that, at high levels of disturbance, for example where environmental variables are extreme—low temperature, high temperature, low rainfall, high

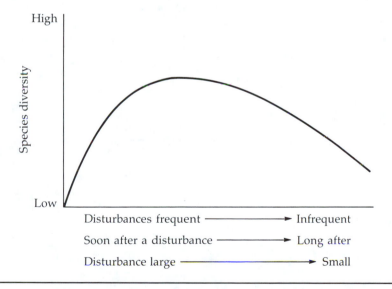

Figure 17.11 *The intermediate-disturbances hypothesis of community organization. The species composition of a community is never in a state of equilibrium, and high species diversity can be maintained only at intermediate levels of disturbances like fires or windstorms. (Redrawn from Connell 1978.)*

wind, high frequency of fire, and the like—only good colonists, *r*-selected species, will survive, giving rise to low diversity. This theory also predicts that, at low rates of disturbance, competitively dominant species will outcompete all other species and only a few *K*-selected species will persist, again giving low diversity. The most diverse communities lie somewhere in between. Natural communities seem to fit into this model fairly well. Tropical rain forests and coral reefs are both examples of communities with high species diversity, and both were used as evidence for the old equilibrium hypothesis. However, Connell (1978) has pointed out that coral reefs maintain their highest diversity in areas disturbed by hurricanes. He has also argued that the richest tropical forests occur where disturbance by storms or people has been documented.

Sousa (1979) has also provided an elegant experimental verification of the intermediate-disturbance hypothesis in a marine intertidal situation. He found small boulders, easily disturbed by waves, to carry a mean of 1.7 species of sessile plant and animal species; large boulders, rarely moved by waves, had a mean of 2.5 species, and intermediate-sized boulders 3.7 species. Sousa then cemented small boulders to the ocean floor to show that these results were a result of rock stability, not rock size. Furthermore, diversity not only changed in response to disturbance but often did so in a predictable way, in a sequence of events known as **succession.**

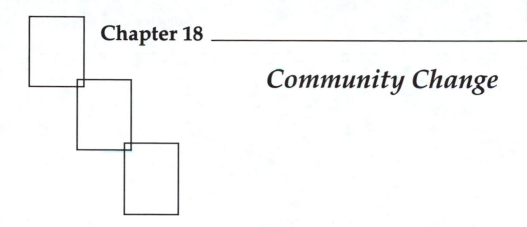

Chapter 18

Community Change

Disturbance in most communities is provided by catastrophic events such as fire, grazing, or erosion. Following such events, bare ground or "light gaps" (patches of clear sky in the canopy) often result in forests, and clear substrate may result in intertidal situations. Species diversity then gradually changes as the community returns to "normal." This change can be predictable and orderly and is termed **succession.**

18.1 Succession

The concept of succession was first developed by botanists who monitored floristic changes along sharp environmental gradients, for example, from seashore and sand-dune halophytes back to scrub and finally mature woodland. The key assumption was that each invading species made the environment a little different (say a little less salty or a little more shady) so that it then became suitable for more K-selected species, which invaded and outcompeted the earlier residents (Clements 1936). This process supposedly continued until the most K-selected species had invaded, when the community was said to be at climax. Retrogression in this sequence was not possible unless another disturbance intervened. The **climax community** for any given region was thought to be determined by climate and soil conditions. Succession in some **ecosystems** seems to fit this pattern well. Over the past 200 years, the glaciers in the Northern Hemisphere have undergone dramatic retreats, up to 100 km in some cases (Fig. 18.1), sure evidence of a climatic change. As glaciers retreat, they leave moraines, deposits of soil and stones brought forward by the ice, whose age has often been determined by direct observation. Cooper (1923) and Crocker and Major (1955) demonstrated distinct soil-type changes and differences in floral diversity with age of moraines. The basic bare soil has a pH of 8.0 and has low nitrogen content. It is first colonized by mosses, fireweed, *Bryas*, willows, and cottonwood. Very

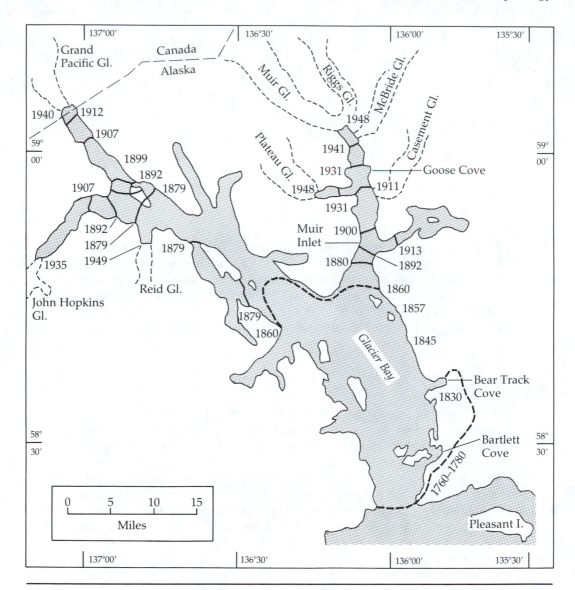

Figure 18.1   *Glacier Bay fjord complex of southeastern Alaska showing the rate of ice recession since 1760.*   (Redrawn from Crocker and Major 1955.)

quickly, alder then invades the area. Because alders can fix atmospheric nitrogen, the nitrogen content of the soil increases dramatically, and pH falls to about 5.0. After about 50 years, the dense stands of alder (*Alnus* sp.) are invaded by Sitka spruce, which after another 120 years may form a dense forest. The alders are shaded out and die. Spruce trees cannot fix nitrogen directly; they take it from the soil, lessening the available nitrogen and stabilizing the pH at about 5.0. Finally, western hemlock and mountain hemlock invade to form the climax spruce-hemlock forest. The only exception to this story is on poorly drained soils, where *Sphagnum* mosses invade the spruce-hemlock forests. They tend to hold yet more water, and eventually the tree roots become waterlogged and too oxygen deficient to survive; the area becomes a *Sphagnum* bog called a *muskeg*. The entire succession sequence from bare substrate to climax is called a **sere**.

Other types of succession do not fit the classical pattern so well. After the retreat of the glaciers from the Great Lakes area, lake levels began to drop and with each drop a raised beach was left. These raised beaches, still evident today parallel to the shores of Lake Michigan, are about 7, 12, and 17 m above the present water level. They are identical in substrate and differ only in age, presenting a good chance to study succession. Cowles (1899) first worked on the system and documented a succession from bare sand to marram grass (*Ammophila Breviligulata*) in six years; sand reed grass (*Calamovilfa longifolia*), little bluestem (*Andropogon scoparius*), sand cherry trees (*Prunus* sp.), and willows (*Salix* sp.) after 20 years; jack pines and white pines after 50–100 years; and black oaks at about 150 years. Cowles predicted the climax community would eventually be beech-maple forest. However, Olson (1958) studied very old sand dunes in this area (about 12,000 years old) and could see no further evidence of succession beyond the black oak stage. Olson also noted that, though soil pH changed from 7.6 at the start of succession to 4.0 after the first 1,000 years, from then on it stabilized, and conditions for the establishment of climax beech-maple forest seemed to diminish (beech and maple require neutral pH and more water). Beech-maple "climax" forest is generally found on moister soils with higher nutrient contents. Black-oak communities are ineffective at nutrient cycling through leaf litter, so on dunes this type of community appears stable. It is also probably true that the actual "target" climax community is likely to keep changing. In New England, old-field successions might take 100–300 years to reach climax, but frequency of major disturbance, fire or hurricane, is every 70 years or so.

Such differences in the patterns of succession in nature prompted Connell and Slatyer (1977) to outline three different models governing succession. The classical view of succession, proposed by Clements (1936), which supposes an orderly progression to a predictable climax community, Connell and Slatyer termed the *facilitation model*, as each species makes the environment more suitable for the next. For the type of succession found on the Michigan sand dunes, they formulated the *inhibition model*, because initial colonists tend to prevent subsequent colonization by other species. Thus, black oaks appear to lessen the chance that beeches and maples will invade. In this model, succession depends on chance events (essentially who arrives first). Succession proceeds as colonists die but it is not orderly. Connell and Slatyer then proposed a third model, which they termed the *tolerance model,* in essence intermediate between the other two. In this model, any species can start the succession, but the eventual climax community is reached in a somewhat orderly fashion. All three models predict that the most likely earlier colonists will be *r*-selected species, weeds in many cases. The key distinction between the models is in how succession proceeds. In the classical facilitation model, species replacement is facilitated by previous colonists, in the inhibition model it is inhibited by the action of previous colonists, and in the tolerance model, it is unaffected (Fig. 18.2). Ten years after the mod-

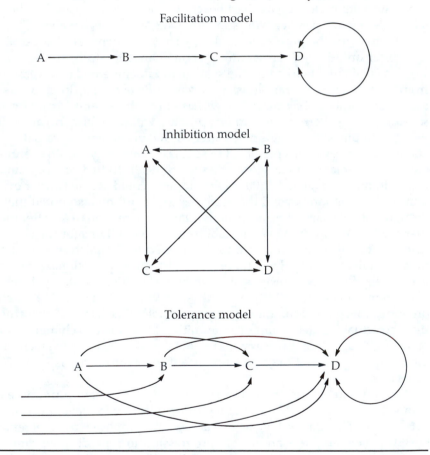

Figure 18.2 *Three models of succession.* Four species are represented by A, B, C, and D. An arrow indicates "is replaced by." The facilitation model is the classic model of succession. In the inhibition model, all replacements are possible, and much depends on who gets there first. The tolerance model is represented by a competitive hierarchy in which later species can outcompete earlier species but can also invade in their absence. (Redrawn from Horn 1976.)

els were proposed and many data later, Connell, Noble, and Slatyer (1987) suggested that these models represent not strict alternatives but rather the ends of a continuum of effects of earlier or later species. In the real world components from each may be prevalent in any one given study system. Thus, for the succession of alder to spruce on Alaskan floodplains, "stochastic, life history, facilitative, competitive and herbivory processes all affect the interaction between alder and spruce during succession and no single successional process or model adequately describes successional change" (Walker and Chapin 1987, p. 131; see also Walker and Chapin 1986). Walker and Chapin (1987) have discussed the relative importance of each of these major factors on communities of a different successional state and have presented their results in a figure, which outlines how major processes determine change in plant-species composition (Fig. 18.3, p. 346). Their conclusion is that succession is a complex process driven by many processes acting simultaneously in any given situation. Thus the effects of such factors as competition, seed-arrival time, insect and mammalian herbivory, and stochastic events vary in importance according to the stage of succession—early, middle, or late. Inouye et al. (1987) provide a detailed comparison of the influence of two factors, nitrogen availability and time since disturbance, on succession in old fields on Minnesota sand plains. Vegetation and soils were sampled in 22 old fields ranging in age from 1 year to 56 years since abandonment. Soil-nitrogen concentration increased significantly with age. Vegetation cover, total above-ground plant **biomass,** and litter cover increased significantly with soil nitrogen. When field age and soil nitrogen were used as independent variables in simple regression and partial correlation analyses, they revealed that, although soil nitrogen was an important determinant of species composition and abundance for some plant groups, dispersal and colonization, which are dependent on field age, are likely to be more important (Fig. 18.4, p. 347).

Connell and Slatyer's original models are also further discussed by Horn (1976), who has provided a simple model of succession that assumes succession is a Markovian replacement process. First, Horn gathered together information on a community, in this case the tree species present in a New Jersey forest. Next, he mapped the seedlings underneath the crowns of the forest trees; these saplings were assumed to replace the mature trees in time. Horn assumed that the higher the number of seedlings of a particular species under a tree, the higher the likelihood that this species would replace the mature tree. Horn thus constructed a series of replacement probabilities like those shown in Table 18.1 (p. 348). For example, there were a total of 837 saplings under large grey birch trees (Horn 1975). Among these, there were no grey birch saplings. Janzen [1970] has argued that this phenomenon, at least in tropical situations, results because seed eaters are often host-specific and congregate

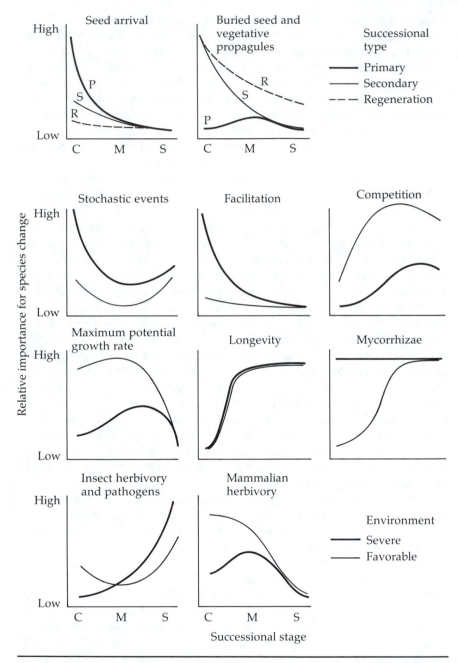

Figure 18.3 *Influence of succession and environmental severity on major successional processes that determine change in species composition during colonization (C), maturation (M), or senescence (S) stages of succession.* (Redrawn from Walker and Chapin 1987.)

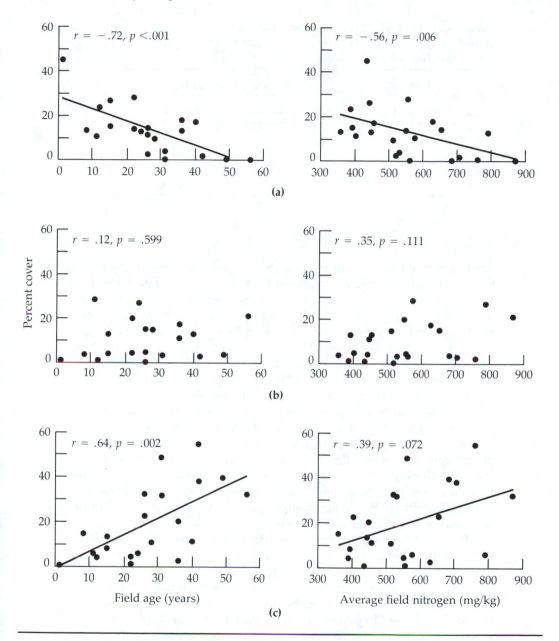

Figure 18.4 *Percent vegetation cover plotted against field age and average field nitrogen concentration for old abandoned fields in Minnesota.* **(a)** *Introduced species.* **(b)** *Native species (but not true prairie species).* **(c)** *True prairie species.* (Redrawn from Inouye et al. 1987.)

Table 18.1 *A 50-year tree-by-tree transition matrix showing the probability of replacement of one individual by another of the same or different species in a New Jersey forest. (From Horn 1975.)*

Present Occupant	Occupant 50 Years Hence			
	Grey Birch	*Blackgum*	*Red Maple*	*Beech*
Grey birch	0.05	0.36	0.50	0.09
Blackgum	0.01	0.57	0.25	0.17
Red maple	0.0	0.14	0.55	0.31
Beech	0.0	0.01	0.03	0.96

under a parent tree, devouring seeds there. Only when seeds happen to fall under different species are they likely to survive. Thus each tree casts a "seed shadow" in which survival of its own kind is reduced—hence the high value of dispersive seeds.) There were 142 red maples, 25 beech seedlings, and a few other species. The replacement probability for grey birch under grey birch is then $0/837 = 0$, for red maple under grey birch $142/837 = 0.17$, and for beech under grey birch $25/837 = 0.03$. These replacement values can be summed to give the total predicted abundance of a species. Beginning with an observed distribution of the canopy species in a stand in New Jersey known to be 25 years old, Horn modelled the change in species composition over several centuries (Table 18.2). Theoretically, red maple is predicted to assume dominance quickly while grey birch disappears. Beech should slowly increase to predominate later. All these predictions agree with what happens in nature. Thus, Horn's model predicts a deterministic successional outcome, independent of the initial forest composition—a prediction very similar to Connell and Slatyer's (1977) tolerance model.

Finally, it is worth remembering that there are some cases in which communities change not in a directional fashion but to a new state of normality. This type of event, more severe and with longer-lasting effects, is often produced by the introduction of a new species into the community. Thus *Endothia parasitica,* the chestnut blight, has forever altered the structure of temperate forests in eastern North America by killing trees, and Dutch elm disease has had the same effect in Europe and the United States. The introduction of weed plants can have similar effects; for example, the arrival of prickly pear cactus in Australia has forever changed the landscape there, though the effects are not so violent now that the cactus has been controlled somewhat by a natural enemy that eats it, the moth *Cactoblastis* (see Chapter 13).

Table 18.2 *The predicted percentage composition of a New Jersey forest consisting initially of 100 percent grey birch. (From Horn 1975.)*

	Age of Forest (Years)						Data from Old Forest
	0	50	100	150	200	∞	
Grey birch	100	5	1	0	0	0	0
Blackgum	0	36	29	23	18	5	3
Red maple	0	50	39	30	24	9	4
Beech	0	9	31	47	58	86	93

18.2 Island Biogeography

Area Effects

Experimental studies on succession were stimulated years ago by observations of succession on offshore islands that had been totally defaunated by volcanic eruptions, such as the one that rocked Krakatau in 1883. The 1980 eruption of Mount St. Helens in Washington state provided another good opportunity to study natural recolonization. In the six years following this eruption, Wood and del Moral (1987) monitored vascular-plant invasion of the barren substrates. They found **dispersal** was limited in many species and that nurse plants played a key role in trapping seeds and promoting seedling establishment. They concluded that the path of early succession depended on spatial position of an area and dispersal abilities of species in the seed pool nearby and that distribution did not reflect environmental gradients. Ultimately, who colonized what depended as much on succession as on the distance to the nearest floral or faunal source. Furthermore, it is important to realize that the ultimate number of species in an area will be dependent not only on distance from source pools but also on their size. A small area is unlikely to support as many different types of colonists as a big area, if only because the range of **habitats** in a small area is smaller. This relationship is best described by the equation

$$S = c A^z$$

or, in logarithmic form,

$$\log S = \log c + z \log A$$

where S = number of species, c = a constant measuring the number of species per unit area, A = area, and z is a constant measuring the slope of the line relating *S* and *A*. An example of a species-area relationship is given in Fig. 18.5, which shows how the number of species of reptiles and amphibians of Caribbean islands increases with island size. Remarkably, the constant *z* tends to be about 0.3 for a variety of island situations, including the amphibians and reptiles of the West Indies, beetles in the West Indies, ants in Melanesia, vertebrates on islands in Lake Michigan, and land plants in the Galapagos (Preston 1962). Human introductions can sometimes upset this balance (see Section Seven and Chaloupka and Domm 1986).

Preston (1962) has compared species-area curves for continental land masses. For example the species-area curve for North American birds is shown in Fig. 18.6. From an area of about 10 acres to about 1 billion acres the relationship between area and species diversity is a straight line of the form

$$S = 40\,A^{0.17}$$

The species-area relationship for flowering plants in England is shown in Fig. 18.7. In general, species-area curves for continental areas or for parts of large islands had *z* values (slopes) from 0.15 to 0.25, generally lower than the 0.3 values found in island studies. Thus, as larger areas are sampled, fewer new species are added on continents than on islands. Many more examples are provided, along with a full discussion, by Connor and McCoy (1979), and for insects by Strong (1979) (Fig. 18.8, p. 352). The rationale here is that each area in the continental studies, except the largest, contains some transient species from adjacent habitats. So the slope

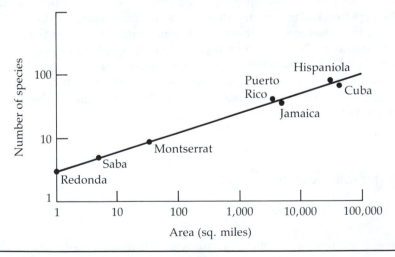

Figure 18.5 *Species-area curve for the amphibians and reptiles of the West Indies.* (Modified from MacArthur and Wilson 1967.)

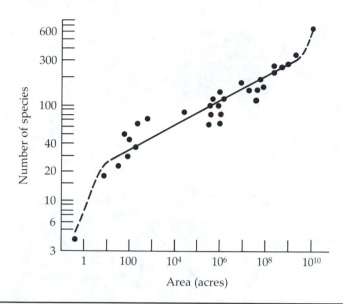

Figure 18.6 *Species-area curve for North American birds.* The points range from a 0.5 acre (0.2 hectare) plot with three species in Pennsylvania to the whole United States and Canada (4.6 billion acres, 1.86 billion hectares) with 625 species. (Redrawn from Preston 1960.)

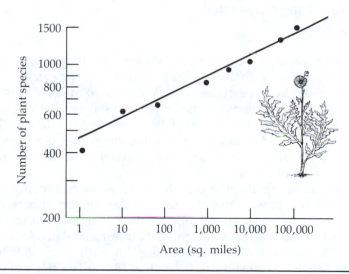

Figure 18.7 *Species-area relationship for flowering plants in England.* The smallest plot is 1 mi^2 (2.6 km^2) in Surrey, and the largest plot is the whole of England, 87,417 mi^2 (227,284 km^2). (Modified from Williams 1964.)

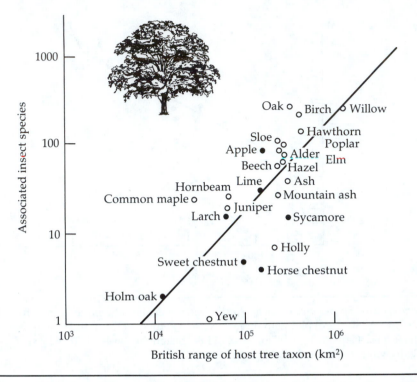

Figure 18.8 *The species-area relationship of the insects of British trees (r = 0.78, P < 0.001).* Closed dots indicate introduced taxa; open dots indicate British natives. (Modified from Strong 1974*a*.)

of the species-areas curve is shallow. Islands are actual isolates with reduced migration rates, so the number of transients in an area is minimal, thus steepening the slope of the curve. What is also interesting is that in many cases the population densities of specific species changes with the size of habitat patches, too. Kareiva (1983) reviewed the data for 19 species of herbivores and found that nine exhibited greater per-plant abundances with increasing patch size, eight revealed no effects, and two were less abundant in large patches.

Not all island-mainland data fit the species-area dichotomy quite so clearly, however. Brown (1978) studied the distribution of boreal mammals and birds in the isolated ranges of the Great Basin in the United States. Each mountaintop was essentially a forest island in a sea of desert (Fig. 18.9a). The species-area relationship for birds (Fig. 18.9b) had a slope of 0.165, that for mammals 0.326 (Fig. 18.9c). The slope of the line for mammals was more like that found on islands. The reason is that the mountain ranges were essentially isolated, and the mammalian fauna was a shrinking relict community of a bygone age when rainfall was

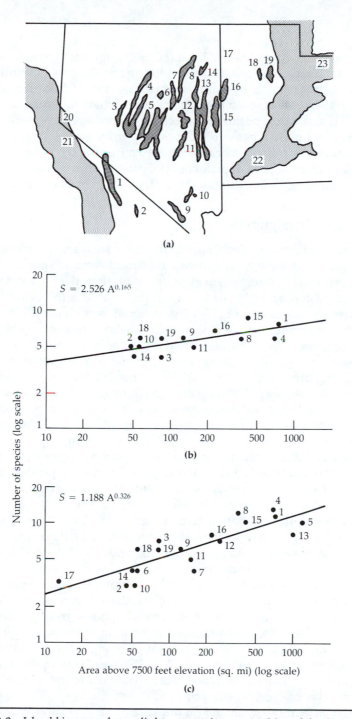

Figure 18.9 *Island biogeography applied to mountaintops.* **(a)** *Map of the Great Basin region of the western United States showing the isolated mountain ranges between the Rocky Mountains on the east (right) and the Sierra Nevada on the west (left).* **(b)** *Species-area relationship for the resident boreal birds of the mountaintops in the Great Basin.* **(c)** *Species-area relationship for the boreal mammal species.* Numbers refer to sample areas on the map. (Redrawn from Brown 1978.)

higher and this type of boreal habitat was contiguous. There is no mammalian migration between mountaintops. In this situation, mountaintops
behave as true islands. By contrast, birds disperse much better, and the z
value in their case is much more in line with a mainland type of relationship. It is also worth noting that, because birds disperse so well, there is
also no effect of distance of a mountaintop from a possible source pool.
This situation is unusual because it goes against the usual belief that colonization of an area is dependent on distance from a source pool. For
most organisms, in most situations, distance is an important parameter.

Rates of Emigration

The **immigration** rates of species into new areas and their **emigration**
rates from them have great influence on succession and species diversity
in a given plot. If immigration exceeds emigration, then the net result is
an increase in the number of species; if the reverse is true, then the area
will lose species over time. Usually, emigration is equated with extinction. For any potential colonist it will probably be easier to colonize islands or areas with an impoverished fauna, because few competitors or
natural enemies exist. Consequently, immigration rates are often thought
of as being highest on islands. With extinction, for a given number of
species, the loss of species from an area (through competitive extinction)
is likely to be higher in smaller areas, where competition is severe, than
in bigger areas, which can support more species. These ideas were expressed by MacArthur and Wilson (1967) (Fig. 18.10). The effects of island size and island distance from a source pool on the expected number
of species on an island can be superimposed on this graph (Fig. 18.11).

Probably the best experimental study of island species diversity,
called species-area effects, and distance from source pools was performed by Daniel Simberloff (Simberloff and Wilson 1970; Simberloff
1976), who fumigated six very small islands, consisting only of mangroves, off the Florida Keys in 1966–1967 and who followed the subsequent patterns of recolonization. Photo 18.1 (p. 356) shows one of the
islands and illustrates the method used to fumigate them. Because of the
simplicity of the islands, pure mangroves, there was no increasing habitat diversity with area. About 4,000 animal species, mostly insects, spiders, and other terrestrial arthropods, constitute the species pool in the
Keys region, and about 500 are resident on mangrove (Simberloff 1978).
The recolonization curves (Fig. 18.12, p. 357) rose to an upper asymptote
in eight-to-nine months and then declined slightly to the original equilibrium number of species observed.

Islands near the mainland (in this case two meters away) did reach a
higher equilibrium level than islands farther (533 m) away. Species **turnover** was indeed rapid, with only around 40 percent of the species recorded prior to a year after fumigation. Unfortunately, the turnover rate

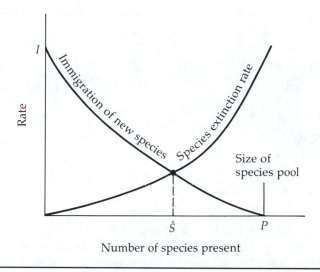

Figure 18.10 *MacArthur and Wilson's equilibrium model of a biota of a single island.* The equilibrial species number (S) is reached at the intersection point between the curve of rate of immigration of new species not already on the island and the curve of extinction of species from the island. (Redrawn from MacArthur and Wilson 1967.)

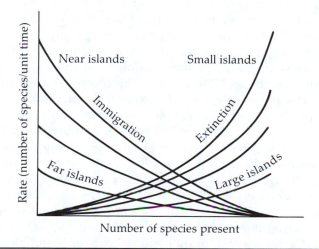

Figure 18.11 *Equilibrium models of biotas of islands with varying distances from the principal source area and of varying size.* An increase in distance (near to far) lowers the immigration curve; an increase in island area (small to large) lowers the extinction curve. (Redrawn from MacArthur and Wilson 1967.)

Photo 18.1 Experimental defaunation of a mangrove islet in the Florida Keys. (a) Construction of a scaffold frame. (b) Installation of a large tent into which insecticide was introduced, killing all life. Tent and scaffold were removed after defaunation, and recolonization was monitored. (Photos by Daniel Simberloff, Florida State University.)

for these small arthropods was so high that species may have colonized an island, gone through a generation, and gone extinct all in the interval between two censuses, so immigration and emigration rates could not be accurately calculated. Furthermore, Simberloff (1978) suggested that most of the observed turnover was in fact due to the "extinction" of species that were only transients in the first place. Nevertheless, the data fit the MacArthur-Wilson model well and probably greatly contributed to its acceptance as dogma. Unfortunately, many subsequent studies have not shown extensive turnover, casting the equilibrium theory of island biogeography into great doubt (Gilbert 1980). This controversy is important because island biogeographic theory has come to be uncritically accepted by conservationists and park planners and is regularly incorporated into refuge design (Boecklen and Simberloff 1986). Island **biogeography** theory was thought to be particularly useful in the continuing debate over the design of wildlife preserves, particularly in the question of whether planners should design many small preserves or a few large ones (given the unlikely prospect of choice). The International Union for Conservation of Nature and Natural Resources (IUCN), a major worldwide conservation group, has stated that refuge design criteria and management practice should be in accord with the equilibrium theory of island biogeography (IUCN 1980). As will be made clear later (Chapter 21), this idea is on shaky ground at best, and more consideration should really be given to the **autoecology** of target species.

It is interesting to note that the equilibrium theory of island biogeog-

Predefaunation surveys

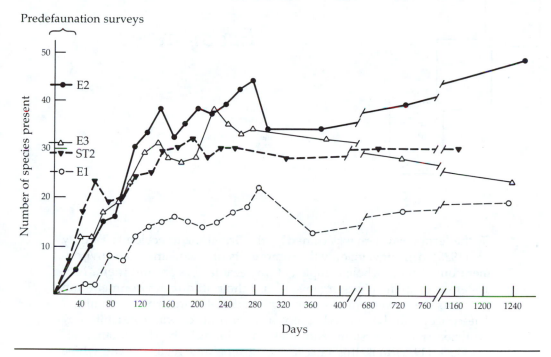

Figure 18.12 Colonization curves for arthropods on mangrove islands near Sugarloaf Key, southern Florida, after defaunation by fumigation. Pre-defaunation species numbers are on the ordinate. (Redrawn from Simberloff 1978.)

raphy, first credited to MacArthur and Wilson (1963), was actually presented 15 years earlier by a British lepidopterist named Munroe, in his doctorial thesis, which dealt specifically with the distribution of butterflies in the West Indies (Munroe, 1948). Munroe noted that the bigger an island, the larger its butterfly fauna, but he never actively pursued this aspect further, instead sticking to taxonomic interests. In rediscovering this work, Brown and Lomolino (1989, p. 1957) note,

> Important as new ideas are in the progress of science, they are often not the unique inspirations of genius that are portrayed in the textbooks. On the one hand, scientific revolutions usually do depend on major conceptual innovations. On the other hand, in order for those insights to have impact they must be promoted cogently at a receptive stage in the development of a discipline. It is not sufficient to have a good idea, it is even more important to develop and publicize it.

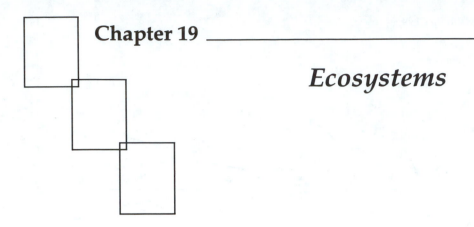

Chapter 19 _____

Ecosystems

The term **ecosystem** was coined by the British plant ecologist Tansley (1935) to include not only the **community** of organisms in an environment but also the whole complex of physical factors around them. Ecosystem ecology involves organisms and their **abiotic** environment and concerns the movement of energy and materials through communities. The concept can be applied at any scale: a drop of water inhabited by protozoa is an ecosystem, and a lake and its biota constitute another. Lovelock (1979) took this idea to its extreme and regarded the whole Earth as one totally interlocked ecosystem, which he named Gaia, after the ancient Greek Earth goddess. In this viewpoint, Gaia was reminiscent of one superorganism, forever regulating temperature, oxygen, and moisture levels to ensure the continuation of life. Lovelock pointed out that levels of such things have not changed appreciably in hundreds of millions of years, whereas on the basis of physical changes alone, such alterations would have been expected. Most ecosystems can never really be regarded as having definite boundaries. Reiners (1986) has argued that, for this reason and others, the ecosystem remains the least coherent of the organization levels of ecology. He suggests it lacks a logical system of interconnected principles and a well-understood and widely accepted focus. Even in a clearly defined pond ecosystem (Fig. 19.1), waterfowl may be moving in and out. The big advantage of ecosystems ecology, however, is the common currency of energy or nutrients, which allows the biology of communities and populations to be compared between and within trophic levels, something no other ecological discipline can boast.

In attempts to determine the importance of individual units within the scheme of an ecosystem, at least three major constituents can be measured. The first is **biomass,** the standing crop of an organism. Attaching too much importance to biomass, however, may lead to erroneous conclusions in some cases. In the applied world of timber technology, a small

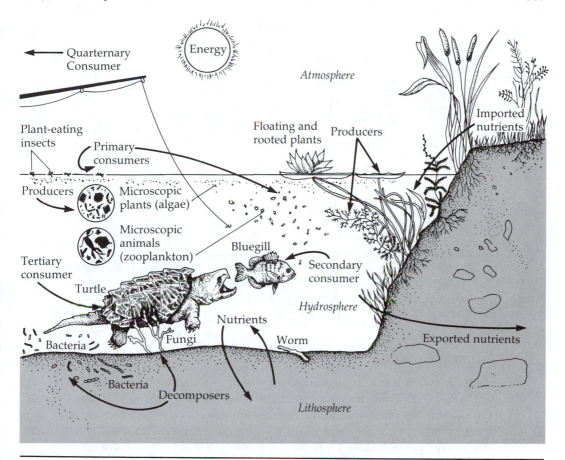

Figure 19.1 *Producer, Consumer, and Decomposer in a pond ecosystem.* Each of these roles is filled by a number of different organisms. For example, additional secondary or tertiary consumers might be water snakes, snapping turtles, and various birds of prey. Although ecosystems are often thought of as closed systems, none really is. Typically, both living and nonliving things are imported and exported.

standing crop may indicate small potential harvest or yield, but this is not necessarily true if the tree species in question has a high growth rate—new biomass will be produced rapidly and a higher rate of harvest could be sustained (Fig. 19.2). In this situation, energy flow may be more critical. Then the community is regarded as an energy transformer; energy flow is the second constituent. Third, the ecosystem may be most limited by the availability of a rare chemical or mineral. In this case, the flow of limiting chemicals through ecosystems becomes the most important factor in understanding how systems work.

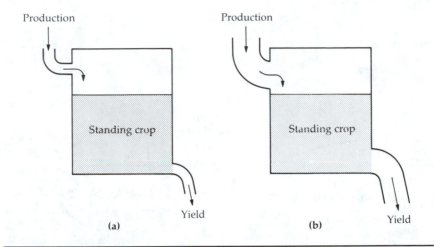

Figure 19.2 *Hypothetical illustration of two equilibrium communities (where input equals output).* **(a)** *Low input, low output, slow turnover.* **(b)** *High input, high output, rapid turnover.* Standing crop is not related to production or yield because turnover time for all systems is not a constant.

19.1 *Trophic Levels, Food Webs, and Guilds*

There are two major ways in which organisms derive energy; **autotrophs** pick up energy from the sun and nonliving sources, whereas **heterotrophs** eat living matter or, in the case of saprophytes and decomposers, dead material originally derived from living autotrophs. The transfer of food energy from the source, in plants, through herbivores to carnivores occurs through the **food chain.** Two examples of food chains are

Example 1	*Example 2*
Sun	Sun
Producer corn	Producer pine tree
Primary consumer/herbivore corn earworm	Herbivore aphid
Secondary consumer/parasite ichneumonid wasp	Predator spider
Tertiary consumer/carnivore warbler	Predator chickadee
Tertiary consumer/carnivore hawk	Predator hawk

Every step in the food chain is termed a **trophic level,** and **herbivores** and **carnivores** feed at different trophic levels. **Omnivores** are unusual in that they feed on at least two trophic levels, plant and animal. Charles Elton (1927) was one of the first to explain the importance of food chains to ecology. He pointed out that most chains are actually fairly short and contain no more than five or six links. Some of the most important links in the food chain are those involving detritivores. It is generally underappreciated that much primary production is not consumed by herbivores but dies and rots on the ground to be consumed by detritivores (see Fig. 19.3). Bacteria and fungi are thus common constituents of many food chains. Often the rate at which mineral resources such as phosphorus and nitrogen, locked up in dead material, are released and again made available for growing plants is controlled by the rate of action of the decomposers.

The classification of organisms by trophic levels is one of function rather than species. For example, male horseflies are herbivores, feeding on nectar and plant juices, whereas females are blood-sucking ectoparasites. Most complete food webs are extremely complicated, even when a simplified subset is considered, as for example the species associated with Wytham Woods, an intensively studied area in England (Fig. 19.4).

In Wytham Woods, and throughout the deciduous forests of Western

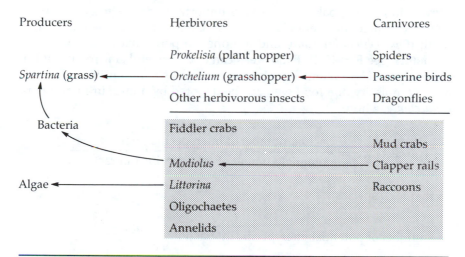

Figure 19.3 Food web of a Georgia salt marsh with groups listed in their approximate order of importance. Often *Spartina* grass is the only plant in this food web. Detritivores and the species that feed on them (under the dotted line) constitute a large fraction of the total species. Note that although newer studies have revealed the presence of additional insect herbivores (Stiling, Brodbeck, and Strong 1982; Stiling and Strong 1983), the general pattern remains unaltered. (Redrawn from Teal 1962.)

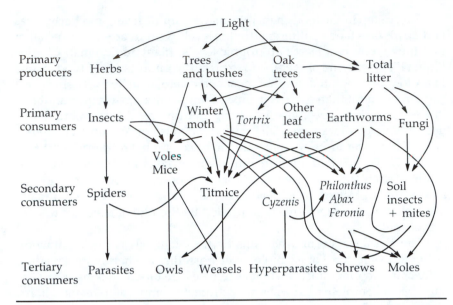

Figure 19.4 *Simplified food web for Wytham Woods, England.* (Redrawn from Varley 1970.)

Europe in general, oak trees are very important and are fed on by over 200 species of lepidopteran larvae alone. One of these is the oak winter moth (*Operophtera brumata*), and it alone supports many parasites and predators (see Fig. 15.1). Food webs can form a useful starting point for the analysis of ecosystem organization (Cohen 1978; Pimm 1982). The relative complexity of a food web can be denoted by a measure known as *connectance*, where

$$\text{connectance} = \frac{\text{actual number of interspecific interactions}}{\text{potential number of interspecific interactions}}$$

so that for *n* species

$$\text{number of potential interspecific interactions} = \frac{n(n-1)}{2}$$

In Fig. 19.4, there are 22 "species," 41 actual interactions, $(22 \times 21)/2 = 231$ potential interactions, and a connectance of 0.18. It must be realized that in such idealized webs higher carnivores are assumed to be able to feed on what may be potentially unrealistic sorts of prey. Auerbach (1979) showed that many food webs in the real world do not lend themselves well to such analyses because linkages may switch according to prey availability. Worse still, Paine (1988) suggests that biologists draw their foods webs in an informal, almost casual manner, omitting many links. Also, many species may be aggregated into larger, convenient groups, difficult to disentangle taxonomically, such as ''spiders'' or even ''invertebrates.'' This aggregation disguises much relevant trophic information. As a result, Paine suggests, mean connectance is not always a valuable thing to measure. For comparative purposes, however, such analyses may be valuable. Pimm (1982) has suggested that, as the complexity of a food web increases, its connectance falls (Fig. 19.5). Some other generalities are shown in Fig. 19.6. Omnivores tend to be rare except in insect communities. Where they occur, omnivores are normally found feeding on species in adjacent trophic levels. Predators tend to outnumber prey by a factor of 1.3 (Cohen 1978; Briand and Cohen 1984; Cohen and Briand 1984). Finally, and perhaps surprisingly, food chains are fairly short, commonly having a median length of four links and generally varying only between three and five links in different environments (Schoener 1989).

In a further level of analysis, each trophic level can be split up into functional units called *feeding guilds*. Herbivores, for example, may form the leaf-chewing guild, the sap-sucking guild, the stem-boring guild, and

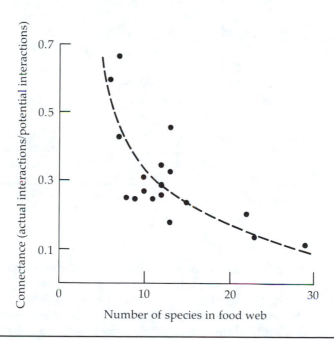

Figure 19.5 *Connectance of 18 food webs in relation to the number of species in the web.* There are fewer interactions in food webs as species richness increases. (Redrawn from Pimm 1982.)

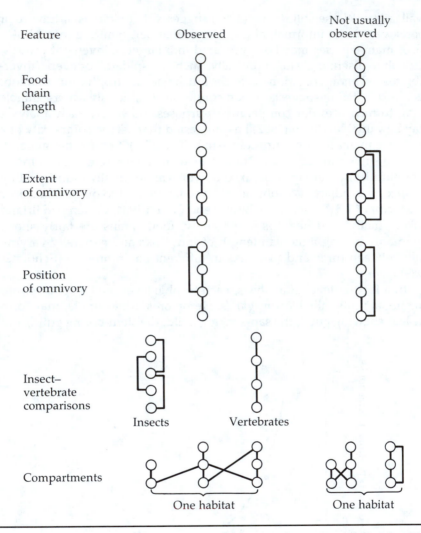

Figure 19.6 *A partial summary of the features observed in natural food webs. (Re-drawn from Pimm 1982.)*

the leaf-mining guild. The stem-boring guild of salt-marsh cord grass, *Spartina alterniflora*, a common salt-marsh plant, is shown in Fig. 19.7. Root (1967) was the first to pioneer this type of analysis during a consideration of **niche** exploitation patterns in birds, and he returned to the same theme in a study of the insect guilds associated with collard plants at Ithaca, New York (Root 1973). Species in a guild may be expected to exhibit certain properties, among the most important of which is that severe competition may occur between members because they are feeding

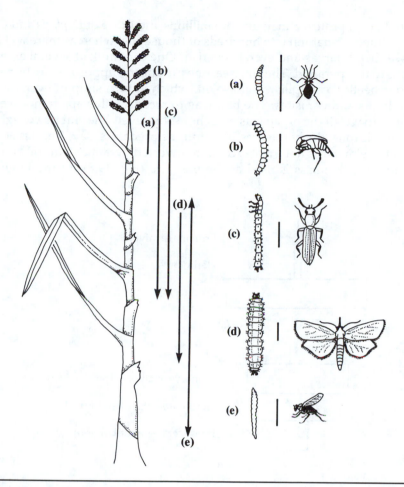

Figure 19.7 *The stem-boring guild associated with salt-marsh cord grass,* Spartina alterniflora, *on the Gulf Coast of North America.* **(a)** Calamomyia alterniflorae *(Diptera).* **(b)** Mordellistena splendens *(Coleoptera).* **(c)** Languria taedata *(Coleoptera).* **(d)** Chilo plejadellus *(Lepidoptera).* **(e)** Thrypticus violaceus *(Diptera).* Arrows indicate where in the stem the larva of each species is found and the direction it bores. Adults, shown at extreme right, are free-living. The length of each scale line between the larva and adult of each species represents 0.5 cm in the drawing.

in the same manner. Of course, according to this line of thinking, competition between conspecific individuals should always be even more intense.

Other widely accepted phenomena evident from a study of food chains are that top predators tend to be rather large and sparsely distributed, whereas herbivores are smaller and more common. This generalization is often termed *the pyramid of numbers*. In a small pond, the numbers of protozoa may run into millions and those of *Daphnia* and *Cyclops* (their predators) into hundreds of thousands, whereas there will be fewer beetle larvae and even fewer fish. One can think of several exceptions to this pyramid. The elm tree, one producer, supports many herbivorous beetles, caterpillars, and so on, which in turn support even more predators and parasites. The best way to reconcile this apparent exception is to weigh the organisms in each trophic level. The elm tree weights 27 metric tons, the herbivores 50 g, and the predators, say, 5 g. It is clear that the elm tree is not a real exception. Inverted pyramids can still occur, even when biomass is used as the measure (Fig. 19.8). In the English

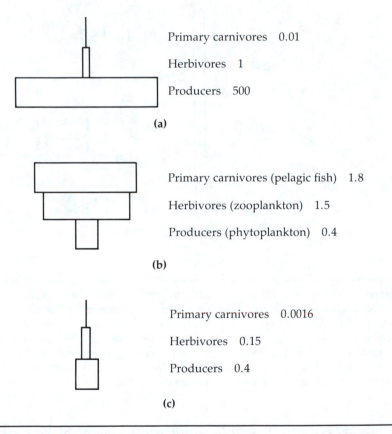

Primary carnivores   0.01

Herbivores   1

Producers   500

(a)

Primary carnivores (pelagic fish)   1.8

Herbivores (zooplankton)   1.5

Producers (phytoplankton)   0.4

(b)

Primary carnivores   0.0016

Herbivores   0.15

Producers   0.4

(c)

Figure 19.8 Biomass pyramids. **(a)** *A pyramid of biomass seen in an old field in Georgia* (g dry wt/m^2). (Odum 1957). **(b)** *An inverted pyramid of biomass in the English Channel* (g dry wt/m^2). Phytoplankton has a lower standing crop than the zooplankton, and the zooplankton biomass is less than that for the fish. **(c)** *Production rates of organisms in the English Channel (g dry wt/m2/day). The inversion in (b) is made* possible by the high rate of production by producers in that system. (Redrawn from Price 1975.)

Channel, the biomass of phytoplankton supports a higher biomass of zooplankton. This is possible because the production rate of phytoplankton is much higher than that of zooplankton, and the small phytoplankton standing crop (biomass at any one point in time) processes large amounts of energy. The most realistic pyramid is thus the energy pyramid, which never becomes inverted (Table 19.1).

Generalizations from Trophic Analyses

There are four major methods of studying trophic relationships between organisms:

- **Direct observations of feeding behavior**

- **Gut analysis:** Comparison of food fragments in the gut with samples of suspected food from the field (difficult on small animals or thorough chewers!)

- **Radionucleotide tracer analysis:** Particularly useful for feeders on fluid food such as plant sap. Plants can be labelled with ^{32}P, heavy phosphorus, and the organisms around the plant later collected and screened for radioactivity. A relative amount of feeding can be obtained if the counts are expressed per minute per milligram of dry weight. Once herbivores have been labelled, predators can be identified in a similar manner. Other useful tracers are heavy metals like rubidium and strontium, nonradioactive and easily detectable by a graphite-furnace atomic-absorption spectrophotometer (Legg and Chiang 1984; Moss and van Steenwyk 1984; Perfect, Cook, and Padgham 1985).

- **Serological techniques:** Based on the precipitin test. Basically, an antigen is prepared from the prey species under study and is injected into a rabbit to produce a specific antibody. The antibody is then used to test the gut contents of potential predators. A positive precipitin reaction demonstrates that the predator had prey parts (antigen) in its gut. Unfortu-

Table 19.1 *Pyramids of energy production in Cedar Bog Lake and Lake Mendota. (From Lindeman 1942.)*

Trophic Level	Cedar Bog Lake	Lake Mendota
Tertiary consumers	—	0.7
Secondary consumers	3.1	6.0
Primary consumers	14.8	144.0
Producers	111.3	428.0

Note: Units are gram calories^{-1} cm^{-2} yr^{-1}.

nately, no quantitative analysis can be performed; this method only gives a positive or negative record. Therefore, although such techniques have been important historically, they are no longer as widely applied as radionucleotide tracer analysis.

From the analysis of many trophic relationships, the following trends have been suggested to hold true as one moves "up" the food chain from herbivores to top carnivores (Price 1975):

- □ Fewer species
- □ Lower population levels
- □ Lower reproductive rates
- □ Increased body size
- □ Increased food searching areas and diversity of habitats utilized
- □ Higher powers of dispersal
- □ Higher searching ability
- □ Higher maintenance cost (energy)
- □ Higher efficiency of food utilization
- □ Food of higher calorific value
- □ Reduced feeding specialization
- □ More complex behavior
- □ Longer life expectancy

As always, there are certain exceptions to the trends. Large herbivores in Africa, such as the rhinoceros, sometimes weigh more than their largest predator, the lion. Perhaps to balance the status quo, social hunting behavior, such as that practiced by hunting dogs, has evolved to exploit large herbivores (see also Chapter 5).

Another interesting trend is that of feeding strategy and size. In general, as animals get large the progression is herbivore → carnivore → omnivore → herbivore. To begin with, carnivores often evolve to be larger than their herbivore prey. Eventually they become too slow and inefficient to catch enough food and omnivory evolves as a strategy. If a species becomes extremely large, it cannot catch animal prey and reverts evolutionarily to herbivory. Among the vertebrates, small birds and mammals are herbivores, canids and felids are predators, bears are omnivores, and elephants are herbivores.

Energy Flow

Viewing species and individuals as energy transformers in a community has a great strength, which is that the calorie is used as a lowest common denominator and species and individuals can be reduced to caloric equivalents. In the search for common theories in communities, an ecosystems approach may provide great insights into how biological systems function by closely examining energy flow.

The first law of thermodynamics states that energy can be neither created nor destroyed. The second law states that, in every energy transformation, potential energy is reduced because heat energy is lost from the system in the process. Thus, as food passes from one organism to another, the potential energy contained in the system is dissipated as heat. Therefore, there is a unidirectional flow of energy through the system, with no possibility of recycling (Fig. 19.9). This is a common explanation for why food chains commonly have only four or five links. The process of energy transfer is so inefficient (about 10 percent is transferred from one level to the next, and 90 percent is lost) that chains remain short. (Pimm [1984] has proposed additional reasons.) In contrast, chemicals are not dissipated and remain in the ecosystem indefinitely unless erosion occurs. Chemicals constantly circulate or recycle in the system, becoming more concentrated in higher organisms (Fig. 19.10), often leading to disastrous results in the case of chemical pollutants.

A relatively complete energy-flow diagram for a Georgia salt marsh (essentially an energetic version of the food web outlined in Fig. 19.3) is shown in Fig. 19.11 (p. 371). As with many ecosystems, most **primary production** goes to the decomposers.

In the salt marsh the **producers** are also the most important **consum-**

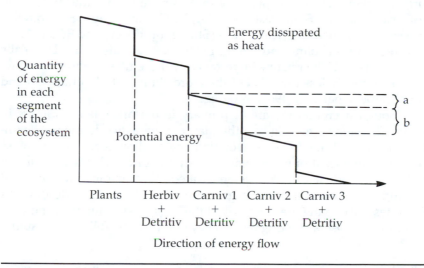

Figure 19.9 *Energy flow through the ecosystem.* At each trophic level maintenance energy is lost as heat (a); energy is also lost as heat in each transformation from one trophic level to the next (b). Ultimately, all energy in the system is dissipated as heat. Thus, the system can be maintained only by an outside supply of energy. Herbiv = herbivore, Carniv = carnivore, Detritiv = detritivore.

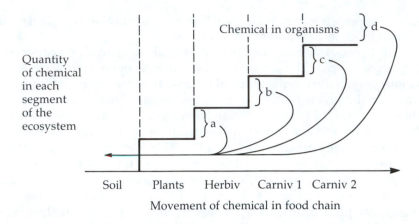

Figure 19.10 *The cycling of a chemical in the ecosystem assuming no erosion.* Chemicals in organisms not consumed by the higher trophic level (a to d) slowly return to the soil as they are released through decay.

ers; in other words, most primary production is used in plant respiration. The bacteria are next in importance; as **decomposers,** they degrade about one-seventh of the energy that plants use. Animal consumers are a poor third in importance, degrading only about one-seventh of the energy the bacteria use, though it must be stressed that, since this study was performed, many additional species of insects have been found to feed on *Spartina* in the U.S. Southeast (Rey 1981; McCoy and Rey 1987 and references therein), including leaf miners (Stiling, Brodbeck, and Strong 1982) and stem borers (Stiling and Strong 1983) not originally noticed in Teal's pioneer studies. The relative importance of these new herbivores, however, is probably less than that of the more abundant planthoppers and grasshoppers that Teal (1962) recognized.

Though decomposers are of primary importance as consumers in many ecosystems, the value of other, larger organisms should not be underestimated. Often breakdown of plant material proceeds in a stepwise fashion reminiscent of traditional succession. A now-classic experiment by Edwards and Heath (1963) demonstrated this phenomenon. They put oak and beech leaves in nylon bags in the soil and examined decomposition rates. By varying the mesh size of the bags, they could vary the sizes of the decomposers entering them. They obtained the following results:

Bag Mesh Size (mm)	Fauna That Could Enter Bags	Percent Oak Leaves Gone in Nine Months
7.0	All	93
0.5	Small invertebrates, microorganisms	38
0.003	Microorganisms only	0

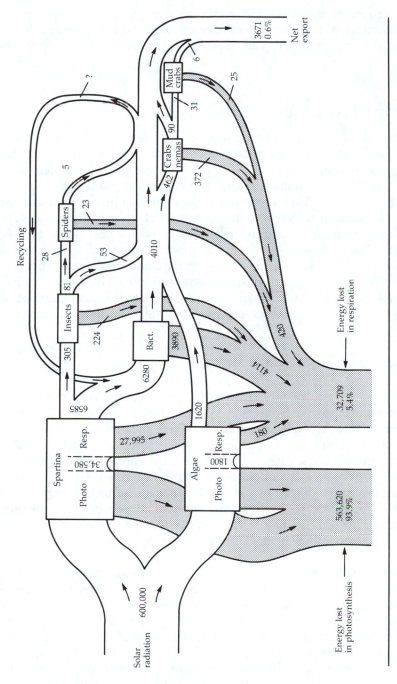

Figure 19.11 *Energy-flow diagram for a Georgia salt marsh.* Nemas = nematodes. (Redrawn from Teal 1962.)

371

Although microorganisms are very important in decay, they cannot begin their work until particle size is reduced by larger organisms. In the soil ecosystem, earthworms are most important in the initial decay process.

19.2 *Primary Production*

The process of **photosynthesis** is the cornerstone of all life and the starting point for studies of community metabolism. The bulk of the Earth's living mantle is green plants (99.9 percent by weight) (Whittaker 1975); only a small fraction of life is animal. **Gross primary production** is equivalent to the energy fixed in photosynthesis, and **net primary production** is gross primary production minus energy lost by plant respiration. The simplest method of measuring net primary production is the harvest method. The amount of plant material produced per unit of time, ΔB, is then equivalent to the biomass change in the community between time 1 and time 2, $\Delta B - \Delta B_1$. Two possible losses must be recognized: L = biomass lost by death of plants, and G = loss due to consumer organisms. Knowing these two elements, one can estimate net primary production as

net primary production $= \Delta B + L + G$

The energy equivalent of the production in biomass can then be obtained if one burns the biomass in a bomb calorimeter, but it must be remembered that this method cannot effectively gauge new root growth, a substantial amount of production. Also, Wallace and O'Hop (1985) showed that, in some cases where grazing by animals is high, net productivity can be grossly underestimated when based solely on estimates of standing-crop biomass. In their study on beetle herbivory on water lilies, larval production of beetles alone, without adjustments for egestion, respiration, or adult feeding, surpassed plant biomass availability. Therefore, rapid macrophyte turnover was going on to support the beetle population.

The best current estimate of global primary production is 110–120 $\times$ 10^9 metric tons dry weight yr^{-1} on land and 50–60 $\times$ 10^9 metric tons in the seas (Lieth 1975; Whittaker 1975). How does primary production vary over the different types of vegetation on the Earth? In general, primary production is highest in the tropical rain forest and decreases progressively toward the poles (Table 19.2). Productivity of the open ocean is very low, approximately the same as that of the Arctic tundra (see Fig. 19.12, p. 374). In agricultural situations, primary production of crops falls far short of that of their more natural counterparts, possibly because of a short growing season. Growth may be increased by the application of fer-

Table 19.2 Primary production and plant biomass for the Earth. (From Whittaker and Likens, cited by Whittaker 1975.)

Ecosystem Type	Area (10^6 km^2)	Mean Net Primary Production (g m^{-2} yr^{-1})	World Net Primary Production (10^9 tons yr^{-1})	Mean Biomass (kg m^{-2})	World Biomass (10^9 tons)
Continental					
Tropical rain forest	17.0	2,200	37.4	45.0	765
Tropical seasonal forest	7.5	1,600	12.0	35.0	260
Temperate evergreen forest	5.0	1,300	6.5	35.0	175
Temperate deciduous forest	7.0	1,200	8.4	30.0	210
Boreal forest	12.0	800	9.6	20.0	240
Woodlands and shrubland	8.5	700	6.0	6.0	50
Savannah	15.0	900	13.5	4.0	60
Temperate grassland	9.0	600	5.4	1.6	14
Tundra and alpine	8.0	140	1.1	0.6	5
Desert and semidesert scrub	18.0	90	1.6	0.7	13
Extreme desert, rock, sand, ice	24.0	3	0.07	0.02	0.5
Cultivated land	14.0	650	9.1	1.0	14
Swamp and marsh	2.0	2,000	4.0	15.0	30
Lake and stream	2.0	250	0.5	0.02	0.05
Total continental	149	773	115	12.3	1,837
Marine					
Open ocean	332.0	125	41.5	0.003	1.0
Upwelling zones	0.4	500	0.2	0.02	0.008
Continental shelf	26.6	360	9.6	0.01	0.27
Algal beds and reefs	0.6	2,500	1.6	2.0	1.2
Estuaries	1.4	1,500	2.1	1.0	1.4
Total marine	361	152	55	0.01	3.9
Grand total	510	333	170	3.6	1,841

tilizers, but it still generally falls short of more natural situations (Table 19.3, p. 375). In an economic sense, care must be taken to ensure that the highest productivity is coupled with high desirability for a particular crop.

Efficiency Measures of Primary Production

How efficient are the vegetations of different communities as energy converters? One can determine the efficiency of utilization of sunlight from the ratio

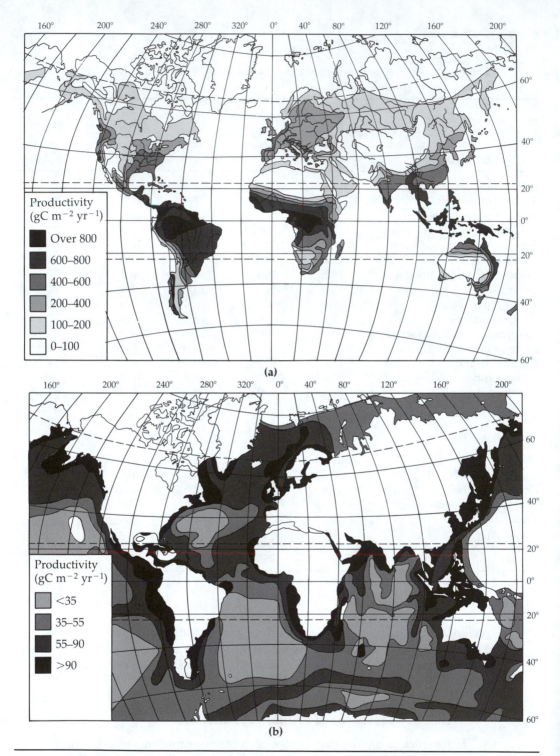

Figure 19.12 **(a)** *Worldwide pattern of net primary productivity on land.* (Redrawn from Reichle, 1970). **(b)** *Worldwide pattern of net primary productivity in the oceans.* (Redrawn from Koblentz-Mishke, Volkounsky, and Kabanova 1970.

Table 19.3 *Year-long net primary productivity of some cultivated and natural ecosystems. (From Odum 1959.)*

System	Net Primary Productivity ($g\ m^{-2}\ yr^{-1}$)
Spartina salt marsh, Georgia	3,285
Desert, Nevada	40
20–35-year-old pine plantation, England	2,190
20–35-year-old deciduous plantation, England	1,095
Wheat, world average	343
Rice, world average	496
Potatoes, world average	400
Sugarcane, world average	1,726
Mass algal culture, outdoors	4,526

$$\frac{\text{efficiency of gross}}{\text{primary production}} = \frac{\text{energy fixed by gross primary production}}{\text{energy in incident sunlight}}$$

Phytoplankton communities have very low efficiencies of usually less than 0.5 percent, herbaceous communities 1–2 percent, and crops generally less than 1.5 percent; the highest values occur in forests—2–3.5 percent (Cooper 1975). Perhaps 50–70 percent of the energy fixed by photosynthesis is lost in **respiration,** so usually less than 1 percent of the sun's energy is actually converted into net primary production. In the temperate zone, this percentage is equivalent to 300–600 calories of primary production cm^{-2}.

The Limits to Primary Production

What limits primary production? In terrestrial systems water is a major determinant, and production shows an almost linear increase with annual precipitation, at least in arid regions (Fig. 19.13). However, the temperature conditions are also important, as there is a much greater range of temperatures over land than over water. Rosenzweig (1968) showed that actual evapotranspiration rate could predict the aboveground production with good accuracy in North America. The relationship between the two could be described by the function (Lieth and Box 1972)

$$\text{primary production} = 3,000\ (1 = e^{-0.0009695(E-20)})$$

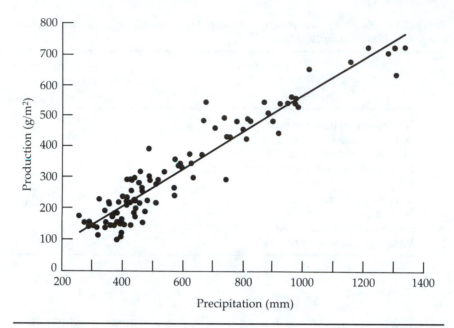

Figure 19.13   *Relationship between mean annual precipitation and mean above-ground net primary production for 100 major land resource areas across the central grassland region, Great Plains, of the United States.* (Redrawn from Sala et al. 1988.)

where E is the annual evapotranspiration (in mm). Lieth (1975) showed that, in addition to evapotranspiration rate, length of growing season was well correlated with net primary production of forests, at least in North America (Fig. 19.14). This correlation might help explain why tropical wet forests, with a very long growing season, are so productive. It also might help explain why conifers are so predominant in northern realms: they effectively extend the short growing season at high latitudes by retaining their leaves (needles) for long durations. One must also take great care, in calculating total productivity in terrestrial systems, to include roots as well, for in some species, especially herbaceous ones, below-ground biomass can be 40 percent that of total biomass. For other species like temperate trees, roots account for only 18 percent of the total biomass, so comparisons of above-ground productivity between these two vegetation types would not provide the most accurate comparisons of total productivity.

Nutrient deficiency, particularly of nitrogen and phosphorus, can limit primary productivity, too, as agricultural practitioners know only too well. For example, the salt marsh plants, *Borrichia frutescens*, in Photo 19.1 (p. 378) were lightly fertilized only twice in the spring of 1990, in February and March. By June, when the picture was taken, the fertilized plants (at right) were dramatically larger than unfertilized plants (at left).

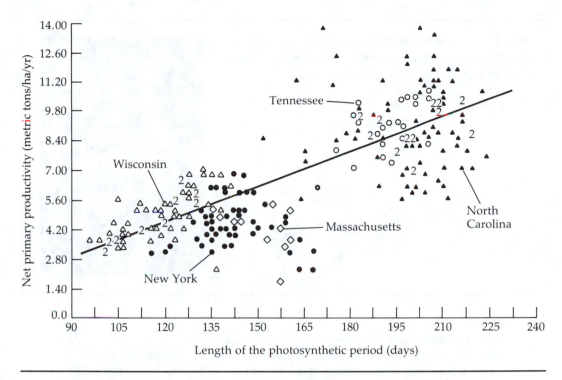

Figure 19.14 *Relationship between net primary production and the length of the growing season for stations in the eastern deciduous forest region of North America. (Redrawn from Lieth 1975.)*

Flowering was also greatly accelerated. Fertilizers are commonly used to boost productivity of annual crops. What is less commonly appreciated is that harvesting timber also removes large amounts of nutrients from the forest ecosystem, and such losses should also be made good by nutrient supplements. Rennie (1957) showed that pines removed fewer nutrients from soil than hardwoods. Some foresters have provided good evidence to show that fertilization increases timber yields. Among the most impressive studies are those involving fertilization of pines (*Pinus radiata*) with phosphate after a fire. Fertilized trees grew to more than twice the size of unfertilized ones after 15 years (Gentle, Humphreys, and Lambert 1965). The basal area of pines on unfertilized plots was only 12 m^2/ha, compared to 27 m^2/ha in fertilized areas. In more natural situations, much less work has been done on the effects of fertilization on climax communities. The reason is probably that the economic stimulus to perform the work is lacking. One exception is that fertilized *Metrosideros polymorpha* trees on Hawaii showed an annual rate of growth up to double that of unfertilized controls (Gerrish, Mueller-Dombois, and Bridges 1988). Some additional work has been performed on grasslands. In Brit-

Photo 19.1 Factors that limit primary production. Gardeners know that fertilizers containing nitrogen and phosphorus can increase the size of many plants. (Photo by P. Stiling.)

ain unfertilized grassland swards yield about 2.5 tons dry matter ha^{-1} yr^{-1}, and grass / legume swards about 6 tons ha^{-1} yr^{-1}. Adding 400 kg ha^{-1} nitrogen increases the yield of both types to about 10 tons ha^{-1} yr^{-1}. Peak grass yields of 12 tons ha^{-1} yr^{-1} require 625 kg ha^{-1} nitrogen, but peak profitability occurs at an application rate of 400 ha^{-1} (Jackson and Williams 1979).

If nutrient availability is so critical to plants, there should be an obvious relationship between soil nutrient content and plant production. Surprisingly, such relationships have not commonly been found (Gessel 1962), which tends to contradict results from fertilization studies. The probable explanation is that in many cases measured nutrients are locked up in a form unavailable to plants. An analogous problem exists in estimating what fraction of plant biomass is available to animals. In many cases, nitrogen availability is critical to animals but measures of total percent nitrogen in plant tissues may not give a good indication of what is available to herbivores, because much nitrogen is locked up in indigestible forms (Bernays 1983) or is unavailable because of the presence of digestibility-reducing substances (Mattson 1980; Wint 1983; Brodbeck and Strong 1987; see also Chapter 13).

Aquatic Systems

Of the factors limiting primary production in aquatic ecosystems, among the most important are available light and available nutrients.

Light is particularly likely to be in short supply because water absorbs light very readily. Even in "clear" water, only about 5–10 percent of the radiation may be present at depths of only 20 m. Too much light can also inhibit the growth of green plants by overheating them. Such a phenomenon can be found in tropical and subtropical surface waters throughout the year, where maximum primary production occurs several meters beneath the surface of the sea.

The most important nutrients affecting primary productivity in aquatic systems again are nitrogen and phosphorus. Important only locally in terrestrial systems, both often limit production in the oceans, where they occur in low concentrations. Few nutrients are tied up in the standing crop, in contrast to the situation in terrestrial systems, especially forests, where large amounts of nutrients occur in plants themselves. Rich, fertile soil contains about 5 percent organic matter and up to 0.5 percent nitrogen. One square meter of soil surface can support 50 kg dry weight of plant matter. In the ocean, the richest water only contains 0.00005 percent nitrogen, and 1 m^2 could support no more than 5 g dry weight of phytoplankton (Ryther 1963). Enrichment of the sea by addition of nitrogen and phosphorus can result in substantial algal blooms. Such enrichment occurs naturally in areas of upwellings, such as the Antarctic or the coasts of Peru and California, where cold, nutrient-rich, deep water is brought to the surface by strong currents, resulting in very productive ecosystems.

Phosphorus is particularly important in limiting productivity in freshwater lakes. During the late 1960's, many lakes in North America became polluted by runoff from the land of rainwater enriched with phosphorus from fertilizer application or from sewage. This new input into the lakes caused huge blooms of blue-green algae, which clogged the lake and increased turbidity, a process termed **eutrophication.** Eventually, the problem was traced to increased phosphorus input (see also Chapter 22). In a dramatic series of experiments conducted in Canada, begun in the late 1960's, lakes were treated with a whole range of chemicals, including phosphorus, and other lakes were left unmanipulated as controls. Only where phosphorus levels were greatly increased was the character of the lake drastically altered by algal blooms and all the other attendant problems of eutrophication (Schindler 1974 and 1977). Carefully designed experiments had clearly implicated phosphorus as the main limiting factor for primary production.

For some systems, aquatic and terrestrial, one particular nutrient may be limiting at first, but if the level of this particular factor is raised, then another factor becomes limiting. This type of sequential progression of limiting factors is illustrated in Fig. 19.15. Such a phenomenon was demonstrated in nature by Menzel and Ryther (1961), who studied primary productivity of the Sargasso Sea. They found that iron levels were

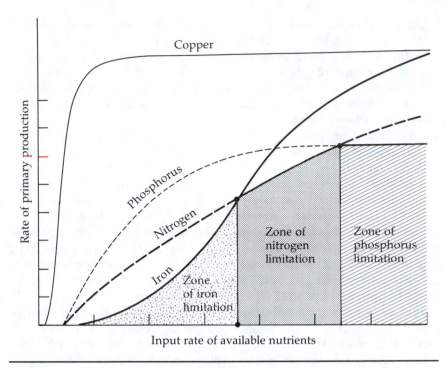

Figure 19.15 *Hypothetical sequence of nutrient factors that limit primary productivity.*
The rate of primary production follows the heavy line and is limited first by iron,
then (as more iron becomes available) by nitrogen, and finally by phosphorus.
Some nutrients, such as copper, may always be present in superabundant
amounts. (Redrawn from Krebs 1985a.)

the most obvious limiting factor, but that nitrogen and phosphorus were
likely to be limiting in the presence of sufficient iron.

19.3 *Secondary Production*

The biomass of plants that accumulates in a community as a result of photosynthesis can eventually go in one of two directions: to herbivores or to
detritus feeders. Most of the biomass rots and is available to detritivores.
Heal and Maclean (1975) calculated that detritivores carried out over 80
percent of the consumption of matter, often "working it over" on a number of occasions to extract the most energy from it. However, it is the
herbivores that feed on the living plant biomass and, as a result, constitute the greatest selective pressure on living plants. Energy flow in such
species will therefore be considered in more detail.

The energy for an individual animal can be seen as a series of dichotomies:

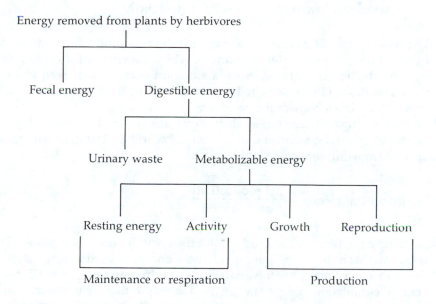

Thus energy can end up in waste products, it can be used in movement, or it can be allocated to growth and reproduction.

Efficiency Measures of Secondary Production

How are the components of **secondary production** in animal communities measured? Metabolizable or assimilation energy can be measured very simply in the laboratory, where food intake (gross energy intake) and fecal and urine production can easily be recorded. In the field such measurements are extremely difficult, and the usual approach is to measure assimilation energy by means of the relation

assimilation energy = respiration + net production

Respiration is measured while an animal is confined in a cage; oxygen consumption, carbon dioxide output, and heat production are monitored directly. Net production can be measured by the growth of individuals and reproduction, the birth of new animals. Production as biomass can again be converted to energy measures by determination of the caloric value of a unit weight of the species. In warm-blooded animals, the resting energy or minimum rate of metabolism, often referred to as the *basal metabolic rate* is a simple function of body size:

$$\text{basal metabolism (kcal / day)} = 70 \times (\text{weight})^{3/4}$$

The basal metabolism is measured under resting conditions, with no food in the stomach and at a temperature at which no energy for cooling or heat production is required. Such a situation is rarely achieved in the field, and Brody (1945) estimated metabolism in the field for maintenance to be approximately twice the basal rate.

In viewing an herbivore as an energy transformer, three main types of efficiency can be measured in the animal world and used to compare species from different communities:

$$\text{growth efficiency} = \frac{\text{net productivity}}{\text{assimilation}}$$

Growth efficiency (or production efficiency) in animals is generally less than that in plants because much more energy is used in respiration. Even within animals, much variation occurs. In mammals, over 95 percent of assimilation energy may be used in respiration, whereas in cold-blooded animals this figure is less (for insects, 60–80 percent), reflecting the energy cost of homeothermy. One consequence is that sparsely vegetated deserts can support healthy populations of snakes and lizards where mammals might easily starve. The large monitor lizard known as the Komodo dragon eats the equivalent of about a pig a month, its own weight every two months, whereas a cheetah consumes something like four times its own weight in the same period (Kruuk and Turner 1967). It is also interesting to note that growth efficiencies are higher in young animals than in old (35 percent, as opposed to 3–5 percent)—hence the practice of raising broiler chickens and calves for meat. Further, smaller species often have higher growth rates than larger species. For a given amount of hay, rabbits produce the same quantity of meat as beef cattle, but do so four times as quickly:

	1 Steer	300 Rabbits
Total body weight	1,300 lb	1,300 lb
Food consumption per day	16 2/3 lb	66 2/3 lb
Time to consume 1 ton of hay	120 days	30 days
Heat loss per day	20,000 kcal	10,000 kcal
Gain in weight per day	2 lb	8 lb
Gain in weight from 1 ton of food	240 lb	240 lb

Efficiency between trophic levels can be measured by two other indices:

$$\text{Lindeman's efficiency} = \frac{\text{assimilation at trophic level } n}{\text{assimilation at trophic level } n - 1}$$

a measure of **assimilation efficiency** named after ecologist R.J. Lindeman (1942) in recognition of his classic work on ecological energetics, and

$$\text{consumption efficiency} = \frac{\text{intake at trophic level } n}{\text{net productivity at trophic level } n - 1}$$

Lindeman's efficiency often appears to be a constant, around 10 percent for each set of trophic levels, although some data on marine food chains shows it can exceed 30 percent (Steele 1974). Consumption efficiency measures the relative pressure of one trophic level on the one beneath it and seems generally to fall in the range of 20–25 percent, meaning that 75–80 percent of the net production of each trophic level either goes into the decomposer chain, is lost in the system, or is used to increase population biomass. For carnivores, however, consumption efficiencies may reach 80 percent, showing that animals are much better equipped to handle meat than they are dead material or chemically protected plants.

In summary, Whittaker (1975) has calculated that the percentages of net primary production going to animal consumption in tropical rain forests and in temperate rain forests and grasslands are, respectively, only 7.5 percent and 10 percent. In open oceans, however, the figure is 40 percent, and in zones of upwelling 35 percent, indicating that zooplankton are better croppers of phytoplankton than terrestrial herbivores are of their food plants. In the general scheme of things, birds and mammals, the animals most keenly observed by humans, contribute almost nothing to secondary production and very little to consumption. Insects are a little more important, but soil animals, including nematodes, remain the major factor. Thus, ecosystems ecology, unlike population ecology, does not stress the importance of higher trophic levels, because they are relatively unimportant in energy flow.

The Limits to Secondary Production

What controls secondary production is a complex question, but it is generally thought to be limited to a high degree by available primary production. McNaughton et al. (1989) have documented a tight correlation between primary productivity in a variety of ecosystems and the biomass of and consumption by herbivores. Existence of this correlation is not as obvious as one might think because it means that secondary chemicals that make plants poisonous or distasteful to herbivores are much less important influences on consumption at the ecosystem level than they are at the plant and population levels (see Chapter 13). When host quality and quantity are increased experimentally by an input of fertilizer, herbivore biomass often increases, too. After reviewing 18 laboratory and field nutrition-interaction studies, Onuf (1978) concluded that a general correlation could be found between nitrogen levels and the susceptibility of plants to insect attack. Leaf-feeding insects removed four times as much foliage from mangrove, *Rhizophora mangle,* on a high-fertility site enriched by the droppings from a large colony of egrets and pelicans as from a nearby low-fertility site (Onuf, Teal, and Valiela 1977). Nitrogen concentration in the phloem sap of grasses increases under fertilization, and sucking species like aphids and planthoppers become much more successful and prolific (Prestidge and McNeill 1983). Vince, Valiela, and Teal (1981) found a dramatic increase in insect biomass on fertilized salt marshes of the sort where Teal (1962) had done his pioneering energy-flow studies. But fertilization does not always increase the growth rates of herbivores. Stark (1964), in a review of 15 tree-insect interactions involving nitrogen fertilizers, showed that insect survival was reduced. Auerbach and Strong (1981) could find no increase in growth rates or nitrogen-accumulation rates of two species of tropical hispine beetles feeding on *Heliconia* spp. in Costa Rica. In such cases natural enemies are often more commonly involved in determining the numbers of herbivores in the field, but this type of argument belongs more in the realm of control of population numbers than in an ecosystem chapter. However, Strauss (1987) has shown that fertilization can increase levels of generalist predators on certain insects, which indirectly decreases the abundance of other herbivores, even on fertilized areas. Fertilized *Artemisia ludoviciana* plants in Minnesota supported greater numbers of phloem- and seed-feeding insects and also had a concurrent increase in patrolling by aphid-tending ants. Such high levels of aggressive predators reduced the levels of other foliage feeders like chrysomelid beetles, even though such beetles readily feed on fertilized foliage.

Finally, Odum (1969) has suggested that the energy relations of communities can also be used as a measure of their maturity (Table 19.4). According to Odum, gross and net primary productivity are very low in mature communities, high in developing ones. Energy in a complex, highly mature system is shunted into maintenance of order, and less is used for production of new materials. Food chains are shorter in the mature communities; those in mature communities are more complex. Odum also views the ratio of production to biomass (P/B) as the single most important measure of maturity in an ecosystem; P/B ratios become lower as maturity increases. Of much interest to applied ecologists is that high P/B ratios mean high crop yields, so agricultural practices should be aimed specifically at exploiting immature ecosystems rather than mature ones. Vitousek and Reiners (1975) showed that biomass accumulated rapidly with time during early succession but then declined (Fig. 19.16, p. 386).

Table 19.4 *Odum's (1969) model of ecosystem development—general trends in 24 variables during ecological succession.*

Ecosystem Attributes	Developmental Stages	Mature Stages
Community Energetics		
1. Gross production / community respiration (P/R ratio)	Greater or less than 1	Approaches 1
2. Gross production / standing crop biomass (P/B ratio)	High	Low
3. Biomass supported / unit energy flow (B/E ratio)	Low	High
4. Net community production (yield)	High	Low
5. Food chains	Linear, predominantly grazing	Weblike, predominantly detritus
Community Structure		
6. Total organic matter	Small	Large
7. Inorganic nutrients	Extrabiotic	Intrabiotic
8. Species diversity—variety component	Low	High
9. Species diversity—equitability component	Low	High
10. Biochemical diversity	Low	High
11. Stratification and spatial heterogeneity (pattern diversity)	Poorly organized	Well organized
Life History		
12. Niche specialization	Broad	Narrow
13. Size of organism	Small	Large
14. Life cycles	Short, simple	Long, complex
Nutrient Cycling		
15. Mineral cycles	Open	Closed
16. Nutrient exchange rate, between organisms and environment	Rapid	Slow
17. Role of detritus in nutrient regeneration	Unimportant	Important
Selection Pressure		
18. Growth form	For rapid growth	For feedback control
19. Production	(r selection)	(K selection)
Overall Homeostasis		
20. Internal symbiosis	Undeveloped	Developed
21. Nutrient conservation	Poor	Good
22. Stability (resistance to external perturbations)	Poor	Good
23. Entropy	High	Low
24. Information	Low	High

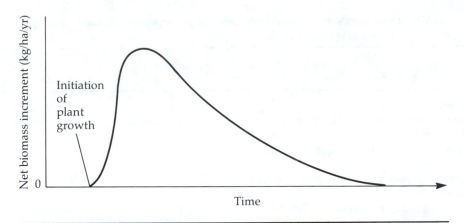

Figure 19.16 *Changes in biomass and nutrient losses during succession.* Biomass should accumulate rapidly during early succession; then the rate of biomass accumulation gradually falls to zero. (Redrawn from Vitousek and Reiners 1975.)

Other predictions about the nature of communities are also listed in Table 19.4. Some of these comparisons are directly comparable to those derived from a consideration of *r*- and *K*-selected organisms in terms of population life styles (see Chapter 11).

19.4 *Nutrient Cycles*

As previously discussed, nutrients such as nitrogen and phosphorus often limit primary or secondary production. For example, McNaughton (1988) reports that the mineral content of foods is an important determinant of the spatial distribution of animals within the Serengeti National Park, Tanzania. Areas of grassland containing higher concentrations of magnesium, sodium, and phosphorus support higher densities of large herbivores than areas of low concentrations of these minerals. It is often argued that ecosystems can best be understood not from the path of energy through them but by the paths of nutrients, the **biogeochemical cycles.** Because chemicals are not dissipated, but remain in the ecosystem indefinitely, they tend to accumulate in individuals or species, which then act as "pools" of nutrients. The rate of nutrient movement between pools is called the *flux rate* and is measured as the quantity of nutrient passing from one pool to another per unit of time. Nutrients cycle between pools through meteorological, geographical, or biological transport mechanisms. Meteorological inputs include dissolved matter in rain and snow, atmospheric gasses, and dust blown by the wind. Geological inputs include elements transported by surface and subsurface drainage, and biological inputs result from the movements of animals, or animal parts, between ecosystems. The turnover times of nutrients in the various compartments of ecosystems seem to get longer with increasing latitude. Jordan and Kline (1972) quote 10.5 years for the cycling time of nutrients in a tropical rain forest and 42.7 years for the taiga in the Soviet Union.

Nutrient cycles have often been studied by means of the introduction of radioactive tracers into ecosystems. For example, Whittaker (1961) followed the introduction of ^{32}P-labelled phosphoric acid into an aquarium. He found a definite sequence of nutrient uptake (Fig. 19.17):

1. ^{32}P was rapidly taken up by phytoplankton and subsequently discharged.
2. Filamentous algae on the sides and bottom slowly picked up ^{32}P.
3. Crustaceans grazing on algae picked up ^{32}P even more slowly.
4. ^{32}P began to accumulate in bottom sediment and was tied up in less active forms.

This type of cycle is broadly similar to that occurring in natural lakes; phosphorus and other nutrients accumulate in the bottom sediment, largely unavailable for use, which is why phosphorus is so often limiting in lake ecosystems.

Nutrient cycles can be divided into two broad types:

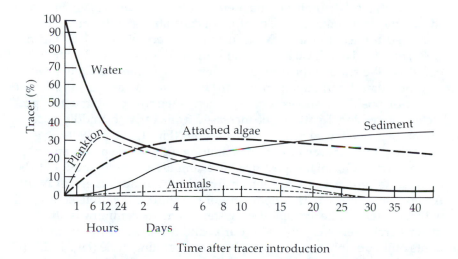

Figure 19.17 *Movement of radiophosphorus in an aquarium microcosm.* Percentage of the tracer present at a given time (after correction for radioactive decay) is on the vertical axis; time after tracer introduction (on a square-root scale) is on the horizontal. (Redrawn from Whittaker 1961.)

□ *Local cycles,* such as the phosphorus cycle just described, which involve elements with no mechanism for long-distance transfer.
□ *Global cycles,* which involve interchange between the atmosphere and the ecosystem and are particularly applicable to elements such as nitrogen, carbon, oxygen, and water. Global nutrient cycles unite all of the world's living organisms into one giant ecosystem called the **biosphere,** the whole Earth system.

The **nitrogen cycle** is a good example of a gaseous global cycle (Fig. 19.18). The quantity of nitrogen tied up in living organisms is very small compared with the total amount in the atmosphere. Almost all nitrogen available for plants comes from **nitrogen-fixing bacteria** or blue-green algae, so there is severe competition between plants for soil nitrogen, which often limits growth (Etherington 1975). Limiting nutrients, such as nitrogen, are usually bound up most tightly in ecosystems, whereas nonlimiting elements, such as sodium, do not accumulate in the food chain. Vitousek and Reiners (1975) showed how, under their "nutrient-retention hypothesis," these types of phenomena would change according to the successional state of the ecosystem (Fig. 19.19, p. 390). In early successional stages, return to the environment of essential elements such as nitrogen are negligible, but, as plant-growth rate levels off, less new tissue is produced and some nitrogen can be released. In support of this hypothesis, they point out that nitrate concentrations are much lower in streams draining successional forests than they are in streams draining old forest stands (Vitousek and Reiners 1975). Lewis (1986) has also shown the nitrogen and phosphorus concentrations of drainages of undisturbed tropical forests in Venezuela to be very high. Nancy Grimm (1987) provided empirical evidence to support Vitousek and Reiners's hypothesis. She studied the nitrogen dynamics of Sycamore Creek, a lowland Sonoran desert stream in Arizona, on seven occasions at different stages of postflood succession. Nitrogen retention exhibited the predicted successional patterns of increases from early to middle successional stages, followed by late-stage declines (Fig. 19.20, p. 390).

Nitrogen cycles can be critical in limiting individual species and population cycles. One of the most striking examples was often thought to be that of the brown lemming, which lives in the tundra areas of North America and Eurasia. The ecosystem of the tundra is simple compared with more temperate or tropical ecosystems, and the lemming is often the major herbivore. Every three-to-four years, numbers of these small rodents build up only to crash again in a never-ending cycle (Fig. 19.21, p. 391). This ecosystem has been studied in some detail on the Arctic coastal-plain tundra near Point Barrow, Alaska. The traditional story is as follows. As lemming numbers increase in winter, their feces and urine stimulate plant growth, which in turn increases lemming production. Eventually, the lemming numbers become so great that the vascular

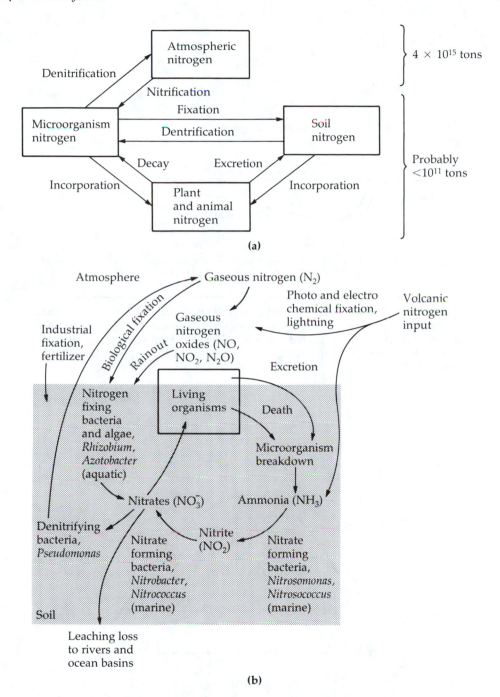

Figure 19.18 *The nitrogen cycle.* **(a)** *Major relationships between the very large atmosphere pool of gaseous nitrogen and the biosphere.* **(b)** *The complex interrelationships of the soil-based portion of the cycle.*

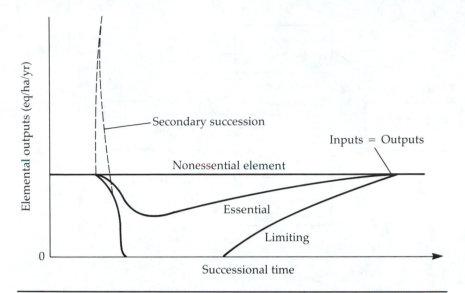

Figure 19.19 *Nutrient retention during succession.* Elements limiting to plant growth are retained most strongly. Losses of nonessential elements should not vary over succession. The dotted line represents very high rates of nutrient loss immediately after a disturbance such as logging. (Modified from Vitousek and Reiners 1975.)

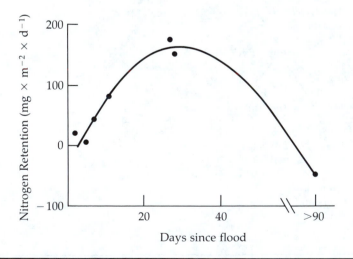

Figure 19.20 *Rates of ecosystem inorganic nitrogen retention as a function of time since disturbance in Sycamore Creek, Arizona.* (Redrawn from Grimm 1987.)

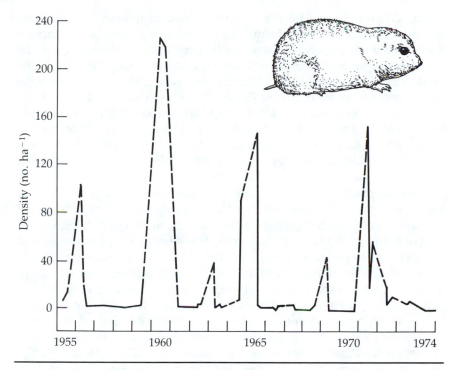

Figure 19.21 *Estimated lemming densities in the coastal tundra at Barrow, Alaska, for a 20 year period.* (Redrawn from Batzli et al. 1980.)

plant cover is thinned out. More sunlight reaches the ground, thawing out the soil overlying the **permafrost,** and plant roots can penetrate a little further. Permafrost lower down prevents water movement, so nutrients are not **leached** out, but accumulate in peat, not available to the plants. The intense grazing and lack of root nutrients soon reduce forage quality. This trend, coupled with high predation rates, social strife, and dramatic emigration reduces lemming numbers drastically. Low nutrient quality also prevents breeding the following year. Gradually, during the next two-to-three years the vegetation slowly recovers, nutrients increase, and the cycle begins again. This whole description of the process that controls population variation was termed the *nutrient-recovery hypothesis* by Pitelka (1964) and Schultz (1964) and although it is an attractive idea, it must be stressed that not all the elements of the cycle have been proven beyond question. Much work remains to be done, and Batzli et al. (1980) have pointed out some inconsistencies between data and theory.

First, food supplementation in times of high densities should prevent population decline, but early experiments failed to produce this result. It is possible, however, that the failure resulted because predators were attracted from nearby unprofitable areas and killed higher numbers of rodents. When Ford and Pitelka (1984) provided supplementary food *and* prevented predation in California, they found that experimental populations declined modestly compared to the natural crash of control populations that summer. It is also worth noting that "intrinsic" changes in the behavior of the rodents themselves may be causing variations in population numbers (see also Chapter 15). As densities of animals increase, aggressiveness between individuals goes up, affecting hormonal levels (Krebs 1985*b*). This process in turn promotes a huge dispersal of animals away from centers of high population density (Stenseth 1983), so the population cycles of lemmings may essentially be caused and regulated by the behavior of the animals themselves.

Similar processes probably occur in other ecosystems as well but are harder to unravel. The ecosystem of the Arctic is simple. There are about 100 species of plant but only about 10 are important, comprising 90 percent of the plant biomass. The lemming is the only major herbivore, and it has two main predators and six minor avian and mammalian enemies. Other ecosystems are much more complex. In addition, much carbon, nitrogen, and phosphorus is held in the soil in the Arctic; only about 2 percent is held in living material. This proportion is in sharp contrast to those in temperate and tropical forests, where between 20 and 70 percent of these nutrients are found in living plants (Fig. 19.22). Finally, it must be emphasized that Arctic systems are commonly not in a state of equilibrium; much organic matter is slowly accumulating as peat. This is not the case in many other ecosystems. Thus, although some of the peculiarities of other ecosystems could probably be explained by studies of nutrient cycling, the answers are likely to be different from those of the Arctic.

Nutrient availability on a large scale can have a severe impact on the type of vegetation supported. In the Northern Hemisphere, where many areas have been glaciated, soils are derived from pulverized bedrock, till, and are very fertile. Those in the Southern Hemisphere, however, have not commonly undergone such a phenomenon. These soils are old, weathered, and infertile, derived from Gondwanaland. Such areas include Australia, South America, and India. Among the most severely limiting nutrients is phosphorus, and it is interesting that *Eucalyptus* species are well adapted to grow on soils with a low phosphorus content. In general, the vegetational communities that grow on such impoverished soils can be termed **oligotrophic** in much the same fashion as crystal-clear lakes (Chapter 16). Nutrient-rich temperate soils are then **eutrophic.** In a natural situation, oligotrophic systems tend to have a large biomass in the humus layer of the soil. Because vegetational types have evolved specifically to cope with local conditions, productivity of natural vegetation differs little between oligotrophic and eutrophic systems. When disturbed, however, for agriculture, the nutrient-poor systems quickly lose their productive potential, whereas the nutrient-rich ones are not so easily perturbed. The reason is that the large stores of nutrients in the humus of an

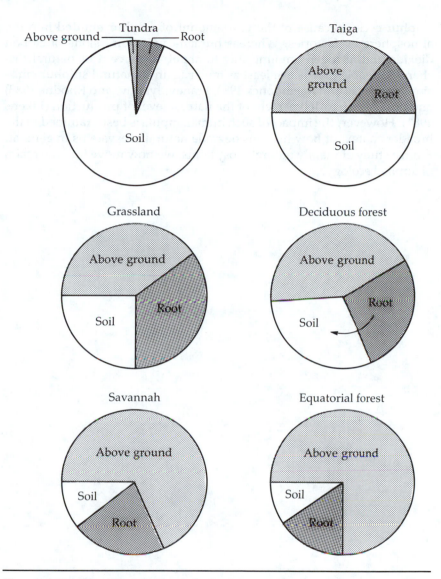

Figure 19.22 *Deposition of nitrogen in the three organic-matter compartments (above ground, root, and soil) for each of six biome types.* (Redrawn from Swift, Heal, and Anderson 1979.)

oligotrophic system are quickly lost under manipulation. This factor must be considered in the search for reasons for the often poor success rates of tropical agriculture.

Many other nutrients cycle, too, like carbon, sulphur, and phosphorus. One of the nutrient cycles under most scrutiny these days is the sulphur cycle, because of the vast amount of sulphur emitted into the atmosphere by factories. Whereas human-caused emissions of carbon dioxide and nitrogen amount only to about 10 percent of natural discharges, people produce at least as much again as natural sulphur emissions (Andreae and Raemdonck 1983; Franey, Ivanov, and Rhodne 1983) and possibly up to 160 percent of the natural level of production (Likens 1981). However, the impact of such human inputs is best examined in the broader context of how humans degrade natural ecosystems in general. From a study of mainly natural ecosystems, we now move into the realm of applied ecology.

Section Six

Applied Ecology

What effect has uncontrolled human population growth had on the environment since this photo of Miami was taken in 1953? How long will Everglades National Park remain unchanged given the growth of nearby Miami? To many people, applied ecology is the study of questions like these. It addresses issues such as the effects of pollutants, loss of wildlife through direct human activity, global patterns of land use, and the effect of exotic species on native biota.
Photos: Florida State Archives.

Applied ecology is the study of people's influence on ecological systems and, concurrently, the management of those systems and resources for society's benefit. Many problems instantly arise when one faces the prospect of ecological management—what constitutes an environmental problem depends partly on public perception. An organism or habitat that has been incorporated into art, poetry, or national symbolism becomes an object of special concern—hence the overriding importance of the bald eagle to the United States. However, some agreement on the important issues in applied ecology have been reached by U.S. national committees (see Table

S6.1). These include the prevention of pollution, the conservation of natural areas, and the preservation of the Earth's genetic diversity. Most ecological problems in today's world are linked to the adverse effects people have on the environment. There is, therefore, much interest in knowing how human population levels will change in the future.

Table S6.1 *Important issues in applied ecology. (From National Research Council 1986.)*

Legal requirements
Air and water quality standards
Public health
Rare, threatened, and endangered species
Protected areas or habitats

Aesthetic values
Landscape appeal
Attractive communities
Appealing species (such as large ungulates, colorful birds, cacti, eagles, and
    tigers)
Clean air and water

Economic concerns
Species or habitats of recreational or commercial interest
Ecosystem components not useful in themselves but necessary to the preser-
    vation of other, economically important components

Environmental values and concerns
Ecosystem rarity or uniqueness
Sensitivity of species or ecosystems to stress
Ecosystem ''naturalness''
Genetic resources
Recovery potential of ecosystems
Keystone species

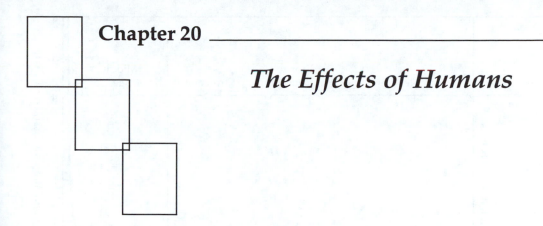

Chapter 20

The Effects of Humans

Though the data are scanty before A.D. 1650, they consistently show populations to have a very slow rate of growth. High infant mortality rates and low longevity meant that increases were small; in Roman times, the average life expectancy for men was about 30 years. Even then, humans severely affected the environment. In Bronze Age Britain, for example, lime trees (*Tilia* sp.) may have been selectively removed because the sod underneath them was agriculturally rich (Turner 1962).

20.1 *Human Population Growth*

The agricultural revolution and modern medicine have allowed our population to skyrocket in recent times (Fig. 20.1 and Table 20.1). Photo 20.1 (p. 399), showing the city of Miami Beach at two points in its history, illustrates just how rapidly urban areas can grow. Over 5,000 million people now inhabit the Earth, and Europe and Asia together contain over 75 percent of them. Whereas the second billion humans were added to the Earth in a period of 130 years, from 1800 to 1930, the fifth billion was added in just 12 years, from 1975 to 1987, and predictions are that the next billion will be added in the succeeding 12 years, by 1999 (Miller 1988). Less than 10 percent of this population lives in the Southern Hemisphere, whereas 50 percent lives in Asia alone, on 10 percent of the habitable area, and 20 percent lives in Europe on 5 percent of the land. Six nations (China, India, Pakistan, Bangladesh, Japan, and Indonesia) have nearly half the world's population (Fig. 20.2, p. 400).

During 1987, a relatively small annual growth rate of 1.7 percent acted on a base population of 5 billion, and 86.7 million new people were added to the planet, an average of nearly 10,000 an hour (Miller 1988). One out of six of these people will be hungry or undernourished, one out of four will lack clean drinking water, and one out of three will not have adequate health care or sewage disposal. The differences in characteris-

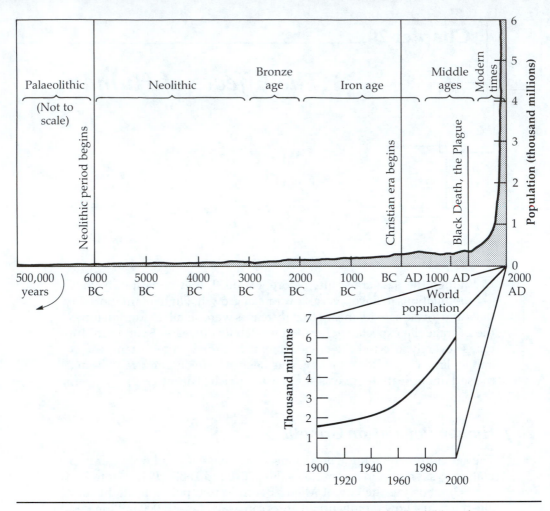

Figure 20.1 *Growth of the world's human population since the Palaeolithic and projected at current rates of increase to* A.D. *2000.* Catastrophes like the Black Death had remarkably little long-term effect. (Redrawn from Trewartha 1969.)

Table 20.1 *Estimates of population growth 1750–1979 (in millions).*

Area	1750	1800	1850	1900	1950	1979
World	791	978	1,262	1,650	2,515	4,321
Asia (except U.S.S.R.)	498	630	801	925	1,381	2,498
Africa	106	107	111	133	222	457
Europe (except U.S.S.R.)	125	152	208	296	327	483
U.S.S.R.	42	56	76	134	180	264
North America	2	7	26	82	166	244
South and Central America	16	24	38	74	162	352
Oceania	2	2	2	6	13	23

(a)   **(b)**

Photo 20.1 Human population growth. (a) Miami Beach in 1925 still had open areas.
(b) By 1953, this area was filled with dwellings. (Photos Florida State Archives.)

tics between the 1.3 billion people living in the more-developed countries
(**MDCs**) and those in the less-developed countries (**LDCs**) are illustrated
in Fig. 20.3 (p. 401). There are 33 MDCs, located mainly in the Northern
Hemisphere, Australia, and New Zealand. They occur in areas of favorable climate and good soil; the people are literate and well fed. There are
133 LDCs, most of which are in the Southern Hemisphere. They generally have less favorable soils and hot climates, and the people are illiterate
and undernourished. Whereas the increase in population of MDCs is essentially zero, the number of people in LDCs is still skyrocketing, and the
world's population growth is not expected to level off until at least 10
billion people are alive, in about 2090.

The best estimate of population growth seems to be about 1.7 percent, which gives a doubling time of 41 years. Of the continents, Europe
has the lowest rate of growth (0.4 percent, doubling time of 150 years)
and Africa has the highest (2.9 percent, doubling time of 25 years). Of the
nations, Germany has the lowest rate (-0.2 percent) and Kenya the
highest (3.5 percent, doubling time of 20 years). Africa has by far the largest rates of natural increase (Fig. 20.4, p. 402). It is immediately obvious
that even at what seem to be low rates of growth, 2–4 percent, doubling
times are only of the order of 20–30 years (Table 20.2, p. 403). At present
rates of increase, each person will have 0.5 m^2 of standing room in A.D.
2600, and in A.D. 4000 the Earth will be a mass of humanity expanding

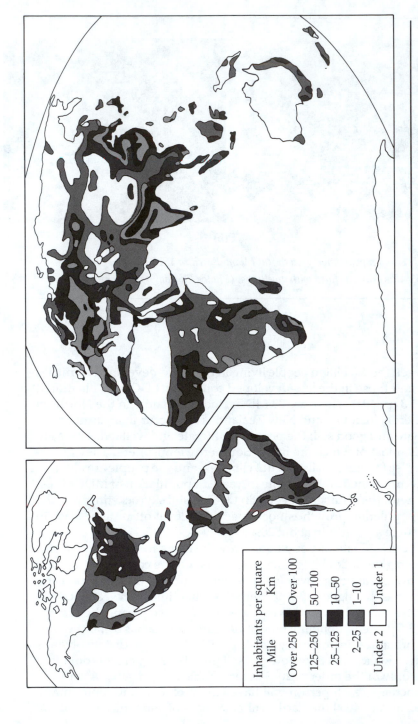

Figure 20.2 World population density. Only in regions of unfavorable climate are low densities found, but high densities are seen both in urban-industrial regions of the world and in rural-agricultural regions, especially in Asia. (Redrawn from Trewartha 1969.)

Inhabitants per square

Mile | Km

Over 250 | Over 100
125–250 | 50–100
25–125 | 10–50
2–25 | 1–10
Under 2 | Under 1

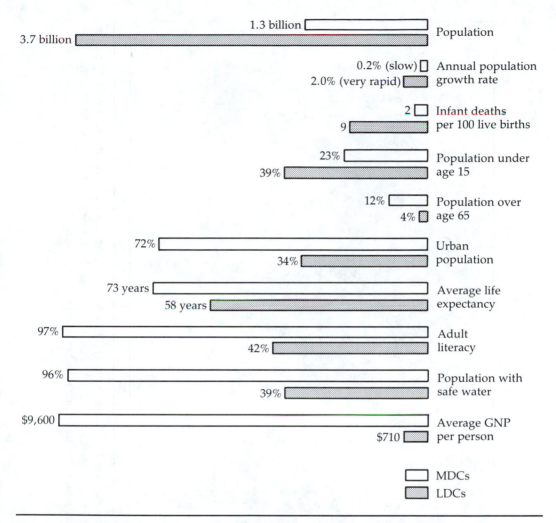

Figure 20.3 *Some characteristics of more-developed countries (MDCs) and less-developed countries (LDCs) in 1986. (Redrawn from Miller 1988.)*

outward at the speed of light. Such comparisons may, of course, seem absurd, but the serious purpose here is to make us aware that absolute limits to human populations exist, much as they do for other animals.

20.2 Loss of Wildlife Through Human Activity

The effect of loss of wildlife due to human activity can be split into two parts—that due to habitat destruction associated with agriculture, forestry, and urban development and that due to direct exploitation. The

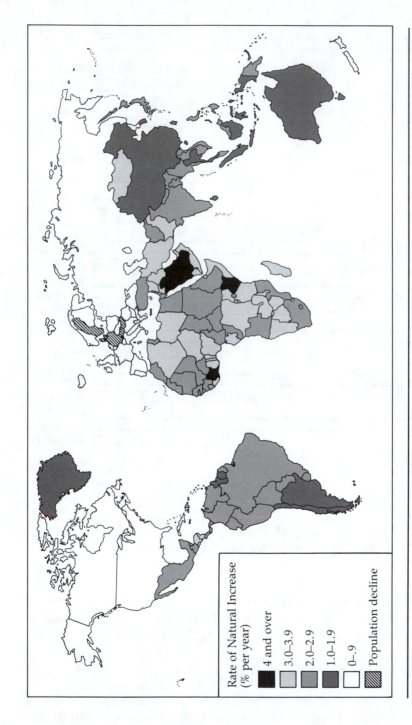

Figure 20.4 Rate of natural increase, the crude birth rate minus the crude death rate. With the crude death rate relatively low in most countries, the rate of natural increase is influenced primarily by the crude birth rate. Several European countries now have negative growth rates, because the crude death rate is higher than the crude birth rate.

Rate of Natural Increase (% per year)

- 4 and over
- 3.0–3.9
- 2.0–2.9
- 1.0–1.9
- 0–.9
- Population decline

Table 20.2 *Growth rates and doubling times.*

Increase Percent per Annum	Years to Double Population	Increase Percent per Annum	Years to Double Population
0.1	693	2.5	28
0.5	139	3.0	23
1.0	70	3.5	20
1.5	47	4.0	18
2.0	35		

Note: Years to double population = 70 divided by percentage annual increase. Percentage annual increase = 70 divided by doubling period.

International Union for Conservation of Nature and Natural Resources (IUCN) has estimated that at least 1,000 animal species and over 25,000 plant species are now **endangered.** After habitat destruction (threatening 67 percent of the species on the IUCN list), the next greatest causes of **extinctions** are the effects of overexploitation (37 percent) and introduced exotic species (19 percent). Why should we care? In truth, even some ecologists find it hard to muster up the enthusiasm to mourn the extinction of another little brown bird (Oddie 1987). Avise (1989) suggests, with good reason, that the conservation of subspecies like the dusky seaside sparrow, the last of which died in Florida in 1987, may be a misdirection of effort. Molecular genetic surveys of the dusky, and other subspecies, show that these populations are not evolutionarily distinct at all, and their preservation would not result in an increase in biodiversity.

But an extinction is forever, and potentially valuable species will not be here to provide us with their unique genes. Though this reasoning may seem like a throwaway line, Myers (1983) and Eltringham (1984) provide thousands of examples of the value of wildlife to people, from new varieties of food to eat and new drugs for medicine to new sources of industrial fuel. For example, 3 million U.S. citizens (and many others worldwide) depend on the foxglove, *Digitalis lanata,* as a source of cardiotonic compounds to keep them alive. Without this wildflower, they would find themselves in terminal heart trouble with as little as 72 hours to live. Commercial sales of digoxin, the drug from these foxgloves, are in the region of $50 million a year. The value of plant-derived prescription drugs in the United States was at least $20 billion in 1982, and many of the plants involved are tropical—hence the value of preserving tropical forests. We never know what new diseases lie around the corner (like AIDS), and new drugs will certainly be needed to treat them. Only five of the 41 higher plants used for plant-derived drugs are native to the United States. Laboratory-synthesized analogs of plant-derived drugs are generally over twice as expensive to produce. Ninety-five percent of raw materials come from natural sources.

Some direct economic values can also be attached to wildlife and habitats. Though it is difficult to assess the aesthetic and philosophical benefits of "nature," one can estimate how much people are willing to pay to keep areas natural (National Research Council 1982). The directexpenditures method seeks to assess the value of wildlife-related activities on the basis of the total amount spent by participants in those activities. A questionnaire might be mailed out to a random subset of the population to elicit information on dollars spent on wildlife-related activities such as entry and license fees, travel costs, food, and lodging. The assumption is that these amounts are equal to the amount spent in enjoying the activity. Nobe and Gilbert (1970) used this method to estimate the economic value of hunting and fishing in Colorado, which they concluded amounted to $250 million in 1968. In many areas, sport fishing has a much greater economic impact than commercial fishing. As an example, the Great Lakes sport fishermen in Michigan annually spend in excess of $157 million. The ex-vessel value of the commercial catch is only $4 million annually (National Research Council 1982).

Habitat Destruction

For convenience, habitats can be grouped according to likely causes of destruction. Each group then serves as a jumping-off point for further studies (Chapter 21). A listing might be

- *Virgin and pristine ecosystems:* These are lands untouched by humans.
- *Lands used for outdoor recreation and nature conservation:* They may include "unnaturally" protected areas, where, for example, forest fires that would normally occur are stopped or herds of game that would occur naturally at reduced densities as a result of lowered carrying capacity are held at higher densities. National parks, where nest boxes and winter feed may be provided, are a good example.
- *Water-catchment areas:* In these areas, although terrestrial systems may be replaced by aquatic systems as a result of damming, the ecological effects are generally beneficial, including increased irrigation potential.
- *Forestry:* The effects of forestry are harder to quantify, because forestry in, say, Europe would not be much different from the "farming" of natural forests, whereas in other areas, especially the tropics, forestry is involved in devastating changes causing the thinning of forests and leaching of nutrients from soil.
- *Agriculture:* Farm lands lie toward the highly manipulated end of the scale. Crop monocultures require huge energy inputs in the form of fertilizers, pesticides, and herbicides. Furthermore, the systems are usually prone to epidemics.

□ *Marine environments:* Marine habitats are not changed as such but are often "mined" for fish and marine mammals.

□ *Urban environments and derelict land.*

In many cases, of course, preservation of habitats stands in the way of economic development of an area. When the U.S. Congress passed the Endangered Species Act in 1973, it was probably thinking of furry brown-eyed creatures or soaring birds (Holden 1977). However, it is difficult biologically to justify denying endangered-species status to an obscure invertebrate like a worm if another invertebrate, like a rare butterfly, is under consideration for protection. In 1966 the Tennessee Valley Authority had authorization from Congress to build the Tellico dam and reservoir across the little Tennessee River. Builders had completed half the dam when the snail darter, a tiny fish scarcely bigger than a large paperclip, was discovered (see Photo 20.2). Because the snail darter cannot breed in still waters, the $116 million dam would have annihilated the newly discovered population in one blow. The Tellico dam was 75 percent finished when the secretary of the interior listed the snail darter as an endangered species. Environmentalists sued to have construction stopped, and in 1978, with the dam 90 percent complete, the Supreme Court ruled in favor of the fish. Even with the best intentions for species preservation, Congress, in passing the Endangered Species Act, had "opened up a Pandora's box containing infinite numbers of creeping things they never dreamed existed" (Holden 1977, p. 1427). Approximately 85 other species of snail darter exist in Tennessee, and subsequent to the controversy, additional populations of the snail darter were discovered; it was removed from the endangered species list (Kellert 1985).

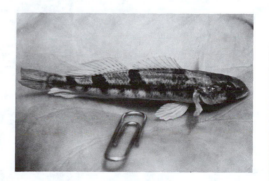

*Photo 20.2 The snail darter.
This is the endangered fish that
stood in the way of the $116 mil-
lion Tellico Dam in Tennessee,
which would have destroyed its
habitat. (Photo from U.S. Department of Agriculture, Fish and
Wildlife Service.)*

Direct Exploitation

In many areas, animals and plants are exploited in their own right. In developing countries, wild game is a principal source of meat, providing 75 percent of the animal protein in Ghana and Zaire and 20 percent along Brazil's Trans-Amazon highway. A large proportion is also taken by hunters and poachers for markets in Europe, North America, or Japan. Wildlife may be traded as hides and skins for fur and leather companies, as exotic meat for exotic dishes, or as products for the perfume, pharmaceutical, or ''aphrodisiac'' industries. Photo 20.3 shows an example, a giraffe killed illegally just for its tail. Alive, wild animals are sold to the pet trade, zoos, and biomedical research. In 1979 the estimated turnover for smuggled Australian wildlife was $30 million. The U.S. Fish and Wildlife Service estimated in 1984 that the illegal traffic in birds was at least as high as the 223,000 that were legally imported each year. Small boatloads of walrus heads have been observed in Alaska, the animals decapitated for their valuable tusks (Allen 1980). The U.S. trade in carved ivory may have accounted for some 32,000 elephants in 1986 (Di Silvestro 1988), and in addition at least 2,000 more may have been legally killed to supply hides to North American markets. The international ivory trade consumes tusks from more than 80,000 African elephants a year, taken from a total population of approximately 625,000 in 1989. Recently, there has been a sharp increase in the numbers killed, which reflects (1) an increase in the price of ivory, (2) deteriorating economic conditions in Africa, and (3) greater availability of automatic weapons. If these exploitation pat-

Photo 20.3   Giraffe poached in Kenya for its tail. The rest of the carcass has been left to rot. (Photo by Thane Riney, Food and Agriculture Organization.)

terns continue, the African elephant population will be halved in 10 years (Pilgram and Western 1986). Happily there are some encouraging signs that that will not necessarily happen. In June 1989, Japan and Hong Kong announced a ban on ivory imports. Japan was the largest ivory importer and consumer of ivory, whereas Hong Kong was heavily involved in making ivory carvings for export. But is a ban on ivory the answer? In southern Africa, numbers of elephants were actually increasing in 1989 in South Africa, Zimbabwe, and Botswana (Armstrong and Bridgland 1989). Careful management and culling of herds allowed numbers to increase and to support a valuable ivory trade. These countries are loath to see this source of revenue denied them, and they rightfully claim that similar success stories in other African countries are possible with good management. Besides, an official ban on ivory may simply drive the whole business underground, which in turn would cause prices to sky-rocket, further encouraging heavy poaching. In Zambia, availability of local hunting licenses drastically reduced poaching, because local poachers were generally willing to apply for licenses. Professional poaching by outsiders continues and is heavily combatted. The United States, as of 1990, continued to threaten to reduce aid to southern African countries that would not institute ivory bans.

The imports of wildlife into Britain during just six months in the 1970's were staggering (Table 20.3) and included species that supposedly should not even have been traded. Taken to its limit, hunting can and does exterminate species in the wild. The Arabian oryx is a case in point. This magnificent animal was virtually snatched from the grave. The last known wild Arabian oryx was reportedly killed in 1972. Luckily, three were taken captive in 1962 in the Rub-al-Khali, the great southern desert of Saudi Arabia, just as a hunting party was preparing to shoot them for fun. These animals, together with a small number of others donated by members of the royal families of Saudi Arabia and Kuwait, were bred in captivity, and a herd of 400 was formed in zoos in the southwestern United States. The story completed its full circle when captively bred animals were released in Oman in 1980.

The Optimal Yield Problem

How many deer can be shot before a marked effect on population numbers occurs? How can people harvest a large part of a population without causing long-term changes in its equilibrium numbers? Removing too many young individuals may seriously impair population age structure and hence population size of future generations, and removal of a large fraction of the dominant reproductive individuals would also seriously affect reproductive output. Unfortunately, this scenario is at loggerheads with an economic viewpoint. Current profits (which can be

Table 20.3   *Selected imports of wildlife into Britain (January–July 1976). (From Burton 1976.)*

Species	Number or Amount Imported
*Leopard (*Panthera pardus*)	661 skins
*Jaguar (*Panthera onca*)	279 skins
Margay (*Felis wiedii*)	9,277 skins
Geoffroy 's cat (*Felis geoffroyi*)	14,169 skins
Ocelot (*Felis pardalis*)	16,619 skins
Tiger cat (*Felix tigrina*)	9,000 skins
*Polar bear (*Thalarctos maritimus*)	101 skins
Malayan bear (*Ursus malayanus*)	100 live
Elephant (*Loxondonta africana*)	42,871 kg ivory
Estuarine crocodile (*Crocodilus porosus*)	100 live
Estuarine crocodile (*Crocodilus porosus*)	523 skins
Caiman (*Caiman crocodilus*)	310 live
Caiman (*Caiman crocodilus*)	5,021 skins
Tortoises (*Testudo graeca*)	245,026 live
(*Testudo horsefieldi*)	25,000 live
*Hawksbill turtle (*Eretmochelys imbricata*)	322 kg tortoiseshell
Monitor lizards *(*Varanus flavescens*)	646,389 skins
*(*Varanus bengalensis*)	462 live
*(*Varanus griseus*)	245,254 skins
(*Varanus exanthematicus*)	25,000 skins
(*Varanus niloticus*)	25,002 skins
(*Varanus salvator*)	22,008 skins
Iguana (*Iguana* spp.)	25,000 skins
Tegu (*Tupinambis* spp.)	40,013 skins

*Species that should not be legally traded.

invested at a favorable rate of interest) are more valuable than future profits. Economically it is best to overexploit a population now.

Maximum Sustainable Yield

In theory, if its environment is constant, every population has a maximum sustainable yield (MSY), the largest number of individuals that can be removed from the population without causing long-term changes. In practice, the MSY is often difficult to determine, let alone achieve. The MSY is a result of both the reproductive rate and the number of individuals reproducing. Both yields and population sizes fluctuate more as the MSY is approached, the effect being most pronounced in large mammals such as whales (May 1980).

In debates over the effects of harvesting or hunting animal populations, two schools of thought have emerged. In one, hunting is seen as an extra strain on the population in addition to those already acting on it.

This view is referred to as the *totally additive mortality hypothesis*. In the other, hunting is seen as a source of mortality that merely replaces other sources of mortality; in other words, shooting game merely removes individuals that would be killed in some other way. This view is known as the *compensatory mortality hypothesis* (Anderson and Burnham 1976). Surprisingly, the data seem to support the compensatory hypothesis rather than the totally additive mortality hypothesis (Nichols et al. 1984) at least for populations of mallards, *Anas platyrhynchos*. Hunting waterfowl is probably the single most important hunting activity in North America (Martin, Pospahala, and Nichols 1979), so the data on this issue are among the best. Nearly 700 species of migratory birds occur in North America north of Mexico, and nearly 100 of these are shot as game. It has been estimated that the fall population of ducks is over 100 million, and hunting them provides over 17 million days of hunting recreation for 2.4 million hunters annually. Of course the acid test for comparing the totally additive mortality and the compensatory mortality hypotheses would be to ban hunting over a large experimental area and then to examine subsequent mortality rates. This experiment is unlikely to be politically or socially acceptable. Data like these are necessary, however, to answer the pressing question of how much game it is acceptable to take; already hunting is responsible for between 35 and 66 percent of known mortality in U.S. black ducks (*Anas rubripes*) and mallards (Krementz et al. 1988). Additional studies on other organisms seem to indicate that mortality of populations due to shooting is "made up" by lessened mortality in other areas. This was true for grey squirrel populations in the United States (Mosby 1969) (often the game animal most hunted by sportsmen) and for wood pigeons in the United Kingdom, shot both for the pot and as an agricultural pest (Murton, Westwood, and Isaacson 1974). This is not to say that harvesting cannot seriously affect population numbers. If the harvesting technique is more thorough, for example trawling for fish, then overexploitation can cause population stocks to crash.

Calculating Yields

Out of all applied fields, it is in fisheries that the most work on the optimal yield problem has been done. Commercial fisheries generate a vast amount of money, and the problem of overfishing has been addressed since at least the 1920's, when commercial stocks of many species began to dwindle.

For a harvested population the important measurement is the *yield* expressed in terms of either weight or numbers over a particular time period to give a catch per unit effort. Catch per unit effort can then be compared year after year to determine how well a particular managed resource is doing. Remember that the maximum yield of a population is related to the maximum population increase. Greatest population increase, dN/dt, occurs according to the equation

$$\frac{dN}{dt} = rN\frac{(K - N)}{K}$$

at the midpoint of the logistic curve, Fig. 9.5. Thus, maximum yield is obtained from populations at less than maximum density, when they are constantly trying to expand their own population densities into unutilized resource areas. Adjusted for fishing losses, dN/dt becomes

$$\frac{dN}{dt} = rN\frac{(K - N)}{K} - qXN$$

where q is a constant and X equals the amount of fishing effort, so qX equals fishing mortality rate. As fishing and hence fishing losses increase, qXN can begin to affect the total catch (yield) severely. This relationship is clearly shown in Fig. 20.5, which details the relationship

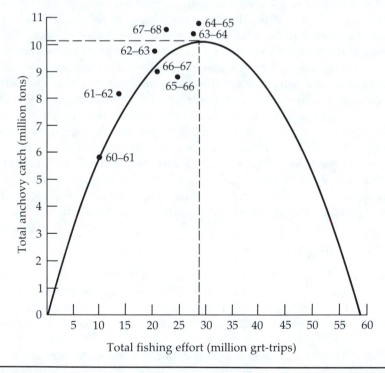

Figure 20.5 *Relation between total fishing effort and total catch for the Peruvian anchovy fishery, 1960–1968.* The effects of humans and seabirds are combined in these data. The parabola represents the logistic model fitted to these data. Arrows indicate maximum-sustained-yield-appropriate fishing effort. (Redrawn from Boersma and Gulland 1973.)

between fishing effort and total catch for the Peruvian anchovy. The Peruvian anchovy fishery was the largest fishery in the world until 1972, when it collapsed. In 1972 12.3 million metric tons were harvested, and this one species alone comprised 18 percent of the world's total harvest of fish. Note that from 1964 to 1971 the catch was close to the supposed maximum. People *may* not have been the primary cause of population collapse. In 1972, the phenomenon known as El Niño, a major climatic change, perhaps caused by volcanic activity, occurred, permitting warm tropical water to move into the normally cool, nutrient-rich upwellings near the Peruvian coast. The anchovies failed to spawn, and adult fish moved south to cooler waters. Whether or not the fishery might have been saved if fishing had ceased or its intensity had been reduced to allow recovery is a matter for speculation. From the logistic equation, MSY can be estimated as

$$MSY = Nr$$

where r is the rate of natural increase and N is the average number of animals present throughout the year. However, this approach assumes a constant rate of cropping throughout the year. If there is one "season," then MSY is best estimated by the equation

$$MSY = N(1 - e^{-r})$$

which is likely to be a lower value. If $r = 0.14$ and $N = 10,000$, then

$$MSY = 10,000(1 - e^{-0.14}) = 1,306$$

But if the harvest is spread out over the year, then N is likely to be reduced by natural deaths. N might be reduced by 10 percent so that $N = 9,000$. Then

$$MSY = 9,000 \times 0.14 = 1,260$$

In practice, cropping is often limited to seasons if only to give animals some respite during breeding. Several techniques are available for estimating MSY of a population (Eltringham 1984).

In a previously uncropped population, one proceeds largely by trial and error, by guessing the MSY and taking a constant number of animals, year after year. It is necessary to count the population before and after harvest (Caughley 1977). For each year, compute

$$\log_e \frac{N_{t+1}}{N_{t-c}}$$

where N_t is the number at time t and C is the crop. Plot this value against N_t, and the y intercept is the rate of natural increase r. Now,

$$MSY = \frac{r}{2} \times \frac{K}{2}$$

where K is the original size of the population.

If a population has been harvested for some time, then other methods can be applied to estimate MSY. For example, r can be estimated from the equation

$$r = \frac{KH}{(K - N)}$$

where H is the harvest in past years and N is the size of the population from which it has been taken. MSY is now calculated as before.

Finally, Caughley (1976) has suggested

$$MSY = \frac{CK^4}{4N^2 (K^2 - N^2)}$$

where, as before, N is the standing crop (size of cropped population), C is the annual crop, and K is the original, unharvested population size.

Economic Yield

The ultimate yield of any harvest is, of course, not in biomass but in dollars. Surprisingly it is often poor business to operate a fishery, or any other type of harvest, at maximum yield (Gordon 1954). Consider Fig. 20.6, a simple economic model for a fishery. Total costs are assumed to be proportional to fishing effort, and revenue is directly proportional to yield, so that the yield curve of Fig. 20.5 is identical to the revenue curve of Fig. 20.6. The important point is that maximum economic revenue occurs at a lower fishing density than does maximum yield. It might seem that sound economic practice would also ensure a safe biological management strategy. Of course if one country's total costs rose at a lower rate than another's, perhaps because of cheaper labor or more efficient gear, then the maximum yield and maximum economic returns for the two would differ. This is a common source of conflict in international waters; what is sound economic practice for one group is not good management for the other.

Some basic principles still emerge from a consideration of MSYs. First, the most difficult (yet most critical) aspect of population management is undoubtedly accurate population censuses. This still remains one

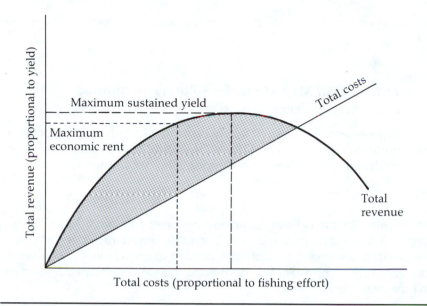

Figure 20.6 *A simple economic model of a fishery in which costs are directly related to fishing effort and revenue is directly related to yield.* The shaded zone is the area in which revenue exceeds costs. Maximum economic return is achieved when the revenue-cost difference is maximal, and this is below maximum sustained yield in this model. (Redrawn from Krebs 1985*a*.)

of the most difficult tasks in wildlife biology. Accuracies of greater than 90 percent are very unusual, and accuracies are often as low as 20–30 percent (Eltringham 1984). Second, in the absence of other information, a good rule of thumb is not to harvest more than 10 percent of the population at carrying capacity.

20.3 *How to Solve Ecological Problems*

It is very difficult to provide a blueprint for success in solving ecological problems. Each case must be dealt with largely on its own merits. There are so many unusual and unique ecological variables associated with each case study that broad generalizations are of little use. Useful predictions are more often based on local field experience than on formal theory. Strong (1980) has argued that predictability in biology will never be as advanced as it is in atomistic sciences such as chemistry and physics. He suggests we cannot compare organisms to atoms; whereas there are only a relatively small number of atom types (105 at last count), there are many millions of organisms (see Chapter 2). Thus, although scientists can discover certain laws linking similar atoms, the profusion of different and independent organisms suggests there will be few universal ecological laws.

Practices That May Aid in the Solution of Applied Ecological Problems

Although ecological problem solving cannot be reduced to a formula, the Committee on the Applications of Ecological Theory to Environmental Problems (National Research Council 1986) recommended that nine general guidelines be followed:

- **Involve scientists from the beginning:** Scientists can help to identify nonobvious goals, can translate environmental goals into scientific objectives, can show what information is needed to answer major questions, and are usually important in determining the values attached by society to ecosystem components.

- **Treat projects as experiments:** Pilot-scale experiments were conducted to study the potential effects of the Alaska pipeline on caribou movements (Cameron et al. 1979). In the process a great deal more was learned about caribou movements than would have been possible if the pipeline had not been constructed or if it had been constructed without prior experimentation. Frequently, a major project itself can function as a large-scale experiment if careful baseline studies are done beforehand.

- **Publish information so that others can learn from any mistakes:** Larkin (1984) believes that literature review can provide 50 percent of the information needed in most initial impact assessments and as much as 75 percent when coupled with brief reconnaissance surveys. Looking for analogous studies is extremely useful; even if some are not perfect, reading them can be useful in "scoping out a project."

- **Set proper boundaries on projects:** Well-intentioned efforts to conserve animals in reserves can be undermined if they are planned for too short a term or too small an area (Soulé and Wilcox 1980).

- **Use natural-history information:** Nonrandom harvests of animals, in which unequal numbers of the sexes are taken (for example, the preferential hunting of male deer [Law 1979] or the unwitting selection of [larger] male fish through size selectivity of fishing gear [Moav, Brody, and Hulata 1978; Richer 1981]) can lead to uneven and catastrophic sex ratios. The carefully designed Garki malaria-control project in Africa and its attendant mathematical predictions were thrown into turmoil because incorrect assumptions were made about the biology of the mosquito vectors, which did not land inside buildings (and thus were not exposed to contact insecticides) as often as predicted (Molineaux and Gramiccia 1980; May 1986).

- **Be aware of interactions:** When **DDT** was first introduced, no one imagined it would eventually show up in marine fish (Buckley 1986). And in northern Australia, when water buffalo were introduced, they brought with them their own blood-sucking fly, a species that bred in cattle dung and transmitted an organism sometimes fatal to cattle. Australia's native dung beetles were accustomed only to the small pellets of grazing marsupials and could not tackle the large pats of the buffalo. The blood-sucking flies proliferated. Eventually, African dung beetles, used to large pats, had to be brought in to compete with the flies for dung, which they successfully did, reducing fly populations (Moon 1980).

- **Be alert for possible cumulative effects:** Recent awareness of such serious cumulative effects as **acid rain,** loss of tropical forests, and the threatened extinction of many species has brought increasing political and scientific attention to these effects. Increasing the intensity of harvesting can lead to population collapse if a critical threshold is passed, especially in conjunction with other environmental changes (Beverton 1983).

- **Plan for heterogeneity in space and time:** Plankton communities undergo seasonal turnover, and any population changes must be viewed in the light of natural variations in population densities.

- **Prepare for uncertainty and think probabilistically:** Attempts to achieve MSYs in many fisheries can result in more variable yields, greater population fluctuations, and a greater likelihood of population collapse (May 1980).

Some of the recent reasons for failure in **ecological impact** assessment are given in Table 20.4.

Sometimes it is simply too hard for ecologists to move from pure science to applied science. As Wigglesworth (1955) explained, pure scientists are used to attempting to solve problem A but, if that proves too difficult, to switching to problem B and even coming across a solution to problem C. In applied cases, however, there is no escape from problem A.

Indicator Species

Living organisms, so-called *biological indicator species* can be used, in a process called **bioassay,** to monitor movements, accumulations, and modifications of materials in their environments and to monitor the biological effects of those materials. Chemical and physical monitoring can detect the quantity of toxic substances in the environment, but only bio-

Table 20.4 *Sources of failure in ecological impact assessment. (From National Research Council 1986.)*

Overelaborate guidelines and requirements that are too diverse

Time and money constraints not recognized

Unreasonable expectations of decision makers

Tendency to start gathering baseline data immediately, at the expense of careful planning

Failure to formulate objectives clearly and to develop a study strategy

Unwarranted belief that ecological principles used in managed systems are as appropriate to unmanaged systems

Failure to recognize the value of early input from those who might later be involved in review, leading to an adversarial process

Failure to define project boundaries

Failure to consider cumulative effects

Failure to state the bases of value judgments

Lack of scientific standards for impact assessment

Lack of respect in academe for impact assessment

Vague and unverifiable predictions

Lack of a rigorous, quantitative approach, especially in monitoring

Lack of continuity in studies conducted during planning, developmental, and operational phases of a project

Failure to follow actions with adequate monitoring studies

Use of impact assessment for disclosure, rather than for learning

Failure to recognize the scientific value of experimentation and monitoring

Failure to consider the recovery potential of species and ecosystems

Poorly written reports in which major points are buried in enormous amounts of information

Inordinate expenditure of effort on descriptive studies with little potential for predictive value

Inaccessibility of reports and results of studies, making them difficult to evaluate and learn from

---

logical monitoring can divulge what such materials are doing to organisms. Living organisms can also accumulate records of past conditions in their tissues and increase the detail of records. Pollen from honeybees provided a much more complete record of the distribution of several pollutants in Puget Sound than was obtained by physical monitoring (Bromenshenk et al. 1985).

The choice of indicator species depends on the type of environmental changes to be monitored. Short-lived organisms respond quickly to environmental changes; long-lived ones might integrate stress over a number of years. Sessile organisms are exposed to all the contaminants within an area because they cannot easily escape; mobile animals can leave the area. Many organisms may stop reproducing under stress, so changes in fecundity can be good signals of environmental change. Some plants, however, are stimulated to reproduce and produce seed under poor conditions—a last-ditch effort at continuation of their genes. Vascular plants in particular are good detectors of air pollution, which often causes lesions on leaves. Sulphur dioxide in acid rain is a case in point. Lichens

and mosses are even more effective at indicating airborne pollution because they absorb water and nutrients directly from the air (Hawksworth
1971; Lawrey and Hale 1979). A measure of pollution is indicated by a
lichen species-diversity index, in which abundance is based on number
of individuals, extent of coverage, growth form, and degree of luxuriance
(Fig. 20.7). It has been calculated that at least one-third of Britain has lost
its epiphytic lichen flora as a result of SO_2 pollution (Rose 1970). In
aquatic environments, **eutrophication** can be assessed from increases in
the biomass of certain algae like *Spirogyra, Cladophora, Oedogonium,* and
Stigeoclonium. Diatoms have been known to concentrate heavy metals
and radioactive substances by a factor of several thousand. Mussels have
also been used to monitor pollution in a scheme known as "Mussel
Watch" (Goldberg et al. 1978). Pollutants can be concentrated in mussel
shells and preserved for long periods. In marine situations, sea bird populations are often used as monitors because they are much easier to assess than fish or aquatic invertebrates. In addition, different species that
nest in a single location may forage at widely differing distances from the
breeding colony, and identifying these patterns can help locate contaminated areas (Boersma 1986; Gilman et al. 1979). The herring gull (*Larus
argentatus*) has been chosen to monitor pollution in the Great Lakes of
Canada and the United States because it is resident year-round and because it accumulates pollutants only from the lakes, not from the surrounding land. Mineau et al. (1984) showed just how effective the species
was in monitoring the fate of persistent organochlorine pesticides after
control measures were established.

In terrestrial systems, top carnivores often concentrate materials in

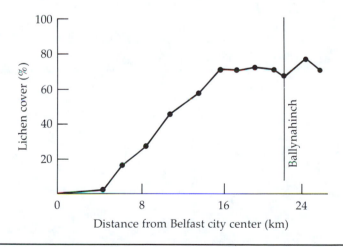

Figure 20.7 *The increase of lichen cover on trees outside the city of Belfast, Northern
Ireland.* (Redrawn from Mellanby 1967.)

their bodies (Chapter 19), especially fat-soluble pesticides. Being wide-ranging, they are often exposed to materials with a patchy distribution, which a fixed monitoring station might miss. Other species may react to pollutants with specific changes in behavior or phenotype. Remember the familiar case of industrial melanism in the peppered moth (*Biston betularia*) (Chapter 3; Kettlewell 1973; May and Dobson 1986).

More recently, preserved organisms are being kept in storage (so-called environmental specimen banking) so as to provide long-term records of pollution. This practice ensures that, as new chemicals become of concern, banked specimens can be examined retrospectively for them, providing the history of the pollutant's buildup in the environment. Furthermore, as new analytical techniques become available, they can be tested on old, known specimens. Nonbiological materials such as sediments can also be used for this type of work but generally cannot provide the range of information available from biological specimens. For example, it was the thinning of eggshells that first provided evidence of the widespread accumulation and destructiveness of DDT (Cooke 1973; Buckley 1986). The methods of preservation and the decisions about what to preserve are well-treated by Lewis, Stein, and Lewis (1984).

Of course, in some situations, merely monitoring an indicator species would not provide a good overview of what is happening in the entire community. Some species in a guild use different microhabitats; thus, monitoring habitat suitability for a guild indicator species might not reflect habitat suitability for other species in the guild. Block, Brennan, and Gutiérrez (1986) found that it was more economical and statistically less variable to monitor a whole guild of ground-foraging birds than it was to monitor the population of any single species.

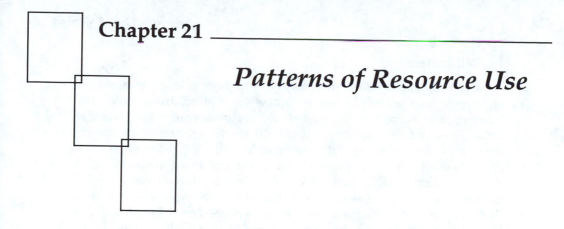

Chapter 21

Patterns of Resource Use

Habitats and patterns of resource use can be grouped according to degree of degradation by humans. Such a grouping would include wilderness areas as the least-affected habitats and areas such as quarries or dumps as the most heavily degraded. In between lies a variety of habitats altered by humans to intermediate extents, such as forests and agricultural lands. A listing in order from least to most degraded might include wilderness areas, national parks, water catchment areas, forests, agricultural land, and derelict land. Each of these categories serves as a useful jumping-off point for discussion.

21.1 Wilderness Areas

The virgin or unaltered land areas of the world can be usefully categorized according to the reason for their unaltered state, as follows (after Simmons 1981):

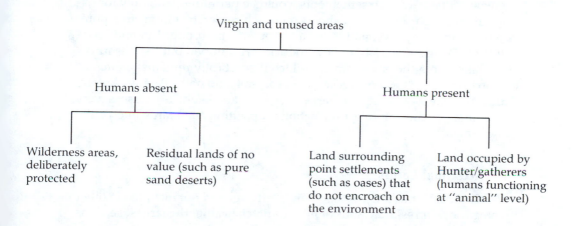

Wilderness

Designated **wilderness** areas may be thought of as large, unmanaged regions containing whole sets of **biotas** (as distinct from parks and reserves, which are usually smaller and are preserved with particular landscapes, recreations, or communities in mind). One rationale for wilderness preservation is that it is everyone's spiritual right to enjoy such areas. Scientifically, it is proposed that they maintain a gene pool of wild organisms (for possible future economic gain) and provide natural communities for scientific research into behavior and ecology. Seventy-seven thousand km^2 of wilderness areas had been preserved by 1970 in the United States (following the signing of the National Wilderness Act of 1964). Recreational and aesthetic arguments were the catalyst to their preservation (Simmons 1981). Of course, designating an area a wilderness is often likely to attract more people, whose numbers make management necessary. Between 1970 and 1986, the amount of land in the National Wilderness Reservation System increased almost tenfold (Miller 1988), most of it in Alaska. In the United States, debate rages over the wilderness-area issue. Many economists were upset that legislation that created the Alaska Wilderness in 1980 cut off the oil industry from at least $1.5 trillion in energy reserves. President Bush's 1989 inaugural address called for the opening of Alaska's coastal plain to exploratory oil drilling, and on March 16, 1989, the Senate Energy Committee approved the plan. One week later, the *Exxon Valdez* accident occurred (see chapter section 22.3), and Congress is expected now to be less hasty in its approval for new drilling in Alaska.

Point Settlements

Even the largest wilderness area in the world, Antarctica, contains point settlements. The Antarctic treaty of 1959 among "user" nations ensured that only such settlements would be permitted, and only for scientific research. Photo 21.1 shows one such research station. In a point settlement, *everything*, all foods and supplies, are brought in and *everything*, all waste materials and byproducts, are shipped out. No resources are taken from the environment, which theoretically remains unaltered. However, the discovery of reserves of coal, oil, and natural gas in Antarctica and the possibility of harvesting krill are likely to cause severe economic discord among user nations, upsetting the status quo of the agreement.

Residual Lands

Residual lands are just that—lands left over and not included in other categories. Of residual lands of no commercial value, the largest sections

Photo 21.1 Point settlement. The Chilean research station in Antarctica, Whaler's Bay, Deception Island. (Photo supplied by Carolina Biological Supply Company.)

are in the Arctic and subarctic zones of Eurasia and North America, between the sparse point settlements that characterize these vast tracts of tundra. Recently the advent of industrialism and the activities of mining and petroleum extraction have begun to alter these lands. Vehicles erode tundra, and waste attracts scavenging polar bears. Because these ecosystems are so simple, it takes little to upset their balance.

Some boreal forest zones of Canada and the Soviet Union also lack density of settlement. Again, minimal disturbance can be far-reaching; wolf control allows herbivorous moose to build up to high, forest-damaging levels. A very large area of "unused" land is the desert, where primary production is very low and exploitation is confined to mineral extraction at point settlements. In the United States, however, recreational use by dune buggies may severely affect stability when binding plant roots in partially vegetated dunes are destroyed and the dunes become subject to severe wind erosion.

Hunter-Gatherer Areas

The hunter-gatherer way of life was once the dominant mode of existence on Earth, but by A.D. 1500 only 15 percent of the world's land surface was inhabited by such peoples, at a time when most major biomes were more or less fully occupied. Today, hunters and gatherers form perhaps 0.001 percent of the population and include such tribes as the Bushmen of southern and southwestern Africa, the Pygmies of central Africa, the Australian Aborigines, and the Indians of "interior" Brazil and Venezuela. Some North American Indian and Eskimo groups also live largely on hunted food, although contact with western culture is high. Manipulation of the environment in these groups is minimal, mainly because such groups do not populate their territories up to the carrying capacity of the land and also because there is some restriction of overexploitation;

for example, some groups do not take female animals. Mortality from starvation, disease, accidents, and predation are generally low, and social mortality such as infanticide has been important in the population-resources equation. Only in extremely rigorous environments do starvation and, more likely, thirst affect population sizes.

21.2 *Protected Ecosystems: National Parks*

National parks differ from wilderness areas in that, although there is protection from certain kinds of environmental alteration, change is acceptable, and management is a necessity, the aim often being to preserve a species or assemblage of species.

Sometimes preserves are proposed for single species. An example is the Wood Buffalo National Park in Canada, erected for the American bison, perhaps more than any other animal the symbol of the native North American grassland. Ranging from small prairies of the East to the Rockies, bison numbered 60–75 million in the 18th century. Following the arrival of European settlers, every year through the early 1870's millions of bison were killed, in part to subjugate the Indians by removing their food base. By 1878 the southern herd was gone, and by 1883 the last of the Dakota bison had disappeared. Only 150 animals were estimated to be alive by 1889, but under careful management their numbers have increased since then. In northern Alberta, the Wood Buffalo National Park preserves one of the few remaining herds of North American bison, shown in Photo 21.2. The herd is culled regularly in order to keep it within the carrying capacity of the park and to try to keep down diseases

Photo 21.2 American bison. (Photo from U.S. Department of Agriculture, Fish and Wildlife Service.)

to which the herd is subject. The park is one of the few remaining areas where the once-plentiful bison now exist (Fig. 21.1). In northeast India, the Kaziranga Sanctuary preserves the great Indian rhinoceros. Animal migration often presents a problem, for what is sacred in one country is fair game to the next. Even in Europe the EEC (European Economic Community) cannot agree on the conservation of migratory birds. Italian

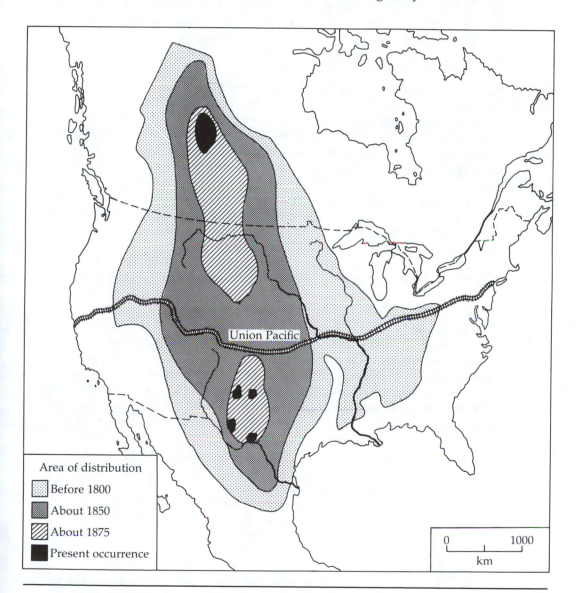

Area of distribution

- Before 1800
- About 1850
- About 1875
- Present occurrence

Union Pacific

0 1000 km

Figure 21.1 *The former and present distribution of the bison in North America.* (Redrawn from Illies 1974.)

hunters regularly kill 150–200 million birds per year of all types together with a few of their own species (three in 1975) on the opening day of the hunting season (Simmons 1981).

Rather more common are reserves that protect an assemblage of species or a whole set of habitats. For example, in the southeastern United States one cannot simply preserve red-cockaded woodpeckers, *Picoides borealis*. One must first preserve the old and dying longleaf and loblolly pines that they use for nesting. Such pines cannot persist without the intermittent fires that eliminate the seedlings of other species that might otherwise outcompete them. Such fires favor the emergence of wiregrass, *Aristida stricta*, and other flora that was predominant before the arrival of Europeans. Thus, maintenance of red-cockaded woodpeckers involves fire management of the whole coastal-plain community.

Larger communities are more often given national park status. Examples include fragile island **ecosystems** with a high proportion of endemics such as Aldabra, in the Indian Ocean, and the Galapagos islands off the coast of Ecuador. Wetlands, often sought after for land reclamation, are among the most productive of the world's ecosystems and should more often be preserved. (They also protect coasts from storm damage and are nurseries for fish.) Finally, of course, an area may be preserved for a fine view, especially in mountainous regions, or for other aesthetic or spiritual reasons.

Few reserves in small countries are big enough to qualify for national park status. In Britain, 10 "national parks" (Brecon Beacons, Dartmoor, Exmoor, Lake District, Northumberland, North Yorkshire Moors, Peak District, Pembrokeshire Coast, Snowdonia, and Yorkshire Dales) have been designated, totalling 1,360,000 ha (MacEwen and MacEwen 1983), but by world standards these are better referred to as "protected landscapes," because each has been materially altered, and over 250,000 people live in them. In larger countries such as the United States, however, there are many national parks, like the Everglades in Florida, which consists of a complex of wetland areas, coastal mangroves, tropical sawgrass marshes, and "hammock" forests on the slightly drier areas a few centimeters above flood level (see Photo 21.3). The 4856 km^2 park supports a wide variety of animal life, including the rare Everglades kite, the roseate spoonbill, alligators, and manatees, and legislation is currently being approved that would increase the park even more.

On an even larger scale are some of the national parks and game reserves of Africa. There the desire to ensure the survival of the fauna is reinforced by two strong economic considerations:

1. Tourism yields foreign currency.
2. Wild game can yield more protein per unit area than domesticated animals and with less damage from overgrazing.

Photo 21.3 River of grass, Everglades National Park, Florida. (Photo from Florida State Archives.)

The latter, however, is a touchy subject, since for some groups like the Masai of Kenya, domestic animals are directly equated with wealth, besides providing meat and milk.

The destructive influences of fire and flood often pose problems for managers of protected ecosystems. Fires, initiated by lightning, may be part of the natural ecology—many species of conifers have cones that open and release their seeds only after being subjected to high temperatures. The giant redwood, *Sequoiadendron giganteum,* is a case in point, and state and national park services must follow a program of controlled burning to initiate new seedlings. In southern England, a rare species of orchid thrives only on chalk grasslands where competing grasses are grazed down. Management has therefore reintroduced sheep into the area.

The Theory of Park Design

In theory, many nature reserves are ''islands'' in a sea of more intensely developed land, and MacArthur and Wilson's (1967) predictions of immigration and extinction rates on islands have been drawn into the debate about reserve size and shape. As early as the mid 1970's, Wilson and Willis (1975) proposed that certain geometrical shapes were better for refuge designs than others. Their arguments were based on the fact that ecological conditions in peripheral areas of reserves—that is, areas near the edges—are never the same as those central areas, simply because peripheral areas are bound to undergo some influence from whatever is on the other side of the edge. For example, at the edge of a forest, conditions may be too sunny for shade-loving plants or bird densities may be higher because of the overlap between forest species that venture onto open ground and grassland species that penetrate a short distance into the forest. All these subtle differences, both known and unknown, are referred to collectively as edge effects, and because of their influence refuge designers cannot be certain that peripheral areas of a refuge are as effective as central areas, which are away from the influence of other habitats. Wilson and Willis (1975) therefore thought that rounded or contiguous areas were better than fragmented or elongated ones because they had less edge per unit area and thus minimized edge effects (see Fig. 21.2). Reserves close to one another were recommended over isolated reserves, to reduce extinctions, and corridors connecting reserves were considered even better. The debate over the advantages of a single large refuge over several small ones became known as the SLOSS (single large or several small) debate (Simberloff 1986a). Large reserves were often argued to be able to harbor more species and to be more diverse than many smaller reserves of equal total area. Ultimately the experimental studies done show little support for the idea of a single large reserve (Simberloff and Abele 1976 and 1982; Simberloff and Gotelli 1984), and of course an advantage of smaller areas is that more of them can be purchased for the

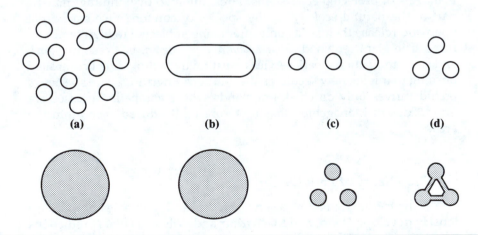

Figure 21.2 Geometrical rules of reserve design based on theories proposed in the literature from the study of island faunas. The upper and lower figures have the same total area and represent reserves in a homogeneous setting, but the lower configuration supposedly leads to a lower extinction rate than the upper. Rounded and continuous areas are thought by many to be preferable to fragmented and elongated areas (a and b); distance effects are minimized by clumping rather than linear arrangements (c); and corridors between isolated areas are even better (d). (Redrawn from Wilson and Willis 1975.)

same money, and each can be chosen to preserve a particularly interesting fauna.

What probably happens is that, even when islands and mainland habitats appear similar, habitat diversity and hence floral and faunal diversity will actually be higher on a series of smaller islands simply because parts of the smaller sites are physically more distant from one another than are parts of a single large site (Simberloff 1986*a*). Studies of herbs in English woodlands (Game and Peterken 1984) and of lizards in the Australian wheatbelt (Kitchener et al. 1980) bear out the idea of higher diversity in island series. Recently, Quinn and Harrison (1988) have reviewed data from over 30 censuses for 15 island groups where all data were reported, island by island, species identities of individuals were confirmed, and at least six islands were included in each survey. The resultant species-area relationships for vertebrates, land plants, and insects show that collections of small islands generally harbor more species than do comparable areas composed of one or a few large islands. National park faunas were shown to be richer in collections of small parks than in the larger parks. This finding has an important bearing on future land purchases for conservation purposes. Furthermore, in 29 out of the 30 cases, the small-islands curve saturated with species more rapidly than did the large-island curve, and in 18 out of the 30 cases, the two curves never intersected, indicating that smaller, more numerous islands always had more species than fewer large ones. It is also worth noting that a large site, or a series of smaller sites connected by corridors, leaves species more vulnerable to extinction by a single catastrophic event or by the spread of disease or introduced predators (Simberloff and Cox 1987). The hordes of rats and cats brought to the Seychelles by Europeans in the late 18th century devastated the local avifauna, but some species were saved because a few birds bred on smaller outlying islands (Simberloff and Cox 1987). Small areas will not work in all cases, however, and Temple (1986) has shown that bird diversity in North American woodland plots is best predicted by "core" area, the area of a plot more than 100 m from the edge. Many birds will only use the nondisturbed area of a habitat, so long, thin refuges or small, isolated ones are of little value. The survival of the spotted owl (*Strix occidentalis*) is a case in point because logging in the northwestern United States is reducing and fragmenting its only habitat, old-growth forest (Dixon and Juelson 1987; Salwasser 1987; Simberloff 1987*b*). Unfortunately, such old-growth forest is extremely valuable to the logging industry, which provides, essentially, the economic backbone of the Pacific Northwest. In the Amazon, pioneering studies on the minimum critical size of ecosystems have suggested that buffer zones hundreds of meters in extent will be needed to alleviate edge effects (Lovejoy et al. 1983 and 1986). Trees fall down more rapidly at edges, where they are more exposed to wind and erosion, microclimate changes, and densities of some organisms decrease. The diversity of some insects, like butterflies, actually increases in disturbed areas in the tropics, but even there, fragmentation can have deleterious effects. In one of the first scientific studies on the effects of fragmentation in the tropics, Klein (1989) showed that dung- and carrion-beetle communities in 1 ha and 10 ha forest fragments differed drastically from those in contiguous forest. Intact forested areas had more species, denser populations, and larger beetles than did forest fragments. These results may help explain why dung decomposes at lower rates in forest fragments, and these effects may ramify through other related community and ecosystem processes as well. Too small a reserve is obviously of little value in conservation.

It is not always remembered in such theoretical debates that species are not simply numbers in an equation and cannot be treated as such. Some plants and animals are of more interest to conservationists than others, and island biogeographic theory fails because it does not draw any distinctions between desirable and undesirable species. Some areas such as oligotrophic lakes are species-poor yet are preserved because of their pristine appearance. Other areas like agriculturally improved grassland or even old railway sidings can be very rich in plant species, yet we do not wish to preserve these areas. Richness seems to be important only when it suits us; otherwise we ignore it. Emphasis should return to autecological studies to determine the habitat requirements of endangered species (Owen 1975; Soulé and Simberloff 1986).

Finally, Mabey (1980) and others have argued that conservation has catered principally to an elite minority and that government agencies and trusts select preserves on pseudoscientific grounds for ecological specialists. What does the public really want? Goldsmith (1983) suggests that their principal interest is near to home, not at some distant reserve. They would like the "feeling" of nature in their immediate neighborhoods. They want to see pretty flowers, birds, and butterflies in the countryside. Swallowtails and bluebells may do just as well as panthers and bobcats, which are very secretive anyway. At the moment, conservation specializes on rare species and untouched habitat. Local conservation projects, closer to urban centers, may also be worthwhile. Photo 21.4 of Central Park in New York City illustrates the importance even of relatively small urban parks in conserving undeveloped land for recreational purposes. Such parks provide the only access some urban dwellers have to wooded or open areas. Goldsmith suggests that most conservation projects focus on ecological measures and assume that these can be equated directly with economic or social measures of value. This is simply untrue. The training of biologists makes the recording of plants and animals easy, but the answers obtained from such data do not give a measure of other values. It might be argued that the importance of different taxonomic groups could be assessed from the membership of the organizations that study them. For example, in Britain the Royal Society for Protection of Birds has

Photo 21.4 *Central Park, New York City, 1905. Before the advent of the automobile, city parks were even more heavily used than today because of the lack of access to other open areas. (Photo from Library of Congress, photo no. LC–D401–9285.)*

about 250,000 members, and the Botanical Society of the British Isles has 2,300. But of course membership is affected by the marketing efficiency of the organization as well as the size, color, and attractiveness of the organisms concerned, and without sound habitat management for plants the suitability of an area for birds would certainly deteriorate.

The Practice of Park Design

In the United States the national park system encompasses acreage acknowledged to be outstanding examples of the national terrain—Yellowstone, Glacier, and Rocky Mountain national parks in the Rockies, for example, Shenandoah and the Great Smoky Mountains in Appalachia, Grand Canyon, Bryce Canyon, Zion, and Canyonlands in the arid Southwest, Yosemite, King's Canyon-Sequoia, Lasser, Crater Lake, and Mount Rainier in the far western ranges, and many more (Fig. 21.3). Strong management is needed in these areas; hunting is not allowed, and government ownership is essential. The popularity of these parks is such that visitor numbers rose by about 10 percent each year during the 1960's.

The tradition of federal park lands in the United States began in 1872, when Congress created a park in the Yellowstone region of Wyoming. The seeds of an idea for national parks had perhaps been planted eight years previously, when the Yosemite Act had donated 44 square miles (113 km^2) of U.S. public land to the state of California to preserve for the public. This gift was largely precipitated by the fact that the lands were for all commercial intents and purposes worthless, with no mineral or usable water resources. The first preserve for animals per se came in 1903 with the establishment of a refuge for birds on Pelican Island off the Flor-

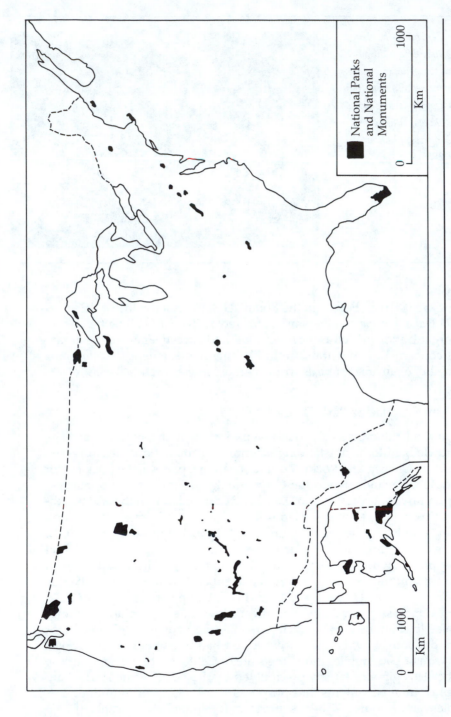

Figure 21.3 *National parks and larger national monuments (such as Death Valley) in the United States, 1990.* Note the concentration of protected areas in the West and the very small proportion of reserves in the nation overall.

ida coast. This park was established at a time when the nation was witnessing, and of course participating in, the wholesale slaughter and extinction of the passenger pigeon.

The preservation of public land in the United States is controlled by many separate bodies, including the U.S. Forest Service, the National Park Service, and the U.S. Fish and Wildlife Service, each of which was originally developed to preserve land for a different purpose (Table 21.1). Clearly, the commonly held belief that national forest and other agencies are in the business of conservation per se is a myth. They are actually in the timber business or some other economically related activity. The largest fraction of public land, held by the Bureau of Land Management, is vast trackless areas never sold privately or donated for preserves, much of it timberland (Fig. 21.4). About 42 percent of U.S. land consists of public lands owned jointly by citizens and managed for them by federal, state, and local government; 35 percent is administered by federal agencies. About 73 percent of that is in Alaska and 22 percent in the West. Most of this land is held under the "multiple use" concept and, under the Multiple Use Sustained Yield Act of 1960, can be managed for a sustained yield of forest products. The Reagan administration, particularly under Secretary of the Interior James Watt in the early 1980's, was itself very eager to explore the possibility of the "development" of such land and even national park lands for mineral resources. Mr. Watt, a born-again Christian, based his policies on his belief that "we must occupy (use) the land until Jesus returns." Opponents, unable to combat his arguments with logic, had themselves to resort to Biblical quotation: "Beware of false prophets . . . in sheep's clothing" (Reed 1981).

Inherent in the concept of a national park is use by the public, yet too many tourists can profoundly alter an area. The boot-clad feet of visitors to Rocky Mountain National Park can wear away in a few hours tundra that takes up to five years to grow a few centimeters. John and Ann Edington (Edington and Edington 1986) document many more cases. For

Table 21.1 *Land areas administered by state and federal agencies. (From Revelle and Revelle 1984.)*

Agency/Office	Name of Land Holding	Millions of Acres		
		"Lower 48"	Alaska	Total
Bureau of Land Management	National resource lands	174	70	244
U.S. Forest Service	National forests	168	23	191
National Park Service	National park system	27	52	79
U.S. Fish and Wildlife Service	National wildlife refuges	13	76	89
State and local parks and forests	———	about 25		

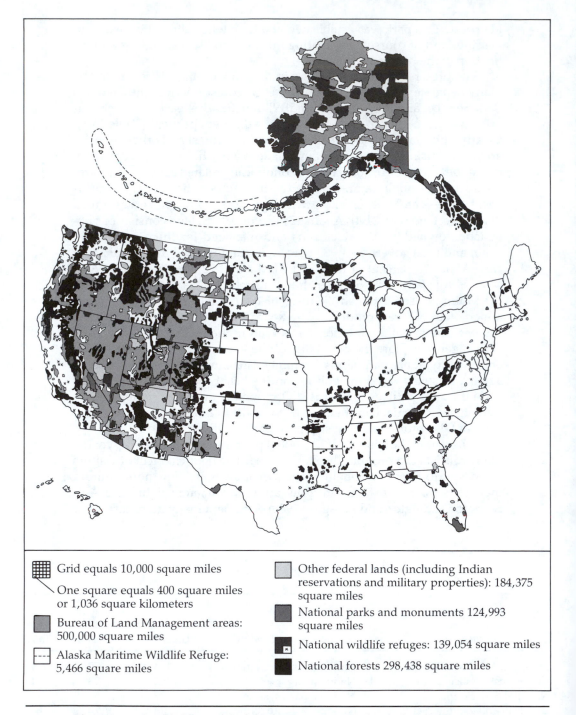

Grid equals 10,000 square miles

One square equals 400 square miles
or 1,036 square kilometers

Bureau of Land Management areas:
500,000 square miles

Alaska Maritime Wildlife Refuge:
5,466 square miles

Other federal lands (including Indian
reservations and military properties): 184,375
square miles

National parks and monuments 124,993
square miles

National wildlife refuges: 139,054 square miles

National forests 298,438 square miles

Figure 21.4   *Land holdings of the federal government.*   (Redrawn from Miller 1988.)

example, cliff-nesting birds can be disturbed by climbers, and water birds on lakes are easily threatened by sailboats. Even the seemingly harmless act of offering tidbits to animals can cause problems of habituation to human foodstuffs. In a game reserve in Zimbabwe, elephants fed oranges by tourists have adopted the interesting habit of uprooting tents in search of these and other morsels. Some grizzly bears (*Ursus arctos horribilis*) have occasionally preyed on the visitors themselves! These and other effects of users on a park system are outlined in Fig. 21.5. Questions regarding public use need to be addressed quickly because leisure time is on the increase. In 1900, 26.5 percent of the time available to U.S. citizens could be classed as leisure; by 1950, leisure was 34 percent, and by the year 2000 it is predicted to reach 40 percent. Expenditure in western nations on tourism runs at 4–7 percent across most income levels. With such time and expenditure available to the tourist, it is evident that tourism itself needs much management. Six million visitors per year over the 272 km² New Forest in southern England seems like saturation level, though many visitors stay by the road and the area, for the moment, remains ecologically rich. More park land is definitely needed. The process of acquiring land for recreational purposes is, however, very expensive. As the United States gave the world the national park idea, it seems only fitting that this country should remain in the forefront of conservation science and solve these types of problems. Newmark (1987) has already proposed that extinctions have occurred in many U.S. parks simply because they are too small. Only the largest western North American parks (such as Kootenay-Banff-Jasper-Yoho, at over 20,000 km²) are large enough to retain an intact mammalian fauna. In some cases, specific reasons can be found. The population of grizzly bears in Yellowstone National Park is declining rapidly as a result of the closure of garbage dumps (Knight and Eberhardt 1985). However, the actual details and general conclusions of Newmark's paper are hotly contested. Following a similar claim that East African parks were rapidly losing large mammals, Boecklen and Gotelli (1984) were able to demonstrate, statistically, that the projections of loss of species were completely incorrect. Leader-Williams and Albon (1988) have presented the interesting alternative that in Africa smaller parks are better than larger ones, because smaller parks can be protected more effectively against poaching, the main threat to the species within. Using data on rhinos and elephants in Zambia, Leader-Williams and Albon were able to show a direct relationship between the rate of decrease of rhino and elephant numbers and the patrol effort. They went on to show that rhino numbers increased most, between 1980 and 1984, in countries that spent the most money protecting them. Clearly, preservation of large areas alone is not a sufficient safeguard against extinction.

Worldwide, some 3,500 protected areas, covering some 4.25 million km² (1050 million acres) have been set aside for conservation (Fig. 21.6,

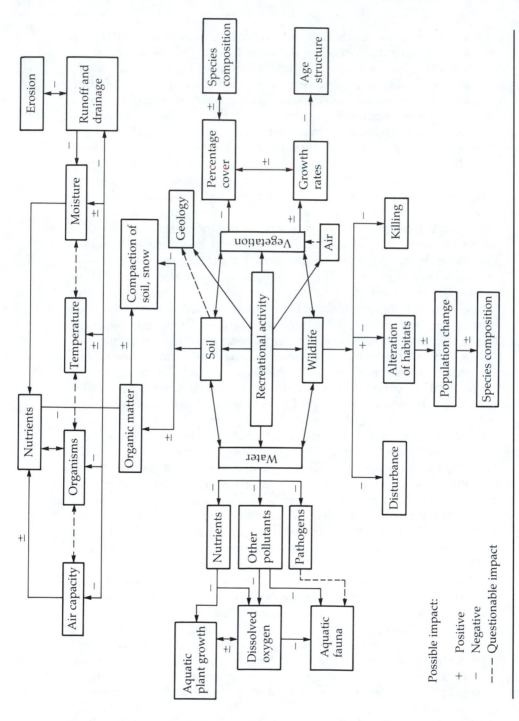

Figure 21.5 A schematic diagram of the interrelationships between recreation and its ecological impact. (Redrawn from Wall and Wright 1977.)

Possible impact:

+   Positive

−   Negative

– – – Questionable impact

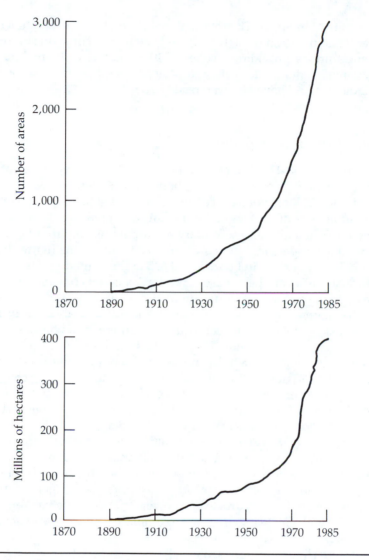

Figure 21.6 *Growth of the global network of protected areas, 1890–1985.* (Redrawn from International Union for Conservation of Nature and Natural Resources 1985.)

p. 435). Before 1970 most protected areas were located in industrial countries, but for the past 15 years LDCs have led in both numbers added and rates of establishment. Of course, the degree of protection against illegal or unmanaged hunting and fishing, farming, and logging varies considerably. Each country allows these activities to a greater or lesser extent. For example, "national park," which already means different things in the United States and in the United Kingdom, is equivalent to "national hunting area'' in Spain. The International Union for Conservation of Nature and Natural Resources (IUCN) has tried to overcome this problem of nomenclature by providing a series of 10 distinct and defined management categories (see McNeely and Miller 1984), but these seem vague and indistinct and have not been readily adopted.

21.3 *Water Catchment Areas*

An organism's most basic metabolic need is for water; people die of thirst long before they die of hunger. Our bodies are 60 percent water, and we need about 2.25 liters per day. To this amount must be added the residential requirements of cooking, washing, and waste disposal. In LDCs, for example in Karachi, these secondary requirements amount to only 90 liters per day per person, but in industrialized nations the figure skyrockets to 263 liters per day in London and 635 in the United States.

Industry is a very heavy user of water, as many consumer goods require up to 20,000 liters per ton of product. For oil, iron, and steel production, the figure may be even higher. Agriculture is also a very heavy user—0.45 kg (1 lb) of dry wheat requires 227 liters for its production; the same amount of rice requires 757–946 liters; and a liter of milk requires 4,000 liters (Simmons 1981). In animal production, water is also important; 1 pig (50 kg) needs 20 kg of water per day, and a lactating cow needs 45 kg of water per day to produce 12 kg of milk. (In contrast, sheep are abstemious, needing perhaps only 5 kg per day on dry range and virtually none on good pasture.) Water is also used as a flotation device. The St. Lawrence seaway carried over 60 million metric tons of cargo in 1966, and the Rhine in Europe carried 230.6 million tons in 1965. Such use is theoretically nonconsumptive, but in reality contamination by shipping generally renders water unusable for other purposes. A genuinely nonconsumptive use is for hydroelectric power. In Peru 68 percent of all electricity comes from this source, in Colombia 63 percent, in Canada 81 percent, and in Norway 99.8 percent!

Of the various conditions in which free water exists, salt water in the oceans claims 97 percent = 1.357×10^6 km^3. Of the remaining 3 percent that is fresh water, 75 percent is immobilized as glaciers and ice caps. Of the last quarter, most is groundwater, so surface fresh water in lakes, streams, and rivers accounts for only 1.5 percent of fresh water and the atmosphere 0.16 percent (van der Leeden 1975) (Fig. 21.7). The average water content of the atmosphere, if rained out all at once, would provide a global fall of 2.5 cm and about 10 days' supply of rainfall (Barry 1969). Ten days is also the average residence time of a water molecule in the atmosphere. This figure points to a rapid turnover in the cycle of evaporation, runoff, and precipitation. The movement of water through the atmosphere, biosphere, and lithosphere (soil) is called the *hydrological cycle*

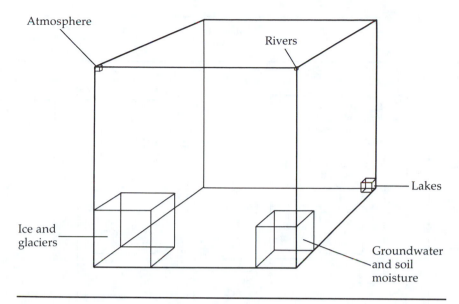

Figure 21.7 Waters of the world. The large cube represents all the waters of the world. The smaller ones represent the quantities occupied by different kinds of water. The rest of the large cube represents ocean water and saline lakes (Redrawn from van Hylckama 1975.)

and is illustrated in Fig. 21.8. Water normally spends 10–100 days on land, unless it enters the groundwater circulation, in which case it stays much longer. In the great Artesian basin of Australia, some water in the aquifers has been there for 20,000 years (Barry 1969).

Tapping into the Water Supply

In order to divert water from the natural hydrological cycle, people must intervene at those places where technology makes it most feasible and where the ratio of benefits to costs is deemed to be favorable. The most usual phases of intervention are the runoff and storage phases. Seeding clouds to produce rain has not yet proven to be statistically effective.

Runoff characteristics of catchment areas on land are affected by many variables such as slope, soil type and depth, and vegetation cover, of which the last is the easiest to alter. In general, the higher the vegetative biomass, the higher the transpiration and hence the greater the water ''loss.'' **Deforesting** a **watershed** would increase runoff. Of course a deforested watershed may yield greater quantities of unwanted silt, polluting the water, so an optimal balance between water yield and quality

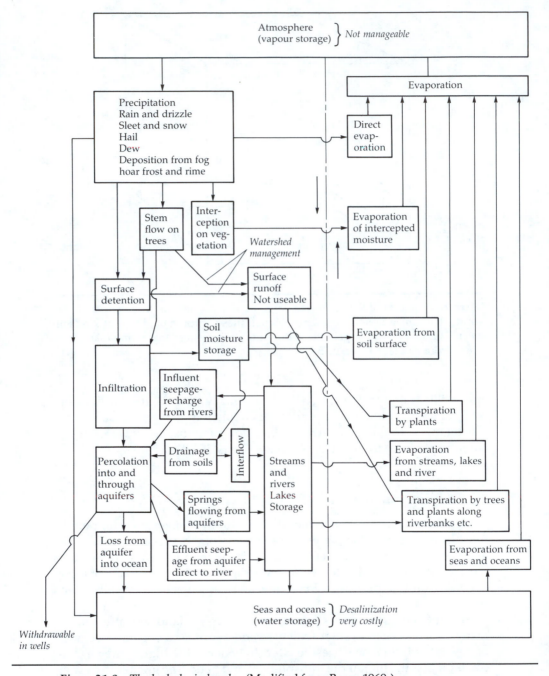

Figure 21.8 *The hydrological cycle.* (Modified from Barry 1969.)

must be sought, and a belt of trees may be left standing in order to trap silt from the watershed (Karr and Schlosser 1978). Another major economic difficulty in deforestation techniques intended to increase supply is that benefits often accrue not to the landowner but to the downstream user; thus the technique is more suitable for land in public ownership. Nevertheless, the technique does work; in southern California a flow of $86.3 \times 10^3 \text{ m}^3$ per day was produced during a wet season instead of $49.3 \times 10^3 \text{ m}^3$ per day before the removal of **riparian** vegetation (Hopkins and Sinclair in Simmons 1981).

More feasible for management are the surface storage phases of fresh water in lakes and rivers. The lakes of Africa store 30 percent of the world's liquid fresh surface water, the lakes of North America store 25 percent, Lake Baikal (U.S.S.R.) stores 18 percent, and the world's smaller lakes and rivers the remaining 27 percent. Rivers contain only 1,200 km³ of the world's fresh surface water, compared with 125,000 km³ in lakes. As a result, manipulation of river systems by deepening of channels increases storage capacity only marginally, although at local scales, such amounts may be very important. Rivers are often the immediate source of water for agricultural irrigation. Manipulations of rivers for irrigation are dominated by China, followed by India, the United States, Pakistan, and Asia.

Dams: Their Advantages and Disadvantages

The most popular device for controlling water supplies on a large scale is the dam. Its advantages include

- □ *Controlled discharge to produce year-round water:* The Shasta Dam on the Sacramento River in California's central valley has facilitated treble and even quadruple cropping in the southern parts of the valley.
- □ *Fisheries or recreation benefits to offset land cost.*
- □ *Control of floods.*
- □ *Hydroelectric power:* This resource can be sold to industry to pay for the dam but is often underexploited in LDCs because there is no market for electrical power. Roughly 40 percent of Brazil's total energy needs are provided by hydroelectricity (Barrow 1988).

Dams also have serious drawbacks, which include

- □ *Explosive growth of producer organisms:* These can transpire water away, as did, for example, the water hyacinth on the White Nile (Holm, Weldon, and Blackburn 1969).
- □ *Disturbance during the construction phase that affects wildlife:* Usually, however, species reinvade the area within a few years of the subsequent decrease in human activity. Such was the case in Newfound-

land, where construction of a hydroelectric dam lessened movement by caribou through the area until after the dam was completed (Kiell, Hill, and Mahoney 1986).

☐ *Silting:* Accumulation of silt, an extreme case of which is illustrated in Photo 21.5, may reduce the life span of an impoundment and represents a loss of nutrients downstream, which must be replaced by fertilizers (see discussion of the Aswan Dam, Chapter 1).

☐ *Interference with fish migration, unless fish passes are built:* In tropical areas, the movements of aquatic mammals, turtles, crocodiles, and alligators may also be curtailed (Barrow 1988).

☐ *Triggering of seismic movements in highly faulted regions:* An earthquake near the Konya dam in India cost 200 lives in 1967 (Rothé 1968).

☐ *Drowning of agricultural settlements and scenic or naturally beautiful areas:* Egypt's Abu Simbel temples would have been drowned under Lake Nasser if they had not been removed by international action.

☐ *Unpredictable effects:* At Southern Indian Lake in northern Manitoba, an impoundment raised water levels a mere 3 m. The commercial fishery based on whitefish, *Coregonus elupaeformis*, was reduced by two-thirds as the fish migrated away from traditional fishing grounds in response to different flow patterns (Lehman 1986*b*). In 1982, Manitoba Hydro provided a cash settlement of $2.5 million Canadian to commercial fishermen for all future losses. Furthermore, the increased lake levels quickly cut into the new permafrost banks, whereas the old lake had been bedrock controlled. Erosion was rampant, and the flooded soils released so much mercury (a common occurrence in reservoir creation [Cox et al. 1979]) that mercury concentrations in fish exceeded Canadian marketing standards.

Photo 21.5 Silting, a process that destroys the usefulness of many dams. (Photo from U.S. Department of Agriculture.)

Alternative Supplies

As a result of the drawbacks, interest has focused on alternative water supplies. Groundwater is often pumped out for irrigation or for urban development. Recovery, however, is often uneven, unpredictable, and of varying quality. The desalinization of seawater, most commonly used in the Middle East, is usually prohibitively expensive. Not only is the chemical treatment not cheap, but the energy requirements to "raise" water from sea level to required areas are astronomical. Icebergs, of course, hold a great volume of fresh water. The flat-topped Antarctic bergs prove best for towing, and a berg of dimensions 3,000 × 3,000 × 250 m could be towed to Australia in one month. Even if half of it melted in transit, there would still be 1×10^9 m^3 of water, enough to supply 4 million people for one year at an average cost of $1.50 per annum (Simmons 1981). Even better figures might be obtained for the dry west coast of South America, where ocean currents would aid in transportation. Finally, transfer of water from one river basin to another has been proposed, usually involving a tunnel or ascent stage necessitating pumping. Some of the proposed schemes are conceptually massive. The North American Water and Power Alliance (NAWAPA) has outlined a scheme that would divert water from northern Canada and Alaska through the Rocky Mountain trench to supply southern California and northern Mexico (Province of Alberta 1968). In Russia, several schemes have been put forward to divert normally northward-flowing rivers into the more arid south. The ecological soundness of both these schemes is of course minimal with regard to effects on wildlife and even the climate itself; as usual, the needs of the urban masses are deemed of paramount importance.

21.4 Forestry

About one-third of the world's land surface is covered with forests, whose continental distribution is outlined in Table 21.2 and Fig. 21.9 (p. 443). Ecologically, the leaves of trees are the fundamental element in the system, as they are the site of **photosynthesis** and are the main food source for animal life. Wood contains few nutrients and is physically intractable to many animals as food. Defoliating insects are thus highly important in primary production. Spruce trees (*Picea abies*) in Switzerland need 2,300 kg (fresh weight) of leaves to produce 1 m^3 of wood. In a similar setting, deciduous trees appear to be more efficient, requiring only 880 kg to produce 1 m^3 (Ovington 1962). Nevertheless, hardwood forests will continue to be replaced by conifers because of better economic returns (National Research Council 1982). Wildlife is certain to suffer. Harris, Hirth, and Marion (1979) estimate that hardwood forests support double the density of game that coniferous forests do. Lowland equato-

Table 21.2 *Forest lands of the world in 1985. (From Food and Agriculture Organization 1986.)*

Location	Thousands of Hectares	World Total in Percent
Europe	155,237	3.8
Soviet Union	935,000	22.9
North and Central America	659,517	16.1 (Canada, 8; United States, 6)
South America	916,889	22.4 (Brazil, 13)
Asia	562,531	13.7 (Indonesia, 3; China 3)
Africa	697,575	17.1 (Zaire, 4)
Oceania	159,887	3.9
World	4,086,636	

rial forests are the most productive forests; their annual growth is double that of any other category of tropical or temperate forests (Table 21.3, p. 444). However, the number of species per unit area is so high that selective harvesting is costly. In contrast, the immense stands of one or two species of pine and spruce in Eurasia and North America provide a uniform product, much in demand for paper and paper products.

Though leaves are important ecologically in forests, in economic terms wood production is of paramount value. Of the annual growth potential of 4,500 million tons, 1,626 million are estimated to be harvested. Among the high number of direct uses, use as fuel has historically been by far the greatest and even today accounts for about half the lumber cut; virtually any species can be utilized. The problem of maintaining a sufficient supply of wood for fuel has become so serious in some developing nations that wood-fuel development programs have been set up (Bradley 1988; Ngugi 1988). The greatest destruction occurs, however, when forested lands are cut for agricultural use. Most of the farmland in any nation today was once at least moderately tree covered, except for the relatively small areas of natural prairies that existed. In industry, lumber is still very important in the construction of houses and furniture, and converted wood (as pulp) is used at over 1 million tons per year for paper. The consequences of removing the forests, either by selective logging (elimination of designated trees) or by **clearcutting** (removal of whole stands), are many and varied (Farnsworth et al. 1981) but clearcutting causes much more severe consequences (see Photo 21.6, p. 444). The following is a partial list of consequences.

- □ *Soil erosion:* Photo 21.7 (p. 445) shows the consequences of uncontrolled logging in one location.
- □ *Acceleration of salt and nutrient loss to streams, often bringing about **eutro-***

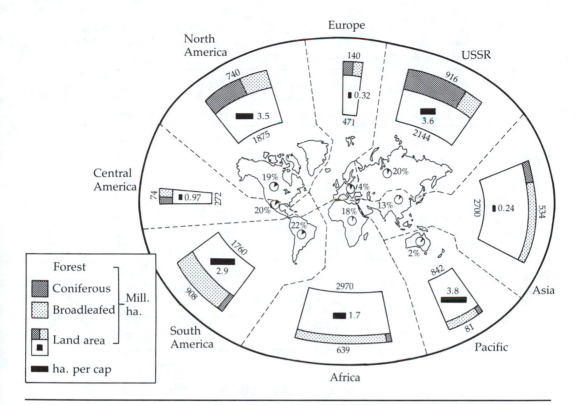

Figure 21.9 *A diagram of forest area and distribution in the world.* The divided circle located in each major region indicates the proportion of the land area devoted to forest. The outer segments show (reading outwards) the total land area (million ha), the number of hectares of forest per person, the division of forest between conifers and broad-leaved types, and the total forest area. The area of the segment itself is proportional to total land area, subdivided into forest and nonforest sectors. The predominance of the USSR and North America in terms of the coniferous trees so much in demand for paper and paper products is evident. The dependence of Europe on outside resources in this field (as in so many others) can easily be inferred (Redrawn from Simmons 1981.)

phication and algal blooms: High floods on the Indus Plains of Pakistan occurred more frequently in the 25 years between 1950 and 1975 than over the previous 65 years, as a result of deforestation (Eckholm 1976).

◻ *Removal of wildlife habitats:* The old-growth forests of the northwestern United States in Washington, Oregon, and northern California provide the only home of species such as red tree voles, western salamanders, flying squirrels, Vaux's swift, and perhaps most importantly, spotted owls. Arguments between lumberjacks and environmentalists rage about how much forest is needed to preserve

Table 21.3   *Estimated forest area and estimated growth potential (in tons of dry matter production per hectare and per year) for different formation classes of the forests of the world. (Based on data from Weck and Wiebecke 1961.)*

Formation Class	Estimated Area			Estimated Growth	
	Million ha	Percent	t/ha/hr	Total Million t/yr	Percent
Equatorial rain forest, lower range	440	18	3.5	1,540	35
Equatorial rain forest, mountain range	40	2	3.0	144	3
Monsoon forests and humid savannah	263	11	1.8	474	11
Dry savannah and dry mountain forests in tropics	530	21	1.0	530	12
Temperate rain forests and laurel; precipitation below 1,000 mm	20	1	7.2	143	3
Sclerophyllous forests	177.5	7	1.0	178	4
Summergreen forests and mountain conifers	393	16	2.2	865	19.5
Boreal conifers	605.5	24	0.9	556	12.5
Total	2,469			4,430	

populations of such species. Ultimately, logging has to stop. The question is whether it will stop now, while some old growth and spotted owls are left, or later, when all the forest and its inhabitants have disappeared. And if *our* forests are going fast, how can we tell other nations, like Brazil, to stop cutting theirs?

Photo 21.6 Clearcutting, the wholesale removal of every living tree in an area. (Photo from U.S. Department of Agriculture.)

Photo 21.7 Severe soil erosion, one of the side effects of uncontrolled logging. (Photo from U.S. Department of Agriculture.)

□ *Removal of recreation areas, of great aesthetic value in themselves.*

□ *Enhancement of desert encroachment in arid areas, for example in the Sahara area and Sudan:* On a global scale, deserts are often argued to be expanding. (Allen [1980] gives a rate of almost 60,000 km^2 a year, an area twice the size of Belgium.) Much of the bad news, however, is based on very scanty data. In fact, it is more likely that deserts merely expand in years of low rainfall; estimates of the spread of deserts taken at such times, and there have been many, are likely to reflect this trend. Further evidence suggests a reversal of desertification in years of good rainfall (Forse 1989). Planting trees in deserts is unlikely to halt desertification and is likely to have low cost-effectiveness; as seems apparent from Photo 21.8.

□ *Reduced oxygen production and carbon dioxide absorption for the planet:* In 1988 two congressional committees held hearings on a proposal that areas the size of Zaire or Australia need to be reforested to prevent excessive CO_2 buildup (see Chapter 22).

Although a major goal of forestry in the past has been the successful production of wood by prevention of insect outbreaks and forest fires, the problems outlined above have become of paramount importance in foresters' minds. At present rates of clearance, the remaining area of unlogged productive forests will be halved by the end of this century. An area the size of Portugal is cleared each year in Asia alone.

One may take the view that ''common sense'' will surely prevail in the end and that ''they'' will surely stop short of the ultimate destruction of forests—but will they? Mount Lebanon, the backbone of the country of the same name, was once richly carpeted in stately cedar trees, much

Photo 21.8 Dune fixation and stabilization, Mauritania, 1984. (Photo by T. Fenyes, Food and Agriculture Organization.)

prized for their strength and utility in a largely scrub zone. From cultures as early as the Phoenicians', 3000 B.C., through Solomon's time and the empire of Rome, right up to World War II and the British, the cedars have been cut. All that remain today are 12 pockets of cedars, protected over the millennia usually by their inaccessibility or by their proximity to religious sites. The rest of the area is a scrubby wasteland. China, for long the world's most populous country, has undergone forest destruction for millennia, and perhaps the most serious deforestation in the world is in that country (Richardson 1966). Many of the rocky, infertile vistas of Greece and other Mediterranean areas are due to the early Greek and Roman civilizations, and even Greek philosophers bemoaned the loss of natural forests. Perhaps we should be grateful, however, that the Industrial Revolution and the adoption of coal as a fuel saved what does remain of Europe's once-rich forests. Reforestation programs in Europe in the 20th century have pushed the area of forested land back up by more than 25 percent.

Yet in the LDCs there seems to be hope. In some countries, patterns of land use will lead to continued and spectacular reductions in forest cover. In others, it now seems unlikely that the needs of the people will require the conversion of all forest over the whole of their territory, given some modest improvement in agricultural techniques (Bunting 1988). In Africa, this seems to be the case in at least 17 of the nations that include humid tropical areas. The corresponding numbers for Central America, South America, and Asia are 5, 10, and 5, the problems in Asia being a bit more severe because of the denser populations. Thus, 37 of the 61 nations that have tropical forest are unlikely to use it all. In Brazil, about which there has been so much concern, needs are more likely to be met more cheaply from the savannahs now that soil acidity and phosphate problems have been more fully investigated (Goedert 1983). Pimentel et al. (1986) suggest that most deforestation could be prevented by doubling of agricultural yields on existing productive land. They point out that agricultural expansion into about 10 million ha of new land per year accounts for most of the 11.6 million ha of forest land being lost annually.

One must realize, however, that U.S. national forests are actually managed as tree farms, not as nature reserves. Yet, stunningly, two-thirds of the national forests lost money in 1987 and 1988 (Rice 1989). The loss is due mainly to below-cost timber sales, from which the government does not even recover the costs of preparing timber for sale. The Forest Service claims that logging provides additional benefits like access for recreation, increased water flows, wildlife diversity, and protection of forests from insects, fire, and disease. However, each reason is on very shaky ground. There exist 356,000 miles of road in national forests, eight times as many as in the interstate highway system and far more than needed for recreation. Any increased flow of water generally occurs in the spring, when it is not needed. Although increased "edges" created by roads may increase diversity, the new species they encourage are often undesirable (*r*-selected species of weeds and pest insects rather than large endangered vertebrates). Such species may indeed be more threatened by habitat fragmentation and increased poaching where roads provide easy access. The fire-break effects of roads may also be disadvantageous; fire has now been recognized as a necessary and integral part of forest history. Virtually all forests in North America have been burned repeatedly for thousands of years. Similarly, insect outbreaks are normally common, creating a patchwork of natural forest openings and increasing the growth and vigor of remaining trees (Rice 1989). Unfortunately, management plans call for increased harvest in most national forests for the years between 1990 and 2040.

21.5 Agricultural Land

The source of much of the food consumed by people is terrestrial agriculture, and agricultural land represents the most manipulated of all the nonurban ecosystems. The scouring of the land to plant agricultural crops creates soil erosion, increased flooding, declining soil fertility, silting of the rivers, deforestation, and desertification. Even though acid rain, pollution, nuclear waste, eutrophication, and the preservation of land for parks and reserves are pressing issues, they seem to pale in comparison with the prospect of potential ruination of the land on which the fabric of our society is constructed. Not only is the agricultural system

human-made, but the plant and animal components have also been genetically engineered. Over half of human energy comes from food grains;
herbivory on other food sources accounts for about another 30 percent;
and carnivory is a distant third (Fig. 21.10). The area of land under cultivation in different parts of the globe is illustrated in Table 21.4. Europe
gives over by far the greatest percentage of its land to crops and pasture
lands, and Africa the least, which helps to explain why Europe often has
food surpluses and Africa often suffers famine. Energetically, the most
efficient agricultural systems include wet rice culture and shifting agriculture (Fig. 21.11, p. 450), hence their popular and, in the case of shifting
agriculture, unfortunate appeal.

The two basic agricultural systems today are *shifting agriculture*—the
total manipulation of natural systems, but only for one to five years—and
sedentary agriculture—permanent replacement of natural systems.

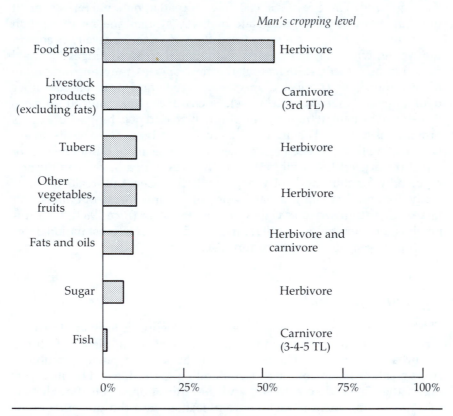

Figure 21.10 Human cropping level for food. Harvest is dominated by grains (TL
= trophic level). (Redrawn from Brown 1970.)

Table 21.4 *Agricultural and grazing lands, 1985, in thousands of hectares. (From Food and Agriculture Organization 1986.)*

Location	Continental Area Excluding Inland Water	Arable Land and Permanent Crops*	Percent	Permanent Pastures and Meadows	Percent
World	13,324,027	1,476,483	11.1	3,170,822	23.8
Africa	2,964,595	184,869	6.2	788,841	26.6
North and Central America	2,242,075	274,626	12.2	367,062	16.4
South America	1,781,851	140,638	7.9	458,364	25.7
Asia	2,757,252	454,253	16.5	644,669	23.4
Europe	487,067	139,625	28.7	84,260	17.3
Oceania	850,967	50,285	5.9	453,026	53.2
Soviet Union	2,240,220	232,187	10.4	374,600	16.7

*Permanent crops includes tree crops such as rubber, citrus, and cocoa.

Shifting Cultivation

Shifting cultivation—often called **slash-and-burn cultivation** is today largely confined to tropical forests and savannahs, where the burning of natural vegetation provides minerals for crop uptake. Crops are planted so as to provide as complete cover as possible, reducing the **leaching** of the soil. Plots are normally abandoned when nutrient drain reduces fertility or when weeds reach an unacceptable level. Breakdown of this system occurs when plots are recultivated too soon, because mineral nutrient cycles never build up again, and fertility is not renewed. Photo 21.9 (p. 451) shows the deforestation that can result when populations that use these techniques grow too dense. Thus, the system cannot cope with intensive crop production in populated areas but suffices in forested land where regeneration eventually occurs.

Sedentary Cultivation

Sedentary cultivation represents the permanent manipulation of an ecosystem; the natural biota is removed and replaced with domesticated plants and animals. The soil is extremely important in this case, for it is the long-term reservoir of all nutrients, although it can be supplemented by chemical fertilizers or organic excreta (the latter is preferable, because it helps maintain the crumb structure of the soil). The paddy culture of rice is an exception—it is essentially an aquariam system with earth boundaries instead of glass. The soil is largely a rooting medium, and water supplies the essential nutrients. Thus, paddy rice can be grown on

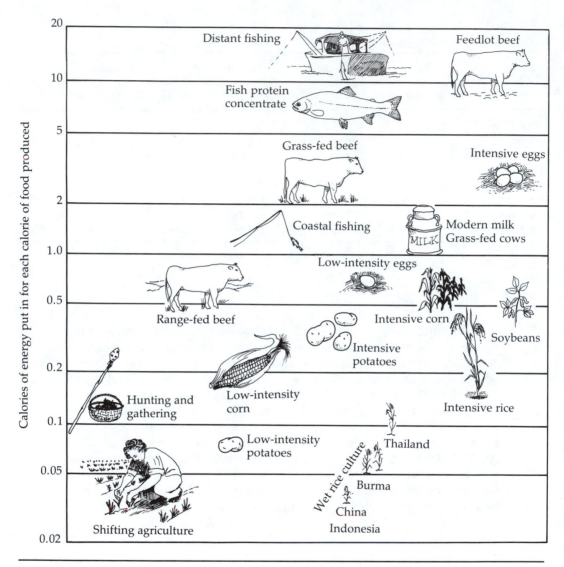

Figure 21.11 *Comparative energy intensiveness of food crops in the United States.*

poor soils. Paramount among the world's crops are rice, wheat, and maize, which comprise 7/9 of the world grain crop. Meat is a much scarcer product, being inefficient to produce. It has the advantage, however, of being nutritionally superior to many plants. Furthermore, animals are often adaptable to climate, and the proteins are not locked away behind a cell wall of cellulose not easily broken down by human stomachs.

Photo 21.9 Shifting agriculture, Sumatra. High population growth and a reluctance to use modern farming techniques has resulted in deforestation. (Photo by H. Hull, Food and Agriculture Organization.)

Improving Crop Yields

The performance of the world's agroecosystems appears so far to have been good. It has permitted not just the survival of the human population but its growth at overall rates of up to 2 percent per annum. Unfortunately, human ingenuity and the seemingly permanent availability of quick-fix solutions to hunger problems has led many economists to think along similar lines for all ecological situations—if there's a problem, it can be fixed. However, the absolute number of undernourished people increases from year to year, and to reduce the number is unlikely to be easy, rapid, or cheap. There is a complete lack of unanimity in future forecasts, which range from famine in a few years to surpluses. The surpluses of Europe and North America might now amount to 10 percent of the world food production and could be used to make good some of the deficiencies in Asia, but given the rise in populations of 2 percent a year, such a contribution can only make an impact for a limited time. Despite the horrors of recent famines, especially in Ethiopia during 1984–1985, aid is not the solution; the answer must come from within. Free food undercuts the local market. In effect, local farmers must compete economically with very low-cost or free imported food. Obviously, they cannot turn a profit themselves, so they stop producing and join the ranks of the poor. This continuing cycle has sometimes been referred to as the *utterly dismal theorem.*

All the world has adequate sunlight and CO_2 for some form of photosynthesis, but when other necessities like temperature, rainfall, topography, and soil are considered, the proportion of the terrestrial surface suited to agriculture falls to magnitudes not far beyond the present total of 12.5 percent. Most authorities agree that intensification, getting more out of the same land area, holds most promise of improving food yields. Future developments include better irrigation, flood control, drainage, erosion control, mechanization, fertilizer and biocide use, and the raising of improved varieties of plants and animals. Unfortunately, a doubling of agricultural output per unit area requires a tenfold increase in the inputs of fertilizers and pesticides (Fig. 21.12). Many have argued for improved strains of wheat and other crops. The so-called *green revolution* has dramatically improved wheat varieties (at the International Corn and Wheat Improvement Center in Mexico) and rice strains (at the Institute of Rice Research in the Philippines) (Harrar and Wortman 1969). Of these, the development in 1966 of IR-8, "miracle rice," was the most famous. Its top yield averages 1,067 kg/ha, compared with around 300 kg/ha for most local varieties in the Philippines. Of course, in any breeding program, care must be taken not to lose the genetic diversity present in old, wild stocks. The advent of a new strain of pathogen or biotype of insect pest that could successfully attack the miracle strains could mean crop failure over huge areas and even entire nations, and the older strains often prove more resistant. With the advent of genetic engineering, the promise of better varieties of crops is an attractive prospect (Chapter 4), but it is worth remembering that, although biotechnology can produce a hog with rapid weight gains and leaner meat (by means of a spliced-in gene for human growth hormone), the animal is also cross-eyed and arthritic, an object of pity unless one views living creatures as nothing but potential food. Forcing livestock into pain and misery should not be considered a solution to the problem of feeding the world (Jackson and Piper 1989).

Future Foods

Other possibilities for increasing yields include developing new "wild" sources of food such as fungi, algae, and wild game. Such possibilities, however, may well remain theoretical ideas rather than practical solutions. Algal culture requires a good deal of technical know-how, sophisticated machinery, and energy input (Pirie 1969), even though 5 m^2 devoted to algae production could feed one person 10^6 kcal yr^{-1}, whereas it would take 1,200 m^2 of grain cultivation and 4,000 m^2 of pork-raising area to do the same thing. Fungal growth on waste straw can also be highly efficient. The common cultivated mushroom, *Agaricus bisporus*, can convert 100 kg of fresh wheat straw into 50 kg of fruiting bodies, though up to 95 percent of the fresh weight is water. Mushrooms, however, are sometimes thought of as luxury food; when the economy drops, so does mushroom production (as the author can testify, having worked on a mushroom farm for many years).

What about the prospect of using more wild game as food? Wild game is often more catholic in its use of forage than domestic varieties

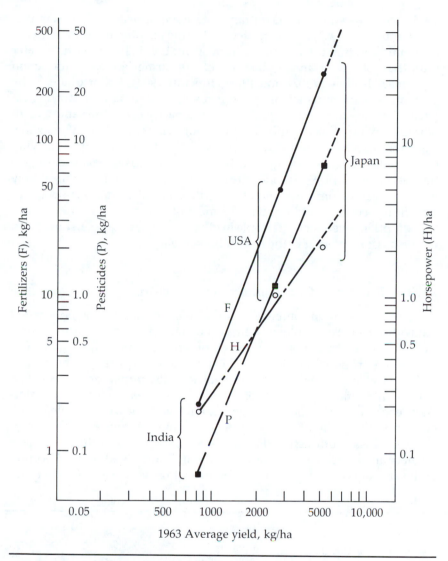

Figure 21.12 *Relationships between yield of food crops and requirements for fertilizer (F), pesticides (P), and horsepower (H) used in cultivation and harvest of crops. Doubling the yield of food requires a tenfold increase in the use of fertilizers, pesticides, and animal or machine power.* (Redrawn from Odum 1971.)

and may eat plants that currently go unharvested. A case in point is a large South American rodent, the capybara, which feeds on aquatic weeds. Attempts to validate the idea that game ranching should be more economically sound than cattle raising have been made since the 1960's, especially in Africa, but have usually proved unsuccessful (Caughley

1976). More recently some contrary data have appeared. For example, Freudenberger (1982) has suggested that game farming in South Africa is economic and that the net return from springbok is $0.25 per ha greater than that for sheep, mainly because of the farmer's lower production costs (see also Lewis 1977 and 1978; Hopcraft 1982). Unfortunately, the economics of the situation are not directly related to natural **carrying capacity.** For one thing, ecological carrying capacity is not the same as economic carrying capacity, the density of animals that will allow maximal sustained harvesting, and later reports showed that economic carrying capacity does not necessarily differ between wild ecosystems and domestic ones (Walker 1976). Second, the quality and desirability of meat may well be greater from cattle than from wild game. By analogy, British or American people would not buy the meat of horses or dogs for human consumption, even though people in other countries may consume them with a passion, at least in the case of horsemeat.

The Ecological Consequences of Agriculture

The extension and intensification of food production is not carried out without environmental side effects and large volumes of farm wastes. Runoff water often contains high levels of nitrogen, phosphorus from fertilizers, and residual pesticides, which contaminate streams. In the last 200 years, at least one-third of the **topsoil** on U.S. croplands has been lost to erosion. Some 4 billion tons of sediments are washed into U.S. waterways each year. The most spectacular and well-known example of soil erosion in the United States was the creation of a huge dustbowl from Kansas across southwest Oklahoma to western Texas in the 1930's as a result of an extensive monoculture of wheat. Photo 21.10 shows a portion of the aftermath. An expensive lesson in ecology was learned on May 12,

Photo 21.10 The result of severe wind erosion, Cimarron County, Oklahoma, 1936. (Photo from U.S. Deparment of Agriculture, Soil Conservation Service.)

1934, when high winds whipped topsoil from as far away as Montana into cities such as Chicago, New York, and Washington, D.C., causing dusty foglike conditions that left a film of dust—actually topsoil—on every exposed surface. As in many cases, however, people may not have been entirely to blame. Early in the 19th century, the Great Plains were known as the Great American Desert—not a strictly accurate designation, but correct in that patterns of rainfall could be very erratic and almost desertlike at times. The term *Great Plains* came into general use after the 1860's as a result of propaganda by the railway companies, who wished to see the plains settled, creating business for the trains. In good years of relatively high rainfall, farms thrived, but in bad years—not uncommon— the plains were much better suited to grazing than to farming. A series of bad years brought on wind-blown erosion of topsoil and economic disorder (Eckholm 1976). The intertwined roots of hardy native grasses normally shielded the soil from these droughts and winds, but in the 1930's, the worst period of drought, after several warning signs in the 1890's and 1910, snowlike drifts of sand began to smother fields. People left in droves, trekking westward to another land of mythical bounty, California, as John Steinbeck so eloquently chronicled in *The Grapes of Wrath*. Ecological lessons travel slowly, however, and, incredibly, virtually the same thing happened in the Soviet Union as late as 1963.

Loss of topsoil can cause severe silting of rivers. It costs Argentina $10 million a year to dredge silt from the River Plate estuary and to keep Buenos Aires open to shipping (Allen 1980). The capacity of India's Nizamsagar Reservoir has been more than halved by silting, from 900 million m^3 to 340 million m^3. Continued irrigation for agriculture can also cause salting of the land, as is now being discovered in California. Even ''fresh'' water contains minute quantities of dissolved salts, and as water evaporates or is used by plants, the salt is left behind and accumulates in the soil. Mesopotamia, probably the world's oldest irrigated area, on the plains of the Tigris and Euphrates, had a fertility legendary throughout the Old World. Today, Iraqi peasants eke out a living on some of the world's lowest crop yields. Some areas of southern Iraq glisten like fields of freshly fallen snow (Eckolm 1976).

Severe grazing by livestock can also play a large part in desertification. Cutting herd numbers is the only answer, and, paradoxically, a reduction of 50 percent could double meat output. The reason lies in the physiological nature of growth. Of all the food consumed by a grazing animal, roughly the first half is used for maintenance, the next quarter for reproduction, and the final quarter for milk production, growth, and fat storage. Any cutback in feeding as a result of overgrazing results in a reduction of these final functions rather than the others. It must be instilled in this society that the benefits of productivity are more important than pure herd numbers.

Finally, agriculture can result in a severe loss of wildlife through land reclamation drainage, hedgerow removal, and accumulation of toxic residues. Although the answer to the question ''Can we feed the populations of the future?'' is probably ''Yes,'' the more relevant question becomes ''What are the ecological consequences of doing so?'' (Holdgate 1978). ''The ghost of Malthus, which Western economists had long thought to have chased away, stalks the planet as ever before'' (Pawley 1976, p. 13).

21.6 *Derelict Land*

Destruction

Apart from affecting the environment through agriculture and by damming rivers, people alter the landscape or geomorphology in many other ways (Table 21.5). Without question, strip mining devastates the land more severely than any other form of physical excavation. Photo 21.11 shows a strip mine in Lewis County, Washington. The world's largest excavation is the Brigham Canyon copper mine in Utah, which has involved the removal of 3,355 million tons of earth over an area of 7.21 km^2 to a depth of 774 m, seven times the amount of material moved to build the Panama Canal. In Britain, as early as 1922, it was realized that ''man is many times more powerful as an agent of denudation than all the atmospheric denuding forces combined'' (Sherlock 1922, p. 333).

Table 21.5   Some anthropogenous landforms. (From Goudie 1986.)

Feature	Cause
Pits and ponds	Mining
Broads (lakes in southeast England)	Peat extraction
Spoil heaps (mine tailings)	Mining
Terracing	Agriculture
Ridge and furrow	Agriculture
Cuttings	Transport
Embankments, impoundments	Transport, river and coast management
Dikes	River and coast management
Mounds	Defense, memorials
Craters	War, bombing
City mounts (*tells*)	Human occupation
Canals	Transport, irrigation
Reservoirs	Water management
Subsidence depressions, sink holes	Mineral and water extraction
Moats	Defense

Photo 21.11 Strip mining for coal, Lewis County, Washington. A 50 ft coal seam is seen in the lower part of the photo. This cut shows the 150 ft of overburden that had to be removed to expose the coal. (Photo from U.S. Department of Agriculture, Soil Conservation Service.)

Ecosystem Restoration

Reclamation of derelict land is of increasing importance to society both because the quantities of land so affected are getting bigger and because the value attached to restoring natural vegetations on these lands is going up. Bradshaw and Chadwick (1980) estimated the amount of derelict land in Great Britain to be over 55,000 ha in 1971, and the total has certainly increased since then. In the United States, an estimated 1,295 km² of land was disturbed by surface mining as of January 1965. It is in Western Europe, however, that most land has been modified by people, and it is there also that reclamation research is at its peak. Most of the derelict land in urban environments, though certainly not all, is built upon. It is in rural settings that the problem is more acute, largely as a result of mining activities, which create open wounds in the countryside and also huge spoil-disposal sites. More recently, in response to huge crop surpluses, some governments, for example in Britain, are proposing that farmers be compensated for taking marginal farmlands out of cereal production in a policy termed ''set-aside'' (Mitchley 1988). Such land is then suitable for restoration projects.

Reclamation research usually focuses on three main objectives: restoring land so that it can once again be used productively, removing local sources of toxic materials, and making sites visually attractive again. General reclamation has usually involved establishing a vegetation cover

Photo 21.12   Ecosystem restoration. A former gravel pit in Norfolk County, Massachusetts, seeded to grasses and planted with trees, 1965. (Photo from U.S. Department of Agriculture, Soil Conservation Service.)

that prevents further weathering of the land. An example is shown in Photo 21.12. Doing so requires knowledge of the environmental tolerances of plants and of **succession** after establishment. Less attention has been paid to the restoration of animal communities, in part because animals are more mobile and are expected to invade these sites naturally and in part because less is known about manipulating animal communities (Wathern 1986). The main method to date for reclamation of land has been to produce a green cover as quickly as possible, irrespective of the appropriateness as a long-term answer. This is the "quick-fix" solution. Often no attention is paid to the use of native flora; alien species are often cheaper and more commonly used. No exotic species used for rehabilitation are more controversial than the hundreds that compose the genus *Eucalyptus*. These trees produce at least twice as much biomass per hectare as other tree species, though their overall consumption of water is high.

The average cost of seeding for reclamation of surface mines in seven western U.S. states has been estimated at $620 per ha (in 1977), but use of earth–moving equipment pushed the figure up to $10,000 per ha (National Research Council 1981); it is no wonder that many areas remain unrestored. Amazingly, in Iowa, Illinois, and Indiana, highly agricultural states, only 8 percent, 11 percent, and 18 percent, respectively, of original terrestrial ecosystems remain. Of course, in some instances there is no need to reclaim derelict areas, as an interesting flora will develop naturally. Some abandoned mine workings are now biologically very interesting, and their attendant vegetation is often adapted to cope with high levels of certain heavy metals (Davies 1976; Lee and Greenwood 1976; Johnson 1978; Johnson, Putwain, and Holliday 1978; Holliday and Johnson 1979). Because mining and quarrying operations have changed over the years, however, it is unlikely that modern sites will acquire the same ecological interest and diversity as an old quarry, with its uneven face, rubble-strewn floor, and scattered waste heaps. New large-scale workings produce uniform conditions over large areas (Wathern 1986), and these new workings are not well tolerated by today's more aware society, unless the public can see that industry is trying to rehabilitate its sites.

Often nitrogen-fixing plants such as legumes can be used successfully to revegetate nutrient-poor soils (Bradshaw et al. 1978), though in some instances such a productive sward would appear completely out of place. Repeated application of fertilizers to reclaimed areas is expensive, and reclamations should ideally result from a single simple treatment.

In cities, though derelict land is most often acquired by developers, the value of obtaining land for public parks cannot be overestimated, expensive though it is. The value of open spaces such as Boston Common, Central Park in New York, and the Royal Parks of London is probably best judged from the fact that they are not built on.

21.7 The Sea

The Physical Nature of the Oceans

Seventy-one percent of the globe's surface is covered by the oceans and seas, an important feature of which is their depth. Whereas only 2 percent of the land is over 300 m above the sea, 77 percent of the ocean floor is more than that depth below sea level. The great trenches of the Philippines and the Marianas have a depth of 10,700 m, deeper than the height of the highest terrestrial mountain. There are three main ocean zones: the continental shelf, extending from the coastline to about 200 m below sea level; the continental slope, falling steeply from the shelf to 2,500 m; and the deep ocean. The **continental shelf** is most often the site of human activity because of its moderate depths and easy access.

The salt nature of the oceans is derived from runoff from the land masses and is relatively steady at about 35 parts per thousand. The chemical elements that produce this salinity are endlessly varied, from chlorine as sodium chloride at 1.9×10^7 μg/liter to gold at 0.2 μg/liter (Goldberg and Bertine 1975). Only where concentrations exceed one part per million does commercial extraction seem likely; candidates are salt, magnesium, and bromine. Replenishment must be taking place naturally at an equal or greater rate, since the salinity of the oceans is always about 35 parts per thousand. Thus the sea does seem to be an "inexhaustible" supply of these minerals. By far the greatest resource in the oceans, however, is not the minerals but the organisms that live in them.

Improving Yields from the Sea

It is commonplace to see calls for greater use of marine biological resources for food, yet there are severe limiting factors on biological productivity in the sea. Photosynthesis can occur only in a band about 60 m
deep, the **euphotic zone,** because from there to 520 m there is only blue
light, the other wavelengths having been absorbed. As in fresh water,
CO_2 tends to be limiting; the amount dissolved is dependent on mixing at
the interface between water and the atmosphere.

Of all ocean areas, the open oceans are the least productive and are
equivalent to something of a biological desert (Table 21.6). The coastal
zone has a higher productivity because of its proximity to the sources of
minerals, and **upwelling** zones share similar characteristics.

Harvesting by people lower down the food chain, where production
is greater (see Chapter 19), is technically difficult. **Phytoplankton** is productive but has a rapid turnover, and the standing crop at any one time is
low. Given the energy involved in processing large volumes of seawater,
larger fish can concentrate protein from plankton more efficiently than
humans (Morris 1970), a fact that might save us from plankton dinners.
Most of our crop comes at the third **trophic level**—desired species of
fish—and at each trophic level biomass is lost to competing taxa such as
sharks or seabirds. A further harvest comes from the detritus feeders that
scavenge the sea floor; many flatfishes and crustaceans belong to this
group. Detritus feeders and bacteria are important food sources, because
there are 10 g of dead and 100 g of dissolved carbon for every gram of
living carbon in the water column.

Estimates of the sustained annual yield of fish from the seas is around
100 million tons per year (Ryther 1969) split almost equally between upwellings and coastal areas; the open ocean produces less than 1 percent.
Only the fish occurring at depths less than 200 m are usually available to

Table 21.6 *Primary productivity in the oceans. (From Whittaker and Likens 1975.)*

	NPP Dry Matter			Biomass Dry Matter	
Ocean Area	Area $10^6 km^2$	Mean $g/m^2/yr$	Total $10^9 t/yr$	Mean kg/m^2	Total $10^6 t$
Open ocean	332.0	125	41.5	0.0003	1.0
Upwelling zones	0.4	500	0.2	0.02	0.008
Continental shelf	26.6	360	9.6	0.001	0.27
Algal beds and reefs	0.6	2,500	1.6	2.0	1.2
Estuaries (excluding marsh)	1.4	1,500	2.1	1.0	1.4
Totals and averages	361.0	155	55.0	0.01	3.9

people, and actual harvests have steadily increased, reaching 90 million metric tons in 1986 (Anonymous 1988). Of this catch, Japan and the Soviet Union reap the most, with about 10 million tons per year, followed by China, Peru, and Norway. The United States and United Kingdom are down the list at 3 and 1 million tons per year, respectively.

Despite the absolute constraints of food production in the seas, extension of fisheries could significantly improve yields. Such extension could utilize untapped species—the anchoveta in Peru, Alaskan pollock, Bering Sea flatfishes, and grenadiers in the northwest Atlantic. Alternatively, krill could be harvested. Since Antarctic whaling has reduced whale stocks to about 10 percent of their former size, the uneaten food of the whales theoretically is available to man. This food is krill, the shrimp *Euphausia superba*, 6–7 cm long and 1 g in weight.

The captive breeding or **aquaculture** of selected species for food is less than a panacea. It requires a complex of buildings, stores, hatcheries, tidal areas, tanks, ponds, and lagoons. One advantage is that waste heat (for example, from nuclear reactors) could be used to maintain warm water temperatures, which in turn promotes faster fish growth. Often more than one species can be cultured together. For example, sewage might be used as a basis for algal production, which forms the food of oysters, which then filter the water as well. The oyster droppings are eaten by worms, which are the prey of bottom-living fishes, whose nitrogenous excretions nourish water weeds, which oxygenate the water. A major disadvantage here is the great technical skill needed for success.

Overfishing

Overfishing is the bane of modern fisheries technology. The annual world catch is 15–20 million tons (20–24 percent), lower than it might be if the resource were not overfished. Particularly favored fish species have shown dramatic declines, included among which are the Asian sardine, the California sardine, the northwest Pacific salmon, the Atlantic-Scandian herring, and the Barents Sea cod. Photo 21.13 shows the vast fish-meal stocks in Peru in the 1960's before collapse of the Peruvian anchovy populations, as a result of overfishing, in the 1970's. Others are showing considerable signs of strain, such as Newfoundland cod, North Sea herring, and yellowfin tuna. Worse still is the accidental capture and killing of nontarget animals. In shrimp fishing in the Gulf of Mexico, the ratio of fish discards to shrimp caught ranges from 3:1 to 20:1. The problem of *incidental intake*, as it is known, is particularly acute in marine environments (Allen 1980) and can lead to controversy. An example is the debate in Florida in the late 1980's and early 1990's over the use of turtle exclusion devices, designed to allow sea turtles to escape from shrimp nets. They save the lives of turtles but reduce the efficiency of the nets in catching shrimp.

Photo 21.13 Fish-meal stocks in Peru. This photo conveys the enormous scale of the Peruvian fishery, which collapsed a few years later. (Photo by R. Coral, Food and Agriculture Organization.)

In the United States, management of marine fisheries has not been distinguished by many great successes. Striped bass (*Roccus saxatilis*), haddock (*Melanogrammus aeglefinus*), and Atlantic halibut (*Hippoglossus hippoglossus*) were once abundant on the East Coast, as were Pacific sardines (*Sardinops sagax*) and chub mackerel (*Scomber joponicus*) in California.

International regulatory measures intended to reduce catches, such as net size and catch limitations, are difficult to enforce, and often once a species has been overfished, it may not be possible for it to regain its place in the energy pathways of the ecosystem. The Pacific sardine of the California Current system was overfished in the 1930's and replaced by a competitor, the anchovy *Engraulis mordox* (Fig. 21.13). At the end of the 1950's, the latter's biomass was similar to that of the sardine 30 years before, and it has clearly ousted the former species (Ehrlich and Ehrlich 1970). Sometimes overfishing alone cannot be given the blame for species decline. For example, it is not yet clear whether changes in ocean conditions due to climatic events such as El Niño can also devastate fishes (Barber and Chavez 1983; Steele and Henderson 1984). Climatic changes can also occur over a wide time span, such as hundreds of years, with concurrent gradual effects on wildlife (Bryson and Murray 1977).

A definite example of overexploitation involves the whaling industry (or rather involved it). In 1982 the International Whaling Commission (IWC) resolved to discontinue commercial whaling within three years, but such action often merely results in long objections from whaling countries or their simple withdrawal from the IWC (Cherfas 1986). A biologically depletive program of cropping, mainly by Japan and the Soviet Union, continues despite falling yields and dwindling stocks (Fig. 21.14) under the transparent excuse of scientific whaling. If managed properly, this natural renewable resource could make an important contribution,

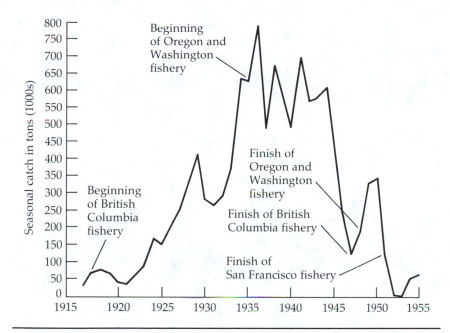

Figure 21.13 *History of the Pacific Coast sardine fishery, at one time the first-ranking fishery in North America in weight of fish landed and third-ranking in value (after tuna and salmon fisheries).* The decline is attributed to overfishing, and there has been no recovery between 1955 and 1970. (Redrawn from Dasmann 1976.)

amounting to about 10 percent of the total yield of marine products on a sustainable basis. At the moment, whales are being mined rather than harvested. Eltringham (1984) gives a good account of the history of this folly. Much environmentalist support advocates the elimination of all whaling, even though some species, like the Antarctic minke, could be safely harvested to a moderate degree. This policy has resulted in a distrust of environmentalist groups by countries such as Japan and the Soviet Union; cooperation by these countries in future discussions over other marine resources is thus jeopardized.

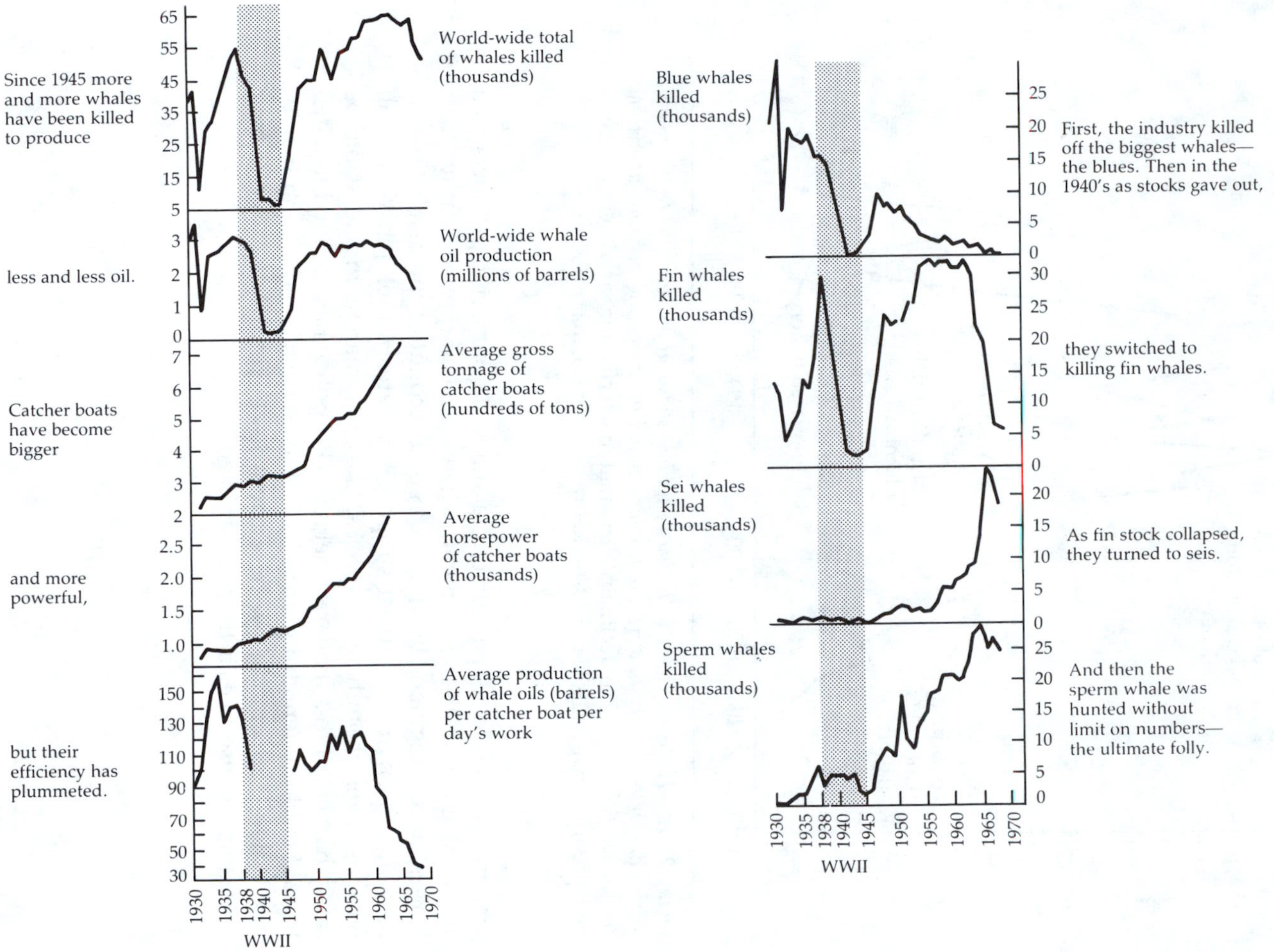
Since 1945 more and more whales have been killed to produce
World-wide total of whales killed (thousands)
less and less oil.
World-wide whale oil production (millions of barrels)
Catcher boats have become bigger
Average gross tonnage of catcher boats (hundreds of tons)
and more powerful,
Average horsepower of catcher boats (thousands)
but their efficiency has plummeted.
Average production of whale oils (barrels) per catcher boat per day's work
Blue whales killed (thousands)
First, the industry killed off the biggest whales— the blues. Then in the 1940's as stocks gave out,
Fin whales killed (thousands)
they switched to killing fin whales.
Sei whales killed (thousands)
As fin stock collapsed, they turned to seis.
Sperm whales killed (thousands)
And then the sperm whale was hunted without limit on numbers— the ultimate folly.
1930
1935
1938
1940
1945
1950
1955
1960
1965
1970
WWII

Figure 21.14 *The decline in whale numbers, related to the inputs of the industry.* (Redrawn from Payne 1973.)

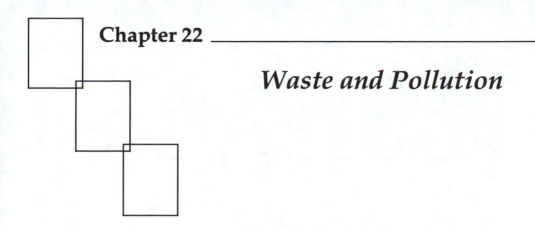

Chapter 22

Waste and Pollution

J ust about everything made or harvested by people turns to waste sooner or later, and an incredible amount is generated (Fig. 22.1). Waste may either be treated as a source of raw materials and **recycled** or it may be ignored as too costly or unsuitable for such a process and be spewed out to do great harm in the environment. Unfortunately, bagged fertilizer is often cheaper than dried sludge, new steel is more convenient than that obtained from reprocessed old cars, and agricultural land is easier to convert to housing than are levelled spoil heaps. However, global scarcity of raw materials may eventually change these relative costs. Choosing sites for the dumping of domestic garbage creates definite problems, but there are also many more serious implications of industrial wastes and pollutants. For purposes of discussion, an attempt is made to classify these **pollutant** types. As a point of interest, among the MDCs, Poland is seen as the most polluted, even by its own scientists, because a very low fraction of the GNP (gross national product) is spent on pollution control (Miller 1988).

22.1 *Wastes Emitted into the Atmosphere*

Most gaseous wastes, of which CO_2, CO, SO_2, and H_2S comprise the biggest volume, are the by-products of fossil fuel combustion. Carbon monoxide, mainly from the internal combustion engine, is the largest of the latter three, and in the United States its volume equals the sum of all other industrial contaminants (about 150 million tons per year), although this amount is small compared to natural emissions (5–10×10^9 tons per year). Carbon monoxide competes with oxygen for hemoglobin and is a poison to humans (and other animals) rather than a factor of ecological change. Even in highly polluted cities, however, CO levels are below fatal doses, and removal of the pollutant allows a rapid return to equilibrium.

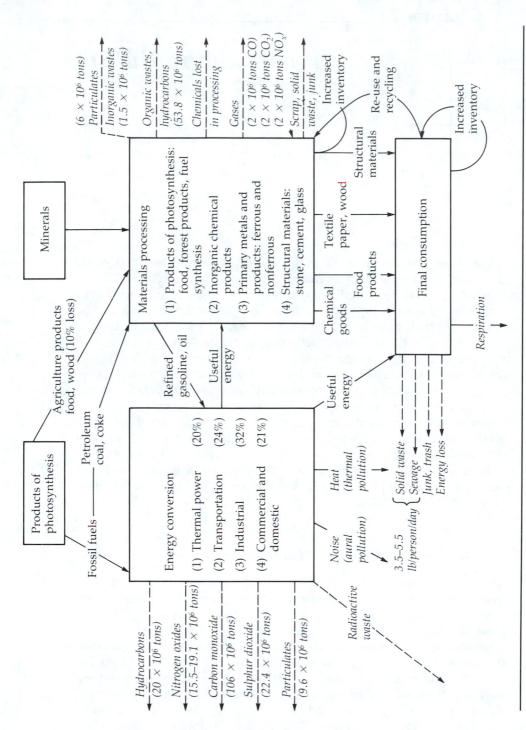

Figure 22.1 A scheme of material flow through the resource processes of the United States, showing how inputs of energy and materials are converted to various kinds of wastes. (Based on data from Simmons 1981.)

Sulphur Dioxide

Sulphur dioxide, SO_2, comes from the combustion of coal and fuel oils, sulphuric acid plants, and the processing of metal ores containing sulphur. In the atmosphere, it lasts an average of 43 days, being converted to SO_3 and reacting with water to produce an aerosol form of sulphuric acid, which is toxic to plants at 0.2 ppm and is highly corrosive to iron, steel, copper, and nickel. Urban-industrial areas emit large quantities—New York City emits 2 million tons per year from coal and Great Britain 5.9 million tons per year. Concentrations of 1–5 ppm usually produce a detectable response in people in the form of inflammation of the upper respiratory tract, and plants are also injured at such levels. The downwind transport of SO_2 and its reaction with water to form **acid rain** (H_2SO_4) are a major ecological headache. Industrial SO_2 from the northeast United States is acidifying the freshwater bodies of Canada and killing the fish; a similar problem has been identified in Scandinavia from acid rain originating in Britain.

A relative measure of acidity of rain is the number of hydrogen ions in a liter of water. In pure water there would be about one hydrogen ion for every 10^7 molecules of water; in rainwater about 1 per 10^6 water molecules, because rainwater comes into contact with carbon dioxide in the atmosphere, and carbonic acid, a very weak acid, is produced. Rain falling in New England, on the other hand, has 1 hydrogen ion for every 10^4 water molecules—that is 1 in 10,000, a 100-fold increase above normal. The reason is that the area is downwind from the industrial centers of the Northeast, where fossil fuels are burned in huge quantities. Acidity is usually measured, however, by **pH,** the negative logarithm of the hydrogen-ion concentration. Thus, for pure water, $-\log_{10}(0.0000001) = 7$, but for the rain in New England, pH $= -\log_{10}(0.0001) = 4$. Of the lakes of the Adirondack Mountains of the northeastern United States, 51 percent had a pH value of less than 5 in 1975, and 90 percent of these had no fish. In total, 45 percent of the high-altitude lakes had no fish—compared to only 4 percent in the 1930's. In southern Norway, one-third of the lakes have been devoid of fish since the 1940's. In all cases, it is usually the young fish that are most susceptible. Acid rain is clearly a serious problem, and the **air pollution** that crosses international borders to cause it can be a subject of intense political debate.

Besides acidifying lakes, acid rain has also been implicated in declining growth rates and increased tree mortality in eastern U.S. forests, particularly where red spruce (*Picea rubens*) grows. However, such modern trends have also been viewed as a natural consequence of a long-term (200-year) warming trend, which has been a major driving force behind the decrease in red spruce populations from 1800 to the present (Hamburg and Cogbill 1988) and an increase in hardwoods over roughly the same period. A 2°C warming trend, as has actually been recorded in for-

ests in New England, would be equivalent to an over 400 m upward altitudinal displacement of the coniferous-deciduous boundary, sufficient to
cause the observed changes.

In retrospect, society has at least made some progress in sulphur-
dioxide control since the 1960's, the days of minimal environmental monitoring. A 1979 convention on long-range, transboundary air pollution
required participating nations to reduce by 1993 national sulphur emissions or their transboundary flows by 30 percent from 1980 levels. A 1982
study showed that pollution reductions brought about by the Clean Air
Act between 1970 and 1978 saved about $14 billion per year in costs asso-
ciated with early death and another $2 billion in costs associated with ill-
ness (Luoma 1988).

Carbon Dioxide

Carbon dioxide has been produced from the burning of **fossil fuels** at
an accelerating rate since the Industrial Revolution. CO_2 levels have
shown a steady increase over the last 30 years (Fig. 22.2). The preindustrial concentration of CO_2 is thought to have been 265–270 ppm, and the
1976 level was 325 ppm. Concentrations by the mid-21st century may
well be double those of pre–Industrial Revolution times. The concern
here is that a doubling of CO_2 will cause a **greenhouse effect,** raising the
global temperature. In 1987, French and Soviet scientists published analyses of the Vostok, Antarctica, ice core. This core revealed an environmental record dating back 160,000 years in the form of the variation in
proportions of deuterium, a heavy form of hydrogen, in the ice. Carbon
dioxide levels and temperature showed an obvious, strong linkage,
though it is still not clear whether rising CO_2 levels caused or followed
rising temperatures. Natural CO_2 levels over this 160,000 year period
ranged from 180 to 300 ppm (Barnola et al. 1987). Even small changes, of
the order of 1.5–2.5°C, could cause shifts in the boundaries of deserts, in
the limits of cultivation, and in the melt rates of polar icecaps, causing
minor but costly rises in sea levels (Wigley, Jones, and Kelly 1980). Although this trend may not be a uniformly bad thing—for example, it is
certainly preferable to a new Ice Age—it will result in expensive alterations in agricultural plans, designs for reservoirs, and so on. Furthermore, increased global temperatures may also promote more intense
storms. The years 1987 and 1988 were the warmest on record, and in September 1988, Hurricane Gilbert, the most powerful hurricane ever recorded, cut across the Caribbean.

Specifically, individual plant and animal species would be much
affected, particularly in their areas of distribution. Beech trees would
probably go extinct in the southeastern United States, except at high elevations, and suitable habitats in the United States would be limited to the
northern Great Lakes. Of course, new habitats would open up 500 to

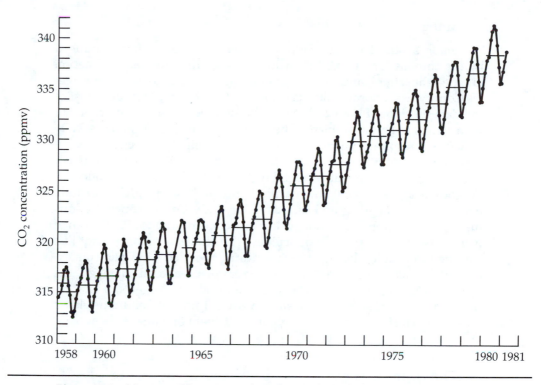

Figure 22.2 *Mean monthly concentrations of atmospheric CO_2 at Mauna Loa (an intermittently active volcano on the island of Hawaii).* The yearly oscillation is explained mainly by the annual cycle of photosynthesis and respiration of plants in the Northern Hemisphere. The steady increase is indicative of an increase in pollutants. (From National Academy of Sciences, 1983.)

1,000 km north in Ontario and Quebec. The sex ratio of lizards and alligators would change, because sex ratio is determined by nest temperature. Warmer temperatures would perhaps mean more males of these species. Turtles, on the other hand, would produce mainly females. For insects, warmer temperatures might allow an extra generation to fit into a longer summer. In agriculture, more insect pests might lead to lower yields.

It is clear that humans can affect the amount of CO_2 in the atmosphere not only by cutting down on fossil-fuel use (ironically, nuclear power is a cleaner alternative in this regard) but also by reforestation—or **deforestation.** Just how many trees would have to be planted to absorb the excess CO_2? The Environmental Protection Agency has looked seriously at the numbers, and by the end of 1988 two congressional committees had held hearings that touched on the subject of massive reforestation. Estimates of the required area needed range from areas equal to the size of Australia to those equal to the size of Zaire (Booth 1988*b*).

Smog

Smog is caused mainly by nitrous oxides in the atmosphere produced by the combustion of coal, oil, natural gas, or the internal-combustion engine—the latter being the most important by a factor of six. Los Angeles County, with normal nitrous oxide levels of 2 ppm, provides most of the data on smog. Photo 22.1 shows Rio de Janeiro, Brazil, another city with a severe smog problem. The major constituent of smog is NO_2, which is a respiratory irritant and absorbs sunlight, especially in blue wavelengths, so it appears as a yellow-brown gas. At concentrations of 8–10 ppm, NO_2 can reduce visibility to 1 mile. Nitrogen dioxide undergoes *photolysis*, breakdown with light, and the products include ozone, formaldehydes, and nitrous oxide, NO. Ozone concentrations may reach 0.65 ppm on a smoggy day, and concentrations of 1.25 ppm of ozone can cause respiratory difficulties. The effects of smog are not just confined to people; damage to crops during the 1960's in California exceeded $8 million per year, and on the eastern seaboard, damage was estimated to be $18 million. The forests around Los Angeles, a vital reserve of recreational space, also suffer various forms of pollution-related die-off, of which ozone is probably the chief agent of damage.

The Ozone Layer

Chlorofluorocarbons (CFCs) are used in refrigerator cooling systems and as aerosol propellants. Between 1930 and 1980, about 8 million tons were made and subsequently released into the atmosphere, where the compound reacted with the ozone layer. A depletion of ozone would

Photo 22.1 Atmospheric pollution. The beautiful city of Rio de Janeiro, Brazil, blanketed in its own smog. (Photo by P. Stiling.)

allow more UV light to penetrate to the Earth and result in an increase in skin cancer. CFC production has doubled the amount in the atmosphere since 1970, though production is now declining. In 1987, 30 nations of the European community signed the Montreal Protocol, which calls for a 50 percent reduction of ozone-depleting chlorofluorocarbons by 1990 (Anonymous 1988). Vermont is the first U.S. state to announce a ban on CFCs for automotive air conditioners. The state legislature voted in 1989 to enforce the measure beginning with new vehicles in 1993.

In summary, the effects of gaseous waste are seen at three scales:

□ *At the **point source:*** An example is the release of the volatile liquid methyl isocyanate from the Union Carbide plant at Bhophal, India, which caused over 2,000 deaths in 1984. Methyl isocyanate, made by combining methylamine and phosgene, had been used there since 1980 in the production of the pesticide carbaryl for the domestic market.

□ *Regionally:* The local climate may be affected and existing illnesses exacerbated, especially respiratory troubles. Acid rain is a good example.

□ *Globally:* The world's climate can be changed. The greenhouse effect is of paramount importance.

22.2 *Economic Poisons*

Many poisons are applied to agricultural ecosystems to eliminate or diminish insect herbivores on crops. Ideally, these toxic substances would be **biodegradable,** that is, would break down into biologically insignificant compounds, but in reality many poisons do not exhibit this quality, and some remain toxic for long periods, contaminating the soil and the rivers by means of runoff and accumulating in food chains, a process called **biological magnification,** often with fatal effects for top predators. Precautions against sheep warble fly have prevented the reproduction of golden eagles, because DDT inhibits calcium-carbonate deposition in the oviducts of birds, resulting in thin-shelled eggs, which are usually crushed by the brooding parents (Cooke 1973). Egg-shell thinning caused by accumulations of pesticides is a problem in many other species of birds as well. Photo 22.2 shows an example. Anti-gnat measures in lakes have caused the death of grebes (Hunt and Bischoff 1960). Two popular contact hormone weed killers, 2,4-D (nonpersistent) and 2,4,5-T (highly persistent), have been criticized because spray drift is an indiscriminate killer of vegetation, and an uncomfortable amount of evidence is coming to light suggesting that teratogenicity to human fetuses can result from certain levels of exposure to any 2,4,5-T that contains the contaminant dioxin. It was dioxin (TCDD) that contaminated the soil around Seveso in

Photo 22.2 Duck's nest with clutch of eggs illustrating effects of DDT. DDT accumulates in food chains and causes thin egg shells, which break under the weight of the parent birds. (Photo from U.S. Department of Agriculture.)

Lombardy, Italy, after an explosion at a chemical factory there in 1976. The area had to be evacuated and much of the vegetation and topsoil scraped up and burned (Bonaccorsi, Fanelli, and Tognoni 1978). High levels of dioxin contamination have also been reported recently from the United States.

DDT (dichlorodiphenyltrichloroethane), first synthesized in 1874, became prominent in the early 1940's as an insecticide desirable because of its lethal nature and long **persistence** in the environment. Peak production reached 80 million pounds in 1959 and decreased thereafter until cancellation of its registration in 1972. Public concern over the use of DDT was highlighted with the release of journalist Rachel Carson's (1962) highly dramatic accounts of pesticides and pollution, *Silent Spring*. During these 20-odd years, DDT and its metabolites appear to have been distributed over the entire globe, including the far oceans, even showing up in Antarctic penguins (Sladen, Menzie, and Reichel 1966). The ability of DDT to build up in food chains was well documented; for example, on the East Coast of the United States, a concentration in water of 0.0005 ppm is accompanied by concentrations of 0.33 ppm in *Spartina* grass, 0.4 ppm in zooplankton, 2.07 ppm in needlefish, 3.57 ppm in herons, 22.8 ppm in fish-eating mergansers, and 77.5 ppm in gulls (Woodwell, Warster, and Isaacson 1967). Quite low dosages of 5 ppm of DDT affect the behavior and reproduction of fish and amphibians. Unfortunately, many of the insect pests for which DDT was applied have now become resistant. Even though the more persistent pesticides are being phased out, their less persistent, more specific "offspring" are still often lethal.

The cost of development of new, more specific pesticides makes them more expensive than their old counterparts, and introduction is therefore often slow, especially in LDCs. It is interesting to note that levels of certain pesticides in the human body, for example the persistent organochlorines, are typically 12–16 ppm in MDCs—in the United States most people are now legally unfit for human consumption. Though such a statement may sound absurd in a noncannibalistic society such as ours, it does raise questions about the desirability of breast-feeding babies. There are, however, still no authenticated cases of clinical damage to people from *approved* uses of insecticides, so they continue to appear on the market. Years ago, Food and Drug Administration workers are said to have attested to DDT's harmlessness by putting a spoonful in their coffee (Buckley 1986).

22.3 Contamination by Organic Substances

Oil

The dominant organic contaminant is crude oil. Incidents of loss from refining, transhipments, tanker cleansing, and accidental spills numbered 595 around the shores of Britain alone during 1976, and 642 in 1977. Such accidental spills easily run into the thousands on a global scale. Even more acute losses occur after accidents like blowouts, such as that at Santa Barbara or Ekofisk B (North Sea) or the wrecking of large tankers, like the *Torrey Canyon* (the first major spill to occur, in 1967), *Tampico Maru*, *Arrow*, and *Amoco Cadiz*, to name but a few. Somewhere in the world a large wreck like this occurs once a week.

Oil can kill directly through coating and asphyxiation, especially acute for intertidal life, or by poisoning by contact or ingestion, as in plants and preening birds, respectively. Water-soluble fractions can be lethal to fish and invertebrates and may disrupt the body insulation of birds, resulting in their death from hypothermia. The jackass penguin, endemic to South Africa, is gradually being extirpated by oil lost by tankers rounding the Cape of Good Hope.

Though natural seepage of oil into the oceans, mainly at junctions of tectonic plates, occurs at the rate of 600,000 tons per year, one large tanker can carry 200,000 tons of oil (Goldberg and Bertine 1975). The *Amoco Cadiz*, wrecked off France in 1978, was a 230,000 ton vessel. Oyster beds along the Brittany coast were totally ruined, fishing and resort industries were affected, and in tidal flats and salt marshes effects lasted for seven years. Estimates of the total damages at the time were on the order of $30 million. It is important to remember that clean-up operations with detergents can prove more devastating to marine life than the oil itself, which is what happened in Britain after the *Torrey Canyon* disaster (Walsh 1973). In 1968 an explosion at a Santa Barbara offshore oil well released oil and poisonous gases, and the crew abandoned the platform. Oil poured out at 200,000 gallons per day and ruined the beautiful beaches. At least 3,600 seabirds died as a result. In 1979, the Mexican oil well Ixtoc, in the Gulf of Mexico, caused a spill of more than 3 million gallons. Oil and dead birds were washed up on the beaches of south Texas more than 6,000 km (3,600 miles) away.

A particularly severe example was the huge spill resulting from the wreck of the *Exxon Valdez* in Alaska at 12:04 A.M. on Good Friday, March 24, 1989 (see Photo 22.3). The rocks at Bligh Island tore five huge gashes in the hull of the ship, one 6 feet high by 200 feet long. The result was one of the worst oil spills in U.S. waters; 10.7 million gallons spilled, most of it in the first 12 hours. The accident took place in fine weather, clear visibility, and no traffic and was clearly the result of human error. A week after the spill, the resultant slick covered nearly 900 square miles. Hundreds of miles of shoreline were covered with oil, in places as much as 6 inches deep. Officially, 27,000 birds, 872 sea otters, and untold numbers

(b)

(a)

Photo 22.3 Aftermath of a massive oil spill. (a) The oil tanker Exxon Valdez, *run aground off the Alaska coast, 1989, showing remaining oil being offloaded to another tanker and a boom placed in an attempt to minimize the amount of oil washed up on the shore. (b) The adjacent shoreline, covered with oil despite containment efforts. Cleaning efforts were slow and laborious. Steam cleaning was sometimes employed. Although the hot water removed the oil more effectively than cold, it killed the animals that lived on the shore. (Photos by U.S. Coast Guard.)*

of fishes died, although the true numbers are probably higher because many dead birds and otters probably sank and were not recovered. These deaths resulted because birds and otters depend on the insulation provided by their feathers and fur to help them maintain proper body temperature. A coating of oil destroys that insulating property, and the animals literally die of exposure. The effects also carried over into the terrestrial ecosystem when bears, otters, and bald eagles feasted on the oily carrion washed up on the beach. Sitka black-tailed deer ate kelp on the beaches. Few of these animals are expected to be found dead on the beaches, because they generally return to their normal habitats before the effects become apparent. Still, the Fish and Wildlife Service found over 100 dead eagles, and most pairs in the area failed to produce young that year. Even more important may be the long-term effects of the accumulation of oil in the bottom sediments, for this will determine the resiliency of the system, how long it takes to recover. The long generation times of the Arctic fauna will slow this process. Even in the tropics, recovery from oil spills is slow. In 1986 more than 8 million liters of crude oil spilled into a complex region of mangroves, seagrasses, and coral reefs near the eastern entrance to the Panama Canal. Areas of such high traffic are likely to experience more than their fair share of disasters. In the case of the 1986 spill, even after a year and a half, only some organisms in areas very exposed to the open sea had recovered.

Sewage

Another common organic contaminant, especially of fresh waters and shallow offshore seas, is **sewage.** Water carriage of sewage from urban areas dates only from the 1840's, before which the contamination of water supplies led to epidemics of cholera, typhoid, and dysentery. Such outbreaks are still common in LDCs, and treatment with chlorine to kill the causal organisms is an important part of processing in MDCs. Even in New York City, some half-million dogs used to leave 20,000 tons of pollutant feces and up to 3.8 million liters of urine in the city streets each year, all of which was flushed into storm sewers (Goudie 1986). Mercifully, since the "pooper scooper" law was passed, pet owners must collect and dispose of this waste (but how do they dispose of it?). Contamination of water by untreated sewage is denoted by the presence of the bacterium *Escherichia coli*, which is not infectious in itself but is an indicator of the presence of the agents of typhoid and dysentery. Treatment of sewage consists of the separation of the organic matter from the water and its conversion to a biologically inactive sludge, a valuable fertilizer, but one that is currently more expensive than factory-produced material. Even among MDCs many large cities do not treat their sewage properly. In Canada, for example, as of 1981, 100 percent of the sewage of Toronto got a two-stage treatment, whereas practically all of the production of the population of Montreal and Quebec was poured untreated into the St. Lawrence River (Simmons 1981).

Agriculture also produces a lot of contaminants, among the most important of which is runoff containing increased levels of nitrogen and phosphorus from fertilizers not taken up by the plants. The main difficulty associated with these wastes is eutrophication (see below), but there is also the problem of **nitrates** entering groundwater, where they may cause metabolic disorders in domestic animals and in children if used for drinking. An example is the human condition known as methemoglobinemia—the blue-baby syndrome—in which the binding action of the contaminant prevents the hemoglobin from taking up oxygen.

22.4 *Eutrophication*

The introduction of large quantities of nitrogen and phosphorus from sewage and fertilizer—so called **agricultural pollution**—into freshwater systems is one of the major problems of pollution today. Even phosphorus from household detergents constitutes an input; "enzyme" detergents are in fact phosphate presoaks. Phosphorus is normally a scarce element and the limiting step in many ecosystems. The large quantities made available by human activity lift this limit, and organisms such as blue-green algae can then flourish. **Algal blooms** appear all over lakes, clogging them up and greatly reducing their value as water supplies or recreation areas. Photo 22.4 shows the results of an experiment in which these elements were deliberately added to lakes so that the effects could

Photo 22.4 Lake 226 in the Experimental Lakes area of northwestern Ontario, showing the role of phosphorus in eutrophication. The far basin, fertilized with phosphorus, nitrogen, and carbon, is covered with a bloom of the blue-green alga Anabaena spiroides. The near basin, fertilized with nitrogen and carbon, showed no changes in algal abundance. Photo taken September 4, 1973. (Photo by D.W. Schindler.)

be studied. **Eutrophication** is a natural phenomenon, and most lakes become eutrophic in time (Chapter 16), but human activity drastically speeds up this process. Eutrophication has probably been best documented in the Great Lakes of North America, although numerous European examples also exist. In Lakes Erie and Ontario, the total dissolved solids has risen by 50 ppm in the last 50 years, and fish populations have changed almost completely. In summer, a serious depletion of dissolved oxygen occurs, blue-green algal scums are formed, and offensive heaps of *Cladophora* pile up on the shore. In Britain, nitrate levels in rivers are 50 to 400 percent higher than they were 20 years ago (Royal Society Study Group 1983), and groundwater is also seriously affected. These increases both correlate well with the increase in fertilizer use. Fortunately, the process of eutrophication is reversible, provided the relevant inputs are curtailed and remedial treatment is undertaken. Lake Washington near Seattle and Lake Tahoe in California are good examples, the latter being now "gin-clear."

The complete case study of the eutrophication of Lake Washington and its reversal to an unpolluted condition has been well documented by Lehman (1986a). Lake Washington at Seattle is a moderately deep (65 m) warm basin that discharges into Puget Sound via a system of locks and canals built in 1916. Seattle began discharging raw sewage into Lake Washington at the beginning of the 20th century, but in 1926 this trend began to change as sewage was diverted into treatment plants and thence into Puget Sound. By 1941 the last sewer outfall into the lake was removed, but thereafter the suburbs of Seattle began to expand, and by 1953, 10 new treatment plants had sprung up around the lake and were discharging 80 million liters into it daily. Alternative options were not as readily available to the suburbs as they had been to Seattle itself. By this time both the scientists at the University of Washington and the lay public were aware that algae had increased in the lake, and a species indicative of classical lake deterioration, *Oscillatoria rubescens*, was found in the lake for the first time. This alga had been connected with the early stages of decline in water quality in the classical examples from Europe, particularly Lake Zurich, and in Madison, Wisconsin (Hasler 1947). Scientists monitoring the Lake Washington situation were able to predict with great certainty the demise of the Lake (Edmondson 1979), and the newspapers were quick to pick up these predictions. Such attention by the media paid off, because by 1958 local politicians voted money to clean up the lake. Ground-breaking ceremonies for the clean-up campaign were held in 1961, and not a moment too soon, as this body of water had already been christened "Lake Stinko" by the local press. Visibility in lake water had declined from 4 m in 1950 to less 1 m in 1962. One by one, the waste-treatment plants around the lake had their effluent diverted. The trend of lake deterioration stopped in 1964, and by 1965 algal abundance was decreasing and water transparency increasing. Surprisingly, *Oscillatoria* persisted into the 1970's, but by 1975 it was gone and phosphorus concentrations had leveled off. The lake was clean again; visibility was as high as 12 m at times! The whole process was later repeated in Canada, this time in a controlled environmental setup, by limnologists in a large-scale series of experimental lakes in Ontario. The results showed beyond a shadow of doubt that phosphorus was the master controlling agent of eutrophication (Schindler 1977).

22.5 *Radioactive Wastes*

Ionizing Radiation

The major concern about radioactive wastes is that they generate ionizing radiation—radiation with sufficient energy for its interactions with matter to produce an ejected electron and a positively charged ion (Whicker and Schultz 1982). The biological problem is that such interactions in the cells of living organisms can cause genetic and physiological damage and even death. Of course cosmic rays and radionuclides in the Earth's crust have ensured that life has always evolved with a background level of radiation, but this natural level is usually low and of little danger to present life. In the case of atmospheric radioactive fallout from tests or explosions, however, a big danger is concentration of pollutants in food chains just as occurs with other toxic substances. Liden and Gustafson (1967) and Nevstrueva et al. (1967) demonstrated that cesium-137, an emitter of gamma radiation with a half-life of 30 years, was adsorbed on the surface of lichens in Scandinavia and became concentrated in the reindeer that browsed on the lichens and then in the tissues of Lapplanders who ate them. Cesium-137 is metabolized in a fashion similar to that of potassium; strontium-90 is similar to calcium and is accumulated in bones. The buildup of Sr-90 in water, breadfruit, and land crabs on Bikini Atoll has meant that the islanders, allowed back in the early 1970's after atomic tests, were moved out again in 1978.

Studies of the effects of ionizing radiation on ecosystems have been carried out in oak-pine forests (Woodwell and Rebuck 1967), in tropical rain forests (Odum et al. 1970), in hardwood forests in the northern (Murphy, Sharitz, and Murphy 1977) and southern (Cotter and McGinnis 1965) United States, and in a shortgrass prairie (Fraley and Whicker 1971). Photo 22.5 shows an example in North America. One useful generalization to emerge was that the sensitivity of plants depends on the ratio of photosynthetic tissue to total tissue (Woodwell 1967; 1970). Thus, the most sensitive plants are trees, which have a relatively low ratio of photosynthetic mass (leaves) to nonphotosynthetic mass (stem and root), then herbs, grasses, and finally algae, which are usually quite resistant. Among trees, pines, which are evergreen, are more sensitive than deciduous hardwoods, in which the investment in leaves is smaller. A gener-

Photo 22.5 Effects of ionizing radiation on a forest ecosystem. This forest was irradiated deliberatly as part of an experiment. (Photo by G.M. Woodwell, Woods Hole Marine Biological Laboratory.)

alization that applies to both plants and animals is that sensitivity to radiation is correlated with size, the largest species being the most sensitive (Woodwell 1967), as they are to stress in general (Woodwell 1970).

Ionizing radiation was the first radioactive pollutant given national attention. Environmental concerns were raised soon after the first test of an atom bomb at Trinity, New Mexico, July 16, 1945. Problems resulting from nuclear power plants are not simply going to go away; 1 kg of uranium-235 produces as much commercial energy as 2,500 tons of coal, and nuclear power is thus a very attractive energy source. In 1972, the World Energy Conference predicted a steady rise in consumption, whereby in 1990 nuclear power would represent 15 percent of the world industrial energy consumption, a figure actually reached in 1986.

A **fission** reactor works when nuclei of unstable atoms such as uranium-235 or plutonium-239 are split apart by neutrons, energy is released and converted to heat, and more neutrons are released, which bombard further uranium or plutonium atoms in a self-sustained chain reaction. Most phases of the fuel cycle of a fission reactor produce radioactive isotopes. Some are very short-lived and can be released in carefully controlled quantities to the atmosphere or bodies of water, but others can have long half-lives and must be kept away from the environment and from people for a very long time. Strontium-90 has a half-life of 28 years, which means that it would take 500 years to drop to one-millionth of its original activity and the recommended level of safety. Plutonium has a half-life of 24,400 years. Fortunately, these wastes are not produced in large quantities: one ton of spent nuclear fuel when reprocessed gives 500 liters of waste. After 25 years, a **burner reactor** has yielded 25 tons of liquid waste and a **breeder reactor** 44 tons. Twenty-five tons of liquid wastes can be condensed to 2.5 m^3 of solid waste and 44 tons to 4.4 m^3 (Ringwood et al. 1979). The disposal of these wastes, however, causes many problems, because they must be isolated from the biosphere for 250,000 years, a high societal commitment. Suggestions for disposal have included sinking them into the Antarctic ice cap, firing them into the sun, or placing them on interplate subductions to be carried into the Earth's core. The most popular, however, is geological disposal in rock formations without any groundwater circulation; salt deposits (especially abandoned salt mines) are favored.

Nuclear Accidents

The greatest risk of nuclear power, however, may lie not in waste production but in accidents at the plant. No full **meltdown** has yet occurred in the United States, but there have been a large number of minor accidents, the most famous of which was the Three Mile Island incident on March 28, 1979, in Hamsburg, Pennsylvania. A faulty water pump caused overheating, which was followed by a whole series of human errors, which in turn allowed a small amount of radioactive water to be released into a nearby river. No deaths or serious injuries were reported, but the clean-up cost at least $1.5 billion and lasted more than 10 years. Other locations have not been so lucky; the Chernobyl disaster in Russia occurred on April 26, 1986. Two gas explosions inside one of the four water-cooled reactors blew the roof off the reactor building and set the graphite core on fire. As usual, human error was to blame; engineers had turned the automatic safety systems off for an experimental test. Thirty-one plant workers died from exposure to radiation, 200 others suffered acute radiation sickness, and a land area of 2,590 km^2 (1,000 square miles) around the reactor was contaminated with radioactive fallout. Winds carried radioactive material over much of Europe (Fig. 22.3).

Even in the event of automatic shutdown of a plant, residual decay from the fuel rods would generate vast amounts of heat. The resultant molten mass could be hot enough to burn through the bottom of the reactor vessel, through the containment building, and into the ground below, where radioactive gases could escape into the atmosphere. Where the molten mass would stop is not really known, and the phenomenon has been dubbed the **China syndrome.** At Three Mile Island the accident failed to breach the containment building. In 1978 the U.S. Nuclear Regulatory Commission received word of 2,835 "reportable occurrences." Every nuclear plant in operation that year had at least one unscheduled shutdown.

Not surprisingly, the nuclear power plant industry has pretty much ground to a halt. After a good start in the early 1970's, no new orders for

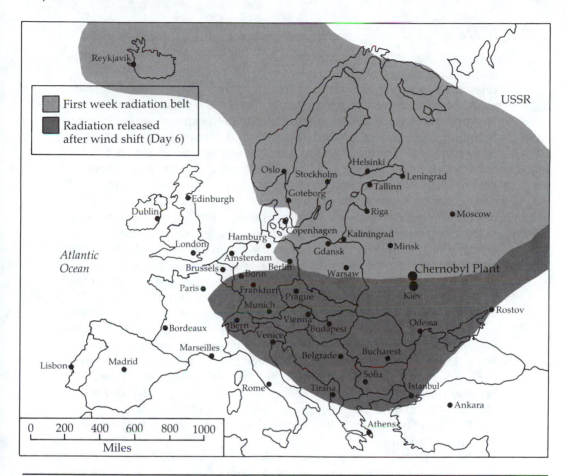

Figure 22.3 *Spread of radioactive fallout over parts of the Soviet Union and much of eastern and western Europe after the Chernobyl accident.* (Redrawn from Miller 1988.)

nuclear reactors have been placed since 1979, and many will probably remain forever half finished (Kupchella and Hyland 1986). On the other hand, of course, with breeder reactors, the world would have a major and almost inexhaustible supply of energy (Zaleski 1980). In a breeder reactor, new fuel, plutonium-239, is actually made or bred within the reactor from nonfissionable (stable) uranium-238, an isotope that is in plentiful supply. The actual costs of nuclear plant development, waste storage, potential accident prevention, and clean-up costs have become astonomical, however, so nuclear power has become economically inviable except in countries like France, where nuclear energy production is heavily subsidized by government (Miller 1988).

22.6 *Metals*

Although metals are part of the "natural" environment and are essential to many organisms, humans have mobilized these elements at far higher rates than those found in nature, often with serious implications. Lead, mostly tetraethyl lead, is used as an additive to automobile fuels at the rate of 4×10^8 kg yr^{-1} and is emitted into the atmosphere where it has a fallout time of about one month (Boggess and Wixson 1979). Young animals and children seem most at risk, especially around urban centers such as freeway exchanges, where lead poisoning from automobile exhaust may impair mental faculties. Mercury, used as an agricultural fungicide, can be even more toxic. A load of seed wheat dressed with a mercury fungicide was made into bread in Iraq in 1972, and deaths were in the hundreds. A more common pathway of exposure is by agricultural runoff into streams and rivers, where it is taken up and concentrated in the flesh of fish. In this way over 3,500 people were poisoned at Minamata Bay in Japan during the late 1950's and early 1960's. One would think such lessons would be learned quickly, but they are not. In the United States, the full power of mercury as a poison hit home in 1969 in Alamogordo, New Mexico. In September of that year, Ernest Huckleby obtained floor-sweepings from a granary to use as feed for his hogs. The sweepings contained millet seed treated with methyl mercury fungicide. Some of the hogs fell ill, but Huckleby chose an apparently healthy one and butchered it. He fed the meat to his family, and three children fell ill and suffered irreparable neurological damage (Grant 1973). Nowadays, treated seed is mixed with a red dye to help prevent such tragic mix-ups.

Not all pesticides, fungicides, herbicides, or "natural" high levels of heavy metals are deleterious, however. Abnormally high levels of cadmium in soils and food at Shipham, England, near the village where the author grew up, do not appear to have caused any ill effects. The general level of hysteria about heavy metals, and indeed all synthetic chemicals, has unfortunately contributed to a state of chemophobia in modern society (Cochran 1987). The truth is that pesticides caused only 10 accidental deaths out of 104,000 (less than 0.01 percent of the total) in 1978. (Motor vehicles caused 51,000 [49 percent] and handguns 17,000 [16.3 percent]). Even among poisonings, pesticides accounted for only 31 out of 6,663 in 1978; drugs and medicines accounted for 3,797. Among **carcinogens**—cancer-causing agents—no pesticide even gets into the top 25 (Cochran 1987). The head of that list is taken up by the latoxins, chemicals produced naturally by certain molds. In other words, naturally occurring substances are not always "good" and man-made ones "bad" (Ames 1983).

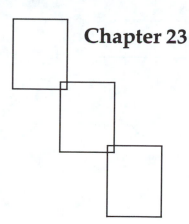

Chapter 23

Introductions of Exotic Species

Not only can people radically alter the environment around them by agricultural activities and forest clearcutting, they can also be important agents in the spread of exotic plants and animals by deliberate and accidental introductions (Bates, 1956).

23.1 Deliberate Introductions

Many plants have been introduced deliberately, for example as new fodder for sheep to bolster the wool trade in England or as ground cover to prevent erosion in the United States (see Photo 23.1 for a spectacular example). Well over 100 species of tree have also been imported into Britain, the bulk of them being conifers imported during the 19th century to satisfy the lumber industry. The Sitka spruce (*Picea sitchensis*), present

Photo 23.1 Kudzu, Pueraria lobata, *a Japanese vine that was deliberately introduced to combat erosion but that now chokes native vegetation in the southeastern United States. (Photo by P. Stiling.)*

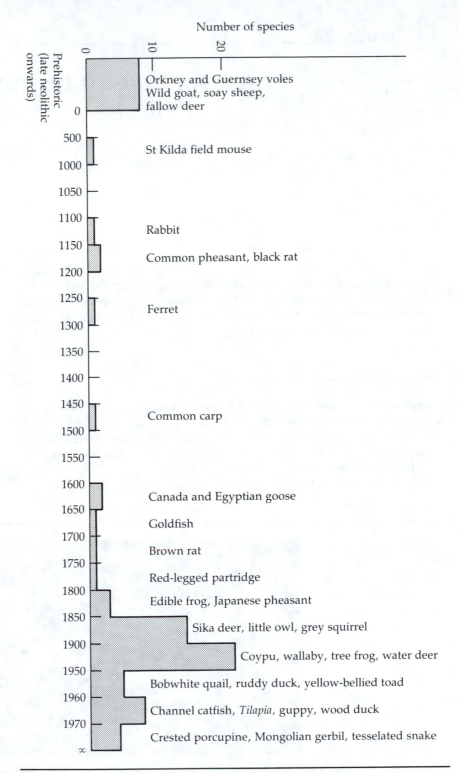

Figure 23.1 *The chronology of animal species now naturalized in Britain with some named examples. Vertical axis is number of species.* (Redrawn from Lever 1977.)

mainstay of the Forestry Commission's plantations, was introduced (Simmons 1979). A more or less complete picture of animal introductions into Britain from the early Bronze Age has been made available by Lever (1977) (Fig. 23.1).

In the British Isles most organisms introduced before the 16th century had some sort of economic merit, but later introductions were motivated by curiosity or decorative value (Jarvis 1979). In the early 1900's, whole rooms full of exotic birds were released in suburban gardens in the hope that they might adapt to local conditions and provide exotic bird song. Such introductions were usually a result of the activity of the Acclimatisation Society, a group formed in England in 1860 specifically to promote the introduction of exotic species, ostensibly to increase the available supply of food (Lever 1977); meat at the time was an expensive luxury. Similar groups have existed in other parts of the world. Nineteenth-century New Yorkers wanted all the birds in the works of Shakespeare, and in Hawaii a group called the Hui Manu collected money from schoolchildren to produce a lowland avifauna that is totally exotic.

Some introduced species, such as bananas and breadfruit, have become utterly dependent on people for reproduction and dispersal, having lost the capacity for producing viable seeds. Other introductions, however, do quite well on their own in an exotic environment. The Irish potato, native to South America, grows unaided in the mountains of Lesotho, and the peach in New Zealand, the guava in the Philippines, and coffee in Haiti are all perennials that have established themselves as wild-growing poulations (Goudie 1986). In Paraguay, oranges (originating in southeast Asia and the East Indies) do well in the native vegetation.

23.2 Accidental Introductions

Many other introductions are not deliberate but inadvertent; they can nevertheless have drastic results. Chaloupka and Domm (1986) have shown that human visitors to islands (called ceys) in Australia's Great Barrier Reef transport many alien diaspores (in the so-called process of anthropochory). They analyzed the number of alien species in the flora on each of 10 ceys and found they were dependent not on island size (as might have been predicted by island biogeography theory) but on the amount of human visitor traffic to that cey (Fig. 23.2). The actual mechanism is proposed to be external attachment to clothing and footwear or in soil adhering to footwear. Heatwole and Walker (1989) have presented new evidence to show that some dispersal of alien plants is aided by gulls and that the density of these birds should also be taken into account. However, this general relationship, an increase in alien species with

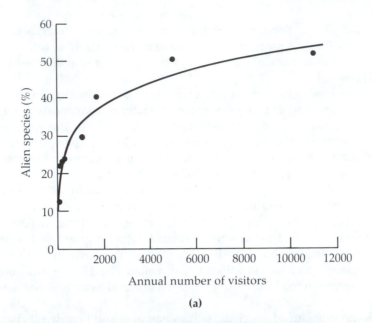

(a)

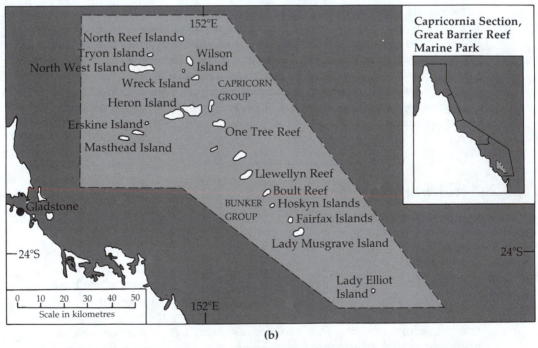

(b)

Figure 23.2 *Humans as agents in the spread of exotic species.* **(a)** *Percentage of alien plants recorded from each of 10 Australian ceys in the Great Barrier Reef as a function of the annual numbers of human visitors.* **(b)** *Location of study area.* (Redrawn from Chaloupka and Domm 1986.)

increased visitor load, has been found to hold for 11 out of 13 nature reserves in North America and South Africa (MacDonald et al. 1989). Furthermore, this overall pattern is one that tends to hold up in a variety of situations. This is something of a dilemma in terms of management—too few visitors cuts down on income; too many and the very nature of the desired preserve is altered. Plants have been dispersed accidentally by moving objects such as vehicles or people themselves, by crop seed movements, and among minerals (such as ballast or road metal); over 50 percent of the 190 major U.S. weed species are invaders from outside the United States (Pimentel 1986).

Perhaps the most frequently introduced taxa are those most easily introduced accidentally—insects, small and inconspicuous and easily overlooked at borders and at international customs. Fire ants, introduced from Argentina into the southern United States in ships' ballast, first landed at Mobile, Alabama, in 1918 and have spread rapidly throughout the Southeast, drastically altering the appearance of old fields with their prominent nests. Their northward spread may ultimately be limited by their inability to overwinter in colder environments. At last count 1,554 species had been introduced into the United States (Sailer 1983), representing 1.7 percent of the continental fauna, and about 40 percent of the pest species of crops in the United States have been introduced (Pimentel 1986), against which many control agents, biological and chemical, have been released. In Hawaii, the situation is even worse. There, 1,476 insect species have been introduced, which represent 29 percent of the islands' total. On the remote island of Tristan da Cunha in the South Atlantic, 38 percent of the insects are not native; 32 species are introduced and 84 are native (Holdgate 1960).

23.3 The Effects of Introductions

Introductions can have drastic influences on the native flora and fauna, often outcompeting it or eating it. The Solomon Islands brown tree snake (*Bioga irregularis*) was introduced onto Guam during World War II to control another introduced species, the Norway rat; it has already eliminated two birds from the island's avifauna and threatens three more (Savidge 1987) as well as bats and lizards. (It has also caused more than 500 power blackouts because pylons and electrical substations are a favorite resting place.) Two decades after its introduction in the late 1950's, the Nile perch, *Lates niloticus*, had taken over Lake Victoria in Africa (Payne 1987). It now comprises 80 percent of all the fish in the lake and has wiped out many species. A voracious fish, weighing as much as 250 kg, it has thrived on the population of small fish that originally inhabited the lake. The species was originally an inhabitant of the Nile and nearby Lake Albert and Lake Turkana, and big waterfalls such as Kabelaga and Owen

Falls probably prevented its colonizing Lake Victoria naturally. Schneider (1927) has estimated that the invasion of the pampas of Argentina by European plants has taken place on such a large scale that at present only 10 percent of the plants growing wild there are native. A similar situation occurs in other parts of the world, particularly on islands (Table 23.1).

In California, the establishment of alien species proceeds rapidly (Fig. 23.3), but it is worth remembering that number of species is not always a good measure of the extent of invasion. Numerous examples are known of single species that, on their own, have a huge effect, such as *Casuarina litorea* in the Bahamas, which creates such a dense shade and has such a toxic effect that few native plants can grow under it. In less than 100 years, *Andropogon pertusus* has become the commonest grass in lowland Jamaica, and bracken fern, *Pteridium aquilinum*, is rampant almost globally.

The economic damage caused by introduced insects alone in the United States has been assessed at about $70 billion (Pimentel 1986). Also, insects can act as vectors of diseases, and mosquitoes can have rad-

Table 23.1 *Percentages of introduced plant species in selected floras. (From data of Moore 1983 and Heywood 1989.)*

Country or Area	Number of Native Species	Number of Alien Species	Percent of Alien Species in Flora
New Zealand	1,790	1,570	46.7
Campbell Island	128	81	38.8
South Georgia	26	54	67.5
Kerguelen	29	33	53.2
Tristan da Cunha	70	97	58.1
Falklands	160	89	35.7
Tierra del Fuego	430	128	23.0
Barbuda	900	180	16.67
Guadeloupe	1,668	149	8.2
Hawaii	12–1,300	228	14.9–16.0
Java	4,598	313	6.4
Australia	15–20,000	1,500–2,000	9.1
Sydney	1,500	4–500	21–25
Victoria	2,750	850	23.6
Austria	3,000	300	9.1
Canada	3,160	881	21.8
Ecuador (Rio Palenque)	1,100	175	13.7
Finland	1,250	120	8.8
France	4,400	500	10.2
Spain	4,900	750	13.2

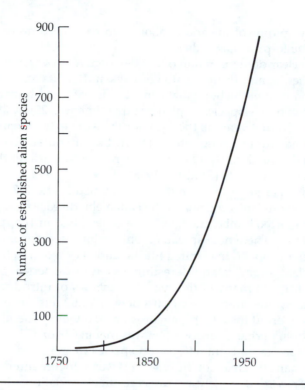

Figure 23.3 *Estimation of the establishment of alien plant species in California since 1750.* (Redrawn from Frenkel 1970.)

ical effects in this way when introduced into a new country. This was a major cause for concern when the tiger mosquito, a vector of dengue fever, was found in the midwestern United States in 1986, imported in old tires from Asia! Luckily, the cold northern winters provided an effective means of eradication. Hawaii was not so lucky. Originally, it was the true island paradise—no mosquitoes. In 1826 a party from Captain Cook's expedition went ashore there to fill the ship's water tanks. They emptied out the dregs of the old water, collected on Mexico's west coast and full of mosquito larvae. These night-flying mosquitoes provided the perfect vector for avian diseases, which spread from migratory bird species to full-time residents. By the late 1800's, great reductions in the native land-bird fauna were noted (Brewer 1988). At this stage, such a die-off was unlikely to be due to malaria, but by the 1920's the disease had had a great negative effect on the population dynamics of the native forest birds (van Riper et al. 1986).

The small island of Laysan harbors one species of honeycreeper, the Laysan finch, and no mosquitoes. When Laysan finches were brought to Honolulu and put in cages that allowed mosquitoes to enter, all the birds died with symptoms of malaria in about two weeks. None of the birds kept in mosquito-proof cages died.

Another clear demonstration of the destructive power of introduced animals comes, again, from the atoll of Laysan. Rabbits and hares were introduced in 1903 in the hope that a meat cannery could be established. The number of native species of plants at that time was 25; by 1923 it had fallen to four. In that year, all the lagomorphs were exterminated to prevent the island from turning into a desert, but of course recovery was slower than destruction. By 1930, there were nine species on the island, and by 1961 there were 16 (Stoddart 1968).

In Britain, a large proportion of elm trees died in the 1970's, as a result of an accidental introduction of the Dutch elm disease fungus, which arrived on imported lumber; hedges, street sidewalks, and coppices were left bare of trees. The whole spread of Dutch elm disease from its native Asia through Europe to the United States, and back again in the form of a new, virulent strain, is an interesting lesson in dispersal. Simberloff (1981) has analyzed many of the available reviews of introductions and lists their disastrous effects. From a list of over 1,500 references on introductions, Simberloff focused his analysis on 10 review papers. The most striking feature, perhaps surprisingly, was that in 678 of 854 instances an introduced species had "no effect" on species in the resident community. In less than 10 percent of the cases (71) was a documented extinction of a resident recorded, largely as a direct result of predation. Herbold and Moyle (1986) took Simberloff to task over this reported lack of effect of introduced species. They pointed out that his report of "no effect" meant no further mention of the effects of the introduced species, no available data on the effects on native species, or no scientific data on actual effects, even though anecdotal evidence clearly suggested major effects to the original authors.

23.4 *Combatting the Spread of Exotics*

How do we combat the problem of introduced species? Simberloff (1986b) suggests it is nearly impossible to predict which invaders will be successful and which will not. For insects, probably the more of a particular taxon's host plants are introduced, the higher the chances it will succeed. Indeed, it is generally a futile exercise to predict which members of any taxon will be successful invaders (Williamson and Brown 1986), though in Britain at least, it has been predicted that the probability an established invader will become a pest is around 10 percent, and it is thought that at least half of the thousand-odd species of recorded invaders have become established.

The problem has certainly been compounded by the efficiency of modern jet travel, which provides transport for invaders from one habitat

Photo 23.2 Signboard on Long Key in the Florida Keys. Combatting introduced species sometimes involves tedious manual removal as was necessary here. (Photo by P. Stiling.)

to another in a matter of hours. Increased vigilance by detection and quarantine stations may be the key. The U.S. Department of Agriculture funds APHIS, the Animal and Plant Health Inspection Service, and individual states may have comparable local agencies. For example, the California Department of Food and Agriculture is unusually vigilant in its pursuit of invading species. In 1985, it intercepted 459 gypsy moth individuals at border stations. From October 1, 1978, through September 30, 1979, APHIS intercepted 18,644 plant pests (Anonymous 1981); of these, 14,002 (75 percent) were insects. What proportion of the invaders are actually caught? Capture of 10 percent of the organisms coming in is considered a success (Dahlsten 1986). Besides, the action of these agencies probably acts as a useful deterrent to individuals who might import millions more illegal arthropod aliens. Only rarely is it possible to eradicate an introduced species, even among vertebrates. The painstaking physical removal of individuals is the usual solution for plants (see Photo 23.2). A mammal called the coypu (*Myocastor coypus*) was introduced in Britain in 1929 and probably escaped from a fur farm in 1937. It was confined to the marshy broads of two counties, Norfolk and Suffolk. Although coypu numbers got as high as 200,000 in the 1960's, government control policies have now eradicated this animal; in January 1989 the government announced that the coypu control organization had completed its task. Coypus had become pests, undercutting river banks and encouraging further erosion and floods as well as feeding on an assortment of crops. The government set itself the target of eradication of coypus within 10 years of the start of the campaign in April 1981 (Baker and Clarke 1988).

Section Seven —

Epilogue

How can problems such as the loss of forests be solved? How can pollution be reduced? (Photos: U.S. Department of Agriculture.)

I t is at once obvious that human population control holds the key to the ecological balance of planet Earth. Only rarely do humans have a positive impact on species welfare, as for example when urban populations feed birds, which, because of increased food intake, lay more eggs and fledge young at a higher rate (Powell, Powell, and Paul 1988). There are also the so-called *synanthropic species*, which actually profit from inadvertent human activity. Rats, mice, sparrows, pigeons, and gulls fall into this category and perhaps the blackbird (*Turdus merula*), too, once a shy forest bird and now a regular and bold inhabitant of European gardens. By and large, however, human ac-

493

tivity correlates well with general extinction. It has been suggested
(Martin and Klein 1984*a*) that global patterns of extinctions of large
land mammals were caused by rapid and substantial changes of climate at the termination of the last glacial period. Evidence is strong,
however, that humans are more to blame (Fig. S7.1). In 1600 there
were approximately 4,226 living species of mammals; since then 36
have become extinct (0.75 percent), and a further 120 (2.84 percent) are
believed to be in great danger of extinction. Martin (1982) has argued
that the global pattern of extinction of large land mammals follows the
footsteps of Palaeolithic settlements, with Africa and Southeast Asia
being affected first (Table S7.1).

Unfortunately, the governments of some countries have been concerned that their populations are not large enough (Romania was one)
and have offered financial inducements to their people to have more
children. Biblical passages also give the impression that people should
dominate the Earth: "Be fruitful and increase, fill the earth and subdue it, rule over the fish in the sea, the birds of the heaven and every
living thing that moves on earth" (Genesis 1:28–29). This idea is essentially disastrous because the population level that can be supported
without reliance on agricultural techniques is 200 million, a figure surpassed by the Middle Ages (though close to the 500 million suggested

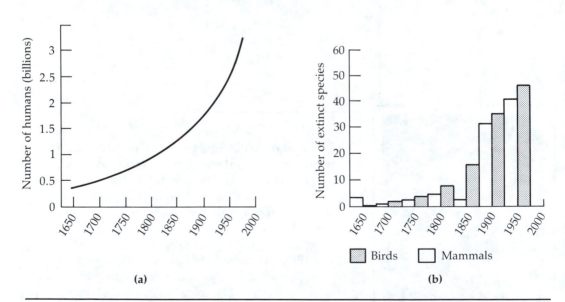

(a)   (b)

Figure S7.1 *Population growth and animal extinctions.* **(a)** *Geometric increase in the
human population.* **(b)** *Increasing numbers of extinctions in birds and mammals.* These
figures suggest that, as human numbers increase, more and more living species
are exterminated. (Redrawn from Goudie 1986.)

Table S7.1   *Dates of major episodes of generic extinction in the late Pleistocene, probably due to human influence. (From data of Martin 1967.)*

Area	Date (Years Before Present)
North America	11,000
South America	10,000
West Indies	Mid-Post Glacial
Australia	13,000
New Zealand	900
Madagascar	800
Northern Eurasia	13,000–11,000
Africa and Southeast Asia (?)	50,000–40,000

by Keyfitz [1975] as ideal for the planet in cultural terms). Our aims should be to reduce population growth so that organic agriculture does not rely so heavily on fossil fuels, to direct resources from weapon production to a more life-enhancing end, and to avoid all waste of energy until the large potential available from solar or fusion sources becomes omnipresent and cheap. There is now one soldier for every 43 people and one doctor for every 1,030; 40 percent of our research expenditure goes to defense, when there are already enough atomic weapons to kill everyone on Earth 67 times over (Miller 1988). It takes the world's military only two and a half hours to spend the entire annual budget of the World Health Organization and two days to spend the cost of the UN plan to conquer desertification. More money should be channeled into conservation of natural habitats. The World Bank's new wildlands policy may help; because of it, biological diversity in developing nations may stand a better chance of being preserved (Goodland 1987). In 1987 World Bank president Barber Conable announced sweeping reorganizations, including the addition of about 40 new environmental positions (Holden 1988). One of the most important additions to the staff was Herman Daly, a rare breed of economist who advocates the concept of "sustainability" rather than the pure mining of biological reserves. The World Bank's policy is now not to fund projects that would convert the most important wildlands. It should encourage development on already degraded lands. If conversion of ecologically valuable lands is justified, then equivalent lands within the same country must be protected. In reality, deforestation was still rampant in World Bank-financed projects in the Amazon after the policy change (Fearnside 1987). Highway development leads to deforestation and low-diversity cattle pastures—the cheapest way to secure land claims. Profits from land sales are increased by road development and by titling. This is a positive-feedback mechanism. Previous commitments to preserving natural habitats and tribal areas are frequently reneged on. Environmental measures are often merely symbolic actions serving to tranquilize public concern during the key period when the development is not yet a *fait accompli* (Fearnside 1987).

It is not widely appreciated that, even in hard economic terms, tropical forests may be worth more standing than logged. Peters, Gentry, and Mendelsohn (1989) presented data on inventory, production, and current market value of all the commercial tree species found in one hectare of species-rich Amazonian forest near Iquitos, Peru. Of the total, 350 individuals (41.6 percent) of 72 tree species (26.2 percent) yielded products with a market value in Iquitos. Edible fruits were produced by 11 species, 1 species produced rubber, and 60 species produced commercial lumber. In addition, many medicinal plants were to be found, but these were not included in the sample. Based on current market prices in Iquitos, the total value of the fruit and rubber was U.S. $698 annually, $422 net. The value of the timber, after logging and transportation, was $1,000 total. Using a simple discounting procedure, Peters, Gentry, and Mendelsohn calculated the net current value of the forest, for sustainable yield of fruit and rubber year after year, to be $6,330 per ha, over six times the value of the timber. The value of the forest in its unchanged state was also over twice the worth of an equal area of fully stocked cattle pasture, figured at $2,960 annually by Peters, Gentry, and Mendelsohn, even before deduction of the costs of weeding, fencing, and animal care, which would lower the value of the pasture substantially. Unfortunately, tropical timber generates foreign exchange, whereas nonwood resources are collected and sold in local markets and are therefore hard to track in national accounting schemes. Even under the World Bank's new policy, worldwide aid to preserve forests will not be spent on protecting forests from timber companies; it will be spent on forestry (Pearce 1989). The World Bank believes, perhaps correctly, that only by taking charge of tropical forests and allowing companies to make money out of them can the world save the rain forests. With the exception of reserves set aside for unique biological features, the forests must cease to be a common resource. They must be fenced, owned, and managed for profit if they are to survive. Such a strategy has worked to save the elephants in southern Africa, and herds are increasing (see Chapter 20). A total ban on logging may lead to uncontrolled destruction of forests by farmers, ranchers, and loggers alike. In the Philippines, on islands where logging is barred, like Mindanao and Marinduque, rates of deforestation have been higher (3–5 percent per year) than in areas that are managed for timber (0.5–2 percent) (Pearce 1989).

The key ecosystems for preservation have been outlined as, first, the tropical forests and, second, aquatic habitats such as estuaries and wetlands. More than half the vertebrate species threatened with extinction are concentrated in 10 areas: tropical forests of southeast Asia, Madagascar, and South America and freshwater habitats of North America and Mexico, western and central Africa and South Africa, islands in the Caribbean, western Indian Ocean (for example, Mauritius, Reunion, Seychelles), and South Pacific (especially New Caledonia), and Hawaii (Allen 1980). Ultimately, preservation of habitats will have to proceed on a finer scale than this one, but recognizing that habitat preservation is critical is a step in the right direction. Often it is a better protective measure to preserve habitat than simply to protect the species within it by law. The British Wildlife and Countryside Act of 1981 placed many species on the endangered list, including sand lizards and smooth snakes, but both are less numerous today than they were then, because destruction of their heathland habitat has gone on unabated.

A method that has been proposed for preserving tropical forests is the "debt for nature" swap, advanced by the Goldsmith brothers, one a financier, the other a rabid environmentalist. This idea, supported vigorously by the World Wildlife Fund in the United States, suggests that poor nations be relieved of part of their foreign debt in return for promoting conservation in their own countries. So far, the deals struck have been small and ad hoc. For example, in 1989, on the initiative of Thomas Lovejoy of the World Wildlife Fund, Costa Rica agreed to spend $17 million on its Guanacaste National Park in return for being relieved of $24 million in debt, but many in the Third World have reacted angrily to these ideas, seeing them as an attack on their national sovereignty, an attempt to annex the rain forests. They argue that the debts were illegitimately incurred in the 1970's, when banks were desperate to lend their huge reserves of petrodollars.

Many experts believe that current rates of deforestation in the tropics could cause the extinction of as many as a million species within the next half century, many of them before they have even been named and described (Lovejoy 1979; Myers 1979 and 1980; Ehrlich and Ehrlich 1981). This trend is disturbing, and something must be done about it. A start has been made in ending the trade in endangered species in the form of the Convention on International Trade in Endangered Species of Wild Fauna and Flora (CITES). This is a young organization formed in Washington, D.C., in 1973 by 10 founding nations. Membership had increased to over 60 by 1986, including the United States. Despite good intentions, the enforcement of existing policies remains the most crucial issue. In January 1979 a Hong Kong magistrate imposed the maximum fine of $1,000 on the Hong Kong Fur Factor Ltd. for smuggling 319 cheetah skins from Ethiopia. Market value of the furs? Officially $40,000; unofficially $1,000,000!

Perhaps the tide of public opinion about conservation and the degradation of the environment is turning. In one survey, conducted by National Wildlife Magazine in 1987, 91 percent of the respondents said they would rather pay higher taxes than have the government cut back on important pollution clean-up programs. This is the type of caring attitude that is necessary to preserve the Earth's biotic diversity in perpetuity. Of course, this figure comes from a biased readership. However, surprisingly perhaps, studies of the U.S. public at large have repeatedly demonstrated that, in various socioeconomic conflict situations, most people are willing to forego diverse social benefits to protect certain endangered species (Kellert 1985). For example, a majority of citizens said they would accept substantial increases in energy development costs to protect such endangered or threatened species as the bald eagle, mountain lion, Agassiz trout, American crocodile, and even silverspot butterfly. Other species like snakes, spiders, and obscure plants fared less well.

S7.1 *Population Reduction*

Ultimately, some form of birth reduction is the best answer to our ecological problems. It has been argued that 20 Somali do not require as much energy to sustain them as does one Swiss (Allen 1980). Although this was true at the time of the study, it may not always be that way in the future. There is no denying that all people want to better themselves, and doing so is generally equated with acquiring more material goods. Some people might argue that for Asians to have a right to outnumber Americans 20 to 1, none of them should aspire to owning a large house or a car. It seems much more likely, however, that people from LDCs will want to drive cars and possess western goods. Therefore, birth reduction should be seriously considered. This solution seems to be the only humanitarian one to the problem of overpopulation, although how we should go about it in western nations, short of giving economic incentives to small families or of promoting complete social and religious reforms, is unclear. Population control, after all, flies in the face of the selfish genes that are in all of us, and it is also contrary to many religious and political beliefs. According to some communists and a few western economists, the world has sufficient resources to eliminate hunger as long as they are shared equally. Luckily, most growth rates in the MDCs have stabilized, and it is the LDCs that contribute most to the problem. Surveys show billions of human beings there are being denied access to family-planning services, despite a discrepancy between desired and actual family size (Eckholm 1976). Most women in developing countries would like fewer children than they already have. Even the desired family size, however, is far above replacement level in many developing countries (Fig. S7.2).

India spent over $200 million on birth control in 1982. Clinics were

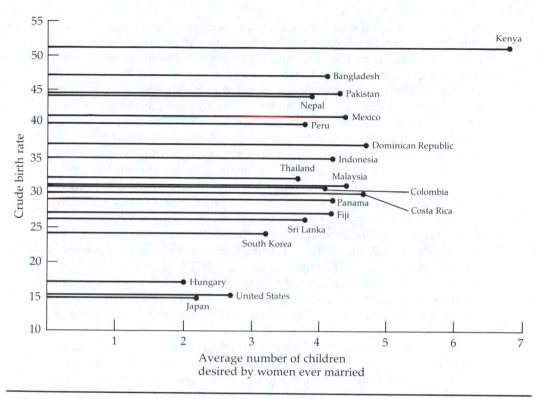

Figure S7.2   *Average desired family size for selected countries.* (Redrawn from Clapham 1981.)

established, and information and birth-control devices were distributed free of charge or at subsidized prices. The government estimated that 37 million births were averted in the 1970's as a result of family planning (Rubenstein and Bacon 1983). In China, the government has decreed that men may not marry until they are age 27 and women until they are 25. Delay in marriage can greatly reduce rates of natural increase. One can easily verify this phenomenon with simple mathematics—over 100 years, a lineage that produces 2 children every 20 years produces 32 offspring, compared to only 27 offspring from parents that produce 3 children but only every 33 years. Population growth in China declined from approximately 2.09 percent per annum in the 1950's to 1.2 percent in 1981—in addition to delay in marriage, people also receive free contraceptives, abortions, and sterilizations. Unfortunately, recent data showed an upsurge in the birth rate to 2 percent in 1986, which was maintained in 1987. When the 1987 figures came out, the family-planning minister, Wang Wei, was sacked. In Peking, couples who pledge to have only one child receive a certificate guaranteeing the child education, medical care, and a

job. All families in the city are allocated housing designed to accommo-
date four people. In addition, families with more than two children must
pay a 10 percent tax on income, and they receive no promotions at work
until the youngest child is 14 years old (Rubenstein and Bacon 1983).
Such measures seem on the face of things very harsh—better perhaps to
emphasize voluntary reduction in birth rates and the advantages it may
bring. Nevertheless, prosperous peasants now have enough money to
pay the government fines for breaking the one-child limit. Fines generally
amount to about $270 per year for five years, a not-prohibitive amount if
several family members contribute. The state of Kerala in India has a per-
capita income below the average for all of India, yet it devotes 39 percent
of its budget to education and 16 percent to health and family-planning
services. Birth rates have dropped from 37 per thousand in 1966 to 25 per
thousand in 1978. The latter figure was the goal of the government in the
mid 1970's, and Prime Minister Gandhi instituted virtually mandatory
sterilization for men to achieve it (Gulhati 1977). In early 1977, Mrs.
Gandhi's government fell, and sterilizations in 1978 fell to 11 percent of
the previous year's total. Future aid to LDCs might take the form of providing birth-control devices, and the services of trained medical staff and
sociologists to teach the correct use of them.

Setting up failproof social security programs should also help reduce
population, because many people raise children to provide security for
themselves in old age. If the state provides good care for the elderly, then
large families will not be necessary for such purposes.

S7.2 *The Economic Viewpoint*

Of course, many economists view the present system as a cornucopia; to
them, things have never been better. The population is growing and with
it the economy. Although this may be true economically for some, such
conclusions are not based on sound ecological principles. Surprisingly,
the words *economics* and *ecology* come from the same Greek root, *oikos*,
economics literally meaning household management. Economics is most
usually based on increasing consumption by a growing population.
Straight away there is a difference of opinion with ecological thinking. In
most economies, economic growth is equated with progress, and **average
GNPs** (the per-capita average of the gross national product) loom large in
equations involving standards of living, but these are a poor measure of
the quality of life. High GNPs may promote more automobiles, and
hence more accidents, injuries, smog, congestion, and pollution. Low
GNPs in LDCs may still allow a good standard of living if the prices of
firewood, food, and housing are low. Some economists even believe that
such environmental problems as pollution will largely tend to be solved
purely on a cost-effective scheme (see Fig. S7.3)—the harmful effects of

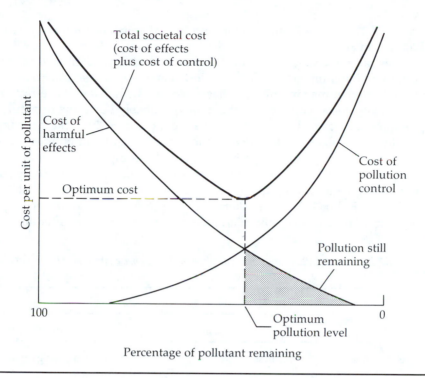

Figure S7.3 *The ineffectiveness of economic solutions in removing pollution from the environment.* Cost-effectiveness analysis involves comparing the costs from the harmful effects of pollution with the costs of pollution control; the object is to minimize the total costs of pollution control while still reducing pollution to an acceptable level. The shaded area shows that some harmful effects remain, but removing these residual damages would make the costs of pollution control too high.

pollution will reduce economic efficiency and will have to be cleaned up or controlled. This type of a model ensures that some pollution will always remain (the shaded area in Fig. S7.3), because it will be uneconomical to control (an ecologically disappointing situation). More controversy arises when other costs and benefits have to be plugged into economic models. An ecologist would probably put a high value on wilderness, wild animals, and beautiful flowers. To an economist, they might have little value. Miller (1988) cites the case of an oil company representative in Montana who, when protesting clean-air standards for his company, argued that some of the people likely to be affected by or die from air pollution would be unemployed and therefore of no economic value. A typical tree has been estimated to be worth $590 in timber, but in a typical 50 year life span it provides $31,250 worth of oxygen, $62,500 in air-pollution reduction, $31,250 in soil fertility and erosion control, $37,500 in water recycling, $31,250 in wildlife habitat, and $2,500 worth of protein—nearly $200,000 in all.

Despite the undercurrent of doom and despair present in this section, there is definitely hope for mankind in an ecological sense. To begin with, though human interference has undoubtedly had a severe effect on many ecosystems, changes in others have probably been due to natural phenomena, particularly climatic change (Table S7.2), and humans should not be blamed. Deciphering the exact cause of a phenomenon is often an extremely difficult problem, given the complexity of ecological changes and lack of baseline data. In an applied situation, few ecologists start taking data until a problem has become well established. Finally, Simon and Kahn (1984) have taken a fairly optimistic view of the ecological future. There is no doubt that much bad news is phony; after all, ex-

Table S7.2 *Human influence or natural phenomena? Some examples. (From Goudie 1986.)*

Change	Causes	
	Natural	*Anthropogenous*
Late Pleistocene animal extinction	Climate	Overhunting by humans
Death of savannah trees	Soil salinization through climatically induced groundwater rise	Overgrazing
Desertification in semi-arid areas	Climatic change	Overgrazing, collection of trees for firewood
Holocene peat-bog development in highland Britain	Climatic change and progressive soil deterioration	Deforestation and ploughing
Holocene elm and linden decline	Climatic change	Feeding and stalling of animals
Tree encroachment into alpine pastures in United States	Temperature amelioration	Cessation of burning
Gully development	Climatic change	Land-use change, deforestation
Early 20th century climatic warming	Changes in solar emission and volcanic activity	CO_2-generated greenhouse effect
Increasing coast recession	Rising sea level	Disruption of sediment supply
Increasing coastal flood risk	Rising sea level, natural subsidence	Pumping of aquifers creating subsidence
Increasing river flood intensity	Higher-intensity rainfall	Creation of drainage ditches, lessening of vegetation cover
Ground collapse	Karstic process	Dewatering by overpumping

aggerated, inflammatory, and generally negative press is news, good news is not (Simon 1980). For example, Avery (1985, p. 408) has written that "the world is not on the brink of famine or ecological disaster brought on by desperate food needs." World agricultural production in the 1980's was at an all-time high and is climbing fast, especially in the developing countries. Even Africa has ample land and technology to feed its population, given more effective national policies. Higher agricultural output in the world, Avery argues, has been stimulated dramatically by new technology (but see also Wiles 1986; Kellogg 1986; and Pimentel and Pimentel 1986, who attack this article on a variety of grounds).

Simon (1986) has shown how estimates of disappearing species and deforestation are based on the flimsiest of data. Wild extrapolations from these data ensue, and the resulting predictions are usually flatly wrong. Simon and Kahn (1985) predicted a world in the year 2000 less crowded (though more populated), less polluted, more stable ecologically, less vulnerable to resource-supply disruption, less precarious economically, and with better food—in short, with a population much richer in most ways than they are today. Their argument is largely based on the technological wizardry of mankind, which they claim will be able to pull us out of any environmental fix. No oil, no problem; we'll simply invent something else. No whales; we'll use synthetic materials. Of course, this attitude views whales as being useful strictly on an economic basis. It should be obvious by now that economics is a useful tool but a poor master. Why should we have to destroy the whales before finding a technological fix? The pro's and con's of these arguments have been presented in a book by Southwick (1985). Anne and Paul Ehrlich (Ehrlich 1985; Ehrlich and Ehrlich 1987) have concluded that Simon's policies are based on faulty logic and that there cannot possibly be an inexhaustible supply of resources on a finite Earth (see also Daly 1982). The year 2000 looms close on the horizon, and the truth of these predictions can then be assessed.

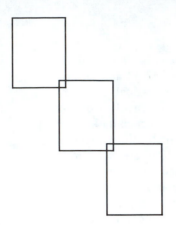

Glossary

Abiotic factors: environmental influences produced other than by living organisms, for example temperature, humidity, pH, and other physical and chemical influences; contrast with "biotic factors."

Acclimation: changes by an organism, often biochemical, subjected to new environmental conditions that enable it to withstand those conditions.

Acid rain: rainfall acidified by contact with sulphur dioxides (a by-product of the burning of fossil fuels) in the atmosphere.

Acquired character: a character not inherited but acquired by an individual organism during its lifetime.

Adaptation: a change in an organism's structure or habits that produces better adjustment to the environment; a genetically determined characteristic that enhances the ability of an organism to cope with its environment.

Adaptive radiation: evolutionary diversification of species derived from a common ancestor into a variety of ecological roles.

Aerobic: pertaining to organisms or processes that require the presence of oxygen.

Aggressive mimicry: resemblance of predators or parasites to harmless species that causes potential prey or hosts to ignore them.

Agricultural pollution: contamination of the environment with liquid and solid wastes from all types of farming, including pesticides, fertilizers, runoff from feedlots, erosion and dust from plowing, animal manure and carcasses, and crop residues and debris.

Air pollution: the presence of contaminants in the air in concentrations that overcome the normal dispersive ability of the air and that interfere directly or indirectly with human health, safety, or comfort or with the full use and enjoyment of property.

Algal bloom: a proliferation of living algae in a lake, stream, or pond.

Allele: one of two or more alternative forms of a gene located at a single point (locus) on a chromosome.

Allelopathy: the negative chemical influence of plants, exclusive of microorganisms, upon one another.

Allen's (1878) rule: among homeotherms, the tendency for limbs and extremities to become shorter and more compact in colder climates than in warmer ones.

Allochronic speciation: separation of a population into two or more evolutionary units as a result of reproductive isolation arising from a difference in mating times.

Allopatric speciation: separation of a population into two or more evolutionary units as a result of reproductive isolation arising from geographic separation.

Altruism: in an evolutionary sense, enhancement of the fitness of an unrelated individual by acts that reduce the evolutionary fitness of the altruistic individual.

Anadromous: living in salt water but breeding in fresh water.

Anaerobic: pertaining to organisms or processes that occur in the absence of oxygen.

Apatetic coloration: coloration of an animal that causes it to resemble physical features of the habitat.

Apomixis: parthenogenetic reproduction in which offspring develop from unfertilized eggs or somatic cells.

Aposematism: conspicuous appearance of an organism warning that it is noxious or distasteful; warning coloration.

Apostatic selection: selective predation on the most abundant of two or more forms in a population, leading to balanced polymorphism (the stable occurrence of more than one form in a population).

Apterous: wingless.

Aquaculture: farming of aquatic or marine systems; rearing of organisms such as fish, algae, or shellfish under controlled conditions.

Aquifer: a layer of rock, sand, or gravel through which water can pass; an underground bed or stratum of earth, gravel, or porous stone that contains water; the place in the ground where groundwater is naturally stored.

Assimilation efficiency: the percentage of energy ingested in food that is assimilated into the protoplasm of an organism.

Association: a group of species occurring in the same place.

Assortative mating: nonrandom mating; the propensity to mate with others of like phenotypes.

Autecology: study of the individual in relation to environmental conditions; contrast ''synecology.''

Autogamous: able to produce offspring sexually by the fusion of gametes from the same individual, for example by fusion of pollen and ovules from the same plant.

Autosome: a chromosome other than a sex chromosome.

Autotroph: an organism that obtains energy from the sun and materials from inorganic sources; contrast with ''heterotroph.''

Average GNP: the gross national product of a country divided by its total population.

Balanced polymorphism: the stable occurrence of more than one form in a population.

Barrier island: a narrow, elongated, sandy island paralleling the coast, separated from the mainland by a bay or lagoon.

Batesian mimicry: resemblance of an edible (mimic) species to an unpalatable (model) species to deceive predators.

Benthic: pertaining to aquatic bottom or sediment habitats.

Benthos: bottom-dwelling aquatic organisms (for example, burrowing worms, molluscs, and sponges).

Bergmann's (1847) rule: among homeotherms, the tendency for organisms in colder climates to have larger body size (and thus smaller surface-to-volume ratio) than those in warm climates.

Bioassay: the use of living organisms to measure the biological effect of some substance, factor, or condition.

Biochemical oxygen demand (BOD): the amount of oxygen that would be consumed if all the organic substances in a given volume of water were oxidized by bacteria and other organisms; reported in milligrams per liter.

Biodegradable: capable of being decomposed quickly by the action of microorganisms.

Biogeochemical cycle: the passage of a chemical element (such as nitrogen, carbon, or sulphur) from the environment into organic substances and back into the environment.

Biogeography: the branch of biology that deals with the geographic distribution of plants and animals.

Biological control: use of natural enemies (diseases, parasites, predators) to regulate populations of pest species.

Biological magnification: the concentration of a substance as it ''moves up'' the food chain from consumer to consumer.

Biomass: dry weight of living material in all or part of an organism, population, or community; commonly expressed as weight per unit area, biomass density.

Biome: a major terrestrial climax community; a major ecological zone or region corresponding to a climatic zone or region; a major community of plants and animals associated with a stable environmental life zone or region (for example, northern coniferous forest, Great Plains, tundra).

Biosphere: the whole Earth ecosystem.

Biota: all the living organisms occurring within a certain area or region.

Biotic factors: environmental influences caused by living organisms; contrast with ''abiotic factors.''

Boreal: occurring in the temperate and subtemperate zones of the Northern Hemisphere.

Breeder reactor: a nuclear reactor that produces more fissionable material than it consumes.

Burner reactor: a nuclear reactor that consumes more fissionable material than it produces.

Calcareous: in soil terminology, rich in calcium carbonate and having a basic reaction.

Canonical distribution: a lognormal distribution of the numbers of individuals or species according to the mathematical formulation of Preston (1962).

Carbon-14: a radioactive isotope of carbon (atomic weight 14) that can be used for dating organic materials.

Carcinogen: a chemical or physical agent capable of causing cancer.

Carnivore: an animal (or plant) that eats other animals; contrast with ''herbivore.''

Carrying capacity: the amount of animal or plant life (or industry) that can be supported indefinitely on available resources; the number of individuals that the resources of a habitat can support.

Catadromous: living in fresh water but breeding in sea water.

Character displacement: divergence in the characteristics of two otherwise similar species where their ranges overlap, caused by selective effects of competition between the species in the area of overlap.

Character divergence: evolution of differences between similar species occurring in the same area.

Chimera: a piece of DNA incorporating genes from two different species; an organism whose cells are not all genetically alike.

China syndrome: a popular term for the consequence of core meltdown in a nuclear reactor in which a molten mass of intensely radioactive material plummets through vessel and containment and into the Earth beneath, in the direction (from the Western Hemisphere) of China.

Clade: the set of species descended from a particular ancestral species.

Clearcutting: the practice of cutting all trees in an area, regardless of species, size, quality, or age.

Cleistogamy: self-pollination within a flower that does not open.

Climax community: the community capable of indefinite self-perpetuation under given climatic and edaphic conditions.

Cline: a gradient of change in population characteristics over a geographic area, usually related to a corresponding environmental gradient.

Clone: a lineage of individuals reproduced asexually.

Coadaptation: evolution of characters of two or more species to their mutual advantage.

Coevolution: development of genetically determined traits in two species to facilitate some interaction, usually mutually beneficial.

Coexistence: occurrence of two or more species in the same habitat; usually applied to potentially competing species.

Cohort: those members of a population that are of the same age, usually in years or generations.

Coliform index: a measure of the purity of water based on a count of the coliform bacteria it contains.

Commensalism: an association between two organisms in which one benefits and the other is not affected.

Community: a group of populations of plants and animals in a given place; used in a broad sense to refer to ecological units of various sizes and degrees of integration.

Competition: the interaction that occurs when organisms of the same or different species use a common resource that is in short supply (''exploitation'' competition) or when they harm one another in seeking a common resource (''interference'' competition).

Competitive exclusion principle: the hypothesis, based on theoretical considerations and laboratory experiments, that two or more species cannot coexist and use a single resource that is scarce relative to demand for it.

Conspecific: belonging to the same species.

Consumer: an organism that obtains its energy from the organic materials of other organisms, living or dead; contrast with ''producer.''

Continental drift: the movement of the continents, by tectonic processes, from their original positions as parts of a common land mass to their present locations.

Continental island: an island that is near to and geologically part of a continent, for example the British Isles or Trinidad.

Continental shelf: the shallow part of the sea floor immediately adjacent to a continent.

Convergent evolution: the development of similar adaptations by genetically unrelated species, usually under the influence of similar environmental conditions.

Coprolite: fossil excrement.

Core: the heart of a nuclear reactor where energy is released; the region of a reactor containing fuel (and moderator, if any) and within which the fission reaction occurs.

Courtship: any behavioral interaction between individuals of opposite sexes that facilitates mating.

Crop rotation: the farming practice of planting the same field with a different crop each year to prevent nutrient depletion.

Crypsis: coloration or appearance that tends to prevent detection of an organism by others, especially predators.

Cultural change: any modification of characteristics specific to a population that is transmitted by learning rather than by genetic mechanisms.

DDT: 1,1,1-trichloro-2,2-bis(p-chloriphenyl) ethane; the first of the modern chlorinated-hydrocarbon insecticides.

Decomposers: consumers, especially microbial consumers, that change their organic food into mineral nutrients.

Deforestation: removal of trees from an area without adequate replanting.

Deme: a local population, usually small and panmictic.

Denitrification: enzymatic reduction by bacteria of nitrates to nitrogen gas.

Density: the number of individuals per unit area.

Density dependent: having an influence on individuals that varies with the number of individuals per unit area in the population.

Density independent: having an influence on individuals that does not vary with the number of individuals per unit area in the population.

Desalinization: the process of removing salt from water.

Desert: a region receiving very small amounts of precipitation or where (for example, ice caps) the moisture present is unavailable to vegetation.

Deterministic model: a mathematical model in which all relationships are fixed and stochastic processes play no part; contrast "stochastic model."

Diapause: a period of suspended growth or development and reduced metabolism in the life cycle of many insects, during which the organism is more resistant to unfavorable environmental conditions than during other periods.

Dimorphism: the occurrence of two forms of individuals in a population.

Dioecious: characterized by individuals each of which has only male or only female reproductive organs.

Diploid: of cells or organisms, having two sets of chromosomes.

Direct competition: exclusion of individuals from resources, by other individuals, by aggressive behavior or use of toxins.

Dispersal: movement of organisms away from the place of birth or from centers of population density.

Disruptive selection: selection against individuals in a population that have intermediate values of a trait, leading to the divergence of subpopulations with extreme values of the trait.

Distribution: the area or areas (taken together) where a species lives and reproduces.

Diversity: the number of species in a community or region; alpha diversity is the diversity of a particular habitat; beta diversity is diversity of a region pooled across habitats.

Dominance: the influence or control exerted by one or more species in a community as a result of their greater number, coverage, or size.

Ecologic: pertaining to the living environment.

Ecologic efficiency: the percentage of energy in biomass produced by one trophic level that is incorporated into biomass by the next highest trophic level.

Ecological impact: the total effect of an environmental change, whether natural or man-made.

Ecological release: the expansion of habitat or food use by populations in regions of low species diversity, permitted by reduced interspecific competition.

Ecology: the branch of science dealing with the relationships of living things to one another and to their environment.

Ecosystem: a biotic community and its abiotic environment.

Ecotone: the transition zone between two diverse communities.

Ecotype: a subspecies or race that is specially adapted to a particular set of environmental conditions.

Ecumene: the portion of the Earth's surface occupied by permanent human settlement.

Edaphic: pertaining to soil.

Emigration: the movement of organisms out of a population.

Endangered species: a species with so few living members that it will soon become extinct unless measures are taken to slow its loss.

Endemic: an organism that is native to a particular region.

Endogenous: produced from within; originating from or due to internal causes; contrast "exogenous."

Energy resource: a natural supply of energy available for use, for example the Earth's internal heat, fossil fuels, hydropower, nuclear energy, solar energy, and wind.

Enrichment: the addition of nutrients to an ecosystem, for example the addition of nitrogen to waterways by agricultural runoff.

Environment: all the biotic and abiotic factors that affect an individual organism at any one point in its life cycle.

Epidemiology: the study of disease in populations or groups.

Epilimnion: the upper layer of water in a lake, usually warm and containing high levels of dissolved oxygen.

Epiphyte: a plant that lives on another plant but uses it only for support, drawing its water and nutrients from natural runoff and the air.

Epistasis: a synergistic effect whereby the effect of two or more gene loci is greater than the sum of their individual effects.

Equilibrium: a condition of balance, such as that between immigration and emigration or birth rates and death rates in a population of fixed size.

Euphotic zone: that part of the water column that receives sufficient sunlight to support photosynthesis; usually limited to the upper 60 m.

Eutrophication: the normally slow aging process by which a lake fills with organic matter, evolves into a bog or marsh, and ultimately disappears.

Evolution: the gradual accumulation of genetic change that is thought to have given rise, beginning with common ancestors, to the diversity of life.

Exponential growth: the steepest phase in a growth curve, that in which the curve is described by an equation containing a mathematical exponent.

Exponential rate of increase: the rate at which a population is growing at a particular instant expressed as a proportional increase per unit time.

Extant: of a species, currently represented by living individuals.

Extinct: of a species, no longer represented by living individuals.

Fecundity: the potential of an organism to produce living offspring.

Feral: having reverted from domestication to the wild but remaining distinct from other wild species.

Fission: splitting or division; nuclear fission is the splitting of the nuclei of the atoms of certain elements into lighter nuclei and is accompanied by the release of relatively large amounts of energy.

Fixation: attainment by an allele of a frequency of 1 (100 percent) in a population, which in effect becomes monomorphic for that allele.

Food chain: figure of speech describing the dependence for food of organisms upon others, in a series beginning with plants and ending with the largest carnivores.

Forest: a region that, because it receives sufficient average annual precipitation (usually 75 cm [30 inches] or more), supports trees and small vegetation.

Fossil fuels: coal, oil, and natural gas, so-called because they are derived from the fossil remains of ancient plant and animal life.

Fossorial: living in burrows.

Founder effect: the principle that a population started by a small number of colonists will contain only a fraction of the genetic variation of the parent population.

Functional response: a change in the rate of exploitation of prey by an individual predator resulting from a change in prey density (see also ''numerical response'').

Fusion: the combination of two atoms into a single atom as a result of a collision, usually accompanied by the release of energy.

Gene: a unit of genetic information.

Gene flow: the exchange of genetic traits between populations by movement of individuals, gametes, or spores.

Genetic drift: change in gene frequency caused solely by chance, usually unidirectional and more important in small populations.

Genome: the entire genetic complement of an individual.

Genotype: the genetic constitution of an organism or a species in contrast to its observable characteristics; contrast ''phenotype.''

Genus: the taxonomic category above the species and below the family; a group of species believed to have descended from a common direct ancestor.

Geometric rate of increase: the factor by which the size of a population changes over a specified period; contrast ''exponential rate of increase.''

Glacial epoch: the Pleistocene epoch, the earlier of the two epochs comprising the Quaternary period, characterized by the extensive glaciation of regions now free from ice.

Global stability: ability to withstand perturbations of a large magnitude without being affected; contrast "local stability."

Grassland: a region with sufficient average annual precipitation (25–75 cm [10–30 inches]) to support grass but not trees.

Green belts: areas from which buildings and houses are excluded, often serving as buffers between pollution sources and concentrations of population.

Greenhouse effect: the heating effect of the atmosphere upon the Earth, particularly as CO_2 concentration rises, caused by its ready admission of light waves but its slower release of the heat they generate on striking the ground.

Gross production: production before respiration losses are subtracted; photosynthetic production for plants and metabolizable production for animals.

Group selection: elimination of a group of individuals with a detrimental genetic trait, caused by competition with other groups lacking the trait.

Guyot: a flat-topped submarine volcano.

Habitat: the sum of the environmental conditions where an organism, population, or community lives; the place where an organism normally lives; the environment in which the life needs of an organism are supplied.

Half-life: the time it takes for one-half the atoms of a radioactive isotope to decay into another isotope; the time it takes certain materials, such as persistent pesticides, to lose half their strength.

Haplodiploidy: the presence of haploid males and diploid females in the same species, for example in the Hymenoptera.

Haploid: containing one set of chromosomes.

Harem: a group of females controlled by one male.

Herbivore: an organism that eats plants; contrast "carnivore."

Heredity: genetic transmission of traits from parents to offspring.

Heterotroph: an organism that obtains energy and materials from other organisms; contrast "autotroph."

Hierarchy: a rank order; the pecking order, leadership, or dominance patterns among the members of a population.

Holotype: the single specimen chosen by the original author of a species as the archetypical example of that species and which any revised description of the species must include; contrast "lectotype" and "neotype," terms for such an archetypical specimen when it is chosen, not by the original author, but by a later author in the absence of a holotype.

Home range: the area in which an individual member of a population roams and carries on all of its activities.

Horizon: in a soil, a major stratification or zone, having particular structural and chemical characteristics.

Host: the organism that furnishes food, shelter, or other benefits to an organism of another species.

Hybridization: breeding (crossing) of individuals from genetically different strains, populations, or sometimes species.

Hypolimnion: the layer of cold, dense water at the bottom of a lake.

Immigration: the movement of individuals into a population.

Inbreeding: a mating system in which adults mate with relatives more often than would be expected by chance.

Inclusive fitness: The total genetic contribution of an individual by way of its sons, daughters, and all other relatives combined.

Independent assortment: the separate inheritances, without mutual influence, of genes occurring on different chromosomes.

Indirect competition: exploitation of a resource by one individual that reduces the availability of that resource to others.

Innate capacity for increase (r): measure of the rate of increase of a population under controlled conditions (also referred to as intrinsic rate of increase).

Interdemic selection: group selection of populations within a species.

Interspecific: between species; between individuals of different species.

Intraspecific: within species; between individuals of the same species.

Isolating mechanism: any condition, for example a genetically determined difference or a mechanical or geographic separation, that prevents gene flow between two populations.

Iterative evolution: the repeated evolution of similar phenotypic characteristics at different times during the history of a clade.

Kin selection: a form of genic selection in which alleles differ in their rate of propagation because they influence the survival of kin who carry the same alleles.

LDC: less-developed country, typically with low GNP, high population growth, low literacy, and low industrialization.

Landfill: a waste-disposal site in which layers of solid waste are laid down in alternation with layers of soil.

Leaching: the process by which soluble materials in the soil, such as nutrients, pesticides, or contaminants, are washed into a lower layer of soil or are dissolved and carried away by water.

Lek: a communal courtship area on which several males hold courtship territories to attract and mate with females; sometimes called an arena.

Lentic: pertaining to standing freshwater habitats (ponds and lakes).

Life table: tabulation presenting complete data on the mortality schedule of a population.

Ligase: an enzyme that joins DNA together.

Limiting resource: the nutrient or substance that is in shortest supply in relation to organisms' demand for it.

Linkage: occurrence of two loci on the same chromosome; functional linkage occurs when two loci do not segregate independently at meiosis.

Locus: the site on a chromosome occupied by a specific gene.

Logistic equation: a model of population growth described by a symmetrical S-shaped curve with an upper asymptote.

Lognormal distribution: a frequency distribution of species abundance in which abundance, on the x axis, is expressed on a logarithmic scale.

Lotic: pertaining to running freshwater habitats (streams and rivers).

MDC: more-developed country, typically with high GNP, low population growth, high literacy, and a strong economy.

Malthusian theory of population: the theory of English economist and religious leader Thomas Malthus that populations increase geometrically (2, 4, 8, 16) while food supply increases arithmetically (1, 2, 3, 4), leading to the conclusion that humans are doomed to overpopulation, misery, and poverty and that population levels will be reduced by disease, famine, and war.

Meiotic drive: a preponderance (generally a frequency greater than 50 percent) of one allele among the gametes produced by a heterozygote.

Melanism: occurrence of black pigment, usually melanin.

Meltdown: of a reactor core, the consequence of overheating that allows part or all of the solid fuel to reach the temperature at which cladding and possibly fuel and support structure liquify and collapse.

Modifier gene: a gene that alters the phenotypic expression of genes at one or more other loci.

Monoculture: cultivation of a single crop to the exclusion of all other species on a piece of land.

Monoecious: having separate male and female reproductive organs on the same individual, used mainly of plants.

Morph: a specific form, shape, or structure.

Mullerian mimicry: mutual resemblance of two or more conspicuously marked, distasteful species to reinforce predator avoidance.

Multivoltine: having several generations during a single season; contrast "univoltine."

Mutant: an organism with a changed characteristic resulting from a genetic change.

Mutation: a change in the genetic makeup of an organism resulting from a chemical change in its DNA.

Mutualism: an interaction between two species in which both benefit from the association.

Natural selection: the natural process by which the organisms best adapted to their environment survive and those less well adapted are eliminated.

Nektonic: free-swimming in the upper zone of ocean water and strong enough to swim against the current.

Neotenic: exhibiting neoteny, the ability of species to reproduce sexually when still exhibiting juvenile characteristics.

Neritic: pertaining to the shallow, coastal marine zone.

Net production: production after respiration losses are subtracted.

Net production efficiency: percentage of assimilated energy that is incorporated into growth and reproduction.

Net reproductive rate (R): the number of offspring a female can be expected to bear during her lifetime, for species with clearly defined discrete generations.

Niche: the place of an organism in an ecosystem; all the components of the environment with which an organism or population interacts.

Nitrate: a salt of nitric acid; a compound containing the radical (NO_3); biologically, the final form of nitrogen from the oxidation of organic nitrogen compounds.

Nitrogen cycle: the biogeochemical processes that move nitrogen from the atmosphere into and through its various organic chemical forms and back to the atmosphere.

Nitrogen fixing bacteria: bacteria that can reduce atmospheric nitrogen to cell nitrogen.

Nonrenewable resource: a resource available in a fixed amount (such as minerals and oil), not replaceable after use.

Norm of reaction: the set of phenotypic expressions of a genotype under different environmental conditions.

Numerical response: change in the population size of a predatory species as a result of a change in the density of its prey.

Oligotrophic: low in nutrients and organisms; low in productivity.

Omnivore: an organism whose diet includes both plant and animal foods.

Operational sex ratio: ratio of sexually ready males to fertilizable females.

Organic: of biological origin; in chemistry, containing carbon.

Palaeontology: the science that deals with life of past geologic ages, treating fossil remains.

Panmixis: the condition in which mating in a population is entirely random.

Parasite: the organism that benefits in an interspecific interaction in which one organism benefits and the other is harmed.

Parasitoid: a specialized insect parasite that is usually fatal to its host and therefore might be considered a predator rather than a classical parasite.

Particulate matter: in air pollution, solid particles and liquid droplets, as opposed to material uniformly dispersed among the air molecules.

Parthenogenesis: reproduction without fertilization by male gametes, usually involving the formation of diploid eggs whose development is initiated spontaneously.

PCBs: polychlorinated biphenyls, a family of chemicals similar in structure to DDT.

Pelagic: pertaining to the upper layers of the open ocean.

Per-capita rate of population growth (r): rate of population growth per individual, used for species with overlapping, nondiscrete generations.

Permafrost: a permanently frozen layer of soil underlying the Arctic tundra biome.

Persistence: of pesticides, the length of time they remain in the soil or on crops after being applied.

pH: a measure of acidity or alkalinity.

Phenology: study of the periodic (seasonal) phenomena of animal and plant life (for example, flowering time in plants) and their relations to weather and climate.

Phenotype: the physical expression in an organism of the interaction between its genotype and the environment; the outward appearance of an organism.

Phoresy: the transport of one organism by another of a different species.

Photic zone: the surface zone of a body of water that is penetrated by sunlight.

Photoperiodism: seasonal response (for example, flowering, seed germination, reproduction, migration, or diapause) by organisms to change in the length of the daylight period.

Photosynthesis: synthesis, with the aid of chlorophyll and with light as the energy source, of carbohydrates from carbon dioxide and water, with oxygen as a by-product.

Phyletic evolution: genetic changes that occur within an evolutionary line.

Phylogeny: the line, or lines, of direct descent in a given group of organisms; also the study or the history of such relationships.

Phylum: one of the primary divisions of the animal and plant kingdoms; a group of closely related classes of animals or plants.

Phytoplankton: the plant community in marine and freshwater situations, containing many species of algae and diatoms, that floats free in the water.

Plate tectonics: the study of the global-scale movements of the Earth's crust that have resulted in continental drift and the creation of many geological formations.

Pleiotropy: the phenotypic effect of a gene on more than one characteristic.

Pleistocene: a geological epoch, characterized by alternating glacial and interglacial stages, that ended about 10,000 years ago, lasted one-to-two million years, and is subdivided into four glacial stages and three interglacials.

Point source: an individual, stationary source of large-volume pollution, usually industrial in origin.

Pollutant: any natural or artificial substance that enters the ecosystem in such quantities that it does harm to the ecosystem; any introduced substance that makes a resource unfit for a specific purpose.

Polygamy: a mating system in which a male pairs with more than one female at a time (polygyny) or a female pairs with more than one male (polyandry).

Polymorphism: occurrence in a population of more than two different forms, independent of sexual differences.

Population: a group of individuals of a single species.

Primary production: production by autotrophs, normally green plants.

Production: amount of energy (or material) formed by an individual, population, or community in a specific time period; see also "primary production," "secondary production," "gross production," "net production."

Protandry: the condition of an individual that, during the course of its development, changes from male to female.

Proximate factors: the mechanisms responsible for an evolutionary adaptation with reference to its physiological and behavioral operation; the mechanics of how an adaptation operates; contrast "ultimate factors."

Punctuated equilibrium: a model that depicts macroevolution as taking place in the form of short periods of rapid speciation alternating with long periods of relative stasis.

r and *K* selection: alternative expressions of selection on traits that determine fecundity and survivorship to favor rapid population growth at low population density (*r*) or competitive ability at densities near the carrying capacity (*K*).

Recombinant DNA: a single molecule combining DNA from two distinct sources.

Recruitment: addition, by reproduction, of new individuals to a population.

Recycling: the process by which waste materials are transformed into usable products.

Red queen hypothesis: the idea, named for the character in Lewis Carroll's *Through the Looking Glass,* that a species must continually evolve just to keep pace with environmental change and with other species, let alone to get ahead in the coevolutionary struggle.

Resource: a substance or object required by an organism for normal maintenance, growth, or reproduction.

Respiration: the complex series of chemical reactions in all organisms by which stored energy is made available for use and which produces carbon dioxide and water as by-products.

Restriction enzyme: an enzyme that recognizes a specific base sequence on DNA and cuts DNA. Some restriction enzymes cut the DNA at a particular point, others at random.

Riparian: related to, living in, or located on the bank of a natural watercourse, usually a river, sometimes a lake or tidewater.

Runoff: water entering rivers, lakes, reservoirs, or the ocean from land surfaces.

Saprophyte: a plant that obtains food from dead or decaying organic matter.

Secondary production: production by herbivores, carnivores, or detritus feeders; contrast "primary production."

Search image: a behavioral selection mechanism that enables predators to increase searching efficiency for prey that are abundant and worth capturing.

Self-regulation: a process of population regulation in which population increase is prevented by a deterioration in the quality of individuals that make up the population; population regulation by adjustments in behavior and physiology internal to the population rather than by external forces such as predators.

Sere: the series of successional communities leading from bare substrate to the climax community.

Sessile: of animals, attached to an object or fixed in place, as for example barnacles.

Sewage: the organic waste and waste water generated by residential and commercial establishments.

Sewerage: the entire system of sewage collection, treatment, and disposal; all effluent carried by sewers, whether sanitary sewage, industrial wastes, or storm-water runoff.

Sex-linked gene: a gene carried on one of the sex chromosomes (expressible phenotypically in either sex).

Sexual selection: selection by one sex for specific characteristics in individuals of the opposite sex, usually exercised through courtship behavior.

Sibling species: species that are difficult or impossible to distinguish by morphological characters.

Sigmoid curve: an S-shaped curve, for example the logistic curve.

Slash-and-burn cultivation: a primarily tropical practice in which forest vegetation is cut, left to dry, and burned to add nutrients to the soil before the land is planted with crops, then abandoned after two-to-five years as a result of falling yields.

Smog: a term coined from "smoke" and "fog" to describe photochemical air pollution.

Sociobiology: the study of the biological reasons behind animal behavior and sociology.

Species (both singular and plural): organisms forming a natural population or group of populations that transmit specific characteristics from parent to offspring; a group of organisms reproductively isolated from similar organisms and usually producing infertile offspring when crossed with them.

Stability: absence of fluctuations in populations; ability to withstand perturbations without large changes in composition.

Stochastic model: a mathematical model incorporating factors determined by chance and providing not a single prediction but a range of predictions; contrast "deterministic model."

Succession: replacement of one kind of community by another; the progressive changes in vegetation and animal life that tend toward climax.

Supergene: a group of two or more loci between which recombination is so reduced that they are usually inherited together as a single entity.

Symbiosis: in a broad sense, the living together of two or more organisms of different species; in a narrow sense, synonymous with "mutualism."

Sympatric: occurring in the same place.

Sympatric speciation: formation of species without geographic isolation; reproductive isolation that arises between segments of a single population.

Synecology: the study (including population, community, and ecosystem ecology) of groups of organisms in relation to their environment; contrast "autecology."

Synergism: the situation in which the combined effect of two factors is greater than the sum of their separate effects.

Taiga: the northern boreal forest zone; a broad band of coniferous forest south of the Arctic tundra.

Territory: any area defined by one or more individuals and protected against intrusion by others of the same or different species.

Thermocline: the thin transitional zone in a lake that separates the epilimnion from the hypolimnion.

Threatened species: a species not yet endangered but whose population levels are low enough to cause concern.

Timberline: the uppermost altitudinal limit of forest vegetation.

Time lag: delay in response to a change.

Topsoil: the top few inches of soil, rich in organic matter and plant nutrients.

Trophic level: the functional classification of an organism in a community according to its feeding relationships.

Tundra: level or undulating treeless land, characteristic of Arctic regions and high altitudes, having permanently frozen subsoil.

Turnover: rate of replacement of resident species by new, immigrant species.

Ultimate factors: the evolutionary survival values of adaptations; the evolutionary reasons for an adaptation; contrast "proximate factors."

Univoltine: having only one generation per year; contrast "multivoltine."

Upwelling: the process whereby, as a result of wind patterns, nutrient-rich bottom waters rise to the surface of the ocean.

Urban: pertaining to city areas.

Vector: an organism (often an insect) that transmits a pathogen (for example, a virus, bacterium, protozoan, or fungus) acquired from one host to another.

Vicarants: disjunct species, closely related, assumed to have been created when the initial range of their common ancestor was split by some historical event.

Vicariance biogeography: the study of distribution patterns of organisms that attempts to reconstruct events through the study of shared characteristics (cladistics), often with little or no attention to dispersal capabilities or ecological properties.

Watershed: the land area that drains into a particular lake, river, or reservoir.

Wilderness: undisturbed area, as it was before human-made changes.

Wild type: the allele, genotype, or phenotype that is most prevalent in wild populations.

Zoogeography: the study of the distributions of animals.

Zooplankton: the animal community, predominantly single-celled animals, that floats free in marine and freshwater environments, moving passively with the currents.

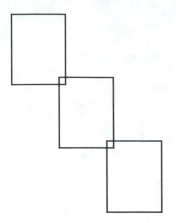

Literature Cited

Abele, L. G., and S. Gilchrist. 1977. Homosexual rape and sexual selection of acanthocephalan worms. *Science* 197:81–83.

Abrahamson, W. G. 1975. Reproductive strategies in dewberries. *Ecology* 56:721–726.

Achiron, M., and R. Wilkinson. 1986. Africa: the last safari? *Newsweek* 108(7):40–42.

Addicott, J. F. 1986. Variation in the costs and benefits of mutualism: the interaction between yuccas and yucca moths. *Oecologia* 70:486–494.

Addicott, J. F., J. M. Aho, M. F. Antolin, D. K. Padilla, J. S. Richardson, and D. A. Soluk. 1987. Ecological neighborhoods: scaling environmental patterns. *Oikos* 49:340–346.

Alcock, J. 1979. *Animal Behavior: An Evolutionary Approach*, 2nd edition. Sinauer Associates, Sunderland, Mass.

Alexander, R. D. 1974. The evolution of social behavior. *Annual Review of Ecology and Systematics* 5:325–383.

Alexander, R. D., and P. W. Sherman. 1977. Local mate competition and parental investment in social insects. *Science* 196:494–500.

Allee, W. C. 1931. *Animal Aggregations. A Study in General Sociology.* University of Chicago Press, Chicago.

Allen, P. H. 1961. Florida longleaf pine fail in Virginia. *Journal of Forestry* 59:453–454.

Allen, R. 1980. *How to Save the World.* Barnes and Noble, Totowa, N.J.

Alvarez, L. W., W. Alvarez, F. Asaro, and H. V. Michel. 1980. Extraterrestrial cause for the Cretaceous-Tertiary extinction. *Science* 208:1095–1108.

Ames, B. H. 1983. Dietary carcinogens and anticarcinogens. *Science* 221:1256–1264.

Andelman, S. J. 1986. Ecological and social determinants of Cercopithecine mating patterns. Pages 201–216 in D. I. Rubenstein and R. W. Wrangham (eds.), *Ecological Aspects of Social Evolution, Birds and Mammals.* Princeton University Press, Princeton, N.J.

Anderson, D. R., and K. P. Burnham. 1976. *Population Ecology of the Mallard: VI. The Effect of Exploitation on Survival.* U.S. Fish and Wildlife Service Publication 128.

Anderson, J. M. 1981. *Ecology for Environmental Sciences: Biosphere, Ecosystems and Man.* Wiley, New York.

Anderson, R. C. 1972. The ecological relationships of meningeal worm and native cervids in North America. *Journal of Wildlife Diseases* 8:304–310.

Anderson, R. C., and A. K. Prestwood. 1979. Lungworms. Pages 266–317 in F. Hayes (ed.), *The Diseases of the White-tailed Deer.* U.S. Department of the Interior, U.S. Fish and Wildlife Service, Washington, D.C.

Anderson, R. M. 1982. Epidemiology. Pages 204–251 in F. E. G. Cox (ed.), *Modern Parasitology.* Blackwell Scientific Publications, Oxford.

Andreae, M. O., and H. Raemdonck. 1983. Dimethyl sulphide in surface oceans and marine atmosphere: a global view. *Science* 221:744–747.

Andrewartha, H. G., and L. C. Birch. 1954. *The Distribution and Abundance of Animals.* University of Chicago Press, Chicago.

Anonymous. 1981. *List of Intercepted Plant Pests from October 1, 1978, through September 30, 1979.* U.S.D.A. APHIS Publication 82-7.

Anonymous. 1988. *World Resources 1988–89.* A report by the World Resources Institute and the International Institute for Environment and Development. Basic Books, Inc., New York.

Antonovics, J., A. D. Bradshaw, and R. G. Turner. 1971. Heavy metal tolerance in plants. *Advances in Ecological Research* 7:1–85.

Applegate, V. C. 1950. *Natural History of the Sea Lamprey,* Petromyzon marinus, *in Michigan.* U.S. Fish and Wildlife Service, Special Scientific Report, Fisheries, No. 55, Washington, D.C.

Armstrong, S., and F. Bridgland. 1989. Elephants and the ivory tower. *New Scientist* 1679:37–41.

Aron, W. I., and S. H. Smith. 1971. Ship canals and aquatic ecosystems. *Science* 174:13–20.

Askew, R. R. 1968. Considerations on speciation in Chalcidoidea (Hymenoptera). *Evolution* 22:642–645.

Askew, R. R. 1971. *Parasitic Insects.* Heineman, London.

Askew, R. R., and M. R. Shaw. 1974. An account of the Chalcidoidea (Hymenoptera) parasitizing leaf mining insects in deciduous trees in Britain. *Biological Journal of the Linnaean Society* 6:289–335.

Attenborough, D. 1979. *Life on Earth.* British Broadcasting Corporation, London.

Auerbach, M. J. 1979. Some real communities are unstable. *Nature* 279:821.

Auerbach, M. J., and D. R. Strong. 1981. Nutritional ecology of *Heliconia* herbivores: experiments with plant fertilization and alternative hosts. *Ecological Monographs* 51:63–83.

Avery, D. 1985. U.S. farm dilemma: the global bad news is wrong. *Science* 230:408–412.

Avise, J. C. 1983. Protein variation and phylogenetic reconstruction. Pages 103–130 in G. S. Oxford and D. Rollinson (eds.), *Protein Polymorphism, Adaptive and Taxonomic Significance.* Academic Press, New York.

Avise, J. C. 1989. A role for molecular genetics in the recognition and conservation of endangered species. *Trends in Ecology and Evolution* 4:279–281.

Avise, J. C., and C. F. Aquadro. 1982. A comparative summary of genetic distances in the vertebrates. *Evolutionary Biology* 15:151–185.

Azim Abul-Atta, E. A. 1978. *Egypt and the Nile After the Construction of the High Aswan Dam.* Department of Irrigation and Land Reclamation, Egypt.

Baker, S. J., and C. N. Clarke. 1988. Cage trapping coypus (*Myocastor coypus*) on baited rafts. *Journal of Applied Ecology* 25:41–48.

Baldwin, N. S. 1964. Sea lamprey in the Great Lakes. *Canadian Audubon Magazine,* November–December: 142–147.

Ballard, R. D. 1977. Notes on a major oceanographic find. *Oceanus* 20:35–44.

Bambach, R. K. 1983. Ecospace utilization and guilds in marine communities through the Phanerozoic. Pages 719–746 in M. J. S. Tevesz and P. L. McCall (eds.), *Biotic Interactions in Recent and Fossil Benthic Communities.* Plenum, New York.

Bambach, R. K., C. R. Scotese, and A. M. Ziegler. 1980. Before Pangaea: the geographies of the Paleozoic world. *American Scientist* 68:26–38.

Barber, R. T., and F. P. Chavez. 1983. Biological consequences of El Niño. *Science* 222:1203–1210.

Barnard, C. J. 1980. Flock feeding and time budgets in the house sparrow, *Passer domesticus* L. *Animal Behavior* 28:295–309.

Barnola, J. M., D. Raynaud, Y. S. Korotkevich, and C. Lorius. 1987. Vostok ice core provides 160,000-year record of atmospheric CO_2. *Nature* 329:410.

Barrons, K. C. 1981. *Are Pesticides Really Necessary?* Regnery Gateway, Chicago.

Barrow, C. 1988. The impact of hydroelectric development on the Amazonian environment, with particular reference to the Tucurui Project. *Journal of Biogeography* 15:67–78.

Barrowclough, G. F. 1980. Gene flow, effective population sizes, and genetic variance components in birds. *Evolution* 34:789–798.

Barry, R. G. 1969. The world hydrological cycle. Pages 11–29 in R. J. Chorley (ed.), *Water, Earth and Man.* Barnes and Noble, New York.

Bartholomew, G. A. 1986. The role of natural history in contemporary biology. *BioScience* 36:324–329.

Barton, A. M. 1986. Spatial variation in the effect of ants on an extrafloral nectary plant. *Ecology* 67:495–504.

Bates, H. W. 1862. Contributions to an insect fauna of the Amazon Valley. *Transactions of the Linnaean Society of London* 23:495–566.

Bates, M. 1956. Man as an agent in the spread of organisms. Pages 788–804 in W. L. Thomas (ed.), *Man's Role in Changing the Face of the Earth.* University of Chicago Press, Chicago.

Bateson, P. P. G., W. Lotwick, and D. K. Scott. 1980. Similarities between the faces of parents and offspring in Bewick's swans and the differences between mates. *Journal of the Zoological Society of London* 191:61–74.

Batzli, G. O. 1983. Responses of arctic rodent populations to nutritional factors. *Oikos* 40:396–406.

Batzli, G. O., R. G. White, S. F. Maclean, Jr., F. A. Pitelka, and B. D. Collier. 1980. The herbivore-based trophic system. Pages 335–340 in J. Brown, P. C. Miller, L. L. Tiezen, and F. L. Bunnell (eds.), *An Arctic Ecosystem.* Dowden, Hutchinson and Ross, Stroudsburg, Pa.

Beattie, A. J. 1985. *The Evolutionary Ecology of Ant-Plant Mutualisms.* Cambridge University Press, Cambridge.

Beattie, A. J., C. Turnbull, R. B. Knox, and E. G. Williams. 1984. Ant inhibition of pollen function: a possible reason why ant pollination is rare. *American Journal of Botany* 71:421–426.

Beaufait, W. R. 1960. Some effects of high temperatures on the cones and seeds of jack pine. *Forest Science* 6:194–199.

Begon, M., J. L. Harper, and C. R. Townsend. 1990. *Ecology: Individuals, Populations and Communities.* Blackwell Scientific Publications, Oxford.

Bell, G. 1982. *The Masterpiece of Nature: The Evolution and Genetics of Sexuality.* University of California Press, Berkeley.

Belovsky, G. E. 1978. Diet optimization in a generalist herbivore: the moose. *Theoretical Population Biology* 14:105–134.

Belsky, A. J. 1986. Does herbivory benefit plants? A review of the evidence. *American Naturalist* 127:870–892.

Belsky, A. J. 1987. The effects of grazing: confounding of ecosystem, community, and organism scales. *American Naturalist* 129:777–783.

Belwood, J. J., and G. K. Morris. 1987. Bat predation and its influence on calling behavior in Neotropical katydids. *Science* 238:64–67.

Bennett, A. F. 1987. The accomplishments of ecological physiology. Pages 1–10 in M. E. Feder, A. F. Bennett, W. W. Burggren, and R. B. Huey (eds.), *New Directions in Ecological Physiology.* Cambridge University Press, Cambridge.

Bennett, F. D., and I. W. Hughes. 1959. Biological control of insect pests in Bermuda. *Bulletin of Entomological Research* 50:423–426.

Bergerud, A. T. 1980. A review of the population dynamics of caribou and wild reindeer in North America. Pages 556–581 in *Second International Reindeer/Caribou Symposium.* Roros, Norway.

Bernays, E. A. 1981. Plant tannins and insect herbivores: an appraisal. *Ecological Entomology* 6:353–360.

Bernays, E. A. 1983. Nitrogen in defence against insects. Pages 321–344 in J. A. Lee, S. McNeill, and I. H. Rorison (eds.), *Nitrogen as an Ecological Factor. 22nd Symposium of the British Ecological Society.* Blackwell Scientific Publications, Oxford.

Bernays, E. A., and E. Graham. 1988. On the evolution of host specificity in phytophagous arthropods. *Ecology* 69:886–892.

Bertram, B. C. R. 1975. Social factors influencing reproduction in wild lions. *Journal of the Zoological Society of London* 177:463–482.

Bertram, B. C. R. 1976. Kin selection in lions and in evolution. Pages 281–301 in P. P. G. Bateson and R. A. Hinde (eds.), *Growing Points in Ethology*. Cambridge University Press, Cambridge.

Bertram, B. C. R. 1979. Serengeti predators and their social systems. Pages 221–248 in A. R. E. Sinclair and M. Morton-Griffiths (eds.), *Serengeti: Dynamics of an Ecosystem*. University of Chicago Press, Chicago.

Beverton, R. J. H. 1983. Science and decision-making in fisheries regulation. Pages 919–938 in G. D. Sharpe and J. Csirke (eds.), *Proceedings of the Expert Consultation to Examine Changes in Abundance and Species Composition of Neritic Fish Resources*. FAO Fishery Report 291. Food and Agriculture Organization, Rome.

Birks, H. J. B. 1980. British trees and insects: a test of the time hypothesis over the last 13,000 years. *American Naturalist* 115:600–605.

Bishop, J. A. 1981. A neoDarwinian approach to resistance: examples from mammals. Pages 37–51 in J. A. Bishop and L. M. Cook (eds.), *Genetic Consequences of Man-made Change*. Academic Press, London.

Bishop, J. A., and L. M. Cook. 1980. Industrial melanism and the urban environment. *Advances in Ecological Research* 11:373–404.

Bishop, J. A., and L. M. Cook (eds.). 1981. *Genetic Consequences of Man-made Change*. Academic Press, London.

Black, F. L. 1975. Infectious diseases in primitive societies. *Science* 187:515–518.

Block, W. M., L. M. Brennan, and R. J. Gutiérrez. 1986. The use of guilds and guild-indicator species for assessing habitat suitability. Pages 109–113 in J. Verner, M. L. Morrison, and C. J. Ralph (eds.), *Wildlife 2000*. University of Wisconsin Press, Madison.

Boecklen, W. J., and N. Gotelli. 1984. Island biogeographic theory and conservation practice: species area or specious-area relationships? *Biological Conservation* 29:63–80.

Boecklen, W. J., and D. Simberloff. 1986. Area-based extinction models in conservation. Pages 247–276 in D. K. Elliott (ed.), *Dynamics of Extinction*. Wiley, New York.

Boersma, L. K., and J. A. Gulland. 1973. Stock assessment of the Peruvian anchovy (*Engraulis ringens*) and management of the fishery. *Journal of the Fisheries Research Board of Canada* 30:2226–2235.

Boersma, P. D. 1986. Seabirds reflect petroleum pollution. *Science* 231:373–376.

Boggess, W. R., and B. G. Wixson. 1979. *Lead in the Environment*. Castle House Publications, Tunbridge Wells, England.

Bonaccorsi, A., R. Fanelli, and A. Tognoni. 1978. In the wake of Seveso. *Ambio* 7:234–239.

Bonnell, M. L., and R. K. Selander. 1974. Elephant seals: genetic variation and near extinction. *Science* 184:908–909.

Bonner, J. T. 1965. *Size and Cycle: An Essay on the Structure of Biology*. Princeton University Press, Princeton, N.J.

Booth, W. 1988*a*. Animals of invention. *Science* 240:718.

Booth, W. 1988*b*. Johnny Appleseed and the greenhouse. *Science* 242:19–20.

Borner, M., C. D. FitzGibbon, M. Borner, T. M. Caro, W. K. Lindsay, D. A. Collins, and M. E. Holt. 1987. The decline of Serengeti Thompson's gazelle population. *Oecologia* 73:32–40.

Boucher, D. H., S. James, and K. Kebler. 1982. The ecology of mutualism. *Annual Review of Ecology and Systematics* 13:315–347.

Bradbury, J. W., and R. M. Gibson. 1983. Leks and mate choice. Pages 109–138 in P. Bateson (ed.), *Mate Choice*. Cambridge University Press, Cambridge.

Bradley, P. M. 1988. Methodology for woodfuel development planning in the Kenyan Highlands. *Journal of Biogeography* 15:157–164.

Bradshaw, A. D., and M. J. Chadwick. 1980. *The Restoration of Land*. Blackwell Scientific Publications, Oxford.

Bradshaw, A. D., R. N. Humphries, M. S. Johnson, and R. D. Roberts. 1978. The restoration of vegetation on derelict land produced by industrial activity. Pages 249–274 in M. W. Holdgate and M. J. Woodward (eds.), *The Breakdown and Restoration of Ecosystems*. Plenum, New York.

Bragg, T. B., and L. C. Hurlbert. 1976. Woody plant invasion of unburned Kansas bluestem prairie. *Journal of Range Management* 29:19–29.

Brett, J. R. 1959. Thermal requirements of fish: Three decades of study, 1940–1960. (Transcript of the second seminar on Biological Problems in Water Pollution, April 1959.) U.S. Public Health Service, Taft Center, Cincinnati, Ohio.

Brewer, R. 1988. *The Science of Ecology*. Saunders, Philadelphia.

Briand, F. 1983. Environmental control of food web structure. *Ecology* 64:253–263.

Briand, F., and J. E. Cohen. 1984. Community food webs have scale-invariant structure. *Nature* 307:264–266.

Brodbeck, B. V., and D. R. Strong. 1987. Amino acid nutrition of herbivorous insects and stress to host plants. Pages 347–364 in P. Barbosa and J. C. Schultz (eds.), *Insect Outbreaks*. Academic Press, San Diego.

Brody, S. 1945. *Bioenergetics and Growth*. Van Nostrand Reinhold, New York.

Bromenshenk, J. J., S. R. Carlson, J. C. Simpson, and J. M. Thomas. 1985. Pollution monitoring of Puget sound with honey bees. *Science* 227:632–634.

Brower, L. P. 1969. Ecological chemistry. *Scientific American* 220:22–29.

Brower, L. P. 1970. Plant poisons in a terrestrial food chain and implication for mimicry theory. Pages 69–82 in K. L. Chambers (ed.), *Biochemical Coevolution*. *Proceedings of the 29th Annual Biological Colloquium*. Oregon State University Press, Corvallis.

Brower, L. P., W. M. Ryerson, L. L. Coppinger, and S. C. Glazier. 1968. Ecological chemistry and the palatability spectrum. *Science* 161:1349–1351.

Brown, A. A., and K. P. Davis. 1973. *Forest Fire Control and Its Use*, 2nd edition. McGraw-Hill, New York.

Brown, C. R. 1988. Enhanced foraging efficiency through information transfer centers: a benefit of coloniality in cliff swallows. *Ecology* 59:602–613.

Brown, C. R., and M. Bomberger Brown. 1986. Ectoparasitism as a cost of coloniality in cliff swallows (*Hirundo pyrrhonota*). *Ecology* 67:1206–1218.

Brown, J. H. 1978. The theory of insular biogeography and the distribution of boreal birds and mammals. *Great Basin Naturalist Memoirs* 2:209–227.

Brown, J. H. 1989. Patterns, modes and extents of invasions by vertebrates. Pages 85–109 in J. A. Drake, H. A. Mooney, F. di Castri, R. H. Groves, F. J. Kruger, M. Rejmanek, and M. Williamson (eds.), *Biological Invasions: A Global Perspective*. Wiley, Chichester, England.

Brown, J. H., D. W. Davidson, J. C. Munger, and R. S. Inouye. 1986. Experimental community ecology: the desert granivore system. Pages 41–61 in J. Diamond and T. J. Case (eds.), *Community Ecology*. Harper and Row, New York.

Brown, J. H., and A. C. Gibson. 1983. *Biogeography*. The C. V. Mosby Company, St. Louis, Mo.

Brown, J. H., and M. V. Lomolino. 1989. Independent discovery of the equilibrium theory of island biogeography. *Ecology* 70:1954–1957.

Brown, J. L. 1969. The buffer effect and productivity in tit populations. *American Naturalist* 103:374–354.

Brown, J. L. 1982. Optimal group size in territorial animals. *Journal of Theoretical Biology* 95:793–810.

Brown, J. L., E. R. Brown, S. D. Brown, and D. D. Dow. 1982. Helpers: effects of experimental removal on reproductive success. *Science* 215:421–422.

Brown, J. L., D. D. Dow, E. R. Brown, and S. D. Brown. 1978. Effects of helpers on feeding and nestlings in the grey-crowned babbler, *Pomatostomus temporalis*. *Behavioral Ecology and Sociobiology* 4:43–60.

Brown, L. R. 1970. Human food production as a process in the biosphere. *Scientific American* 223(3):161–170.

Bruce, H. M. 1966. Smell as an exteroceptive factor. *Journal of Animal Science,* Supplement 25:83–89.

Brunson, M. H. 1939. Influence of Japanese beetle instar on the sex and population of the parasite *Tiphia popilliavora*. *Journal of Agricultural Research* 37:379–386.

Bryson, R. A., and T. J. Murray. 1977. *Climates of Hunger*. University of Wisconsin Press, Madison.

Buckingham, G. R. 1987. Florida's #1 weed: *Hydrilla* vs. biocontrol. Pages 22–25 in *Research* 87, IFAS Editorial Department, University of Florida, Gainesville.

Buckley, J. 1986. Environmental effects of DDT. Pages 358–374 in *Ecological Knowledge and Environmental Problem-Solving, Concepts and Case Studies*. National Academy Press, Washington, D.C.

Bunting, A. H. 1988. The humid tropics and the nature of development. *Journal of Biogeography* 15:5–10.

Burton, J. A. 1976. Illicit trade in rare animals. *New Scientist* 72:168.

Bush, G. L. 1975*a*. Modes of animal speciation. *Annual Review of Ecology and Systematics* 6:334–364.

Bush, G. L. 1975*b*. Sympatric speciation in phytophagous parasitic insects. Pages 187–206 in P. W. Price (ed.), *Evolutionary Strategies of Parasitic Insects and Mites*. Plenum, New York.

Buth, D. G. 1984. The application of electrophoretic data in systematic studies. *Annual Review of Ecology and Systematics* 15:501–522.

Buttel, F. H., and R. Barker. 1985. Emerging agricultural technologies, public policy and implications for third world agriculture: the case of biotechnology. *American Journal of Agricultural Economics* 67:1170–1175.

Bygott, J. D., B. C. R. Bertram, and J. P. Hanby. 1979. Male lions in large coalitions gain reproductive advantage. *Nature* 282:839–841.

Cade, W. 1979. The evolution of alternative male reproductive strategies in field crickets. Pages 343–379 in M. S. Blum and N. A. Blum (eds.), *Sexual Selection and Reproductive Competition in Insects*. Academic Press, New York.

Cain, A. J., and P. M. Sheppard. 1954*a*. Natural selection in *Cepaea*. *Genetics* 39:89–116.

Cain, A. J., and P. M. Sheppard. 1954*b*. The theory of adaptive polymorphism. *American Naturalist* 88:321–326.

Cameron, R. D., K. R. Whitten, W. T. Smith, and D. D. Roby. 1979. Caribou distribution and group composition associated with construction of the trans-Alaska pipeline. *Canadian Field Naturalist* 93:155–162.

Caraco, T., S. Martindale, and T. G. Whitham. 1980. An empirical demonstration of risk-sensitive foraging preferences. *Animal Behaviour* 28:820–830.

Caraco, T., and L. L. Wolf. 1975. Ecological determinants of group sizes of foraging lions. *American Naturalist* 109:343–352.

Carlquist, S. 1974. *Island Biology*. Columbia University Press, New York.

Caro, T. M. 1986. The functions of stotting: a review of the hypotheses. *Animal Behaviour* 34:649–662.

Carothers, J. H. 1986. Homage to Huxley: On the conceptual origin of minimum size ratios among competing species. *The American Naturalist* 128:440–442.

Carson, R. 1962. *Silent Spring*. Houghton Mifflin, Boston.

Caswell, H. 1978. Predator mediated coexistence: a nonequilibrium model. *American Naturalist* 112:127–154.

Cates, R. G., and G. H. Orians. 1975. Successional status and the palatability of plants to generalist herbivores. *Ecology* 56:410–418.

Caughley, G. 1976. Wildlife management and the dynamics of ungulate populations. *Advances in Applied Biology* 1:183–246.

Caughley, G. 1977. *Analysis of Vertebrate Populations*. Wiley, Chichester, England.

Caughley, G., G. C. Grigg, J. Caughley, and G. J. E. Hill. 1980. Does dingo predation control the densities of kangaroos and emus? *Australian Wildlife Research* 7:1–12.

Chadab, R., and C. W. Rettenmeyer. 1975. Mass recruitment by army ants. *Science* 188:1124–1125.

Chaloupka, M. J., and S. B. Domm. 1986. Role of anthropochory in the invasion of coral cays by alien flora. *Ecology* 67:1536–1547.

Charlesworth, B. 1984. The cost of phenotypic evolution. *Paleobiology* 10:319–327.

Chase, C. 1990. Personal communication.

Cherfas, J. 1982. *Man Made Life.* Basil Blackwell, Oxford.

Cherfas, J. 1986. What price whales? *New Scientist* 110(1511):36–40.

Christensen, M. L. 1981. Fire regimes in southeastern ecosystems. Pages 117–136 in H. A. Mooney, T. M. Bonnicksen, M. L. Christensen, J. E. Lotan, and W. A. Reiners (eds.), *Fire Regimes and Ecosystem Properties.* U.S.D.A. Forest Service, General Technical Report WO-26, Washington, D.C.

Christian, J. J. 1971. Population density and reproductive efficiency. *Biological Reproduction* 4:248–294.

Christie, W. J. 1974. Changes in the fish species composition of the Great Lakes. *Journal of the Fisheries Research Board of Canada* 31:827–854.

Clapham, W. B., Jr. 1981. *Human Ecosystems.* Macmillan, New York.

Claridge, M. F. 1985. Acoustic signals in the Homoptera: Behavior, taxonomy, and evolution. *Annual Review of Entomology* 30:297–317.

Clark, B. C. 1962. Balanced polymorphism and the diversity of sympatric species. *Systematics Association Publication* 4:47–70.

Clarke, C. A., G. S. Mani, and G. Wynne. 1985. Evolution in reverse: clean air and the peppered moth. *Biological Journal of the Linnaean Society* 26:189–199.

Clay, K. 1988. Fungal endophytes of grasses: a defensive mutualism between plants and fungi. *Ecology* 69:10–16.

Clegg, M. T., and A. D. H. Brown. 1983. The founding of plant populations. Pages 216–228 in C. M. Schonewald-Cox, S. M. Chambers, B. MacBryde, and L. Thomas (eds.), *Genetics and Conservation.* Benjamin/Cummings, Menlo Park, Calif.

Clements, F. E. 1936. Nature and structure of the climax. *Journal of Ecology* 24:252–284.

Clutton-Brock, T. H. 1974. Primate social organization and ecology. *Nature* 250:539–542.

Clutton-Brock, T. H., and S. D. Albon. 1979. The roaring of red deer and the evolution of honest advertisement. *Behaviour* 69:145–170.

Clutton-Brock, T. H., S. D. Albon, R. M. Gibson, and F. E. Guinness. 1979. The logical stag: adaptive aspects of fighting in red deer (*Cervus elephas* L.). *Animal Behaviour* 27:211–225.

Clutton-Brock, T. H., F. E. Guinness, and S. D. Albon. 1982. *Red Deer: Behavior and Ecology of Two Sexes.* University of Chicago Press, Chicago.

Clutton-Brock, T. H., and P. H. Harvey. 1977. Primate ecology and social organization. *Journal of the Zoological Society of London* 183:1–39.

Clutton-Brock, T. H., and P. H. Harvey. 1984. Comparative approaches to investigating adaptation. Pages 7–29 in J. R. Krebs and N. B. Davies (eds.), *Behavioural Ecology, an Evolutionary Approach*, 2nd edition. Blackwell Scientific Publications, Oxford.

Cochran, D. G. 1987. Our chemophobic society. *Bulletin of the Entomological Society of America* 33:128–133.

Cock, M. J. W. 1985. *A Review of Biological Control of Pests in the Commonwealth Caribbean and Bermuda up to 1982*. Technical Communication of the Commonwealth Institute of Biological Control 9.

Cockburn, A. T. 1971. Infectious diseases in ancient populations. *Current Anthropology* 12:45–62.

Cody, M. L. 1974. *Competition and the Structure of Bird Communities*. Princeton University Press, Princeton, N.J.

Cohen, J. E. 1978. *Food Webs and Niche Space*. Princeton University Press, Princeton, N.J.

Cohen, J. E. and F. Briand. 1984. Trophic links of community food webs. *Proceedings of the National Academy of Sciences of the United States of America* 81:4105–4109.

Coley, P. D. 1983. Herbivory and defensive characteristics of tree species in a lowland tropical forest. *Ecological Monographs* 53:209–233.

Coley, P. D. 1988. Effects of plant growth and leaf lifetime on the amount and type of anti-herbivore defense. *Oecologia* 74:531–536.

Collinson, A. S. 1977. *An Introduction to World Vegetation*. George Allen and Unwin, London.

Colwell, R. K. 1973. Competition and coexistence in a simple tropical community. *American Naturalist* 107:737–760.

Colwell, R. K. 1984. What's new? Community ecology discovers biology. Pages 387–396 in P. W. Price, C. N. Slobodchikoff, and W. S. Gaud (eds.), *A New Ecology: Novel Approaches to Interactive Systems*. Wiley, New York.

Connell, J. H. 1961. The influence of interspecific competition and other factors on the distribution of the barnacle *Chthamalus stellatus*. *Ecology* 42:710–723.

Connell, J. H. 1978. Diversity in tropical rain forests and coral reefs. *Science* 199:1302–1310.

Connell, J. H. 1980. Diversity and the coevolution of competitors, or the ghost of competition past. *Oikos* 35:131–138.

Connell, J. H. 1983. On the prevalence and relative importance of interspecific competition: evidence from field experiments. *American Naturalist* 122:661–696.

Connell, J. H. 1990. Personal communication.

Connell, J. H., I. R. Noble, and R. O. Slatyer. 1987. On the mechanisms of producing successional change. *Oikos* 50:136–137.

Connell, J. H., and R. O. Slatyer. 1977. Mechanisms of succession in natural communities and their role in community stability and organization. *American Naturalist* 111:1119–1144.

Connell, J. H., and W. P. Sousa. 1983. On the evidence needed to judge ecological stability or persistence. *American Naturalist* 121:789–824.

Connor, E., and E. D. McCoy. 1979. The statistics and biology of the species-area relationship. *American Naturalist* 113:791–833.

Connor, E.F., and D. Simberloff. 1979a. You can't falsify ecologic hypotheses without data. *Bulletin of the Ecological Society of America* 60:154–155.

Connor, E., and D. Simberloff. 1979b. The assembly of species communities: chance or competition? *Ecology* 60:1132–1140.

Connor, S. 1988. Genes on the loose. *New Scientist* 118(1614):65–68.

Conover, W. J. 1980. *Practical Nonparametric Statistics,* 2nd edition. Wiley, New York.

Conway, G. R. 1976. Man versus pests. Pages 257–281 in R. M. May (ed.), *Theoretical Ecology: Principles and Applications.* Blackwell Scientific Publications, Oxford.

Cook, E. 1975. Flow of energy through a technological society. Pages 30–62 in J. Lenihan and W. W. Fletcher (eds.), *Energy Resources and the Environment.* Blackie, Glasgow.

Cook, R. E. 1969. Variation in species diversity of North American birds. *Systematic Zoology* 18:63–84.

Cooke, A. S. 1973. Shell thinning in avian eggs by environmental pollutants. *Environmental Pollution* 4:85–152.

Cooper, J. P. (ed.). 1975. *Photosynthesis and Productivity in Different Environments.* Cambridge University Press, London.

Cooper, S. M., and N. Owen-Smith. 1985. Condensed tannins deter feeding by browsing ruminants on a South African savanna. *Oecologia* 67:142–146.

Cooper, S. M., and N. Owen-Smith. 1986. Effects of plant spinescence on large mammalian herbivores. *Oecologia* 68:446–455.

Cooper, W. S. 1923. The recent ecological history of Glacier Bay, Alaska. II. The present vegetation cycle. *Ecology* 4:223–246.

Corayon, J. 1974. Insémination traumatique hétérosexuelle et homosexuelle chez *Xylocoris maculipennis* (Hem. Anthocoridae). *Comptes Rendus de l'Académie des Sciences de Paris D* 278:2803–2806.

Cornell, H. 1974. Parasitism and distributional gaps between allopatric species. American Naturalist 108:880–883.

Cotter, D. J., and J. T. McGinnis, 1965. Recovery of hardwood stands 3–5 years following acute irradiation. *Health Physics* 11:1663–1673.

Cowles, H. C. 1899. The ecological relations of the vegetation on the sand dunes of Lake Michigan. *Botanical Gazette* 27:95–117, 167–202, 361–391.

Cox, C. B., I. N. Healey, and P. B. Moore. 1976. *Biogeography, an Ecological and Evolutionary Approach,* 2nd edition. Blackwell Scientific Publications, Oxford.

Cox, C. B., and P. D. Moore. 1986. *Biogeography, an Ecological and Evolutionary Approach,* 4th edition. Blackwell Scientific Publications, Oxford.

Cox, F. E. G. 1982. Immunology. Pages 173–203 in F. E. G. Cox (ed.), *Modern Parasitology*. Blackwell Scientific Publications, Oxford.

Cox, F. E. G. 1989. Parasites and sexual selection. *Nature* 341:289.

Cox, G. W., and B. J. Le Boeuf. 1977. Female initiation of male competition: a mechanism of mate selection. *American Naturalist* 111:317–335.

Cox, J. A., J. Carnahan, J. D. DiNuazio, J. McCoy, and J. Meister. 1979. Source of mercury in fish in new impoundments. *Bulletin of Environmental Contamination and Toxicology* 23:779–783.

Coyne, J. A. 1976. Lack of genic similarity between two sibling species of *Drosophila* as revealed by varied techniques. *Genetics* 84:593–607.

Coyne, J. A., A. A. Felton, and R. C. Lewontin. 1978. Extent of genetic variation at a highly polymorphic esterase locus in *Drosophila pseudoobscura*. *Proceedings of the National Academy of Sciences of the United States of America* 75:5090–5093.

Crawford, M. 1987. EPA okays field test. *Science* 235:840.

Crawley, M. J. 1983. *Herbivory, the Dynamics of Animal-Plant Interactions. Studies in Ecology, Vol. 10*. Blackwell Scientific Publications, Oxford.

Crawley, M. J. 1985. Reduction of oak fecundity by low-density herbivore populations. *Nature* 314:163–164.

Crawley, M. J. 1987. Benevolent herbivores? *Trends in Ecology and Evolution* 2:167–169.

Crawley, M. J. 1988. Cogene/Scope at Lake Como. *Trends in Ecology and Evolution* 3(4) and *Trends in Biotechnology* 6(4), special combined issue: 2–3.

Cristoffer, C., and J. Eisenberg. 1985. *On the Captive Breeding and Reintroduction of the Florida Panther in Suitable Habitats*. Task #1, Report # 2, Florida Game and Fresh Water Fish Commission and Panther Technical Advisory Committee, Tallahassee.

Crocker, R. L., and J. Major. 1955. Soil development in relation to vegetation and surface age at Glacier Bay, Alaska. *Journal of Ecology* 43:427–448.

Cronin, E. W., and P. W. Sherman. 1977. A resource-based mating system: the orange rumped honey guide. *Living Bird* 15:5–32.

Crosby, A. W. 1986. *Ecological Imperialism, the Biological Expansion of Europe 900–1900*. Cambridge University Press, Cambridge.

Crosby, G. T. 1972. Spread of the cattle egret in the Western Hemisphere. *Birdbanding* 43:205–212.

Crow, J. F., and M. Kimura. 1970. *An Introduction to Population Genetics Theory*. Harper and Row, New York.

Currie, D. J., and V. Paquin. 1987. Large-scale biogeographical patterns of species richness of trees. *Nature* 329:326–327.

Daday, H. 1954. Gene frequencies in wild populations of *Trifolium repens* L. I. Distribution by latitude. *Heredity* 8:61–78.

Dahlsten, D. L. 1986. Control of invaders. Pages 275–302 in H. A. Mooney and J. A. Drake (eds.), *Ecology of Biological Invasions of North America and Hawaii*. Springer-Verlag, New York.

Daley, H. E. 1982. Review of *The Ultimate Resource* by J. Simon. *The Bulletin of the Atomic Scientists* 38(1):39–42.

Darlington, P. J., Jr. 1959. Area, climate and evolution. *Evolution* 13:488–510.

Darwin, C. 1859. *On the Origin of Species by Means of Natural Selection.* John Murray, London.

Darwin, C. 1871. *The Descent of Man, and Selection in Relation to Sex.* John Murray, London.

Dasmann, R. F. 1976. *Environmental Conservation,* 4th edition. Wiley, New York and Chichester.

Davies, B. N. K. 1976. Wildlife, urbanisation and industry. *Biological Conservation* 10:249–291.

Davies, N. B. 1978. Territorial defense in the speckled wood butterfly (*Pararge aegeria*): the resident always wins. *Animal Behaviour* 26:138–147.

Davies, N. B., and T. R. Halliday. 1978. Deep croaks and fighting assessment in toads *Bufo bufo. Nature* 274:683–685.

Davies, N. B., and A. I. Houston. 1984. Territory economics. Pages 148–169 in J. R. Krebs and N. B. Davies (eds.), *Behavioural Ecology, an Evolutionary Approach,* 2nd edition. Blackwell Scientific Publications, Oxford.

Dawkins, R. 1976. *The Selfish Gene.* Oxford University Press, Oxford.

Dawkins, R. 1979. Twelve misunderstandings of kin selection. *Zeitschrift für Tierpsychologie* 51:184–200.

Dawkins, R. 1980. Good strategy or evolutionarily stable strategy? Pages 331–367 in G. W. Barlow and J. Silverberg (eds.), *Sociobiology: Beyond Nature/Nurture.* Westview Press, Boulder, Col.

Dawkins, R. 1986. *The Blind Watchmaker.* Norton, New York.

Dawkins, R., and J. R. Krebs. 1979. Arms races between and within species. *Proceedings of the Royal Society of London Series B* 205:489–511.

Dayton, P. K., and M. J. Tegner. 1984. The importance of scale in community ecology: a kelp forest example with terrestrial analogs. Pages 457–481 in P. W. Price, C. N. Slobodchikoff, and W. S. Gaud (eds.), *A New Ecology: Novel Approaches to Interactive Systems.* Wiley, New York.

Debach, P. S., and H. S. Smith. 1941. The effect of host density on the rate of reproduction of entomophagous parasites. *Journal of Economic Entomology* 34:741–745.

Debach, P. S., and R. A. Sundby. 1963. Competitive displacement between ecological homologues. Hilgardia 43:105–166.

den Boer, P. J. 1968. Spreading of risk and stabilization of animal numbers. *Acta Biotheoretica* 18:165–194.

den Boer, P. J. 1981. On the survival of populations in a heterogeneous and variable environment. *Oecologia* 50:39–53.

Denno, R. F., L. W. Douglass, and D. Jacobs. 1985. Crowding and host plant nutrition: environmental determinants of wing form in *Prokelisia marginata. Ecology* 66:1588–1596.

Denno, R. F., L. W. Douglass, and D. Jacobs. 1986. Effects of crowding and host plant nutrition on the development and body size of the wing-dimorphic planthopper, *Prokelisia marginata. Ecology* 67:116–123.

Desowitz, R. S. 1981. *New Guinea Tapeworms and Jewish Grandmothers: Tales of Parasites and People.* Norton, New York.

De Vos, A., R. H. Manville, and G. Van Gelder. 1956. Introduced mammals and their influence on native biota. *Zoologica* 41:163–194.

Diamond, J. 1984. "Normal" extinctions of isolated populations. Pages 191–246 in M. H. Nitecki (ed.), *Extinctions.* University of Chicago Press, Chicago.

Diamond, J. 1986a. The environmentalist myth. *Nature* 324:19–20.

Diamond, J. 1986b. Overview: laboratory experiments, field experiments, and natural experiments. Pages 3–22 in J. Diamond and T. J. Case (eds.), *Community Ecology.* Harper and Row, New York.

Dickerson, R. E. 1978. Chemical evolution and the origin of life. *Scientific American* 239(3):70–86.

Di Silvestro, R. L. 1988. U.S. demand for carved ivory hastens African elephant's end. *Audubon* 90(2):14.

Diver, C. 1929. Fossil records of Mendelian mutants. *Nature* 124:183.

Dixon, A. F. G., P. Kindlmann, J. Leps, and J. Holman. 1987. Why there are so few species of aphids, especially in the tropics. *American Naturalist* 129:580–592.

Dixon, K. R., and T. C. Juelson. 1987. The political economy of the spotted owl. Ecology 68:772–776.

Dobzhansky, T. 1936. Studies on hybrid sterility. II. Localization of sterility factors in *Drosophila pseudoobscura* hybrids. *Genetics* 21:113–135.

Dobzhansky, T. 1950. Evolution in the tropics. *American Scientist* 38:209–221.

Dobzhansky, T. 1970. *Genetics of the Evolutionary Process.* Columbia University Press, New York.

Downhower, J. F., and K. B. Armitage. 1971. The yellow-bellied marmot and the evolution of polygamy. *American Naturalist* 105:355–370.

Dunbar, M. J. 1980. The blunting of Occam's Razor, or to hell with parsimony. *Canadian Journal of Zoology* 58:123–128.

Dyson, F. J. 1988. *Infinite in All Directions.* Harper and Row, New York.

Eckholm, E. P. 1976. *Losing Ground.* Norton, New York.

Edington, J. M., and M. A. Edington. 1986. *Ecology, Recreation and Tourism.* Cambridge University Press, Cambridge.

Edmondson, W. T. 1979. Lake Washington and the predictability of limnological events. *Archiv für Hydrobiologie, Beiheft* 13:234–241.

Edmunds, M. 1974. *Defence in Animals.* Harlow, Essex.

Edwards, C. A., and G. W. Heath. 1963. The role of soil animals in breakdown of leaf material. Pages 76–84 in D. Doiksen and J. van der Pritt (eds.), *Soil Organisms.* North-Holland, Amsterdam.

Edwards, P. J., and S. P. Wratten. 1985. Induced plant defenses against insect grazing: fact or artifact? *Oikos* 44:70–74.

Ehler, L. E. 1979. Assessing competitive interactions in parasite guilds prior to introduction. *Environmental Entomology* 8:558–560.

Ehler, L. E., and R. W. Hall. 1982. Evidence for competitive exclusion of introduced natural enemies in biological control. *Environmental Entomology* 1:1–4.

Ehrlich, A., and P. R. Ehrlich. 1987. *Earth*. Franklin Watts, New York.

Ehrlich, P. R. 1975. The population biology of coral reef fishes. *Annual Review of Ecology and Systematics* 6:213–247.

Ehrlich, P. R. 1985. Human ecology for introductory biology courses: an overview. *American Zoologist* 25:379–394.

Ehrlich, P. R., and A. Ehrlich. 1970. *Population Resources Environment: Issues in Human Ecology*, 2nd edition. Freeman, San Francisco.

Ehrlich, P. R., and A. Ehrlich. 1981. *Extinction: The Causes and Consequences of the Disappearance of Species*. Random House, New York.

Ehrlich, P. R., A. Ehrlich, and J. P. Holden. 1977. *Ecoscience: Population, Resources, Environment*. Freeman, San Francisco.

Ehrlich, P. R., and P. H. Raven. 1964. Butterflies and plants: a study in coevolution. *Evolution* 18:586–608.

Ehrlich, P. R., and J. Roughgarden. 1987. *The Science of Ecology*. Macmillan, New York.

Eisner, T., and D. J. Aneshansley. 1982. Spray aiming in bombardier beetles: jet deflection by the Coanda effect. *Science* 215:83–85.

Eisner, T., and J. Meinwald. 1966. Defensive secretions of arthropods. *Science* 153:1341–1350.

Elner, R. W., and R. N. Hughes. 1978. Energy maximization in the diet of the shore crab, *Carcinus maenas*. *Journal of Animal Ecology* 47:103–116.

Elton, C. 1927. *Animal Ecology*. Sidgwick and Jackson, London.

Elton, C. 1958. *The Ecology of Invasions by Animals and Plants*. Methuen, London.

Elton, C., and M. Nicholson. 1942. The ten-year cycle in numbers of the lynx in Canada. *Journal of Animal Ecology* 11:215–244.

Eltringham, S. K. 1984. *Wildlife Resources and Economic Development*. Wiley, Chichester, England.

Erickson, E., and S. L. Buchmann. 1983. Electrostatics and pollination. Pages 173–184 in C. E. Jones and R. J. Little (eds.), *Handbook of Experimental Pollination Biology*. Van Nostrand Reinhold, New York.

Erlinge, S., G. Göransson, G. Högstedt, G. Jansson, O. Liberg, J. Loman, I. N. Nilsson, T. von Schantz, and M. Sylvén. 1984. Can vertebrate predators regulate their prey? *American Naturalist* 123:125–133.

Erwin, T. L. 1982. Tropical forests: their richness in Coleoptera and other arthropod species. *Coleopterists Bulletin* 36:74–75.

Erwin, T. L. 1983. Beetles and other insects of tropical forest canopies at Manaus, Brazil, sampled by insecticidal fogging. Pages 59–75 in S. L. Sutton, T. C.

Whitmore, and A. C. Chadwick (eds.), *Tropical Rain Forest: Ecology and Management*. Blackwell Scientific Publications, Oxford.

Estes, J. A., R. J. Jameson, and E. B. Rhode. 1982. Activity and prey selection in the sea otter: influence of population status on community structure. *American Naturalist* 120:242–258.

Etherington, J. R. 1975. *Environmental Plant Ecology*. Wiley, New York.

Faeth, S. H. 1985. Quantitative defense theory and patterns of feeding by oak insects. *Oecologia* 68:34–40.

Faeth, S. H. 1988. Plant-mediated interactions between seasonal herbivores: enough for evolution or coevolution? Pages 391–414 in K. C. Spencer (ed.), *Chemical Mediation of Coevolution*. Academic Press, New York.

Fagerström, T. 1987. On theory, data and mathematics in ecology. *Oikos* 50:258–261.

Fahim, H. M. 1981. *Dams, People and Development, the Aswan High Dam Case*. Pergamon, New York.

Farnsworth, E. G., T. T. Tidrick, C. F. Jordan, and W. M. Smathers. 1981. The value of natural ecosystems: an economic and ecological framework. *Environmental Conservation* 8:275–282.

Fearnside, P. M. 1987. Deforestation and international economic development projects in Brazilian Amazonia. *Conservation Biology* 1:214–221.

Feeny, P. 1970. Seasonal changes in the oak leaf tannins and nutrients as a cause of spring feeding by winter moth caterpillars. *Ecology* 51:565–581.

Feeny, P. 1976. Plant apparency and chemical defense. *Recent Advances in Phytochemistry* 10:1–40.

Feinsinger, P. 1983. Coevolution and pollination. Pages 283–310 in D. J. Futuyma and M. Slatkin (eds.), *Coevolution*. Sinauer Associates, Sunderland, Mass.

Felsenstein, J. 1974. The evolutionary advantage of recombination. *Genetics* 78:737–756.

Felsenstein, J. 1982. Numerical methods for inferring evolutionary trees. *Quarterly Review of Biology* 57:379–404.

Felsenstein, J. 1985. Phylogenies from gene frequencies: a statistical problem. *Systematic Zoology* 34:300–311.

Fenchel, T. 1974. Intrinsic rate of natural increase: the relationship with body size. *Oecologia* 14:317–326.

Fenner, F., and F. Ratcliffe. 1965. *Myxamatosis*. Cambridge University Press, Cambridge.

Firey, W. J. 1960. *Man, Mind and Land: Theory of Resource Use*. Free Press, Glencoe, Ill. Greenwood Press, London.

Fischer, J., N. Simon, and J. Vincent. 1969. *The Red Book—Wildlife in Danger*. Collins, London.

Fisher, R. A. 1930. *The Genetical Theory of Natural Selection*. Clarendon Press, Oxford.

Fisher, R. A., A. S. Corbet, and C. B. Williams. 1943. The relation between the number of species and the number of individuals in a random sample of an animal population. *Journal of Animal Ecology* 12:42–58.

Fleiss, J. L. 1981. *Statistical Methods for Rates and Proportions.* Wiley, New York.

Flessa, K. W., and D. Jablonski. 1985. Declining Phanerozoic background extinction rates: effects of taxonomic structure? *Nature* 313:216–218.

Food and Agriculture Organization. 1986. *Production Yearbook.* FAO, Rome.

Foose, T. J. 1983. The relevance of captive populations to the conservation of biotic diversity. Pages 374–401 in C. M. Schonewald-Cox, S. M. Chambers, B. MacBryde, and L. Thomas (eds.), *Genetics and Conservation.* Benjamin/Cummings, Menlo Park, Calif.

Ford, J. 1988. Sheep will grow woolier on a bioengineered diet. *New Scientist* 117(1603):24.

Ford, R. G., and F. A. Pitelka. 1984. Resource limitation in the California vole. *Ecology* 65:122–136.

Forse, B. 1989. The myth of the marching desert. *New Scientist* 121:31–32.

Foster, G. M., and B. G. Anderson. 1979. *Medical Anthropology.* Wiley, New York.

Fowler, H. G. 1987. Field confirmation of the phonotaxis of *Euphasiopteryx depleta* (Diptera: Tachinidae) to calling males of *Scapteriscus vicinus* (Orthoptera: Gryllotalpidae). *Florida Entomologist* 70:409–410.

Fowler, S. V., and J. H. Lawton. 1985. Rapidly induced defenses and talking trees: the devil's advocate position. *American Naturalist* 126:181–195.

Fowler, S. V., and M. MacGarvin. 1985. The impact of hairy wood ants, *Formica lugubris*, on the guild structure of herbivorous insects on birch, *Betula pubescens. Journal of Animal Ecology* 54:847–855.

Fraley, L., and F. W. Whicker. 1971. Response of a native shortgrass plant stand to ionizing radiation. Pages 999–1006 in D. J. Nelson (ed.), *Radionuclides in Ecosystems. Proceedings of the Third National Symposium on Radioecology.* U.S. Atomic Energy Commission, Washington, D.C.

Franey, J. R., M. V. Ivanov, and H. Rhodne. 1983. The sulphur cycle. Pages 56–61 in B. Bolin and R. B. Cook (eds.), *The Major Biogeochemical Cycles and Their Interactions. Scientific Committee on Problems of the Environment Publication 21.* Wiley, Chichester, England.

Frankel, O. H., and M. E. Soulé. 1981. *Conservation and Evolution.* Cambridge University Press, Cambridge.

Frenkel, R. E. 1970. *Ruderal Vegetation Along Some California Roadsides. University of California Publications in Geography 20,* Berkeley.

Fretwell, S. D. 1972. *Populations in a Seasonal Environment.* Princeton University Press, Princeton, N.J.

Fretwell, S. D., and H. L. Lucas. 1970. On territorial behaviour and other factors influencing habitat distribution in birds. *Acta Biotheoretica* 19:16–36.

Freudenberger, D. 1982. Southern Africa. Pages 3–192 in D. Yerex (ed.), *The Farming of Deer.* Agricultural Associates, Wellington, New Zealand.

Fricke, H., and S. Fricke. 1977. Monogamy and sex change by aggressive dominance in coral reef fish. *Nature* 266:830–832.

Fryer, G., and T. D. Iles. 1972. *The Cichlid Fishes of the Great Lakes of Africa.* T. F. H. Publications, Neptune City, N.J.

Futuyma, D. J. 1983. Evolutionary interactions among herbivores and plants. Pages 207–231 in D. J. Futuyma and M. Slatkin (eds.), *Coevolution.* Sinauer Associates, Sunderland, Mass.

Futuyma, D. J. 1986. *Evolutionary Biology,* 2nd edition. Sinauer Associates, Sunderland, Mass.

Futuyma, D. J., and S. C. Peterson. 1985. Genetic variation in the use of resources by insects. *Annual Review of Entomology* 30:217–238.

Gaertner, F., and L. Kim. 1988. Current applied recombinant DNA projects. *Trends in Ecology and Evolution* 3(4) and *Trends in Biotechnology* 6(4), special combined issue: 4–6.

Gaines, M. S., and L. R. McClenaghan. 1980. Dispersal in small mammals. *Annual Review of Ecology and Systematics* 11:163–196.

Game, M., and F. G. Peterken. 1984. Nature reserve selection strategies in the woodlands of central Lincolnshire, England. *Biological Conservation* 29:157–181.

Gause, G. F. 1932. Experimental studies on the struggle for existence. I. Mixed population of two species of yeast. *Journal of Experimental Biology* 9:389–402.

Gause, G. F. 1934. *The Struggle for Existence.* Macmillan (Hafner Press), New York (reprinted 1964).

Gentle, W., F. R. Humphreys, and M. J. Lambert. 1965. An examination of *Pinus radiata* phosphate fertilizer trial fifteen years after treatment. *Forest Science* 11:315–324.

Gentry, A. H. 1988. Tree species of upper Amazonian forests. *Proceedings of the National Academy of Sciences of the United States of America* 85:156.

George, C. J. 1972. The role of the Aswan High Dam in changing the fisheries of the southeastern Mediterranean. Pages 159–178 in M. Taghi Farvar and J. P. Milton (eds.), *The Careless Technology: Ecology and International Development.* Natural History Press, New York.

Gerrish, G., D. Mueller-Dombois, and K. W. Bridges. 1988. Nutrient limitation and *Metrosideros* forest dieback in Hawaii. *Ecology* 69:723–727.

Gessel, S. P. 1962. Progress and problems in mineral nutrition of forest trees. Pages 221–235 in T. T. Kozlowski (ed.), *Tree Growth.* Ronald Press, New York.

Ghiselin, M. T. 1974. *The Economy of Nature and the Evolution of Sex.* University of California Press, Berkeley and Los Angeles.

Gilbert, F. S. 1980. The equilibrium theory of island biogeography: fact or fiction. *Journal of Biogeography* 7:209–235.

Gill, D. E. 1974. Intrinsic rate of increase, saturation density, and competitive ability. II. The evolution of competitive ability. *American Naturalist* 108:103–116.

Gill, F. B., and L. L. Wolf. 1975. Economies of feeding territoriality in the golden-winged sunbird. *Ecology* 56:333–345.

Gilman, A. P., D. P. Peakall, D. J. Hallett, G. A. Fox, and R. J. Norstrom. 1979. *Animals as Monitors of Environmental Pollutants.* National Academy Press, Washington, D.C.

Gilpin, M. E. 1987. Theory *vs.* practice. *Trends in Ecology and Evolution* 2:169.

Givnish, T. J. 1982. On the adaptive significance of leaf height in forest herbs. *American Naturalist* 120:353–381.

Givnish, T. J., and G. J. Vermeij. 1976. Sizes and shapes of liane leaves. *American Naturalist* 110:743–778.

Godfrey, L. R. 1983. *Scientists Confront Creationism.* Norton, New York.

Goedert, W. J. 1983. Management of the Cerrado soils of Brazil: a review. *Journal of Soil Science* 34:405–428.

Goldberg, E. D., and K. K. Bertine. 1975. Marine pollution. Pages 273–295 in W. W. Murdoch (ed.), *Environment,* 2nd edition. Sinauer Associates, Sunderland, Mass.

Goldberg, E. D., V. T. Bower, J. W. Farrington, G. Harvey, J. H. Martin, P. L. Parker, R. W. Riseborough, W. Robertson, E. Schneider, and E. Gamble. 1978. The Mussel Watch. *Environmental Conservation* 5:101–125.

Goldsmith, F. B. 1983. Evaluating nature. Pages 233–246 in A. Warren and F. B. Goldsmith (eds.), *Conservation in Perspective.* Wiley, Chichester, England.

Goodland, R. J. A. 1987. The World Bank's wildlands policy: a major new means of financing conservation. *Conservation Biology* 1:210–213.

Goodman, D. 1975. The theory of diversity-stability relationships in ecology. *Quarterly Review of Biology* 50:237–266.

Gordon, H. S. 1954. The economic theory of a common property resource: the fishery. *Journal of Political Economics* 62:124–142.

Götmark, F., M. Åhlund, and M. O. G. Eriksson. 1986. Are indices reliable for assessing conservation value of natural areas? An avian case study. *Biological Conservation* 38:55–73.

Goudie, A. 1986. *The Human Impact on the Natural Environment,* 2nd edition. MIT Press, Cambridge, Mass.

Gould, S. J. 1974. The origin and function of "bizarre" structures: antler size and skull size in the "Irish elk," *Megaloceros giganteus. Evolution* 28:191–220.

Gould, S. J., and N. Eldredge. 1977. Punctuated equilibria: the tempo and mode of evolution reconsidered. *Paleobiology* 3:115–151.

Gould, S. J., and R. C. Lewontin. 1979. The spandrels of San Marco and the Panglossian paradigm: a critique of the adaptationist programme. *Proceedings of the Royal Society of London B* 205:581–598.

Goulding, M. 1980. *The Fishes and the Forest: Explorations in Amazonian Natural History.* University of California Press, Berkeley and Los Angeles.

Grant, K. A., and V. Grant. 1964. Mechanical isolation of *Salvia apiana* and *Salvia mellifera* (Labiatae). *Evolution* 18:196–212.

Grant, M. 1973. Mercury in man. Pages 96–105 in W. Jackson (ed.), *Man and the Environment*, 2nd edition. William C. Brown Company, Dubuque, Iowa.

Grant, P. R., and B. R. Grant. 1987. The extraordinary El Niño event of 1982–1983: effects on Darwin's finches on Isla Genovesa, Galapagos. *Oikos* 49:55–66.

Grant, V. 1977. *Organismic Evolution*. Freeman, San Francisco.

Grant, V. 1981. *Plant Speciation*, 2nd edition. Columbia University Press, New York.

Grant, V. 1985. *The Evolutionary Process, a Critical Review of Evolutionary Theory*. Columbia University Press, New York.

Gray, J., and A. J. Boucot (eds.). 1979. *Historical Biogeography, Plate Tectonics and the Changing Environment*. Oregon State University Press, Corvallis.

Greene, J. C. 1959. *The Death of Adam: Evolution and Its Impact on Western Thought*. Iowa State University Press, Ames.

Greenslade, P. J. M. 1983. Adversity selection and the habitat templet. *American Naturalist* 122:352–365.

Greenwood, P. J. 1980. Mating systems, philopatry, and dispersal in birds and mammals. *Animal Behaviour* 28:1140–1162.

Greenwood, P. J., P. H. Harvey, and C. M. Perrins. 1978. Inbreeding and dispersal in the great tit. *Nature* 271:52–54.

Greenwood, P. J., and P. Wheeler. 1985. The evolution of sexual size dimorphism in birds and mammals: a "hot-blooded hypothesis." Pages 287–299 in P. J. Greenwood, P. H. Harvey, and M. Slatkin (eds.), *Evolution, Essays in Honour of John Maynard Smith*. Cambridge University Press, Cambridge.

Grime, J. P. 1977. Evidence for the existence of three primary strategies in plants and its relevance to ecological and evolutionary theory. *American Naturalist* 111:1169–1194.

Grime, J. P. 1979. *Plant Strategies and Vegetation Process*. Wiley, New York.

Grimm, N. B. 1987. Nitrogen dynamics during succession in a desert stream. *Ecology* 68:1157–1170.

Gulhati, K. 1977. Compulsory sterilization: the change in India's population policy. *Science* 195:1300–1305.

Gulland, J. 1988. The end of whaling? *New Scientist* 120(1636):42–47.

Gupta, A. P., and R. C. Lewontin. 1982. A study of reaction norms in natural populations of *Drosophila pseudoobscura*. *Evolution* 36:934–948.

Hadley, J. L., and W. K. Smith. 1986. Wind effects on needles of timberline conifers: seasonal influence on mortality. *Ecology* 67:12–19.

Hairston, N. G., J. D. Allen, R. K. Colwell, D. J. Futuyma, J. Howell, M. D. Lubin, J. Mathias, and J. H. Vandermeer. 1968. The relationship between species diversity and stability: an experimental approach with protozoa and bacteria. *Ecology* 49:1091–1101.

Hairston, N. G., F. E. Smith, and L. B. Slobodkin. 1960. Community structure, population control, and competition. *American Naturalist* 44:421–425.

Haldane, J. B. S. 1953. Animal populations and their regulation. *Penguin Modern Biology* 15:9–24.

Haldane, J. B. S. 1963. The acceptance of a scientific idea. Reprinted in R. L. Weber (ed.), 1982, *More Random Walks in Science*. Institute of Physics, Bristol.

Hall, R. W., L. E. Ehler, and B. Bisabri-Ershadi. 1980. Rates of success in classical biological control of arthropods. *Bulletin of the Entomological Society of America* 26:111–114.

Hallam, A. 1983. Plate tectonics and evolution. Pages 367–386 in D. S. Kendall (ed.), *Evolution from Molecules to Man*. Cambridge University Press, Cambridge.

Hallam, A. 1984. Pre-Quaternary sea-level changes. *Annual Review of Earth and Planetary Science* 12:205–243.

Halliday, T. 1978. *Vanishing Birds*. Holt, Rinehart and Winston, New York.

Hamburg, S. P., and C. V. Cogbill. 1988. Historical decline of red spruce populations and climatic warming. *Nature* 331:428–431.

Hamilton, W. D. 1964. The genetical evolution of social behaviour. I, II. *Journal of Theoretical Biology* 7:1–52.

Hamilton, W. D. 1967. Extraordinary sex ratios. *Science* 156:477–488.

Hamilton, W. D. 1971. Geometry for the selfish herd. *Journal of Theoretical Biology* 31:295–311.

Hamilton, W. D., and M. Zuk. 1982. Heritable true fitness and bright birds: a role for parasites? *Science* 218:384–387.

Hansen, A. J. 1986. Fighting behavior in bald eagles: a test of game theory. *Ecology* 67:787–797.

Harcourt, A. H., P. H. Harvey, S. G. Larson, and R. V. Short. 1981. Testis weight, body weight and breeding system in primates. *Nature* 293:55–57.

Hardin, G. 1957. The threat of clarity. *American Journal of Psychiatry* 114:392–396.

Hardin, G. 1960. The competitive exclusion principle. *Science* 131:1292–1297.

Hardin, G. (ed.). 1964. *Population, Evolution and Birth Control*, 2nd edition. Freeman, San Francisco.

Hardin, G. 1968. The tragedy of the commons. *Science* 162:1243–1248.

Harley, J. L., and S. E. Smith. 1983. *Mycorrhizal Symbiosis*. Academic Press, London.

Harrar, J. G., and S. Wortman. 1969. Expanding food production in hungry nations: the promise, the problems. Pages 84–135 in C. M. Hardin (ed.), *Overcoming World Hunger: The American Assembly*, 1969. Prentice-Hall, New York.

Harris, H. 1966. Enzyme polymorphisms in man. *Proceedings of the Royal Society of London Series B* 164:298–310.

Harrison, R. G. 1980. Dispersal polymorphism in insects. *Annual Review of Ecology and Systematics* 11:95–118.

Harvey, P. H., and P. J. Greenwood. 1978. Anti-predator defense strategies: some evolutionary problems. Pages 129–151 in J. R. Krebs and N. B. Davies (eds.), *Behavioural Ecology, an Evolutionary Approach.* Sinauer Associates, Sunderland, Mass.

Harvey, P. H., and A. H. Harcourt. 1982. Sperm competition, testes size and breeding systems in primates. Pages 589–600 in R. C. Smith (ed.), *Sperm Competition and the Evolution of Animal Mating Systems.* Academic Press, London.

Harvey, P. H., M. Kavanagh, and T. H. Clutton-Brock. 1978. Sexual dimorphism in primate teeth. *Journal of Zoology* 186:475–486.

Hasler, A. D. 1947. Eutrophication of lakes by domestic drainage. *Ecology* 28:383–395.

Hassell, M. F., and G. C. Varley. 1969. New inductive population model for insect parasites and its bearing on biological control. *Nature* 223:1133–1137.

Hastings, J. R., and R. M. Turner. 1965. *The Changing Mile.* University of Arizona Press, Tucson.

Haukioja, E. 1980. On the role of plant defenses in the fluctuation of herbivore populations. *Oikos* 35:202–213.

Hawksworth, D. 1971. Lichens as litmus for air pollution: a historical review. *International Journal of Environmental Studies* 1:281–296.

Heal, O. W., and S. F. Maclean. 1975. Comparative productivity in ecosystems—secondary productivity. Pages 89–108 in W. H. van Dobben and R. H. Lowe-McConnell (eds.), *Unifying Concepts in Ecology.* Dr W. Junk, The Hague.

Heaney, L. R. 1986. Biogeography of mammals in S.E. Asia: estimates of rates of colonization, extinction, and speciation. *Biological Journal of the Linnaean Society* 28:127–165.

Heatwole, H., and T. A. Walker. 1989. Dispersal of alien plants to coral cays. *Ecology* 70:787–790.

Heinrich, B. 1976. Flowering phenologies: bog, woodland, and disturbed habitats. *Ecology* 57:890–899.

Heinrich, B. 1979. *Bumblebee Economics.* Harvard University Press, Cambridge, Mass.

Helliwell, P. R. 1974. The value of vegetation for conservation. II. M1 motorway area. *Journal of Environmental Management* 2:75–78.

Hennig, W. 1979. *Phylogenetic Systematics.* University of Illinois Press, Urbana.

Henwood, K., and A. Fabrick. 1979. A quantitative analysis of the dawn chorus: temporal selection for communicatory optimization. *American Naturalist* 114:260–274.

Herbold, B., and P. B. Moyle. 1986. Introduced species and vacant niches. *The American Naturalist* 128–751–760.

Heron, A. C. 1972. Population ecology of a colonizing species: the pelagic tunicate *Thalia democratica. Oecologia* 10:269–293, 294–312.

Hershkovitz, P. 1977. *Living New World Monkeys (Platyrhini), Vol. 1.* University of Chicago Press, Chicago.

Heslop-Harrison, J. 1964. Forty years of genecology. *Advances in Ecological Research* 2:159–247.

Hesse, R., W. C. Allee, and K. P. Schmidt. 1951. *Ecological Animal Geography,* 2nd edition. Wiley, New York.

Hessler, R., P. Lonsdale, and J. Hawkins. 1988. Patterns on the ocean floor. *New Scientist* 117(1605):47–51.

Heywood, V. H. 1989. Patterns, extents and modes of invasions by terrestrial plants. Pages 31–60 in J. A. Drake, H. A. Mooney, F. di Castri, R. H. Groves, F. J. Kruger, M. Rejmanek, and M. Williamson (eds.), *Biological Invasion: A Global Perspective.* Wiley, Chichester, England.

Hibler, C. P., R. E. Lange, and C. Metzger. 1972. Transplacental transmission of *Protostrongylus* spp. in big-horn sheep. *Journal of Wildlife Diseases* 8:389.

Hilden, O. 1965. Habitat selection in birds: a review. *Annales Zoologici Fennici* 2:53–75.

Hocker, H. W., Jr. 1956. Certain aspects of climate as related to the distribution of loblolly pine. *Ecology* 37:824–834.

Hokkanen, H., and D. Pimentel. 1984. New approach for selecting biological control agents. *Canadian Entomologist* 116:1109–1121.

Holden, C. 1977. Endangered species: review of law triggered by Tellico impasse. *Science* 196:1426–1428.

Holden, C. 1987. New questions in the Strobel case. *Science* 237:1097–1098.

Holden, C. 1988. The greening of the World Bank. *Science* 240:1610–1611.

Holdgate, M. W. 1960. The fauna of the mid-Atlantic islands. *Proceedings of the Royal Society of London B* 152:550–567.

Holdgate, M. W. 1978. The balance between food production and conservation. Pages 227–242 in J. G. Hawkes (ed.), *Conservation and Agriculture.* Duckworth, London.

Hölldobler, B. 1977. Communication in social Hymenoptera. Pages 418–471 in T. A. Sebeok (ed.), *How Animals Communicate.* Indiana University Press, Bloomington, Ind.

Holliday, R. J., and M. S. Johnson. 1979. The contribution of derelict mineral and industrial sites to the conservation of rare plants in the United Kingdom. *Minerals and the Environment* 1:1–7.

Holling, C. S. 1959. Some characteristics of simple types of predation and parasitism. *Canadian Entomologist* 91:385–398.

Holling, C. S. (ed.). 1978. *Adaptive Environmental Assessment and Management.* Wiley, Chichester, England.

Holm, L. G., L. W. Weldon, and R. D. Blackburn. 1969. Aquatic weeds. *Science* 166:699–709.

Hopcraft, D. 1982. Wildlife ranching in perspective. *Tigerpaper* 9:17–20.

Horn, H. S. 1968. The adaptive significance of colonial nesting in Brewer's blackbird (*Euphagus cyanocephalus*). *Ecology* 49:682–694.

Horn, H. S. 1971. *Adaptive Geometry of Trees.* Princeton University Press, Princeton, N.J.

Horn, H. S. 1975. Markovian processes of forest succession. Pages 196–213 in M.

L. Cody and J. Diamond (eds.), *Ecology and Evolution of Communities*. Harvard University Press, Cambridge, Mass.

Horn, H. S. 1976. Succession. Pages 187–204 in R. M. May (ed.), *Theoretical Ecology: Principles and Applications*. Saunders, Philadelphia.

Horn, H. S., and R. M. May. 1977. Limits to similarity among coexisting competitors. *Nature* 270:660–661.

Houston, A. I., and N. B. Davies. 1985. The evolution of cooperation and life history in the dunnock, *Prunella modularis*. Pages 471–487 in R. M. Sibly and R. H. Smith (eds.), *Behavioural Ecology, Ecological Consequences of Adaptive Behaviour*. Blackwell Scientific Publications, Oxford.

Houston, A. I., and J. M. McNamara. 1982. A sequential approach to risk taking. *Animal Behaviour* 30:1260–1261.

Howard, R. D. 1978. The evolution of mating strategies in bullfrogs, *Rana catesbeiana*. *Evolution* 32:850–871.

Howard, W. E. 1949. Dispersal, amount of inbreeding, and longevity in a local population of prairie deer mice on the George Reserve, southern Michigan. Contributions from the Laboratory of Vertebrate Biology of the University of Michigan 43:1–50.

Howarth, F. G. 1983. Classical biological control: Panacea or Pandora's box? *Proceedings of the Hawaii Entomological Society* 24:239–244.

Howe, H. F., and J. Smallwood. 1982. Ecology of seed dispersal. *Annual Review of Ecology and Systematics* 13:201–228.

Huffaker, C. B., and C. E. Kennett. 1959. A ten year study of vegetational changes associated with biological control of Klamath weed. *Journal of Range Management* 12:69–82.

Huffaker, C. B., and C. E. Kennett. 1969. Some aspects of assessing efficiency of natural enemies. *Canadian Entomologist* 101:425–440.

Hungate, R. E. 1975. The rumen microbial system. *Annual Review of Ecology and Systematics* 6:39–66.

Hunt, E. C., and A. I. Bischoff. 1960. Inimical effects on wildlife of periodic DDT applications to Clear Lake, California. *California Fish and Game Bulletin* 46:91–106.

Hurlbert, S. H. 1971. The nonconcept of species diversity: a critique and alternative parameters. *Ecology* 52:577–586.

Hutchinson, G. E. 1959. Homage to Santa Rosalia, or why are there so many kinds of animals? *American Naturalist* 93:145–159.

Hutchinson, J. B. 1965. Crop-plant evolution: A general discussion. Pages 166–181 in J. B. Hutchinson (ed.), *Essays on Crop Plant Evolution*. Cambridge University Press, New York.

Iason, G. R., C. D. Duck, and T. H. Clutton-Brock. 1986. Grazing and reproductive success of red deer: the effect of local enrichment by gull colonies. *Journal of Animal Ecology* 55:507–515.

Ichikawa, T., and S. Ishii. 1974. Mating signal of the brown planthopper

Nilaparvata Lugens (Stål) (Homoptera: Delphacidae): vibration of the sub-strate. *Applied Entomology and Zoology* 9:196–198.

Illies, J. 1974. *Introduction to Zoogeography*. Macmillan, London.

Inouye, R. S., N. J. Huntly, D. Tilman, J. R. Tester, M. Stillwell, and K. C. Zinnel. 1987. Old-field succession on a Minnesota sand plain. *Ecology* 68:12–26.

International Union for Conservation of Nature and Natural Resources. 1980. *World Conservation Strategy*. International Union for Conservation of Nature and Natural Resources, United National Environmental Program, World Wildlife Fund. Gland, Switzerland.

International Union for Conservation of Nature and Natural Resources. 1985. *The United Nations List of National Parks and Protected Areas*. Gland, Switzerland.

Iwasa, Y. 1982. Vertical migration of zooplankton: a game between predator and prey. *American Naturalist* 120:171–180.

Jablonski, D. 1984. Keeping time with mass extinctions. *Paleobiology* 10:139–145.

Jablonski, D. 1986a. Evolutionary consequences of mass extinctions. Pages 313–330 in D. M. Raup and D. Jablonski (eds.), *Patterns and Processes in the History of Life*. Springer-Verlag, Berlin.

Jablonski, D. 1986b. Causes and consequences of mass extinctions: a comparative approach. Pages 183–230 in D. K. Elliott (ed.), *Dynamics of Extinction*. Wiley, New York.

Jackson, M. V., and T. E. Williams. 1979. Response of grass swards to fertilizer N under cutting or grazing. *Journal of Agricultural Science*, Cambridge, 92:549–562.

Jackson, W., and J. Piper. 1989. The necessary marriage between ecology and agriculture. *Ecology* 70:1591–1593.

Jacobson, M. 1982. Plants, insects and man—their interrelationships. *Economic Botany* 36:346–354.

Janzen, D. H. 1966. Coevolution of mutualism between ants and acacias in Central America. *Evolution* 20:249–275.

Janzen, D. H. 1970. Herbivores and the number of tree species in tropical forests. *American Naturalist* 104:501–528.

Janzen, D. H. 1979a. How to big a fig. *Annual Review of Ecology and Systematics* 10:13–51.

Janzen, D. H. 1979b. New horizons in the biology of plant defenses. Pages 331–350 in G. A. Rosenthal and D. H. Janzen (eds.), *Herbivores: Their Interaction with Secondary Plant Metabolites*. Academic Press, New York.

Janzen, D. H. 1979c. Why fruit rots. *Natural History Magazine* 88(6):60–64.

Janzen, D. H. 1981. The peak in North American ichneumonid species richness lies between 38° and 42°N. *Ecology* 62:532–537.

Janzen, D. H. 1983. Dispersal of seeds by vertebrate guts. Pages 232–262 in D. J. Futuyma and M. Slatkin (eds.), *Coevolution*. Sinauer Associates, Sunderland, Mass.

Janzen, D. H., and P. S. Martin. 1982. Neotropical anachronisms: the fruits the gomphotheres ate. *Science* 215:19–27.

Jarvis, J. V. M. 1981. Eusociality in a mammal: co-operative breeding in naked mole rat colonies. *Science* 212:241–250.

Jarvis, J. V. M., and J. B. Sale. 1971. Burrowing and burrow patterns of East African mole rats—*Tachyoryctes, Heliophobius,* and *Heterocephalus. Journal of Zoology* 163:451–479.

Jarvis, P. A. 1979. The ecology of plant and animal introductions. *Progress in Physical Geography* 3:187–214.

Jenny, J. 1980. *The Soil Resource: Origin and Behavior.* Springer-Verlag, New York.

Johanson, D. C., and T. D. White. 1979. A systematic assessment of early African hominids. *Science* 203:321–330.

Johnson, C. G. 1969. *Migration and Dispersal of Insects by Flight.* Methuen, London.

Johnson, M. S. 1978. The botanical significance of derelict industrial sites in Britain. *Environmental Conservation* 5:223–238.

Johnson, M. S., P. D. Putwain, and R. J. Holliday. 1978. Wildlife conservation value of derelict metalliferous mine workings in Wales. *Biological Conservation* 14:131–148.

Jones, C. E., and R. J. Little (eds.). 1983. *Handbook of Experimental Pollination Biology.* Van Nostrand Reinhold, New York.

Jones, D. A. 1973. Co-evolution and cyanogenesis. Pages 213–242 in V. H. Heywood (ed.), *Taxonomy and Ecology.* Academic Press, New York.

Jones, D. F. 1924. The attainment of homozygosity in inbred strains of maize. *Genetics* 9:405–418.

Jones, J. S., B. H. Leith, and P. Rawllings. 1977. Polymorphism in *Cepaea:* a problem with too many solutions. *Annual Review of Ecology and Systematics* 8:109–143.

Jordan, C. F., and J. R. Kline. 1972. Mineral cycling: some basic concepts and their application in a tropical rain forest. *Annual Review of Ecology and Systematics* 3:33–49.

Jukes, T. H. 1983. Evolution of the amino acid code. Pages 191–207 in M. Nei and R. K. Koehn (eds.), *Evolution of Genes and Proteins.* Sinauer Associates, Sunderland, Mass.

Karban, R. 1980. Periodical cicada nymphs impose periodical oak tree wood accumulation. *Nature* 287:326–327.

Karban, R. 1982. Experimental removal of 17-year cicada nymphs and growth of host apple trees. *Journal of the New York Entomological Society* 90:74–81.

Karban, R. 1985. Addition of periodical cicada nymphs to an oak forest: effects on cicada density, acorn production and rootlet density. *Journal of the Kansas Entomological Society* 58:269–276.

Karban, R. 1987. Effects of clonal variation of the host plant, interspecific competition, and climate on the population size of a folivorous thrips. *Oecologia* 74:298–303.

Karban, R. 1989. Community organization of *Erigeron glaucus* folivores: effects of competition, predation, and host plant. *Ecology* 70:1028–1039.

Karban, R., and J. R. Carey. 1984. Induced resistance of cotton seedlings to mites. *Science* 225:53–54.

Karban, R., and R. E. Ricklefs. 1984. Leaf traits and species richness and abundance of lepidopteran larvae on deciduous trees in southern Ontario. *Oikos* 43:165–170.

Kareiva, P. 1983. Influence of vegetation texture on herbivore populations: resource concentraton and herbivore movement. Pages 259–289 in R. F. Denno and M. S. McClure (eds.), *Variable Plants and Herbivores in Natural and Managed Systems.* Academic Press, New York.

Karr, J. R., and I. J. Schlosser. 1978. Water resources and the land-water interface. *Science* 201:229–234.

Keith, L. B. 1983. Role of food in hare population cycles. *Oikos* 40:385–395.

Keller, M. A. 1984. Reassessing evidence for competitive exclusion of introduced natural enemies. *Environmental Entomology* 13:192–195.

Kellert, S. R. 1985. Social and perceptual factors in endangered species management. *Journal of Wildlife Management* 49:528–536.

Kellog, W. W. 1986. Letter to the editor. *Science* 230:1491.

Kenward, R. E. 1978. Hawks and doves: factors affecting success and selection in goshawk attacks on wood-pigeons. *Journal of Animal Ecology* 47:449–460.

Kerr, R. A. 1988. Whom to blame for the Great Storm? *Science* 239:1238–1239.

Kethley, J. B., and D. E. Johnston. 1975. Resource tracking patterns in bird and mammal ectoparasites. *Miscellaneous Publications of the Entomological Society of America* 9:231–236.

Kettlewell, H. B. D. 1955. Selection experiments on industrial melanism in the Lepidoptera. *Heredity* 10:287–301.

Kettlewell, H. B. D. 1973. *The Evolution of Melanism.* Clarendon Press, Oxford.

Keyfitz, M. 1975. Population growth: causes and consequences. Pages 39–64 in W. W. Murdoch (ed.), *Environment,* 2nd edition. Sinauer Associates, Sunderland, Mass.

Kiell, D. J., E. L. Hill, and S. P. Mahoney. 1986. Protecting caribou during hydroelectric development in Newfoundland. Pages 205–226 in *Ecological Knowledge and Environmental Problem-Solving, Concepts and Case Studies.* National Academy Press, Washington, D.C.

Kimura, M. 1983a. *The Neutral Theory of Molecular Evolution.* Cambridge University Press, Cambridge.

Kimura, M. 1983b. The neutral theory of molecular evolution. Pages 208–233 in M. Nei and R. K. Koehn (eds.), *Evolution of Genes and Proteins.* Sinauer Associates, Sunderland, Mass.

King, D. A. 1986. Tree form, height growth, and susceptibility to wind damage in *Acer saccharum. Ecology* 67:980–990.

King, M. C., and A. C. Wilson. 1975. Evolution at two levels: molecular similarities and biological differences between humans and chimpanzees. *Science* 188:107–116.

Kitchener, D. J., A. Chapman, J. Dell, B. G. Muir, and M. Palmer. 1980. Lizard assemblage and reserve size and structure in the western Australian wheatbelt—some implications for conservation. *Biological Conservation* 17:25–62.

Klein, B. C. 1989. Effects of forest fragmentation on dung and carrion beetle communities in central Amazonia. *Ecology* 70:1715–1725.

Klein, D. R. 1968. The introduction, increase, and crash of reindeer on St. Matthew Island. *Journal of Wildlife Management* 32:350–367.

Klein, D. R. 1970. Food selection by North American deer and their response to over-utilization of preferred plant species. Pages 25–46 in A. Watson (ed.), *Animal Populations in Relation to Their Food Resources*. Blackwell Scientific Publications, Oxford.

Klomp, H. 1966. The dynamics of a field population of the pine looper, *Bupalus piniarius* (Lepidoptera: Geom.). *Advances in Ecological Research* 3:207–305.

Knight, R. R., and L. L. Eberhardt. 1985. Population dynamics of Yellowstone grizzly bears. *Ecology* 66:323–334.

Knowlton, N. 1979. Reproductive synchrony, parental investment and the evolutionary dynamics of sexual selection. *Animal Behaviour* 27:1022–1033.

Koblentz-Mishke, I. J., V. V. Volkounsky, R. J. B. Kabanova. 1970. Plankton primary production of the world ocean. Pages 183–193 in W. S. Wooster (ed.), *Scientific Exploration of the South Pacific*. National Academy of Sciences, Washington, D.C.

Koehn, R. K., A. J. Zera, and J. G. Hall. 1983. Enzyme polymorphism and natural selection. Pages 115–136 in M. Nei and R. K. Koehn (eds.), *Evolution of Genes and Proteins*. Sinauer Associates, Sunderland, Mass.

Kozhov, M. 1963. Lake Baikal and its life. *Monographs in Biology* 11:1–344.

Krebs, C. J. 1985a. *Ecology, the Experimental Analysis of Distribution and Abundance*, 3rd edition. Harper and Row, New York.

Krebs, C. J. 1985b. Do changes in spacing behaviour drive population cycles in small mammals? Pages 295–312 in R. M. Sibly and R. H. Smith (eds.), *Behavioural Ecology: Ecological Consequences of Adaptive Behaviour*. Blackwell Scientific Publications, Oxford.

Krebs, C. J. 1988. The experimental approach to rodent population dynamics. *Oikos* 52:143–149.

Krebs, J. R. 1971. Territory and breeding density in the great tit, *Parus major* L. *Ecology* 52:2–22.

Krebs, J. R., and N. B. Davies. 1981. *An Introduction to Behavioural Ecology*. Sinauer Associates, Sunderland, Mass.

Krebs, J. R., J. T. Ericksen, M. I. Webber, and E. L. Charnov. 1977. Optimal prey selection in the great tit, *Parus major*. *Animal Behaviour* 25:30–38.

Krementz, D. G., M. J. Conroy, J. E. Hines, and H. F. Percival. 1988. The effects of hunting on survival rates of American black ducks. *Journal of Wildlife Management* 52:214–226.

Kruuk, H. 1964. Predators and anti-predator behaviour of the black headed gull, *Larus ridibundus. Behaviour Supplement* 11:1–129.

Kruuk, H., and M. Turner. 1967. Comparative notes on predation by lion, leopard, cheetah and wild dog in the Serengeti area, East Africa. *Mammalia* 31:1–27.

Kucera, C. L. 1981. Grasslands and fire. Pages 9–111 in H. A. Mooney, T. M. Bonnicksen, M. L. Christensen, J. E. Lotan, and W. A. Reiners (eds.), *Fire Regimes and Ecosystem Properties.* U.S.D.A. Forest Service General Technical Report WO-26, Washington, D.C.

Kupchella, C. E., and M. C. Hyland. 1989. *Environmental Science,* 2nd edition. Allyn and Bacon, Newton, Mass.

Kuris, A. M., A. R. Blaustein, and J. J. Alio. 1980. Hosts as islands. *American Naturalist* 116:370–386.

Kurtén, B. 1963. Return of a lost structure in the evolution of the felid dentition. *Societas Scientiarum Fennica Arsbok-Vuosikirja* 26:3–11.

Lack, D. 1933. Habitat selection in birds with special references to the effects of afforestation on the Breckland avifauna. *Journal of Animal Ecology* 2:239–262.

Lack, D. 1968. *Ecological Adaptations for Breeding in Birds.* Methuen, London.

Lamarck, J. B. P. de. 1809. *Philosophie Zoologique.* Paris.

Lande, R. 1976. The maintenance of genetic variability by mutation in a polygenic character with linked loci. *Genetic Research* 26:221–235.

Lank, D. B., L. W. Oring, and S. J. Maxson. 1985. Mate and nutrient limitation of egg-laying in a polyandrous shorebird. *Ecology* 66:1513–1524.

Larkin, P. A. 1984. A commentary on environmental impact assessment for large projects affecting lakes and streams. *Canadian Journal of Fisheries and Aquatic Science* 41:1121–1127.

Law, R. 1979. Optional life histories under age-specific predation. *American Naturalist* 114:399–417.

Lawrey, J. D., and M. E. Hale, Jr. 1979. Lichen growth responses to stress induced by automobile exhaust pollution. *Science* 204:423–424.

Laws, R. M., I. S. C. Parker, and R. C. B. Johnstone. 1975. *Elephants and Their Habitat.* Clarendon Press, Oxford.

Lawton, J. H. 1984. Non-competitive populations, non-convergent communities, and vacant niches: the herbivores of bracken. Pages 67–100 in D. R. Strong, D. Simberloff, L. G. Abele, and A. B. Thistle (eds.), *Ecological Communities: Conceptual Issues and the Evidence.* Princeton University Press, Princeton, N.J.

Lawton, J. H., and S. McNeill. 1979. Between the devil and the deep blue sea: On the problem of being a herbivore. Pages 223–244 in R. M. Anderson, B. D. Taylor, and L. R. Taylor (eds.), *Population Dynamics.* Blackwell Scientific Publications, Oxford.

Lawton, J. H., and D. R. Strong. 1981. Community patterns and competition in folivorous insects. *American Naturalist* 118:317–338.

Le Boeuf, B. J. 1974. Male-male competition and reproductive success in elephant seals. *American Zoologist* 14:163–176.

Le Boeuf, B. J., and S. Kaza (eds.). 1981. *The Natural History of Año Nuevo*. Boxwood Press, Pacific Grove, Calif.

Leader-Williams, N., and S. D. Albon. 1988. Allocation of resources for conservation. *Nature* 336:533–535.

Lee, J. A., and B. Greenwood. 1976. The colonization by plants of calcareous wastes from the salt and alkali industry in Cheshire. *Biological Conservation* 1:131–149.

Lees, D. R. 1981. Industrial melanism: genetic adaptation of animals to air pollution. Pages 129–176 in J. A. Bishop and L. M. Cook (eds.), *Genetic Consequences of Man-made Change*. Academic Press, London.

Legg, D. E., and H. C. Chiang. 1984. Rubidium marking technique for the European corn borer (Lepidoptera: Pyralidae) in corn. *Environmental Entomology* 13:579–583.

Lehman, J. T. 1986a. Control of eutrophication in Lake Washington. Pages 301–316 in *Ecological Knowledge and Environmental Problem-Solving, Concepts and Case Studies*. National Academy Press, Washington, D.C.

Lehman, J. T. 1986b. Raising the level of a subarctic lake. Pages 317–330 in *Ecological Knowledge and Environmental Problem-Solving, Concepts and Case Studies*. National Academy Press, Washington, D.C.

Lehmkuhl, J. F. 1984. Determining size and dispersion of minimum viable populations for land management planning and species conservation. *Environmental Management* 8:167–176.

Lerner, I. M. 1954. *Genetic Homeostasis*. Oliver and Boyd, Edinburgh.

Lever, C. 1977. *The Naturalized Animals of the British Isles*. Hutchinson, London.

Levin, D. A. 1979. The nature of plant species. *Science* 204:381–384.

Levin, D. A. 1981. Dispersal versus gene flow in plants. *Annals of the Missouri Botanical Garden* 68:233–253.

Levin, D. A. 1983. Polyploidy and novelty in flowering plants. *American Naturalist* 122:1–25.

Levins, R. 1968. *Evolution in Changing Environments*. Princeton University Press, Princeton, N.J.

Lewin, R. 1983. Santa Rosalia was a goat. *Science* 221:636–639.

Lewis, J. G. 1977. Game domestication for animal production in Kenya: activity patterns of eland, oryx, buffalo and zebu cattle. *Journal of Agricultural Science, Cambridge*, 89:551–563.

Lewis, J. G. 1978. Game domestication for animal production in Kenya: shade behavior and factors affecting the herding of eland, oryx, buffalo and zebu cattle. *Journal of Agricultural Science, Cambridge*, 90:587–595.

Lewis, R. A., N. Stein, and C. W. Lewis (eds.). 1984. *Environmental Specimen Banking and Monitoring as Related to Banking*. Martinus Nijhoff, Boston.

Lewis, W. M., Jr. 1986. Nitrogen and phosphorus runoff losses from a nutrient-poor tropical moist forest. *Ecology* 67:1275–1282.

Lewontin, R. C. 1974. The analysis of variance and the analysis of causes. *American Journal of Human Genetics* 26:400–411.

Lewontin, R. C. 1978. Fitness, survival and optimality. Pages 2–31 in D. J. Horn, R. D. Mitchell, and G. R. Stairs (eds.), *Analysis of Ecological Systems.* Ohio State University Press, Columbus.

Lewontin, R. C., and J. L. Hubby. 1966. A molecular approach to the study of genic heterozygosity in natural populations. II. Amount of variation and degree of heterozygosity in natural populations of *Drosophila pseudoobscura. Genetics* 54:595–609.

Liden, K., and M. Gustafsson. 1967. Relationships and seasonal variation of ^{137}Cs in lichen, reindeer, and man in northern Sweden 1961–1965. Pages 193–208 in B. Aberg and F. P. Hungate (eds.), *Radioecological Concentration Processes. Proceedings of an International Symposium.* Pergamon, Oxford.

Lieth, H. 1975. Primary productivity in ecosystems: comparative analysis of global patterns. Pages 67–88 in W. H. van Dobben and R. H. Lowe-McConnell (eds.), *Unifying Concepts in Ecology.* Dr W. Junk, The Hague.

Lieth, H., and E. Box. 1972. Evapotranspiration and primary productivity: C. W. Thornthwaite Memorial Model. Pages 37–44 in J. R. Mather (ed.), *Papers on Selected Topics in Climatology.* Elmer, New York.

Likens, G. E. (ed). 1981. *Some Perspectives on the Major Biogeochemical Cycles.* Wiley, New York.

Lindeman, R. L. 1942. The trophic-dynamic aspect of ecology. *Ecology* 23:399–418.

Lindow, S. E. 1985. Ecology of *Pseudomonas syringae* relevant to the field use of ice-deletion mutants constructed in vitro for plant frost control. Pages 23–35 in H. O. Halvorson, D. Pramer, and M. Rogul (eds.), *Engineered Organisms in the Environment: Scientific Issues.* American Society for Microbiology, Washington, D.C.

Livezey, B. C., and P. S. Humphrey. 1986. Flightlessness in steamer-ducks (Anatidae: *Tachyeres*): its morphological bases and probable evolution. *Evolution* 40:540–558.

Lloyd, J. E. 1966. Studies on the flash communication system in *Photinus* fireflies. *Miscellaneous Publications of the Museum of Zoology of the University of Michigan* 130:1–195.

Lloyd, J. E. 1975. Aggressive mimicry in *Photuris* fireflies: signal repertoires by femme fatales. *Science* 187:452–453.

Lotka, A. J. 1925. *Elements of Physical Biology.* (Reprinted in 1956 by Dover Publications, New York.)

Lovejoy, T. E. 1979. A projection of species extinctions. Pages 328–329 in *Council on Environmental Quality and Department of State, The Global 2000 Report to the President: Entering the Twenty-first Century.* Council on Environmental Quality, Washington, D.C.

Lovejoy, T. E., R. O. Bierregaard, J. M. Rankin, and H. O. R. Schubert. 1983.

Ecological dynamics of tropical forest fragments. Pages 377–384 in S. L. Sutton, T. C. Whitmore, and A. C. Chadwick (eds.), *Tropical Rain Forest: Ecology and Management*. Blackwell Scientific Publications, Oxford.

Lovejoy, T. E., R. O. Bierregaard, A. B. Rylands, J. R. Malcolm, C. E. Quintela, L. H. Harper, K. S. Brown, Jr., A. H. Powell, G. V. N. Powell, H. O. R. Schubert, and M. R. Hays. 1986. Edge and other effects of isolation on Amazon forest fragments. Pages 257–285 in M. E. Soulé (ed.), *Conservation Biology: The Science of Scarcity and Diversity*. Sinauer Associates, Sunderland, Mass.

Lovelock, J. E. 1979. *Gaia: A New Look at Life on Earth*. Oxford University Press, Oxford.

Lucas, G., and H. Synge. 1978. *Red Data Book*. International Union for the Conservation of Nature and Natural Resources, Morges, Switzerland.

Luck, R. F., and H. Podoler. 1985. Competitive exclusions of *Aphytis lignanensis* by *A. melinus:* potential role of host size. *Ecology* 66:904–913.

Luck, R. F., H. Podoler, and R. Kfir. 1982. Host selection and egg allocation behaviour by *Aphytis melinus* and *A. lignanensis:* comparison of two facultatively gregarious parasitoids. *Ecological Entomology* 7:397–408.

Luoma, J. R. 1988. The human cost of acid rain. *Audubon* 90(4):16–27.

Lyell, C. 1969. *Principles of Geology*. 3 volumes. Reprint of 1830–1833 ed. Introduction by M. J. S. Rudwick. Johnson Reprint Collection, New York.

Mabey, R. 1980. *The Common Ground: A Place for Nature in Britain's Future?* Hutchinson, London.

MacArthur, R. H. 1955. Fluctuations of animal populations, and a measure of community stability. *Ecology* 36:538–546.

MacArthur, R. H., and J. W. MacArthur. 1961. On bird species diversity. *Ecology* 42:594–598.

MacArthur, R. H., and E. R. Pianka. 1966. On the optimal use of a patchy environment. *American Naturalist* 100:603–609.

MacArthur, R. H., and E. O. Wilson. 1963. An equilibrium theory of insular biogeography. *Evolution* 17:373–387.

MacArthur, R. H., and E. O. Wilson. 1967. *The Theory of Island Biogeography*. Princeton University Press, Princeton, N.J.

McCann, T. S. 1981. Aggression and sexual activity of male southern elephant seals. *Journal of the Zoological Society of London* 195:295–310.

McCoy, E. D., and J. R. Rey. 1987. Terrestrial arthropods on northwest Florida saltmarshes: Hymenoptera (Insecta). *Florida Entomologist* 70:90–97.

MacDonald, I. A., L. L. Loope, M. B. Usher, and O. Hamann. 1989. Wildlife conservation and the invasion of nature reserves by introduced species: a global perspective. Pages 215–255 in J. A. Drake, H. A. Mooney, F. di Castri, R. H. Groves, F. J. Kruger, M. Rejmanek, and M. Williamson (eds.), *Biologial Invasions: A Global Perspective*. Wiley, Chichester, England.

MacEwen, A., and M. MacEwen. 1983. National parks: a cosmetic conservation system. Pages 391–409 in A. Warren and F. B. Goldsmith (eds.), *Conservation in Perspective*. Wiley, Chichester, England.

McIntosh, R. P. 1987. Pluralism in ecology. *Annual Review of Ecology and Systematics* 18:321–341.

McKaye, K. R. 1983. Ecology and breeding behavior of a cichlid fish, *Cyrtocara eucinostomus*, in a large lek in Lake Malawi, Africa. *Environmental Biology of Fish* 8:81–96.

McKey, D. 1979. The distribution of secondary compounds within plants. Pages 55–133 in G. A. Rosenthal and D. H. Janzen (eds.), *Herbivores: Their Interaction with Secondary Plant Metabolites*. Academic Press, New York.

McNamara, J. M., and A. I. Houston. 1985. Optimal foraging and learning. *Journal of Theoretical Biology* 117:231–249.

McNaughton, S. J. 1986. On plants and herbivores. *American Naturalist* 128:765–770.

McNaughton, S. J. 1988. Mineral nutrition and spatial concentrations of African ungulates. *Nature* 334:343–345.

McNaughton, S. J., M. Oesterheld, D. A. Frank, and K. J. Williams. 1989. Ecosystem-level patterns of primary productivity and herbivory in terrestrial habitats. *Nature* 341:142–144.

McNeely, J. A., and K. R. Miller (eds.). 1984. National parks, conservation, and development: the role of protected area in sustaining society. *Proceedings of the World Congress on National Parks, Bali, Indonesia, October 11–12, 1982*. Smithsonian Institution Press, Washington, D.C.

Mahoney, M. J. 1977. Publication prejudices: an experimental study of confirmatory bias in the peer review system. *Cognitive Therapy and Research* 1:161–175.

Maiorana, V. C. 1978. An explanation of ecological and developmental constants. *Nature* 273:375–377.

Malcolm, S. B. 1986. Aposematism in a soft-bodied insect: a case for kin selection. *Behavioral Ecology and Sociobiology* 18:387–393.

Mallet, J., J. T. Longino, D. Murawski, A. Murawski, and A. Simpson de Gamboa. 1987. Handling effects in *Heliconius*: where do all the butterflies go? *Journal of Animal Ecology* 56:377–386.

Malthus, T. R. 1798. *An Essay on the Principle of Population, as It Affects the Future Improvement of Society, with Remarks on the Speculations of Mr. Godwin, M. Condorcet and Other Writers*. J. Johnson, London.

Marler, C. A., and M. C. Moore. 1988. Evolutionary costs of aggression revealed by testosterone manipulations in free-living male lizards. *Behavioral Ecology and Sociobiology* 23:21–26.

Marshall, L. G. 1988. Land mammals and the Great American Interchange. *American Scientist* 76:380–388.

Martin, F. W., R. S. Pospahala, and J. D. Nichols. 1979. Assessment and population management of North American migratory birds. Pages 187–239 in J. Cairns, Jr., G. P. Patil, and W. E. Waters (eds.), *Environmental Biomonitoring, Assessment, Prediction and Management—Certain Case Studies and Related Quantitative Issues*. International Co-operative Publishing House, Fairland, Md.

Martin, P. S. 1967. Prehistoric overkill. Pages 75–120 in P. S. Martin and H. E.

Wright (eds.), *Pleistocene Extinctions*. Yale University Press, New Haven, Conn.

Martin, P. S. 1982. The pattern and meaning of Holarctic mammoth extinction. Pages 399–408 in D. M. Hopkins, J. V. Matthews, C. S. Schweger, and S. B. Young (eds.), *Paleoecology of Beringia*. Academic Press, New York.

Martin, P. S., and R. G. Klein. 1984*a*. *Pleistocene Extinctions*. University of Arizona Press, Tucson.

Martin, P. S., and R. G. Klein. 1984*b*. *Quaternary Extinctions: A Prehistoric Revolution*. University of Arizona Press, Tucson.

Martinat, P. J. 1987. The role of climatic variation and weather in forest insect outbreaks. Pages 241–268 in P. Barbosa and J. C. Schultz (eds.), *Insect Outbreaks*. Academic Press, San Diego.

Massey, A. B. 1925. Antagonism of the walnuts (*Juglans nigra* L. and *J. cinerea* L.) in certain plant associations. *Phytopathology* 15:773–784.

Mattson, W. J. 1980. Herbivory in relation to plant nitrogen content. *Annual Review of Ecology and Systematics* 11:119–161.

Mattson, W. J., and N. D. Addy. 1975. Phytophagous insects as regulators of forest primary production. *Science* 190:515–522.

Mauffette, Y., and W. C. Oechel. 1989. Seasonal variation in leaf chemistry of the coast live oak *Quercus agrifolia* and implications for the California oak moth *Phrygaridia californica*. *Oecologia* 79:439–445.

May, R. M. 1973. *Stability and Complexity in Model Ecosystems*. Princeton University Press, Princeton, N.J.

May, R. M. 1975. Patterns of species abundance and diversity. Pages 81–120 in M. L. Cody and J. Diamond (eds.), *Ecology and Evolution of Communities*. Harvard University Press, Cambridge, Mass.

May, R. M. 1976. Models for two interacting populations. Pages 49–70 in R. M. May (ed.), *Theoretical Ecology: Principles and Applications*. Saunders, Philadelphia.

May, R. M. 1977. Food lost to pests. *Nature* 267:669–670.

May, R. M. 1978. The dynamics and diversity of insect faunas. Pages 188–204 in L. A. Mound and N. Waloff (eds.), *Diversity of Insect Faunas*. Blackwell Scientific Publications, Oxford.

May, R. M. 1979. Fluctuations in abundance of tropical insects. *Nature* 278:505–507.

May, R. M. 1980. Mathematical models in whaling and fisheries management. American Mathematics Society Lectures. *Mathematics in the Life Sciences* 13:1–64.

May, R. M. 1981. Population biology of parasitic infections. Pages 208–235 in K. S. Warren and E. F. Purcell (eds.), *The Current Status and Future of Parasitology*. Josiah Macy, Jr., Foundation, New York.

May, R. M. 1983. Parasitic infections as regulators of animal populations. *American Scientist* 71:36–45.

May, R. M. 1986. Experimental control of malaria in west Africa. Pages 190–204 in

Ecological Knowledge and Environmental Problem-Solving, Concepts and Case Studies. National Academy Press, Washington, D.C.

May, R. M., and R. M. Anderson. 1979. Population biology of infectious diseases. *Nature* 280:455–461.

May, R. M., and A. P. Dobson. 1986. Population dynamics and the rate of evolution of pesticide resistance. In National Research Council, *Management of Resistance to Pesticides: Strategies, Tactics, and Research Needs*. National Academy Press, Washington, D.C.

May, R. M., and R. H. MacArthur. 1972. Niche overlap as a function of environmental viability. *Proceedings of the National Academy of Sciences of the United States of America* 69:1109–1113.

Maynard Smith, J. 1968. *Mathematical Ideas in Biology*. Cambridge University Press, New York.

Maynard Smith, J. 1976*a*. Group selection. *Quarterly Review of Biology* 51:277–283.

Maynard Smith, J. 1976*b*. Evolution and the theory of games. *American Scientist* 64:41–45.

Maynard Smith, J. 1978. The ecology of sex. Pages 159–179 in J. R. Krebs and N. B. Davies (eds.), *Behavioural Ecology, an Evolutionary Approach*. Sinauer Associates, Sunderland, Mass.

Maynard Smith, J. 1979. Game theory and the evolution of behavior. *Proceedings of the Royal Society of London B* 205:475–488.

Maynard Smith, J. 1982. *Evolution and the Theory of Games*. Cambridge University Press, Cambridge.

Mayr, E. 1942. *Systematics and the Origin of Species*. Columbia University Press, New York.

Mayr, E. 1963. *Animal Species and Evolution*. Harvard University Press, Cambridge, Mass.

Mellanby, K. 1967. *Pesticides and Pollution*. Fontana, London.

Menzel, D. W., and J. H. Ryther. 1961. Nutrients limiting the production of phytoplankton in the Sargasso Sea, with special reference to iron. *Deep-Sea Research* 7:276–281.

Michod, R. E., and B. R. Levin (eds). 1988. *The Evolution of Sex, an Examination of Current Ideas*. Sinauer Associates, Sunderland, Mass.

Milinski, M. 1979. An evolutionarily stable feeding strategy in sticklebacks. *Zeitschrift für Tierpsychologie* 51:36–40.

Milinski, M., and R. Heller. 1978. Influence of a predator on the optimal foraging behaviour of sticklebacks (*Gasterosteus aculeatus*). *Nature* 275:642–644.

Miller, D. 1970. Biological control of weeds in New Zealand 1927–48. *New Zealand Department of Scientific and Industrial Research Information Series* 74:1–104.

Miller, G. T., Jr. 1988. *Living in the Environment*. Wadsworth Publishing Company, Belmont, Calif.

Miller, S. L., and H. Urey. 1953. Organic compound synthesis on the primitive earth. *Science* 130:245–251.

Milne, A. 1961. Definition of competition among animals. *Society for Experimental Biology (Symposia)* 15:40–61.

Mineau, P. G., G. A. Fox, R. J. Norstrom, D. V. Weseloh, D. J. Hallett, and J. A. Ellenton. 1984. Using the herring gull to monitor levels of organochlorine contaminants in the Canadian Great Lakes. Pages 425–452 in J. O. Nriagu and M. S. Simmons (eds.), *Contaminants in the Great Lakes*. Wiley, Toronto.

Mitchell, G. C. 1986. Vampire bat control in Latin America. Pages 151–164 in *Ecological Knowledge and Environmental Problem-Solving, Concepts and Case Studies*. National Academy Press, Washington, D.C.

Mitchley, J. 1988. Restoration of species-rich calcicolous grassland on ex-arable land in Britain. *Trends in Ecology and Evolution* 3:125–127.

Mitter, C., and D. R. Brooks. 1983. Phylogenetic aspects of coevolution. Pages 65–98 in D. J. Futuyma and M. Slatkin (eds.), *Coevolution*. Sinauer Associates, Sunderland, Mass.

Moav, R., T. Brody, and G. Hulata. 1978. Genetic improvement of wild fish populations. Science 204:1090–1094.

Modell, W. 1969. Horns and antlers. *Scientific American* 220:114–122.

Molineaux, L., and G. Gramiccia (eds.). 1980. *The Garki Project*. World Health Organization, Geneva.

Møller, A. P. 1988. Female choice selects for male sexual tail ornaments in the monogamous swallow. *Nature* 332:640–642.

Moodie, G. E. E. 1972. Predation, natural selection and adaptation in an unusual three-spined stickleback. *Heredity* 28:155–167.

Moon, R. D. 1980. Biological control through interspecific competition. *Environmental Entomology* 9:723–728.

Moore, D. M. 1983. Human impact on island vegetation. Pages 237–248 in W. Holzner, M. J. A. Werger, And I. Ikusima (eds.), *Man's Impact on Vegetation*. Dr W. Junk, The Hague.

Moore, J. A. 1961. A cellular basis for genetic isolation. Pages 62–68 in W. F. Blair (ed.), *Vertebrate Speciation*. University of Texas Press, Austin.

Moore, J. A. 1985. Science as a way of knowing—human ecology. *American Zoologist* 25:483–637.

Moran, R. J., and W. L. Palmer. 1963. Ruffed grouse introductions and population trends on Michigan islands. *Journal of Wildlife Management* 27:606–614.

Morris, I. 1970. Restraints on the big fish-in. *New Scientist* 48:373–375.

Morris, R. F. 1959. Single-factor analysis in population dynamics. *Ecology* 40:580–588.

Morris, R. F. 1963. The dynamics of epidemic spruce budworm populations. *Memoirs of the Entomological Society of Canada* 31:1–332.

Morris, R. F., and C. A. Miller. 1954. The development of life tables for the spruce budworm. *Canadian Journal of Zoology* 32:283–301.

Morton, N. E., J. F. Crow, and H. J. Muller. 1956. An estimate of the mutational damage in man from data on consanguineous marriages. *Proceedings of the National Academy of Sciences of the United States of America* 42:855–863.

Mosby, H. S. 1969. The influence of hunting on the population dynamics of a woodlot gray squirrel population. *Journal of Wildlife Management* 33:59–73.

Moss, J. I., and R. A. van Steenwyk. 1984. Marking cabbage looper (Lepidoptera: Noctuidae) with cesium. *Environmental Entomology* 13:390–393.

Muller, C. H. 1966. The role of chemical inhibition (allelopathy) in vegetational composition. *Bulletin of the Torrey Botanical Club* 93:332–351.

Muller, C. H. 1970. Phytotoxins as plant habitat variables. *Recent Advances in Phytochemistry* 3:105–121.

Muller, F. 1879. *Ituna* and *Thyridis,* a remarkable case of mimicry in butterflies (translated from the German by R. Meldola). *Proceedings of the Entomological Society of London* 27:20–29.

Munroe, E. G. 1948. The geographical distribution of butterflies in the West Indies. Dissertation, Cornell University, Ithaca, N.Y.

Murdoch, W. W. 1975. Diversity, complexity, stability and pest control. *Journal of Applied Ecology* 12:795–807.

Murie, A. 1944. *Wolves of Mount McKinley*. Fauna of National Parks, U.S. Fauna Series Number 5, Washington, D.C.

Murphy, P. G., R. R. Sharitz, and A. J. Murphy. 1977. Response of a forest ecotone to ionizing radiation. Pages 43–48 in J. Zavitkovski (ed.), *The Enterprise, Wisconsin, Radiation Forest*. USERDATID-26113-p2. U.S. Energy Research and Development Administration, Washington, D.C.

Murray, B. G., Jr. 1986. The structure of theory, and the role of competition in community dynamics. *Oikos* 46:145–158.

Murton, R. K., N. J. Westwood, and A. J. Isaacson. 1974. A study of wood-pigeon shooting: the exploitation of a natural animal population. *Journal of Applied Ecology* 11:61–81.

Myers, J. H., and K. S. Williams. 1984. Does tent caterpillar attack reduce the food quality of red alder foliage? *Oecologia* 62:74–79.

Myers, J. H., and K. S. Williams. 1987. Lack of short or long term inducible defenses in the red alder-western test caterpillar system. *Oikos* 48:73–78.

Myers, J. P., P. G. Connors, and F. A. Pitelka. 1981. Optimal territory size and the sanderling: compromises in a variable environment. Pages 135–158 in A. C. Kamil and T. D. Sargent (eds.), *Foraging Behavior, Ecological, Ethological and Psychological Approaches*. Garland STPM Press, New York.

Myers, N. 1979. *The Sinking Ark*. Pergamon, Oxford.

Myers, N. 1980. *Conversion of Tropical Moist Forests*. National Academy of Sciences, Washington, D.C.

Myers, N. 1983. *A Wealth of Wild Species: Storehouse for Human Welfare*. Westview Press, Boulder, Col.

National Academy of Sciences. 1984. *Science and Creationism. A View from the National Academy of Sciences*. National Academy Press, Washington, D.C.

National Academy of Sciences. 1987. *Introduction of Recombinant DNA-engineered Organisms into the Environment: Key Issues.* National Academy Press, Washington, D.C.

National Research Council. 1981. *Surface Mining: Soil, Coal and Society.* National Academy Press, Washington, D.C.

National Research Council. 1982. *Impacts of Emerging Agricultural Trends in Fish and Wildlife Habitat.* National Academy Press, Washington, D.C.

National Research Council. 1986. *Ecological Knowledge and Environmental Problem-Solving.* National Academy Press, Washington, D.C.

Neel, J. V. 1983. Frequency of spontaneous and induced ''point'' mutations in higher eukaryotes. *Journal of Heredity* 74:2–15.

Nei, M. 1983. Genetic polymorphism and the role of mutation in evolution. Pages 165–190 in M. Nei and R. K. Koehn (eds.), *Evolution of Genes and Proteins.* Sinauer Associates, Sunderland, Mass.

Neill, S. R. St. J., and J. M. Cullen. 1974. Experiments on whether schooling by their prey affects the hunting behaviour of cephalopods and fish predators. *Journal of the Zoological Society of London* 172:549–569.

Nevstrueva, M. A., P. V. Ramzaev, A. A. Moiseer, M. S. Ibatullin, and L. A. Teplytch. 1967. The nature of ^{137}Cs and ^{90}Sr transport over the lichen-reindeer-man food chain. Pages 209–215 in B. Aberg and F. P. Hungate (eds.), *Radioecological Concentration Processes. Proceedings of an International Symposium.* Pergamon, Oxford.

Newmark, W. D. 1987. A land-bridge island perspective on mammalian extinctions in western North American parks. *Nature* 325:430–432.

Ngugi, A. W. 1988. Cultural aspects of fuelwood shortage in the Kenyan highlands. *Journal of Biogeography* 15:165–170.

Nichols, J. P., M. J. Conroy, D. R. Anderson, and K. P. Burnham. 1984. Compensatory mortality in waterfowl populations: a review of the evidence and implications for research and management. *Transactions of the North American Wildlife and Natural Resources Conference* 49:535–554.

Nicholson, A. J. 1933. The balance of animal populations. *Journal of Animal Ecology* 2:132–178.

Nicholson, A. J., and V. A. Bailey. 1935. The balance of animal populations. Part I. *Proceedings of the Zoological Society of London* 1935:551–598.

Niemelä, P., and J. Tuomi. 1987. Does the leaf morphology of some plants mimic caterpillar damage? *Oikos* 50:256–257.

Niemelä, P., J. Tuomi, R. Mannila, and P. Ojala. 1984. The effect of previous damage on the quality of Scots pine foliage as food for dipronid sawflies. *Zeitschift angewandte Entomologie* 98:33–43.

Niklas, K. J. 1986. Large-scale changes in animal and plant terrestrial communities. Pages 383–405 in D. M. Raup and D. Jablonski (eds.), *Patterns and Processes in the History of Life.* Springer-Verlag, Berlin.

Niklas, K. J., B. H. Tiffney, and A. H. Knoll. 1980. Apparent changes in the diversity of fossil plants. *Evolutionary Biology* 12:1–89.

Nilsson, S. G., and U. Wästljung. 1987. Seed predation and cross-pollination in mast-seeding beech (*Fagus sylvatica*) patches. *Ecology* 68:260–265.

Nobe, K. C., and A. H. Gilbert. 1970. *A Survey of Sportsmen's Expenditures for Hunting and Fishing in Colorado, 1968*. GFP-R-T-24. Colorado Division of Game, Fish, and Parks, Denver.

Noble, G. K. 1931. *The Biology of the Amphibia*. McGraw-Hill, New York.

Numbers, R. L. 1982. Creationism in 20th-century America. *Science* 218:538–544.

Oddie, B. 1987. The tragedy of Tibbles wren. *New Scientist* 116(1590):67.

Odum. E. P. 1957. The ecosystem approach in the teaching of ecology illustrated with simple class data. Ecology 38:531–535.

Odum, E. P. 1959. *Fundamentals of Ecology*, 2nd edition. Saunders, Philadelphia.

Odum, E. P. 1969. The strategy of ecosystem development. *Science* 164:262–270.

Odum, E. P. 1971. *Fundamentals of Ecology*, 3rd edition. Saunders, Philadelphia.

Odum, H. T., P. Murphy, G. Drewry, F. McCormick, C. Schinan, E. Morales, and J. A. McIntyre. 1970. Effects of gamma radiation on the forest at El Verde. Pages D-3–D-75 in H. T. Odum and R. F. Pigeon (eds.), *A Tropical Rain Forest*. U.S. Atomic Energy Commission, Washington, D.C.

Olson, J. S. 1958. Rates of succession and soil changes on southern Lake Michigan sand dunes. *Botanical Gazette* 119:125–170.

Onuf, C. P. 1978. Nutritive value as a factor in plant-insect interactions with an emphasis on field studies. Pages 85–96 in G. G. Montgomery (ed.), *The Ecology of Arboreal Folivores*. Smithsonian Institution Press, Washington, D.C.

Onuf, C. P., J. M. Teal, and I. Valiela. 1977. Interactions of nutrients, plant growth and herbivory in a mangrove ecosystem. *Ecology* 58:514–526.

Oparin, A. I. 1953. *The Origin of Life*. Dover Publications, New York.

Orians, G. H. 1969*a*. The number of bird species in some tropical forests. *Ecology* 50:783–797.

Orians, G. H. 1969*b*. On the evolution of mating systems in birds and mammals. *American Naturalist* 103:589–603.

Oring, L. W. 1981. Avian mating systems. Pages 1–92 in D. S. Farner and J. R. King (eds.), *Avian Biology, Vol. 6*. Academic Press, London.

Osmond, C. H., and J. Monro. 1981. Prickly pear. Pages 194–222 in D. J. Carr and S. G. M. Carr (eds.), *Plants and Man in Australia*. Academic Press, New York.

Oster, G. F., and E. O. Wilson. 1978. *Caste and Ecology in the Social Insects*. *Monographs in Population Biology No. 12*. Princeton University Press, Princeton, N.J.

Ovington, J. D. 1962. Quantitative ecology and the woodland ecosystem concept. *Advances in Ecological Research* 1:103–192.

Owen, D. F. 1966. Polymorphism in Pleistocene land snails. *Science* 152:71–72.

Owen, D. F. 1980. *Camouflage and Mimicry*. Oxford University Press, Oxford.

Owen, D. F., and D. L. Whiteley. 1986. Reflexive selection: Moment's hypothesis resurrected. *Oikos* 47:117–120.

Owen, D. F., and R. G. Wiegert. 1987. Leaf eating as mutualism. Pages 81–95 in

P. Barbosa and J. C. Schultz (eds.), *Insect Outbreaks*. Academic Press, San Diego.

Owen, O. S. 1975. *Natural Resource Conservation, an Ecological Approach*, 2nd edition. Macmillan, New York.

Packer, C. 1977. Reciprocal altruism in *Papio anubis. Nature* 265:441–443.

Pain, S. 1988. Vagrant eagle "recrosses" the Atlantic. *New Scientist* 117(1600):28.

Paine, R. T. 1966. Food web complexity and species diversity. *American Naturalist* 100:65–75.

Paine, R. T. 1988. Food webs: road maps of interactions or grist for theoretical development? *Ecology* 69:1648–1654.

Paine, R. T., and S. A. Levin. 1981. Intertidal landscapes: disturbance and the dynamics of pattern. *Ecological Monographs* 51:145–178.

Palmer, W. L. 1962. Ruffed grouse flight capability over water. *Journal of Wildlife Management* 26:338–339.

Park, T. 1948. Experimental studies of interspecies competition. I. Competition between populations of the flour beetles *Tribolium confusum* Duval and *Tribolium castaneum* Herbst. *Ecological Monographs* 18:265–307.

Park, T. 1954. Experimental studies of interspecies competition. II. Temperature, humidity, and competition in two species of *Tribolium. Physiological Zoology* 27:177–238.

Park, T., P. H. Leslie, and D. B. Metz. 1964. Genetic strains and competition in populations of *Tribolium. Physiological Zoology* 37:97–162.

Parker, G. A. 1978. Searching for mates. Pages 214–245 in J. R. Krebs and N. B. Davies (eds.), *Behavioral Ecology: An Evolutionary Approach*. Blackwell Scientific Publications, Oxford.

Parker, J. 1950. Planting loblolly pine outside its native range. *Journal of Forestry* 48:278–279.

Parker, J. 1955. Survival of some southeastern pine seedlings in northern Idaho. *Journal of Forestry* 58:137.

Parry, G. D. 1981. The meanings of *r*- and *K*-selection. *Oecologia* 48:260–264.

Parsons, K. A., and A. A. de la Cruz. 1980. Energy flow and grazing behavior of canocephaline grasshoppers in a *Juncus roemerianus* marsh. *Ecology* 61:1045–1050.

Partridge, L. 1980. Mate choice increases a component of offspring fitness in fruit flies. *Nature* 283:290–291.

Pawley, W. H. 1976. World picture—present and future. Pages 13–24 in A. N. Duckham, J. G. W. Jones, and E. H. Roberts (eds.), *Food Production and Consumption: The Efficiency of Human Food Chains and Nutrient Cycles*. North-Holland, Amsterdam.

Payne, I. 1987. A lake perched on piscine peril. *New Scientist* 115(1575):50–54.

Payne, R. 1973. Decline of whales. Pages 143 in W. Jackson (ed.), *Man and the Environment*, 2nd edition. William C. Brown Company, Dubuque, Iowa.

Peakall, R., A. J. Beattie, and S. J. James. 1987. Pseudocopulation of an orchid by male ants: a test of two hypotheses accounting for the rarity of ant pollination. *Oecologia* 78:522–524.

Pearce, F. 1989. Kill or cure? Remedies for the rainforest. *New Scientist* 1682:40–43.

Pearl, R. 1928. *The Rate of Living*. Knopf, New York.

Pearl, R., and L. J. Reed. 1920. On the rate of growth of the population of the United States since 1790 and its mathematical representation. *Proceedings of the National Academy of Sciences of the United States of America* 6:275–288.

Pearsall, W. H. 1954. Biology and land use in East Africa. *New Biology* 17:9–26.

Perfect, T. J., A. G. Cook, and D. E. Padgham. 1985. Interpretation of the flight activity of *Nilaparvata lugens* (Stål) and *Sagatella furcifera* (Horvark) (Hemiptera: Delphacidae) based on comparative trap catches and field marking with rubidium. *Bulletin of Entomological Research* 75:93–106.

Peterman, R. M. 1984. Interaction among sockeye salmon in the Gulf of Alaska. Pages 187–199 in W. G. Pearcy (ed.), *The Influence of Ocean Conditions on the Production of Sockeye Salmonids*. Oregon State University Sea Grant Communications, Corvallis.

Peters, C. M., A. H. Gentry, and R. O. Mendelsohn. 1989. Valuation of an Amazonian rainforest. *Nature* 339:655–656.

Pettersson, B. 1985. Extinction of an isolated population of the middle spotted woodpecker *Dendrocops medius* (L.) in Sweden and its relation to general theories on extinction. *Biological Conservation* 32:335–353.

Petts, G. E. 1985. *Impounded Rivers: Perspectives for Ecological Management*. Wiley, Chichester, England.

Pianka, E. R. 1970. On *r*- and *K*-selection. *American Naturalist* 104:592–597.

Pianka, E. R. 1976. Competition and niche theory. Pages 114–141 in R. M. May, (ed.), *Theoretical Ecology: Principles and Applications*. Blackwell Scientific Publications, Oxford.

Pienkowski, M. W., and P. R. Evans. 1985. The role of migration in the population dynamics of birds. Pages 331–352 in R. M. Sibly and R. H. Smith (eds.), *Behavioural Ecology: Ecological Consequences of Adaptive Behaviour*. Blackwell Scientific Publications, Oxford.

Pierce, G. J. and J. G. Ollason. 1987. Eight reasons why optimal foraging theory is a complete waste of time. *Oikos* 49:111–118.

Pilgram, T., and D. Western. 1986. Inferring hunting patterns on African elephants from tusks in the international ivory trade. *Journal of Applied Ecology* 23:503–514.

Pimentel, D. 1986. Biological invasions of plants and animals in agriculture and forestry. Pages 149–162 in H. A. Mooney and J. A. Drake (eds.), *Ecology of Biological Invasions of North America and Hawaii*. Springer-Verlag, New York.

Pimentel, D. 1988. Herbivore population feeding pressure on plant hosts: feedback evolution and host conservation. *Oikos* 53:289–302.

Pimentel, D., D. Andow, R. Dyson-Hudson, D. Gallahan, S. Jacobson, M. Irish, G. Kroop, A. Moss, I. Schreiner, M. Shepard, T. Thompson, and B. Vinzant.

1980. Environmental and social costs of pesticides: a preliminary assessment. *Oikos* 34:126–140.

Pimentel, D., W. Dazhag, S. Eigenbrode, H. Emerson, and M. Korasik. 1986. Deforestation: Interdependence of fuelwood and agriculture. *Oikos* 46:404–412.

Pimentel, D., and M. Pimentel. 1986. Letter to the editor. *Science* 231:1491–1492.

Pimm, S. L. 1979. The structure of food webs. *Theoretical Population Ecology* 16:144–158.

Pimm, S. L. 1980. Food web design and the effect of species deletion. *Oikos* 35:139–147.

Pimm, S. L. 1982. *Food Webs*. Chapman and Hall, London.

Pimm, S. L. 1984. Food chains and return times. Pages 397–412 in D. R. Strong, D. Simberloff, L. G. Abele, and A. B. Thistle (eds.), *Ecological Communities: Conceptual Issues and the Evidence*. Princeton University Press, Princeton, N.J.

Pirie, N. W. 1969. *Food Resources, Conventional and Novel*. Penguin, Harmondsworth.

Pitelka, F. A. 1964. The nutrient-recovery hypothesis for Arctic microtine cycles. I. Introduction. Pages 55–56 in D. J. Crisp (ed.), *Grazing in Terrestrial and Marine Environments*. Blackwell Scientific Publications, Oxford.

Pleszczynska, W. K. 1978. Microgeographic prediction of polygyny in the lark bunting. *Species* 201:935–937.

Podoler, H., and D. Rogers. 1975. A new method for the identification of key factors from life-table data. *Journal of Animal Ecology* 44:85–115.

Popper, K. R. 1972a. *The Logic of Scientific Discovery*, 3rd edition. Hutchinson, London.

Popper, K. R. 1972b. *Objective Knowledge: An Evolutionary Approach*. Clarendon Press, Oxford.

Posey, C. E. 1967. Natural regeneration of loblolly pine within 230 miles of its native range. *Journal of Forestry* 65:732.

Potter, D. A., and T. W. Kimmerer. 1988. Do holly leaf spines really deter herbivory? *Oecologia* 75:216–221.

Powell, G. V. M., A. H. Powell, and N. K. Paul. 1988. Brother, can you spare a fish? *Natural History* 97(2):34–39.

Prescott-Allen, C., and R. Prescott-Allen. 1986. *The First Resource: Wild Species in the North American Economy*. Yale University Press, New Haven, Conn.

Prestidge, R. A., and S. McNeill. 1983. The role of nitrogen in the ecology of grassland Auchenorrhyncha. Pages 257–283 in J. A. Lee, S. McNeil, and I. H. Rorison (eds.), *Nitrogen as an Ecological Factor. 22nd Symposium of the British Ecological Society*. Blackwell Scientific Publications, Oxford.

Preston, F. W. 1948. The commonness and rarity of species. *Ecology* 29:254–283.

Preston, F. W. 1960. Time and space and the variation of species. *Ecology* 41:611–627.

Preston, F. W. 1962. The canonical distribution of commonness and rarity. *Ecology* 43:185–215, 410–432.

Price, P. W. 1970. Characteristics permitting coexistence among parasitoids of a sawfly in Quebec. *Ecology* 51:445–454.

Price, P. W. 1975. *Insect Ecology.* Wiley, New York.

Price, P. W. 1980. *Evolutionary Biology of Parasites.* Princeton University Press, Princeton, N.J.

Prins, H. H. T., and F. J. Weyerhaeuser. 1987. Epidemics in populations of wild ruminants: anthrax and impala, rinderpest and buffalo in Lake Manyara National Park, Tanzania. *Oikos* 49:28–38.

Proctor, J., and S. Proctor. 1978. *Color in Plants and Flowers.* Everest, New York.

Province of Alberta. 1968. *Water Diversion Proposals of North America.* Department of Agriculture, Water Resources Division, Edmonton.

Pullin, A. S., and J. E. Gilbert. 1989. The stinging nettle, *Urtica dioica,* increases trichome density after herbivore and mechanical damage. *Oikos* 54:275–280.

Pursel, V. G., C. A. Pinkert, K. F. Miller, D. J. Bolt, R. G. Campbell, R. D. Palmiter, R. L. Brinster, and R. E. Hammer. 1989. Genetic engineering of livestock. *Science* 244:1281–1288.

Pusey, A. E. 1987. Sex-biased dispersal and inbreeding avoidance in birds and mammals. *Trends in Ecology and Evolution* 2:295–300.

Quinn, J. F. 1983. Mass extinctions in the fossil record. *Science* 219:1239–1240.

Quinn, J. F., and A. E. Dunham. 1983. On hypothesis testing in ecology and evolution. *American Naturalist* 122:602–617.

Quinn, J. F., and S. P. Harrison. 1988. Effects of habitat fragmentation and isolation on species richness: evidence from biogeographic patterns. *Oecologia* 75:132–140.

Rabinowitz, D. 1981. Seven forms of rarity. Pages 205–217 in H. Synge (ed.), *The Biological Aspects of Rare Plant Conservation.* Wiley, London.

Radinsky, L. B. 1978. Evolution of brain size in carnivores and ungulates. *American Naturalist* 11:815–831.

Ralls, K., and J. Ballou. 1983. Extinction: lessons from zoos. Pages 164–184 in C. M. Schonewald-Cox, S. M. Chambers, B. MacBryde, and L. Thomas (eds.), *Genetics and Conservation.* Benjamin/Cummings, Menlo Park, Calif.

Ralls, K., P. H. Harvey, and A. M. Lyles. 1986. Inbreeding in natural populations of birds and mammals. Pages 35–56 in M. E. Soulé (ed.), *Conservation Biology: The Science of Scarcity and Diversity.* Sinauer Associates, Sunderland, Mass.

Rauch, J. 1987. Drug on the market. *National Journal* 4:818–821.

Raup, D. M. 1962. Computer as aid in describing form in gastropod shells. *Science* 138:150–152.

Raup, D. M. 1966. Geometric analysis of shell coiling: general problems. *Journal of Paleontology* 40:1178–1190.

Raup, D. M. 1979. Biases in the fossil record of species and genera. *Bulletin of the Carnegie Museum of Natural History* 13:85–91.

Raup, D. M. 1984. Evolutionary radiations and extinctions. Pages 5–14 in H. D.

Holl and A. F. Trendall (eds.), *Patterns of Change in Earth Evolution.* Springer-Verlag, Berlin.

Raup, D. M., S. J. Gould, T. J. M. Schopf, and D. Simberloff. 1973. Stochastic models of phylogeny and the evolution of diversity. *Journal of Geology* 81:525–542.

Raup, D. M., and J. J. Sepkoski, Jr. 1984. Periodicities of extinctions in the geologic past. *Proceedings of the National Academy of Sciences of the United States of America* 81:801–805.

Real, L. 1983. *Pollination Biology.* Academic Press, Orlando, Fla.

Reed, J. M., P. D. Doerr, and J. R. Walters. 1988. Minimum viable population size of the red-cockaded woodpecker. *The Journal of Wildlife Management* 52:385–391.

Reed, N. P. 1981. In the matter of Mr. Watt. *Sierra Club Bulletin* 66:6–15.

Reichle, D. E. 1970. *Analysis of Temperate Forest Ecosystems.* Springer-Verlag, New York.

Reiners, W. A. 1986. Complementary models for ecosystems. *The American Naturalist* 127:59–73.

Rennie, P. J. 1957. The uptake of nutrients by timber forest and its importance to timber production in Britain. *Quarterly Journal of Forestry* 51:101–115.

Rensch, B. 1959. *Evolution Above the Species Level.* Columbia University Press, New York.

Revelle, P., and C. Revelle. 1984. *The Environment, Issues and Choices for Society,* 2nd edition. Willard Grant Press, Boston.

Rey, J. R. 1981. Ecological biogeography of arthropods on *Spartina* islands in northwest Florida. *Ecological Monographs* 51:237–265.

Reynolds, S. G. 1988. Some factors of importance in the integration of pastures and cattle with coconuts (*Cocos nucifera*). *Journal of Biogeography* 15:31–39.

Rhoades, D. F. 1979. Evolution of plant chemical defense against herbivores. Pages 3–54 in G. A. Rosenthal and D. H. Janzen (eds.), *Herbivores: Their Interaction with Secondary Plant Metabolites.* Academic Press, New York.

Rhoades, D. F., and R. G. Cates. 1976. Toward a general theory of plant anti-herbivore chemistry. *Recent Advances in Phytochemistry* 10:168–213.

Rice, R. E. 1989. *National Forests, Policies for the Future. Vol. 5. The Uncounted Costs of Logging.* The Wilderness Society, Washington, D.C.

Richardson, S. D. 1966. *Forestry in Communist China.* Johns Hopkins University Press, Baltimore, Md.

Richer, W. E. 1981. Changes in the average size and average age of Pacific salmon. *Canadian Journal of Fisheries and Aquatic Science* 38:1636–1656.

Rifkin, J. 1983. *Algeny.* Viking, New York.

Ringwood, A. E., S. E. Kesson, N. G. Ware, W. Hibberson, and A. Major. 1979. Immobilisation of high level nuclear reactor wastes in SYNROC. *Nature* 278:219–223.

Roberts, R. J. 1980. Restriction and modification enzymes and their recognition sequences. *Nucleic Acids Research* 8:r63–r80.

Rockwood, L. L. 1975. Distribution, density, and dispersion of two species of leafcutting ants (*Atta*) in Guanacaste Province, Costa Rica. *Biotropica* 73:176–193.

Rohwer, S., and F. C. Rohwer. 1978. Status signalling in Harris sparrows: experimental deceptions achieved. *Animal Behaviour* 26:1012–1022.

Roland, J. 1988. Decline of winter moth populations in North America: direct versus indirect effect of introduced parasites. *Journal of Animal Ecology* 57:523–531.

Rollo, C. D. 1986. A test of the principle of allocation using two sympatric species of cockroaches. *Ecology* 67:616–628.

Rood, J. P. 1978. Dwarf mongoose helpers at the den. *Zeitschrift für Tierpsychologie* 48:277–287.

Room, P. M., K. L. S. Harley, I. W. Forno, and P. P. D. Sands. 1981. Successful biological control of the floating weed *Salvinia*. *Nature* 294:78–80.

Root, R. 1967. The niche exploitation pattern of the blue-gray gnatcatcher. *Ecological Monographs* 37:317–350.

Root, R. 1973. Organization of a plant-arthropod association in simple and diverse habitats: The fauna of collards (*Brassica oleracea*). *Ecological Monographs* 43:95–124.

Rose, F. 1970. Lichens as pollution indicators. *Your Environment* 5:185–189.

Rosenzweig, M. L. 1968. Net primary productivity of terrestrial communities: prediction from climatological data. *American Naturalist* 102:67–74.

Rosenzweig, M. L. 1971. Paradox of enrichment: destabilization of exploitation ecosystems in ecological time. *Science* 171:385–387.

Rosenzweig, M. L., and R. H. MacArthur. 1963. Graphical representation and stability conditions of predator-prey interactions. *American Naturalist* 97:209–223.

Rothé, J. P. 1968. Fill a dam, start an earthquake. *New Scientist* 39:75–78.

Rothenbuhler, W. C. 1964. Behaviour genetics of nest cleaning in honey bees. IV. Responses of F1 and backcross generations to disease killed brood. *American Zoologist* 4:111–123.

Rotheray, G. E. 1981. Host searching and oviposition behavior of some parasitoids of aphidophagous Syrphidae. *Ecological Entomology* 6:79–87.

Rotheray, G. E. 1986. Colour, shape and defense in aphidophagous syrphid larvae (Diptera). *Zoological Journal of the Linnaean Society* 88:201–216.

Rotheray, G. E., and P. Barbosa. 1984. Host related factors affecting oviposition behavior in *Brachymeria intermedia*. *Entomologia Experimentalis et Applicata* 35:141–145.

Roughgarden, J. 1983. Competition and theory in community ecology. *American Naturalist* 122:583–601.

Royal Society Study Group. 1983. *The Nitrogen Cycle of the United Kingdom*. The Royal Society, London.

Rubenstein, D. I., and R. W. Wrangham (eds.). 1986. *Ecological Aspects of Social Evolution: Birds and Mammals.* Princeton University Press, Princeton, N.J.

Rubenstein, J. M., and R. S. Bacon. 1983. *The Cultural Landscape: An Introduction to Human Geography.* West Publishing Company, St. Paul, Minn.

Russell, P. F., and T. R. Rao. 1942. On the relation of mechanical obstruction and shade to ovipositing of *Anopheles culicifacies. Journal of Experimental Zoology* 91:303–329.

Ryther, J. H. 1963. Geographic variation in productivity. Pages 347–380 in M. N. Hill (ed.), *The Sea,* Vol. 2, 2nd edition. Wiley-Interscience, New York.

Ryther, J. H. 1969. Photosynthesis and fish production in the sea. *Science* 166:72–76.

Sailer, R. I. 1983. History of insect introductions. Pages 15–38 in C. Graham and C. Wilson (eds.), *Exotic Plant Pests and North American Agriculture.* Academic Press, New York.

Sala, O. S., W. J. Parton, L. A. Joyce, and W. K. Lauenroth. 1988. Primary production of the central grassland region of the United States. *Ecology* 69:40–45.

Salisbury, E. J. 1961. *Weeds and Aliens.* Collins, London.

Salt, G. 1970. *The Cellular Defense Reactions of Insects.* Cambridge University Press, New York.

Salwasser, H. 1987. Spotted owls: turning a battleground into a blueprint. *Ecology* 68:776–779.

Salwasser, H., S. Mealey, and K. Johnson. 1984. Wildlife population viability—a question of risk. *Transactions of the North American Wildlife National Research Conference* 49:421–439.

Sanders, H. L. 1968. Marine benthic diversity: a comparative study. *American Naturalist* 102:243–282.

Sanderson, I. T. 1945. *Living Treasure.* Viking, New York.

Sargent, T. D. 1968. Cryptic moths: effects on background selection of painting the circumocular scales. *Science* 159:100–101.

Savidge, J. 1987. Extinction of an island forest avifauna by an introduced snake. *Ecology* 68:660–668.

Schaffer, M. L., and F. B. Samson. 1985. Population size and extinction: a note on determining critical population sizes. *American Naturalist* 125:144–152.

Schaller, G. B. 1972. *The Serengeti Lion.* Chicago University Press, Chicago.

Scheffer, V. B. 1951. The rise and fall of a reindeer herd. *Scientific Monthly* 73:356–362.

Schemske, D. W. 1980. The evolutionary significance of extrafloral nectar production by *Costus woodsonii* (Zingiberaceae): an experimental analysis of ant protection. *Journal of Ecology* 68:959–967.

Schemske, D. W. 1983. Limits to specialization and coevolution in plant-animal mutualisms. Pages 67–109 in M. H. Nitecki (ed.), *Coevolution.* University of Chicago Press, Chicago.

Schery, R. W. 1972. *Plants for Man.* Prentice-Hall, Englewood Cliffs, N.J.

Schindler, D. W. 1974. Eutrophication and recovery in experimental lakes: implications for lake management. *Science* 184:897–899.

Schindler, D. W. 1977. Evolution of phosphorus limitation in lakes. *Science* 195:260–262.

Schneider, O. 1927. Alteration of the Argentine Pampa in the colonial period. *University of California Publications in Geography* 2:303–321.

Schoener, T. W. 1974. Resource partitioning in ecological communities. *Science* 185:27–39.

Schoener, T. W. 1983. Field experiments on interspecific competition. *American Naturalist* 122:240–285.

Schoener, T. W. 1986. Overview: kinds of ecological communities—ecology becomes pluralistic. Pages 467–479 in J. Diamond and T. J. Case (eds.), *Community Ecology*. Harper and Row, New York.

Schoener, T. W. 1987. The geographical distribution of rarity. *Oecologia* 74:161–173.

Schoener, T. W. 1989. Food webs from the small to the large. *Ecology* 70:1559–1589.

Schoonhoven, L. M. 1982. Biological aspects of antifeedants. *Entomologia Experimentalis et Applicata* 31:57–69.

Schultz, A. M. 1964. The nutrient-recovery hypothesis for Arctic microtine cycles. II. Ecosystem variables in relation to Arctic microtine cycles. Pages 57–68 in D. J. Crisp (ed.), *Grazing in Terrestrial and Marine Environments*. Blackwell Scientific Publications, Oxford.

Schultz, J. C., and I. T. Baldwin. 1982. Oak leaf quality declines in response to defoliation by gypsy moth larvae. *Science* 217:149–151.

Schupp, E. W. 1986. *Azteca* protection of *Cecropia:* ant occupation benefits juvenile trees. *Oecologia* 70:379–385.

Schutt, D. A. 1976. The effect of plant oestrogens on animal reproduction. *Endeavour* 35:110–113.

Scott, G. R. 1970. Rinderpest. Pages 20–35 in J. W. Davis, L. H. Karstad, and D. O. Trainer (eds.), *Infectious Diseases of Wild Mammals*. Iowa State University Press, Ames.

Scriber, J. M., and F. Slansky. 1981. The nutritional ecology of immature insects. *Annual Review of Entomology* 26:183–211.

Seghers, B. H. 1974. Schooling behaviour in the guppy *Poecilia reticulata:* an evolutionary response to predation. *Evolution* 28:486–489.

Selander, R. K. 1976. Genic variation in natural populations. Pages 21–45 in F. J. Ayala (ed.), *Molecular Evolution*. Sinauer Associates, Sunderland, Mass.

Sepkoski, J. J., Jr. 1976. Species diversity in the Phanerozoic: species-area effects. *Paleobiology* 2:298–303.

Sepkoski, J. J., Jr. 1978. A kinetic model of Phanerozoic taxonomic diversity, I. Analysis of marine orders. *Paleobiology* 4:223–251.

Sepkoski, J. J., Jr. 1979. A kinetic model of Phanerozoic taxonomic diversity. II. Early Phanerozoic families and multiple equilibria. *Paleobiology* 5:222–251.

Sepkoski, J. J., Jr. 1984. A kinetic model of Phanerozoic taxonomic diversity. III. Post-Paleozoic families and mass extinctions. *Paleobiology* 10:246–267.

Sharples, F. E. 1983. Spread of organisms with novel genotypes: thoughts from an ecological perspective. *Recombinant DNA Technical Bulletin* 6:43–56.

Sherlock, R. L. 1922. *Man as a Geological Agent.* Witherby, London.

Sherman, P. W. 1977. Nepotism and the evolution of alarm calls. *Science* 197:1246–1253.

Silvertown, J. W. 1980. The evolutionary ecology of mast seeding in trees. *Biological Journal of the Linnaean Society* 14:235–250.

Simberloff, D. 1974. Permo-Triassic extinctions: effects of area on biotic equilibrium. *Journal of Geology* 82:267–274.

Simberloff, D. 1976. Experimental zoogeography of islands: effects of island size. *Ecology* 57:629–648.

Simberloff, D. 1978. Colonization of islands by insects: immigration, extinction, and diversity. Pages 139–153 in L. A. Mound and N. Waloff (eds.), *Diversity of Insect Faunas.* Blackwell Scientific Publications, Oxford.

Simberloff, D. 1981. Community effects of introduced species. Pages 53–81 in M. H. Nitecki (ed.), *Biotic Crises in Ecological and Evolutionary Time.* Academic Press, New York.

Simberloff, D. 1983. Competition theory, hypothesis-testing, and other community ecological buzzwords. *American Naturalist* 122:626–635.

Simberloff, D. 1985. Predicting ecological effects of novel entities: evidence from higher organisms. Pages 152–161 in H. O. Halvorson, D. Pramer, and M. Rogul (eds.), *Engineered Organisms in the Environment: Scientific Issues.* American Society for Microbiology, Washington, D.C.

Simberloff, D. 1986a. Design of natural reserves. Pages 316–337 in M. B. Usher (ed.), *Wildlife Conservation Evaluation.* Chapman and Hall, London.

Simberloff, D. 1986b. Introduced insects: a biogeographic and systematic perspective. Pages 3–26 in H. A. Mooney and J. A. Drake (eds.), *Ecology of Biological Invasions of North America and Hawaii.* Springer-Verlag, New York.

Simberloff, D. 1986c. The proximate causes of extinction. Pages 259–276 in D. M. Raup and D. Jablonski (eds.), *Patterns and Processes in the History of Life.* Springer-Verlag, Berlin.

Simberloff, D. 1987a. Calculating probabilities that cladograms match: a method of biogeographical inference. *Systematic Zoology* 36:175–195.

Simberloff, D. 1987b. The spotted owl fracas: mixing academic, applied and political ecology. *Ecology* 68:766–772.

Simberloff, D. 1988a. Effects of drift and selection on detecting similarities between large cladograms. *Systematic Zoology* 37:56–59.

Simberloff, D. 1988b. The contribution of population and community biology to conservation science. *Annual Review of Ecology and Systematics* 19:473–512.

Simberloff, D., and L. G. Abele. 1976. Island biogeography theory and conservation practice. *Science* 191:285–286.

Simberloff, D., and L. G. Abele. 1982. Refuge design and island biogeographic theory: effects of fragmentation. *American Naturalist* 120:41–50.

Simberloff, D., and W. J. Boecklen. 1981. Santa Rosalia reconsidered: size ratios and competition. *Evolution* 35:1206–1228.

Simberloff, D., B. J. Brown, and S. Lowrie. 1978. Isopod and insect root borers may benefit Florida mangroves. *Science* 201:630–632.

Simberloff, D., and J. Cox. 1987. Consequences and costs of conservation corridors. *Conservation Biology* 1:63–71.

Simberloff, D., and N. Gotelli. 1984. Effects of insularisation on plant species richness in the prairie-forest ecotone. *Biological Conservation* 29:27–46.

Simberloff, D., and E. O. Wilson. 1970. Experimental zoogeography of islands: a two year record of recolonization. *Ecology* 51:934–937.

Simmons, I. G. 1979. *Biogeography: Natural and Cultural.* Edward Arnold, London.

Simmons, I. G. 1981. *The Ecology of Natural Resources,* 2nd edition. Edward Arnold, London.

Simon, H. A. 1956. Rational choice and the structure of the environment. *Psychology Review* 63:129–138.

Simon, J. L. 1980. Resources, population, environment: an oversupply of false bad news. *Science* 208:1431–1437.

Simon, J. L. 1986. Disappearing species, deforestation and data. *New Scientist* 110(1508):60–63.

Simon, J. L., and H. Kahn. 1984. *The Resourceful Earth—A Response to Global 2000.* Blackwell Scientific Publications, Oxford.

Simonsen, L., and B. R. Levin. 1988. Evaluating the risk of releasing genetically engineered organisms. *Trends in Ecology and Evolution* 3(4) and *Trends in Biotechnology* 6(4), special combined issue: 27–30.

Simpson, G. G. 1964. Species density of North American Recent mammals. *Systematic Zoology* 13:57–73.

Sinclair, A. R. E. 1986. Testing multi-factor causes of population limitation: an illustration using snowshoe hares. *Oikos* 47:360–364.

Skinner, G. J., and J. B. Whittaker. 1981. An experimental investigation of interrelationships between the wood-ant (*Formica rufa*) and some tree-canopy herbivores. *Journal of Animal Ecology* 50:313–326.

Sladen, W. J. L., C. M. Menzie, and W. L. Reichel. 1966. DDT residues in Adelie penguins and a crab-eater seal from Antarctica: ecological implications. *Nature* 210:670–671.

Sláma, K. 1969. Plants as a source of materials with insect hormone activity. *Entomologia Experimentalis et Applicata* 12:721–728.

Slobodkin, L. B. 1986a. The role of minimalization in art and science. *The American Naturalist* 127:257–265.

Slobodkin, L. B. 1986b. Natural philosophy rampant. *Paleobiology* 12:111–118.

Slobodkin, L. B. 1988. Intellectual problems of applied ecology. *BioScience* 38:337–342.

Smallwood, P. D., and W. D. Peters. 1986. Grey squirrel food preferences: the effects of tannin and fat concentration. *Ecology* 67:168–174.

Smiley, J. 1986. Ant constancy at *Passiflora* extrafloral nectaries: effects on caterpillar survival. *Ecology* 67:516–521.

Smith, J. N. M. 1962. Detoxification mechanisms. *Annual Review of Entomology* 7:465–480.

Smith, J. N. M., C. J. Krebs, A. R. E. Sinclair, and R. Boonstra. 1988. Population biology of snowshoe hares. II. Interactions with winter food plants. *Journal of Animal Ecology* 57:269–286.

Smith, K. L., Jr. 1985. Deep-sea hydrothermal vent mussels: nutritional state and distribution at the Galapagos rift. *Ecology* 66:1067–1080.

Smith, R. L. 1974. *Ecology and Field Biology,* 2nd edition. Harper and Row, New York.

Sober, E., and R. C. Lewontin. 1982. Artifact, cause and genic selection. *Philosophy of Science* 49:157–180.

Soulé, M. E. 1980. Thresholds for survival: maintaining fitness and evolutionary potential. Pages 151–170 in M. E. Soulé and B. A. Wilcox (eds.), *Conservation Biology: An Evolutionary-Ecological Perspective.* Sinauer Associates, Sunderland, Mass.

Soulé, M. E. 1983. What do we really know about extinction? Pages 111–124 in C. M. Schonewald-Cox, S. M. Chambers, B. MacBryde, and L. Thomas (eds.), *Genetics and Conservation.* Benjamin/Cummings, Menlo Park, Calif.

Soulé, M. E. (ed.). 1987. *Viable Populations for Conservation.* Cambridge University Press, Cambridge.

Soulé, M. E., and D. Simberloff. 1986. What do genetics and ecology tell us about the design of nature reserves? *Biological Conservation* 35:19–40.

Soulé, M. E., and B. A. Wilcox (eds.). 1980. *Conservation Biology.* Sinauer Associates, Sunderland, Mass.

Sousa, W. P. 1979. Disturbance in marine intertidal boulder fields: the nonequilibrium maintenance of species diversity. *Ecology* 60:1225–1239.

Southwick, C. H. (ed.). 1985. *Global Ecology.* Sinauer Associates, Sunderland, Mass.

Southwood, T. R. E. 1961. The number of species of insect associated with various trees. *Journal of Animal Ecology* 30:1–8.

Southwood, T. R. E. 1976. Bionomic strategies and population parameters. Pages 26–48 in *Theoretical Ecology, Principles and Applications.* Saunders, Philadelphia.

Southwood, T. R. E. 1977. The relevance of population dynamic theory to pest status. Pages 35–54 in J. M. Cherrett and G. R. Sagor (eds.), *Origins of Pest, Parasite, Disease and Weed Problems.* Blackwell Scientific Publications, Oxford.

Southwood, T. R. E. 1978. *Ecological Methods with Particular Reference to the Study of Insect Populations,* 2nd edition. Methuen, London.

Southwood, T. R. E. 1987. The concept and nature of the community. Pages 3–27 in J. H. R. Gee and P. S. Geller (eds.), *Organization of Communities Past and Present*. Blackwell Scientific Publications, Oxford.

Sparks, J. 1982. *Discovering Animal Behaviour*. BBC Publications, London.

Spear, L. 1988. A naturalist at large, the halloween mask episode. *Natural History* 97(6):4–8.

Spencer, W. P. 1957. Genetic studies on *Drosophila mulleri*. I. Genetic analysis of a population. *Texas University Publication* 5721:186–205.

Stanley, S. M. 1975. A theory of evolution above the species level. *Proceedings of the National Academy of Sciences of the United States of America* 72:646–650.

Stanley, S. M. 1979. *Macroevolution: Pattern and Process*. Freeman, San Francisco.

Stark, R. W. 1964. Recent trends in forest entomology. *Annual Review of Entomology* 10:303–324.

Stearns, S. C. 1980. A new view of life-history evolution. *Oikos* 35:266–281.

Stebbins, G. L. 1950. *Variation and Evolution in Plants*. Columbia University Press, New York.

Stebbins, G. L. 1974. *Flowering Plants: Evolution Above the Species Level*. Harvard University Press, Cambridge, Mass.

Steele, J. H. 1974. *The Structure of Marine Ecosystems*. Blackwell Scientific Publications, Oxford.

Steele, J. H., and E. W. Henderson. 1984. Modeling long-term fluctuations in fish stocks. *Science* 224:985–987.

Steele, R. H. 1986. Courtship feeding in *Drosophila subobscura*. II. Courtship feeding by males influences female mate choice. *Animal Behaviour* 34:1099–1108.

Stenseth, N. C. 1983. Causes and consequences of dispersal in small mammals. Pages 63–101 in I. R. Swingland and P. J. Greenwood (eds.), *The Ecology of Animal Movement*. Oxford University Press, Oxford.

Stephens, D. W. 1981. The logic of risk-sensitive foraging preferences. *Animal Behaviour* 29:628–629.

Stephens, D. W., and J. R. Krebs. 1986. *Foraging Theory*. Princeton University Press, Princeton, N.J.

Stern, C. 1973. *Principles of Human Genetics*, 3rd edition. Freeman, San Francisco.

Sternberg, J. G., G. P. Waldbauer, and M. R. Jeffords. 1977. Batesian mimicry: selective advantage of color pattern. *Science* 195:681–683.

Stevenson-Hamilton, J. 1957. Tsetse fly and the rinderpest epidemic of 1896. *South African Journal of Science* 58:216.

Stewart, A. J. A. 1986*a*. Nymphal colour/pattern polymorphism in the leafhoppers *Eupteryx urticae* (F.) and *E. cyclops* Matsumara (Hemiptera: Auchenorrhyncha): spatial and temporal variation in morph frequencies. *Biological Journal of the Linnaean Society* 27:79–101.

Stewart, A. J. A. 1986*b*. The inheritance of nymphal colour/pattern polymorphism in the leafhoppers *Eupteryx urticae* (F.) and *E. cyclops* Matsumara

(Hemiptera: Auchenorrhyncha). *Biological Journal of the Linnaean Society* 27:57–77.

Stiling, P. D. 1980. Colour polymorphism in some nymphs of the genus *Eupteryx*. *Ecological Entomology* 5:175–178.

Stiling, P. D. 1987. The frequency of density dependence in insect-host-parasitoid systems. *Ecology* 68:844–856.

Stiling, P. D. 1988. Key factors and density-dependent processes in insect populations. *Journal of Animal Ecology* 57:581–593.

Stiling, P. D. 1990 Calculating the establishment rates of parasitoids in classical biological control. *American Entomologist* 36:225–230.

Stiling, P. D., B. V. Brodbeck, and D. Simberloff. 1991. Variation in rates of leaf abscission between plants may affect the distribution patterns of sessile insects. *Oecologia* (in press).

Stiling, P. D., B. V. Brodbeck, and D. R. Strong. 1982. Foliar nitrogen and larval parasitism as determinants of leafminer distribution patterns on *Spartina alterniflora*. *Ecological Entomology* 7:447–452.

Stiling, P. D., and D. Simberloff. 1989. Leaf abscission: induced defense against pests or response to damage? *Oikos* 55:43–49.

Stiling, P. D., and D. R. Strong. 1983. Weak competition among *Spartina* stem borers by means of murder. *Ecology* 64:770–778.

Stoddart, D. R. 1968. Catastrophic human interference with coral atoll ecosystems. *Geography* 58:25–40.

Strassman, J. E. 1989. Altruism and relatedness at colony foundation in social insects. *Trends in Ecology and Evolution* 4:371–374.

Strathmann, R. R. 1978. progressive vacating of adaptive types during the Phanerozoic. *Evolution* 32:907–914.

Strauss, S. Y. 1987. Direct and indirect effects of host-plant fertilization on an insect community. *Ecology* 68:1670–1678.

Strauss, S. Y. 1988. Determining the effects of herbivory using damaged plants. *Ecology* 69:1628–1630.

Strauss, S. Y. 1989. Personal communication.

Strickberger, M. W. 1986. *Genetics,* 3rd edition. Macmillan, New York.

Strong, D. R. 1974a. Nonasymptotic species richness models and the insects of British trees. *Proceedings of the National Academy of Sciences of the United States of America* 71:2766–2769.

Strong, D. R. 1974b. The insects of British trees: community equilibrium in ecological time. *Annals of the Missouri Botanical Garden* 61:692–701.

Strong, D. R. 1979. Biogeographic dynamics of insect-host plant communities. *Annual Review of Entomology* 24:89–119.

Strong, D. R. 1980. Null hypotheses in ecology. *Synthèse* 43:271–285.

Strong, D. R. 1988. Insect host range (Special Feature). *Ecology* 69:885.

Strong, D. R., J. H. Lawton, and T. R. E. Southwood. 1984. *Insects on Plants, Community Patterns and Mechanisms.* Blackwell Scientific Publications, Oxford.

Strong, D. R., E. D. McCoy, and J. R. Rey. 1977. Time and the number of herbivore species: the pests of sugarcane. *Ecology* 58:167–175.

Strong, D. R., and P. D. Stiling. 1983. Wing dimorphism changed by experimental density manipulation in a planthopper, *Prokelisia marginata* (Homoptera: Delphacidae). *Ecology* 64:206–209.

Strong, D. R., L. Szyska, and D. Simberloff. 1979. Tests of community-wide character displacement against null hypotheses. *Evolution* 33:897–913.

Struhsaker, T. T. 1967. Social structure among vervet monkeys (*Cercopithecus aethiops*). *Behaviour* 29:83–121.

Sunquist, F. 1988. Zeroing in on keystone species. *International Wildlife* 18(5):18–23.

Suppe, F. 1977. *The Structure of Scientific Theories*, 2nd edition. University of Illinois Press, Urbana.

Swift, M. J., O. W. Heal, and J. M. Anderson. 1979. *Decomposition in Terrestrial Ecosystems*. Blackwell Scientific Publications, Oxford.

Tansley, A. G. 1935. The use and abuse of vegetational concepts and terms. *Ecology* 16:284–307.

Teal, J. M. 1962. Energy flow in the salt marsh ecosystem of Georgia. *Ecology* 43:614–624.

Temple, S. A. 1977. Plant-animal mutualism: co-evolution with dodo leads to near extinction of plant. *Science* 197:885–886.

Temple, S. A. 1986. Predicting impacts of habitat fragmentation on forest birds: a comparison of two models. Pages 301–304 in J. Verner, M. L. Morrison, and C. J. Ralph (eds.), *Wildlife 2000*. University of Wisconsin Press, Madison.

Templeton, A. R., and B. Read. 1983. The elimination of inbreeding depression in a captive herd of Speke's gazelle. Pages 241–261 in C. M. Schonewald-Cox, S. M. Chambers, B. MacBryde, and L. Thomas (eds.), *Genetics and Conservation*. Benjamin/Cummings, Menlo Park, Calif.

Terborgh, J. 1973. On the notion of favorableness in plant ecology. *American Naturalist* 107:481–501.

Terborgh, J., and B. Winter. 1980. Some causes of extinction. Pages 119–134 in M. E. Soulé and B. A. Wilcox (eds.), *Conservation Biology: An Evolutionary-Ecological Perspective*. Sinauer Associates, Sunderland, Mass.

Thomas, C. D., D. Ng, C. M. Singer, J. L. B. Mallet, C. Parmesan, and H. L. Billington. 1987. Incorporation of a European weed into the diet of a North American herbivore. *Evolution* 41:892–901.

Thompson, D. W. 1942. *On Growth and Form*, 2nd edition. Cambridge University Press, Cambridge.

Thompson, D., and R. Dunbar. 1988. Sex for dragons and damsels. *New Scientist* 117(1601):45–48.

Thompson, R. 1988. What's going wrong with the weather? *New Scientist* 117(1605):65.

Thornhill, R. 1976. Sexual selection and nuptial feeding behavior in *Bittacus apicalis* (Insecta: Mecoptera). *American Naturalist* 110:529–548.

Thornhill, R. 1979. Adaptive female-mimicking behavior in a scorpionfly. *Science* 205:412–414.

Thornton, I. W. B. (ed.). 1987. *1986 Zoological Expedition to the Krakataus.* Preliminary Report, La Trobe University, Department of Zoology, Miscellaneous series no. 3.

Tilman, D. 1987. The importance of mechanisms of interspecific competition. *American Naturalist* 129:769–774.

Tinbergen, L. 1960. The natural control of insects in pine woods. I. Factors influencing the intensity of predation in song birds. *Archives Néerlandaise de Zoologie* 13:265–343.

Tinbergen, N. 1963. On aims and methods of ethology. *Zeitschrift für Tierpsychologie* 20:410–433.

Toksöz, M. N. 1975. The subduction of the lithosphere. *Scientific American* 233(5):88–98.

Townsend, C. R., and P. Calow. 1981. *Physiological Ecology: An Evolutionary Approach to Resource Use.* Blackwell Scientific Publications, Oxford.

Tranquillini, W. 1979. *Physiological Ecology of the Alpine Timberline.* Springer-Verlag, Berlin.

Trewartha, G. T. 1969. *A Geography of Population: World Patterns.* Wiley, New York.

Trivers, R. L. 1971. The evolution of reciprocal altruism. *Quarterly Review of Biology* 46:35–37.

Trivers, R. L., and H. Hare. 1976. Haplodiploidy and the evolution of social insects. *Science* 191:249–263.

Trostel, K., A. R. E. Sinclair, C. J. Walters, and C. J. Krebs. 1987. Can predation cause the 10-year hare cycle? *Oecologia* 74:185–192.

Tschinkel, W. R. 1990. Personal communication.

Turner, J. 1962. The *Tilia* decline: an anthropogenic interpretation. *New Phytologist* 61:328–341.

Turner, J. R. G. 1986. The genetics of adaptive radiation: a neo-Darwinian theory of punctuational evolution. Pages 183–207 in D. M. Raup and D. Jablonski (eds.), *Patterns and Processes in the History of Life.* Springer-Verlag, Berlin.

Turner, J. R. G., C. M. Gratehouse, and C. A. Carey. 1987. Does solar energy control organic diversity? Butterflies, moths and the British climate. *Oikos* 48:195–205.

Turner, J. R. G., J. J. Lennon, and J. A. Lawrenson. 1988. British bird species distributions and the energy theory. *Nature* 335:539–541.

Uhazy, L. S., J. C. Holmes, and J. G. Stelfox. 1973. Lungworms in the rocky mountain bighorn sheep of western Canada. *Canadian Journal of Zoology* 51:817–824.

U.S. Congress Office of Technology Assessment. 1987. *Technologies to Maintain Biological Diversity, OTA-F-330.* U.S. Government Printing Office, Washington, D.C.

U.S. Department of the Interior and U.S. Department of Commerce, U.S. Fish and Wildlife Service, and U.S. Bureau of the Census. 1982. *1980 National Survey of Fishing, Hunting, and Wildlife-Associated Recreation.* Washington, D.C.

Uyeda, S. 1978. *The New View of the Earth: Moving Continents and Moving Oceans.* Freeman, San Francisco.

Vaeck, M., A. Reynaerts, H. Höfte, S. Jansens, M. DeBeuckeleer, C. Dean, M. Zabeau, M. Van Montagu, and J. Leemans. 1987. Transgenic plants protected from insect attack. *Nature* 328:33–37.

van der Leeden, F. 1975. *Water Resources of the World. Selected Statistics.* Water Information Center, Port Washington, N.Y.

van der Schalie, H. 1972. WHO project Egypt 10: a case history of a schistosomiasis control project. Pages 116–136 in M. Taghi Farvar and J. P. Milton (eds.), *The Careless Technology: Ecology and International Development.* Natural History Press, New York.

van Hylckama, T. E. A. 1975. Water resources. Pages 147–165 in W. W. Murdoch (ed.), *Environment,* 2nd edition. Sinauer Associates, Sunderland, Mass.

van Lenteren, J. C. 1980. Evaluation of control capabilities of natural enemies: does art have to become science? *Netherlands Journal of Zoology* 30:369–381.

van Riper, C., III, S. G. van Riper, M. L. Goff, and M. Laird. 1986. The epizootiology and ecological significance of malaria in Hawaiian land birds. *Ecological Monographs* 56:327–344.

Van Valen, L. M. 1973. A new evolutionary law. *Evolutionary Theory* 1:1–30.

Van Valen, L. M. 1984. Catastrophes, expectations, and the evidence. *Paleobiology* 10:121–127.

Varley, G. C. 1970. The concept of energy flow applied to a woodland community. Pages 389–405 in A. Watson (ed.), *Animal Populations in Relation to Their Food Resources.* Blackwell Scientific Publications, Oxford.

Varley, G. C. 1971. The effects of natural predators and parasites on winter moth populations in England. Pages 103–116 in *Proceedings, Tall Timbers Conference on Ecological Animal Control by Habitat Management No. 2.* Tall Timbers Research Station, Tallahassee, Fla.

Varley, G. C., G. R. Gradwell, and M. P. Hassell. 1973. *Insect Population Ecology, an Analytical Approach.* Blackwell Scientific Publications, Oxford.

Verhulst, P. F. 1838. Notice sur la loi que la population suit dans son accroissement. *Correspondence in Mathematics and Physics* 10:113–121.

Vince, S. W., I. Valiela, and J. M. Teal. 1981. An experimental study of the structure of herbivorous insect communities in a salt marsh. *Ecology* 62:1662–1678.

Vitousek, P. M., and W. A. Reiners. 1975. Ecosystem succession and nutrient retention: a hypothesis. *BioScience* 25:376–381.

Volterra, V. 1926. Fluctuations in the abundance of a species considered mathematically. *Nature* 118:558–560.

von Frisch, K. 1967. *The Dance Language and Orientation of Bees.* Harvard University Press, Cambridge, Mass.

von Haartman, L. 1956. Territory in the pied flycatcher *Muscicapa hypoleuca*. *Ibis* 98:460–475.

Waage, J. K. 1979. Dual function of the damselfly penis: sperm removal and transfer. *Science* 203:916–918.

Waage, J. K. 1982. Sib-mating and sex ratio strategies in scelionid wasps. *Ecological Entomology* 7:103–112.

Waage, J. K., and J. A. Lane. 1984. The reproductive strategy of a parasitic wasp. II. Sex allocation and local mate competition in *Trichogramma evanescens*. *Journal of Animal Ecology* 53:417–426.

Wade, D., J. Ewel, and R. Hotstetler. 1980. *Fire in South Florida Ecosystems*. USDA Forest Service General Technical Report SE-17, Asheville, N.C.

Wade, M. 1980. Court says lab-made life can be patented. *Science* 208:1445.

Waldbauer, G. P. 1988. Aposematism and Batesian mimicry. *Evolutionary Biology* 22:261–286.

Walde, S. J., and W. W. Murdoch. 1988. Spatial density dependence in parasitoids. *Annual Review of Entomology* 33:441–466.

Walker, B. A. 1976. An assessment of the ecological basis of game ranching in southern African savannas. *Proceedings of the Grassland Society of South Africa* 11:125–130.

Walker, L. R., and F. S. Chapin III. 1986. Physiological controls over seedling growth in primary succession on an Alaskan floodplain. *Ecology* 67:1508–1523.

Walker, L. R., and F. S. Chapin III. 1987. Interactions among processes controlling successional change. *Oikos* 50:131–135.

Walker, T. J. 1982. Sound traps for sampling male crickets (Orthoptera: Gryllotalpidae: *Scapteriscus*). *Florida Entomologist* 65:13–25.

Wall, G., and C. Wright. 1977. *The Environmental Impact of Outdoor Recreation*. Waterloo, Ontario, University. Department of Geography Publication Series No. 11.

Wallace, A. R. 1878. *Tropical Nature and Other Essays*. Macmillan, New York.

Wallace, J. B., and J. O'Hop. 1985. Life on a fast pad: waterlily leaf beetle impact on water lilies. *Ecology* 66:1534–1544.

Waloff, N., and O. W. Richards. 1977. The effect of insect fauna on growth, mortality, and natality of broom, *Sorothamnus scoparius*. *Journal of Applied Ecology* 14:787–798.

Walsh, J. 1973. The wake of the *Torrey Canyon*. Pages 60–62 in W. Jackson (ed.), *Man and the Environment*, 2nd edition. William C. Brown Company, Dubuque, Iowa.

Walton, S. 1981. Aswan revisited: U.S.-Egypt Nile project studies high dam's effects. *BioScience* 31:9–13.

Ward, P., and A. Zahavi. 1973. The importance of certain assemblages of birds as "information-centres" for food finding. *Ibis* 115:517–534.

Warner, R. R., D. R. Robertson, and E. G. Leigh. 1975. Sex change and sexual selection. *Science* 190:633–638.

Warren, A., and F. B. Goldsmith. 1983. An introduction to nature conservation. Pages 1–15 in A. Warren and F. B. Goldsmith (eds.), *Conservation in Perspective*. Wiley, Chichester, England.

Warren, S. D., H. L. Black, D. A. Eastmond, and W. H. Whaley. 1988. Structure function of buttresses of *Tachigalia versicolor*. *Ecology* 69:532–536.

Wasserman, A. O. 1957. Factors affecting interbreeding in sympatric species of spadefoots (genus *Scaphiopus*). *Evolution* 11:320–338.

Wathern, P. 1986. Restoring derelict lands in Great Britain. Pages 248–274 in *Ecological Knowledge and Environmental Problem-Solving, Concepts and Case Studies*. National Academy Press, Washington, D.C.

Watson, A. 1967. Territory and population regulation in the red grouse. *Nature* 215:1274–1275.

Watson, A., R. Moss, and R. Parr. 1984. Effects of food enrichment on numbers and spacing behavior of red grouse. *Journal of Animal Ecology* 53:663–678.

Weck, J., and C. Wiebecke. 1961. *Weltwirtschaft und Deutschlands Forst- und Holzwirtschaft*. Bayerischer Landwirtschaftsverlag, Munich.

Wegener, A. 1966. *The Origin of Continents and Oceans* (translation of 1929 edition by J. Biram). Dover Publications, New York.

Weismann, A. 1893. *The Germ-Plasm: A Theory of Heredity* (translation). Walter Scott, London.

Werren, J. H. 1980. Sex ratio adaptations to local mate competition in a parasitic wasp. *Science* 208:1157–1159.

Werren, J. H. 1983. Sex ratio evolution under local mate competition in a parasitic wasp. *Evolution* 37:116–124.

West Eberhard, M. J. 1975. The evolution of social behaviour by kin selection. *Quarterly Review of Biology* 50:1–33.

Whicker, F. W., and V. Schultz. 1982. *Radioecology: Nuclear Energy and the Environment, Vols. I and II*. CRC Press, Boca Raton, Fla.

White, G. G. 1981. Current status of prickly pear control by *Cactoblastis cactorum* in Queensland. Pages 609–616 in *Proceedings of the Fifth International Symposium for the Biological Control of Weeds, Brisbane, 1980*.

White, M. J. D. 1978. *Modes of Speciation*. Freeman, San Francisco.

White, T. C. R. 1984. The abundance of invertebrate herbivores in relation to the availability of nitrogen in stressed food plants. *Oecologia* 63:90–105.

Whitham, T. G. 1978. Habitat selection by *Pemphigus* aphids in response to resource limitation and competition. *Ecology* 59:1164–1176.

Whitham, T. G. 1979. Territorial behaviour of *Pemphigus* gall aphids. *Nature* 279:324–325.

Whitham, T. G. 1980. The theory of habitat selection examined and extended using *Pemphigus* aphids. *American Naturalist* 115:449–466.

Whittaker, R. H. 1961. Experiments with radiophosphorus tracer in aquarium microcosms. *Ecological Monographs* 31:157–188.

Whittaker, R. H. 1975. *Communities and Ecosystems,* 2nd edition. Macmillan, New York.

Whittaker, R. H., and G. E. Likens. 1975. The biosphere and man. Pages 305–328 in H. Lieth and R. H. Whittaker (eds.), *Primary Productivity of the Biosphere.* Springer-Verlag Ecological Studies Vol. 14, Berlin.

Whittaker, R. J., M. B. Bush, and F. K. Richards. 1989. Plant recolonization and vegetation succession on the Krakatau Islands, Indonesia. *Ecological Monographs* 59:59–123.

Wickler, W. 1968. *Mimicry in Plants and Animals.* McGraw-Hill, New York.

Wiens, J. A., J. F. Addicott, T. J. Case, and J. Diamond. 1986. Overview: the importance of spatial and temporal scale in ecological investigations. Pages 145–153 in J. Diamond and T. J. Case (eds.), *Community Ecology.* Harper and Row, New York.

Wigglesworth, V. B. 1955. The contribution of pure science to applied biology. *Annals of Applied Biology* 42:34–44.

Wigley, T. M. L., P. D. Jones, and P. M. Kelly. 1980. Scenarios for a warm high-CO_2 world. *Nature* 238:17–21.

Wilcox, B. A. 1986. Extinction models and conservation. *Trends in Ecology and Evolution* 1:46–48.

Wiles, R. 1986. The future of U.S. agriculture. Letter to the editor. *Science* 231:1490–1491.

Wiley, R. H. 1973. Territoriality and non-random mating in the sage grouse, *Centrocerus urophasianus. Animal Behaviour Monographs* 6:87–169.

Wilkinson, P. F., and C. C. Shank. 1977. Rutting-fight mortality among musk oxen on Banks Island, Northwest Territories, Canada. *Animal Behaviour* 24:756–758.

Williams, A. G., and T. G. Whitham. 1986. Premature leaf abscission: an induced plant defense against gall aphids. *Ecology* 67:1619–1627.

Williams, C. B. 1964. *Patterns in the Balance of Nature and Related Problems in Quantitative Ecology.* Academic Press, New York.

Williams, E. D. 1978. *Botanical Composition of the Park Grass Plots at Rothamstead 1856–1976.* Rothamstead Experimental Station, Harpenden, U.K.

Williams, G. C. 1966. *Adaptation and Natural Selection.* Princeton University Press, Princeton, N.J.

Williams, N. H. 1983. Floral fragrances as cues in animal behavior. Pages 50–72 in C. E. Jones and R. J. Little (eds.), *Handbook of Experimental Pollination Biology.* Van Nostrand Reinhold, New York.

Williamson, M. H., and K. C. Brown. 1986. The analysis and modelling of British invasions. *Proceedings and Transactions of the Royal Society of London B* 314:502–522.

Willis, E. O. 1972. Do birds flock in Hawaii, a land without predators? *California Birds* 3:1–8.

Wilson, D. B. (ed.). 1983. *Did the Devil Make Darwin Do It?* Iowa State University Press, Ames.

Wilson, E. O. 1980. Caste and division of labor in leaf-cutter ants (Hymenoptera: Formicidae: *Atta*). I. The overall pattern in *A. sexdens*. *Behavioral Ecology and Sociobiology* 7:143–156.

Wilson, E. O. 1985. The principles of caste evolution. Pages 307–324 in B. Hölldobler and M. Lindauer (eds.), *Experimental Behavioral Ecology and Sociobiology*. Sinauer Associates, Sunderland, Mass.

Wilson, E. O. 1987. Causes of ecological success: the case of the ants. The Sixth Transley Lecture. *Journal of Animal Ecology* 56:1–9.

Wilson, E. O., and W. L. Brown. 1953. The subspecies concept and its taxonomic applications. *Systematic Zoology* 2:97–111.

Wilson, E. O., and E. O. Willis. 1975. Applied biogeography. Pages 522–534 in M. L. Cody and J. Diamond (eds.), *Ecology and Evolution of Communities*. Harvard University Press, Cambridge, Mass.

Wint, G. R. W. 1983. The effect of foliar nutrients upon the growth and feeding of a lepidopteran larva. Pages 301–320 in J. A. Lee, S. McNeill, and I. H. Rorison (eds.), *Nitrogen as an Ecological Factor. 22nd Symposium of the British Ecological Society*. Blackwell Scientific Publications, Oxford.

Wise, D. H. 1981. A removal experiment with darkling beetles: lack of evidence for interspecific competition. *Ecology* 62:727–738.

Wolda, H. 1983. "Long-term" stability of tropical insect populations. *Researches on Population Ecology*, Tokyo, Supplement No. 3:112–126.

Wolda, H. 1986. Spatial and temporal variation in abundance in tropical animals. Pages 93–105 in S. L. Sutton, T. C. Whitmore, and A. C. Chadwick (eds.), *Tropical Rain Forest: Ecology and Management*. Blackwell Scientific Publications, Oxford.

Wolda, H., and E. Broadhead. 1985. The seasonality of Psocoptera in two tropical forests in Panama. *Journal of Animal Ecology* 54:519–530.

Wood, D. M., and R. del Moral. 1987. Mechanisms of early primary succession in sabal pine habitats on Mount St. Helens. *Ecology* 68:780–790.

Wood, T. K., and S. I. Guttman. 1983. *Enchenopa binotata* complex: sympatric speciation? *Science* 220:310–312.

Woodley, J. D., E. A. Chornesky, P. A. Clifford, J. B. C. Jackson, L. S. Kaufman, N. Knowlton, J. C. Lang, M. P. Pearson, J. W. Porter, M. C. Rooney, K. W. Rylaarsdam, V. J. Tunnicliffe, C. M. Wahle, J. L. Wulft, A. S. G. Curtis, M. D. Dallmeyer, B. P. Jupp, M. A. R. Koehl, J. Neigel, and E. M. Sides. 1981. Hurricane Allen's impact on Jamaican coral reefs. *Science* 214:749–755.

Woodwell, G. M. 1967. Radiation and the pattern of nature. *Science* 156:461–470.

Woodwell, G. M. 1970. Effects of pollution on the structure and physiology of ecosystems. *Science* 168:429–433.

Woodwell, G. M., and A. L. Rebuck. 1967. Effects of chronic gamma radiation on the structure and diversity of an oak-pine forest. *Ecological Monographs* 37:53–64.

Woodwell, G. M., C. F. Warster, and P. A. Isaacson. 1967. DDT residues in an east coast estuary. *Science* 156:821–824.

Woolfenden, G. E. 1975. Floridà scrub jay helpers at the nest. *Auk* 92:1–15.

Woolfenden, G. E., and J. W. Fitzpatrick. 1984. *The Florida Scrub Jay: Demography of a Cooperative Breeding Bird.* Princeton University Press, Princeton, N.J.

Wright, D. A. 1983. Species-energy theory: an extension of species-area theory. *Oikos* 41:496–506.

Wright, M. 1988. Mixed blessings of the flooding in Sudan. *New Scientist* 119(1631)44–47.

Wynne-Edwards, V. C. 1962. *Animal Dispersion in Relation to Social Behaviour.* Oliver and Boyd, Edinburgh.

Wynne-Edwards, V. C. 1977. Intrinsic population control and introduction. Pages 1–22 in F. J. Ebling and D. M. Stoddart, (eds.), *Population Control by Social Behavior.* Institute of Biology, London.

Young, A. 1988. Agroforestry and its potential to contribute to land development in the tropics. *Journal of Biogeography* 15:19–30.

Young, T. P. 1987. Increased thorn length in *Acacia depranolobium*—an induced response to browsing. *Oecologia* 71:436–438.

Zahavi, A. 1975. Mate selection—a selection for a handicap. *Journal of Theoretical Biology* 53:205–214.

Zahavi, A. 1977. The cost of honesty (further remarks on the handicap principle). *Journal of Theoretical Biology* 67:603–605.

Zaleski, C. P. 1980. Breeder reactors in France. *Science* 108:137–144.

Zar, J. H. 1984. *Biostatistical Analysis,* 2nd edition. Prentice-Hall, Englewood Cliffs, N.J.

Zaret, T. M. 1982. The stability/diversity controversy: a test of hypotheses. *Ecology* 63:721–731.

Zaret, T. M., and R. T. Paine. 1973. Species introduction in a tropical lake. *Science* 182:449–455.

Zimmerman, B. L., and R. O. Bierregaard. 1986. Relevance of the equilibrium theory of island biogeography and species-area relations to conservation with a case from Amazonia. *Journal of Biogeography* 13:133–143.

Zucker, W. V. 1983. Tannins: does structure determine function? An ecological perspective. *American Naturalist* 121:335–365.

Press. **Figure 7.1,** from J.F. Downhower and K.B. Armitage, "The yellow-bellied marmot and the evolution of polygamy," *American Naturalist* 105(1971), p. 361. Copyright © 1971 University of Chicago Press. **Figure 7.2,** from G.H. Orians, "On the evolution of mating systems in birds and mammals," *American Naturalist* 103(1969), p. 592. Copyright © 1969 University of Chicago Press. **Figure 8.1** and **8.13,** from C.B. Cox, I.N. Healey, and P.B. Moore, *Biogeography: An Ecological and Evolutionary Approach* (Oxford: Blackwell Scientific Publications, 1976, pp. 27, 28). **Figure 8.2,** from A.S. Collinson, *An Introduction to World Vegetation* (London: Unwin Hyman Ltd., 1977, p. 86). **Figure 8.6,** from J. Terborgh, "On the notion of favorableness in plant ecology," *American Naturalist* 107(1973), p. 491. Copyright © 1973 University of Chicago Press. **Figures 8.10** and **8.11,** from T.J. Givnish and G.J. Vermeij, "Sizes and shapes of liane leaves," *American Naturalist* 110(1976), pp. 754, 757. Copyright © 1976 University of Chicago Press. **Figure 8.12,** By permission from J.R. Hastings and R.M. Turner, *The Changing Mile* (Tucson, Ariz.: University of Arizona Press, 1965, p. 20). Copyright 1965. **Figures 8.14, 9.8, 19.15, 20.6,** and **Table 17.1,** From *Ecology: The Experimental Analysis of Distribution and Abundance,* 3rd ed. by Charles J. Krebs. Copyright © 1985 by Harper & Row, Publishers, Inc. Reprinted by permission of the publisher. **Figure 8.15,** from D.J. Currie and V. Paquin, "Large-scale biogeographical patterns of species richness of trees," *Nature* 329(1987), p. 326. Reprinted by permission from *Nature.* Copyright © 1987 Macmillan Magazines Ltd. **Figure 8.16,** From "Certain aspects of climate as related to the distribution of loblolly pine" by H.W. Hocker, Jr., *Ecology,* 1956, **37,** 824–834. Copyright © 1956 by the Ecological Society of America. Reprinted by permission. **Figure 8.17** and **Table 17.2,** Reproduced by permission from J.H. Brown and A.C. Gibson, *Biogeography* (St. Louis, 1983, The C.V. Mosby Co., pp. 55, 498). **Figure 8.18,** from P. Hershkovitz, *Living New World Monkeys (Platyrhini),* Vol. 1 (Chicago: University of Chicago Press, 1977, p. 400). Copyright © 1977 University of Chicago Press. **Figure 8.19,** Figure from *Ecology and Field Biology,* 2nd ed. by Robert Leo Smith. Copyright © 1966, 1974 by Robert Leo Smith. Reprinted by permission of Harper & Row, Publishers, Inc. **Figure 8.20,** from R.J. Whittaker, M.B. Bush, and F.K. Richards, "Plant recolonization and vegetation succession on the Krakatau Islands, Indonesia," *Ecological Monographs,* 1989, **59,** 59–123. Copyright © 1989 by the Ecological Society of America. Reprinted by permission. **Figure 9.2,** from R.F. Morris, "The dynamics of epidemic spruce budworm populations," *Memoirs of the Entomological Society of Canada* 31(1963), p. 19. **Figure 9.6,** from V.B. Scheffer, "The rise and fall of a reindeer herd," *Science* (73(1951), p. 356–362. Copyright 1951 by the AAAS. **Figure 9.9,** from John Tyler Bonner, *Size and Cycle: An Essay on the Structure of Biology* (Princeton, N.J.: Princeton University Press, 1965, p. 17). Copyright © 1965 by Princeton University Press. Fig. 1 reprinted with permission of Princeton University Press. **Figure 9.10,** from A.C. Heron, "Population ecology of a colonizing species: The pelagic tunicate *Thalia democratica,*" *Oecologia* 10(1972), p. 305. **Figure 9.11,** from T. Fenchel, "Intrinsic rate of natural increase: The relationship with body size," *Oecologia* 14(1974), p. 319. **Figure 11.3,** From "The influence of interspecific competition and other factors on the distribution of the barnacle *Chthamalus stellatus*" by J.H. Connell, *Ecology,* 1961, **42,** 710–723. Copyright © by the Ecological Society of America. Reprinted by permission. **Figure 11.4,** from J.H. Lawton, "Non-competitive populations, non-convergent communities, and vacant niches: The herbivores of bracken," in Donald R.

Strong et al., eds., *Ecological Communities: Conceptual Issues and the Evidence*
(Princeton, N.J.: Princeton University Press, 1984, p. 74). Copyright © 1984 by
Princeton University Press. Fig. 6.3 reprinted with permission of Princeton
University Press. **Figure 11.5,** From "Characteristics permitting coexistence
among parasitoids of a sawfly in Quebec" by P.W. Price, *Ecology*, 1970, **51,**
445–454. Copyright © 1970 by the Ecological Society of America. Reprinted by
permission. **Figure 11.6a,** from R.M. May and R.H. MacArthur, "Niche overlap
as a function of environmental viability," *Proceedings of the National Academy of
Sciences of the U.S.A.* 69(1972), p. 1109. **Figure 11.6b,** from E.R. Pianka,
"Competition and niche theory," in R.M. May, ed., *Theoretical Ecology* (Oxford:
Blackwell Scientific Publications, 1976, p. 117). **Figures 12.3, 12.4, 12.6, 12.7,** and
12.10, from M. Begon, J.L. Harper, and C.R. townsend, *Ecology, Individuals,
Populations, and Communities* (Oxford: Blackwell Scientific Publications, 1986, pp.
359, 360, 362, 363, 369). **Figure 12.11,** from G. Caughley et al., "Does dingo
predation control the densities of kangaroos and emus?" *Australian Wildlife
Research* 7(1980), p. 5. **Figure 12.12,** from N.S. Baldwin, "Sea lamprey in the
Great Lakes," *Canadian Aubudon Magazine*, November–December 1964, p. 142.
Figure 13.1, from G.G. White, "Current status of prickly pear control by
Cactoblastis cactorum in Queensland," in *Proceedings of the Fifth International
Sumposium for the Biological Control of Weeds,*Brisbane, 1980 (Melbourne: CSIRO,
1981, p. 610). **Figure 13.2,** from D. Simberloff, B.J. Brown, and S. Lowrie,
"Isopod and insect root borers may benefit Florida mangroves," *Science*
201)1978), p. 630–632. Copyright 1978 by the AAAS. **Figures 13.3, 13.4, 13.5,** and
13.6, from M.J. Crawley, "Herbivory, the dynamics of animal-plant
interactions," *Studies in Ecology*, Vol. 10 (Oxford: Blackwell Scientific
Publications, 1983, pp. 250, 252, 253, 254). **Figures 14.1, 14,2 15.2, 15.3,** and **15.5,**
from G.C. Varley, G.R. Gradwell, and M.P. Hassell, *Insect Population Ecology, an
Analytical Approach* (Oxford: Blackwell Scientific Publications, 1973, pp. 68, 121,
123). **Figure 14.3,** from M.P. Hassell and G.C. Varley, "New inductive
population model for insect parasites and its bearing on biological control,"
Nature 223(1969), p. 1135. Reprinted by permission from *Nature*. Copyright ©
1969 Macmillan Magazines Ltd. **Figure 15.1,** from G.C. Varley, "The effects of
natural predators and parasites on winter moth populations in England," in
*Proceedings of Tall Timbers Conference on Ecological Animal Control by Habitat
Management*, No. 2, Tallahassee, Florida, 1971, p. 105. **Figure 16.1,** Reprinted with
permission of Macmillan Publishing Company from *Communities and Ecosystems*,
2nd ed., by Robert H. Whittaker (p. 167). Copyright © 1975 by Robert H.
Whittaker. **Figure 16.2** and **21.12,** from E.P. Odum, *Fundamentals of Ecology*, 3rd
ed. (Philadelphia: Saunders, 1971). Reprinted with permission of the publisher.
Figures 16.3 and **16.4,** from T.J. Givnish, "On the adaptive significance of leaf
height in forest herbs," *American Naturalist* 120(1982), pp. 358, 359. Copyright ©
1982 University of Chicago Press. **Figure 16.5** and **17.5,** Reprinted with
permission of Macmillan Publishing Company from *The Science of Ecology* by Paul
R. Ehrlich and Jonathan Roughgarden. Copyright © 1987 by Macmillan
Publishing Company. **Figures 17.1, 17.2, 17.3, 17.6,** and **18.7,** from C.B. Williams,
Patterns in the balance of nature and related problems in quantitative ecology (New York:
Academic Press, Inc., 1964, pp. 22, 28, 46, 48, 50). By permission of Academic
Press, Inc. **Figures 17.7** and **17.8a,** from G.G. Simpson, "Species density of
North American Recent mammals," *Systematic Zoology* 13(1964), pp. 60, 63.

Index

Some Major Drugs Derived From Plants

PLANT	DRUG	USE
Amazonian Liana	Curare	Muscle relaxant
Autumn crocus	Colchicine	Antitumor agent
Belladonna	Atropine	Anticholinergic
Coca	Cocaine	Local anesthetic
Foxglove	Digitoxin, digitalis	Cardiotonic
Meadowsweet	Salicylic acid	Analgesic
Mexican yam	Diosgenin	Birth-control pill
Opium poppy	Codeine, morphine	Analgesic
Tea	Caffeine	CNS stimulant
Yellow cinchona	Quinine	Antimalarial

Pacific Ocean

North America

MEDITERRANEAN

Asparagus	Common grape	Oat
Broad bean	Globe	Parsnip
Cabbage	artichoke	Rape
Cauliflower	Lavender	Sugar beet
Celery	Mint	

BELGIUM
Brussel sprouts

NORTH AFRICA
Cattle   Marjoram

MEXICO/CENTRAL AMERICA

Avacado	Papaya	Tobasco
Common bean	Pecan	Pepper
Hemp/sisal	Sweet potato	Vanilla
Maize/corn	Tomato	

NORTH AMERICA

Cranberry	Jerusalem	Muscadine
Sunflower	Artichoke	Grape
		Turkey

CARIBBEAN
Grapefruit

ANDES/SOUTH AMERICA

Cashew	Lima bean	Quinine
Cayenne	Pepper	Rubber
Cocoa	Pineapple	Upland cotton
Groundnut/	Potato	
peanut	Pumpkin	

Wheat: Graminae
Cereals, rice, sugar, fibers
c. 9,000 spp.

Bee orchid: Orchidaceae.
Ornamentals: c.18,000 spp.

Cabbage: Cruciferae.
Cabbage family, forage,
spices c. 3,000 spp.

Crab apple: Rosaceae.
Apple, peach, pear, plum,
herbs: c. 2,000 spp.

Cinchona (Quinine):
Rubiaceae. Coffee,
quinine, ornamentals
c. 7,000 spp.

Globe artichoke:
Compositae. Artichoke,
oil, endive, sunflower
c. 25,000 spp.

Wood spurge:
Euphorbiaceae. Rubber,
cassava, castor oil
c. 5000 spp.

Leek: Liliaceae.
Ornamentals, leeks,
onions c. 3,500 spp.